Instructor's Solutions Manual

Maryanne Clifford
Eastern Connecticut State University

to accompany

Introductory Statistics

Fifth Edition

Prem S. Mann
Eastern Connecticut State University

WILEY

JOHN WILEY & SONS, INC.

To order books or for customer service call 1-800-CALL-WILEY (225-5945).

ISBN 0-471-44813-3

Printed in the United States of America.

10 9 8 7 6 5 4 3 2 1

Printed and bound by Bradford & Bigelow, Inc.

For Georgette, Grace, and Jimmy

ACKNOWLEDGEMENTS

I would like to thank Gerald Geissert for assisting me by checking the solutions for this manual. Many thanks go to my family for the unceasing support and the hours of help they provided all along the way, especially Mary Lord-Clifford, Colleen Clifford, and James Clifford for assisting in preparing this solutions manual. I also want to thank, Deanna Dixon, Stefanie Fraser, Melissa Klar, Marie Pelkey and Beth Reel for their contributions to this manual. Lastly, I thank my colleague Prem Mann for recommending me for this project and answering many of my questions.

Maryanne Clifford

Contents

Solutions

Data Sets

Chapter One

1.1   The word 'statistics' has the following two meanings:

     i.   First, it refers to numerical facts such as the ages of persons, incomes of families, etc.

     ii.   Second, it refers to the field of study. It provides us with techniques that help us to collect, analyze, present, and interpret data and to make decisions.

1.2   There are two types of statistics: descriptive statistics and inferential statistics. Descriptive statistics consists of methods that help us organize, display, and describe data using tables, graphs, and summary measures. Inferential statistics is the part of statistics that helps us make decisions about the population based on sample results.

1.3   **Population:** All elements whose characteristics are being studied.

     Sample: A portion of the population selected for a study.

     Representative sample: A sample that possesses the characteristics of the population as closely as possible.

     Random sample: A sample drawn in such a way that each element of the population has some chance of being included in the sample.

     Sampling with replacement: A sampling procedure in which the item selected at each selection is put back in the population before the next item is drawn.

     Sampling without replacement: A sampling procedure in which the item selected at each selection is not replaced in the population.

1.4   i.   Suppose the employees of a company select one employee from all employees to be on the management board. Then, they select one employee from the same group of all employees to be a representative on the workers union. This is an example of sampling with replacement.

ii. Suppose the employees of a company select one employee from all employees to be on the management board. Then, they exclude the selected employee from the list of all employees and select one employee from the remaining employees to be a representative on the workers union. This is an example of sampling without replacement.

1.5   A census is a survey that includes all members of the population. A survey based on a portion of the population is called a sample survey. A sample survey is preferred over a census for the following reasons:

i. Conducting a census is very expensive because the size of the population is usually very large.

ii. Conducting a census is very time consuming.

iii. In many cases it is almost impossible to identify every member of the target population.

1.6   a. Population   b. Sample   c. Population   d. Population   e. Sample

1.7   a. Population   b. Sample   c. Population   d. Sample   e. Population

1.8   **Element:** A specific subject or object on which the information is collected.

Variable: A characteristic that is being investigated and that can assume different values for different elements.

Observation: The value of a variable for a single element.

Data set: A collection of observations on one or more variables.

1.9   **Member:** Each city included in the table.

Variable: Number of dog bites reported.

Measurement: Dog bites in a specific city. For example, Oakdale's 12 dog bites is a measurement.

Data set: Collection of dog bite numbers for the six cities listed in the table.

1.10   **Member:** Each airport included in the table.

Variable: Number of passengers.

Measurement: Number of passengers at a specific airport. For example, Los Angeles had 34.9 million passengers, which is a measurement.

Data set: Collection of passenger traffic information for the five airports listed in the table.

1.11   a.   Dog Bites   b. Six observations   c. Six elements (cities)

1.12   a.   Number of Passengers   b. Five   c. Five elements (airports)

1.13   a.   **Quantitative variable:** A variable that can assume numerical values.

b.   **Qualitative variable:** A variable that cannot be measured numerically but can be divided into different categories.

c.   **Discrete variable:** A variable whose values are countable.

d.   **Continuous variable:** A variable that can assume any value over a certain interval or intervals.

e.   **Quantitative data:** Data collected on a quantitative variable.

f.   **Qualitative data:** Data collected on a qualitative variable.

1.14   a. Quantitative   b. Qualitative   c. Qualitative   d. Quantitative   e. Quantitative

1.15   a. Quantitative   b. Quantitative   c. Qualitative   d. Quantitative   e. Quantitative

1.16   a. Discrete   d. Continuous   e. Discrete

1.17   a. Discrete   b. Continuous   d. Discrete   e. Continuous

1.18   Data collected on different elements for the same period of time are called cross-section data. Prices for twenty cars of 1997 models is an example of cross-section data.

Data collected on the same element for the same variable for different periods of time are called time-series data. Average prices of houses in Connecticut for the years 1970 to 1997 is an example of time-series data.

1.19   Internal sources of data are a company's own files and records.

External sources of data are the sources that do not belong to a company.

1.20   a.   Time-series data   b. Time-series data   c. Cross-section data   d. Cross-section data

1.21   a.   Cross-section data   b. Cross-section data   c. Time-series data   d. Time-series data

1.22

m	f	f^2	mf	m^2f
5	12	144	60	300
10	8	64	80	800
7	6	36	102	1734
20	16	256	320	6400
25	4	16	100	2500
Sum = 77	46	516	662	11,734

a. $\sum m = 77$ b. $\sum f^2 = 516$ c. $\sum mf = 662$ d. $\sum m^2 f = 11,734$

1.23

m	f	m^2	mf	$m^2 f$
3	16	9	48	144
6	11	36	66	396
25	16	625	400	10,000
12	8	144	96	1152
15	4	225	60	900
18	14	324	252	4536
Sum= 79	69	1363	992	17,128

a. $\sum f = 69$ b. $\sum m^2 = 1363$ c. $\sum mf = 992$ d. $\sum m^2 f = 17,128$

1.24

x	y	xy	x^2	y^2
15	10	150	225	100
22	12	264	484	144
11	14	154	121	196
8	9	72	64	81
5	18	90	25	324
$\sum x = 61$	$\sum y = 63$	$\sum xy = 730$	$\sum x^2 = 919$	$\sum y^2 = 845$

a. $\sum x = 61$ b. $\sum y = 63$ c. $\sum xy = 730$ d. $\sum x^2 = 919$ e. $\sum y^2 = 845$

1.25

x	y	xy	x^2	y^2
4	12	48	16	144
18	5	90	324	25
25	14	350	625	196
9	7	63	81	49
12	12	144	144	144
20	8	160	400	64
$\sum x = 88$	$\sum y = 58$	$\sum xy = 855$	$\sum x^2 = 1590$	$\sum y^2 = 622$

a. $\sum x = 88$ b. $\sum y = 58$ c. $\sum xy = 855$ d. $\sum x^2 = 1590$ e. $\sum y^2 = 622$

1.26   a. $\sum y = 20 + 14 + 57 + 23 + 7 + 102 = \$223$ b. $(\sum y)^2 = (223)^2 = 49{,}729$

c. $\sum y^2 = (20)^2 + (14)^2 + (57)^2 + (23)^2 + (7)^2 + (102)^2 = 14{,}827$

1.27   a. $\sum y = 83 + 205 + 57 + 134 = \$479$ b. $(\sum y)^2 = (479)^2 = 229{,}441$

c. $\sum y^2 = (83)^2 + (205)^2 + (57)^2 + (134)^2 = 70{,}119$

1.28   a. $\sum x = 23 + 35 + 29 + 54 + 18 = 159$ thousand dollars   b. $(\sum x)^2 = (159)^2 = 25{,}281$

c. $\sum x^2 = (23)^2 + (35)^2 + (29)^2 + (54)^2 + (18)^2 = 5{,}835$

1.29   a. $\sum x = 7 + 39 + 21 + 16 + 3 + 43 + 19 = 148$ students   b. $(\sum x)^2 = (148)^2 = 21{,}904$

c. $\sum x^2 = (7)^2 + (39)^2 + (21)^2 + (16)^2 + (3)^2 + (43)^2 + (19)^2 = 4486$ students

1.30   **Variable:** Tax returns filed electronically.

Measurement: The number of tax returns filed electronically for a given year.

Data Set: Collection of the number of tax returns filed electronically for the seven years listed in the table.

1.31   **Variable:** The number of U.S. babies born as triplets and larger sets.

Measurement: The number of U.S. babies born as triplets and larger sets for a specific year.

Data Set: Collection of the number of U.S. babies born as triplets and larger sets for the years listed in the table.

1.32   The data set in Exercise 1.30 is on the same variable (electronically filed tax returns) for different periods of time. Hence, it is a time-series data set. The data set in exercise 1.31 is on the same variable (U.S. babies born as triples and larger sets) for different states, but for the same period of time. Hence, it is a cross-section data set.

1.33   a.   Sample   b. Population for the year   c. Sample   d. Population

1.34   a.   Population   b. Sample   c. Sample   d. Population

1.35   a.   Sampling without replacement, because once a patient is selected, he/she will not be replaced before the next patient is selected. All 10 selected patients must be different.

   b.   Sampling with replacement because both times the selection is made from the same group (consisting of all professors).

1.36   a.   Sampling without replacement, because once a city is selected it will not be replaced before the next city is selected. All seven cities selected must be different.

   b.   Sampling with replacement, because both times the selection is made from the same group (consisting of all high school teachers).

1.37   a. $\sum x = 8 + 14 + 3 + 7 + 10 + 5 = 47$ shoe pairs   b. $(\sum x)^2 = (47)^2 = 2209$
   c. $\sum x^2 = = (8)^2 + (14)^2 + (3)^2 + (7)^2 + (10)^2 + (5)^2 = 443$ shoes

1.38   a. $\sum y = 4 + 12 + 8 + 10 + 5 = 39$ restaurants   b. $(\sum y)^2 = (39)^2 = 1521$
   c. $\sum y^2 = (4)^2 + (12)^2 + (8)^2 + (10)^2 + (5)^2 = 349$

1.39

m	f	f^2	mf	m^2f	m^2
3	7	49	21	63	9
16	32	1024	512	8192	256
11	17	289	187	2057	121
9	12	144	108	972	81
20	34	1156	680	13,600	400
Sum= 59	102	2662	1508	24,884	867

a. $\sum m = 59$ b. $\sum f^2 = 2662$ c. $\sum mf = 1508$ d. $\sum m^2f = 24,884$ e. $\sum m^2 = 867$

1.40

x	y	x^2	xy	x^2y	y^2
7	5	49	35	245	25
11	15	121	165	1815	225
8	7	64	56	448	49
4	10	16	40	160	100
14	9	196	126	1764	81
28	19	784	532	14,896	361
Sum = 72	65	1230	954	19,328	841

a. $\sum y = 65$ b. $\sum x^2 = 1230$ c. $\sum xy = 954$ d. $\sum x^2y = 19{,}328$ e. $\sum y^2 = 841$

Self-Review Test for Chapter One

1. b

2. c

3. a. A sample without replacement   b. A sample with replacement

4. a. Qualitative   b. Quantitative; continuous   c. Quantitative; discrete   d. Quantitative; continuous

5. **Member:** A specific job included in the table. For example, Coach is a member.

Variable: Votes for the best job at the Superbowl.

Measurement: Votes for a specific Superbowl job. For example, Coach received 1,927 votes is a measurement.

Data set: The collection of votes for the five Superbowl jobs listed in the table.

6. a. $\sum x = 8 + 5 + 10 + 6 + 5 + 8 = 42$ rooms   b. $(\sum x)^2 = (42)^2 = 1764$

c. $\sum x^2 = (8)^2 + (5)^2 + (10)^2 + (6)^2 + (5)^2 + (8)^2 = 314$

7.

m	f	m^2	mf	m^2f	f^2
3	15	9	45	135	225
6	25	36	150	900	625
9	40	81	360	3240	1600
12	20	144	240	2880	400
15	12	225	180	2700	144
$\sum m = 45$	$\sum f = 112$	$\sum m^2 = 495$	$\sum mf = 975$	$\sum m^2f = 9855$	$\sum f^2 = 2994$

a. $\sum m = 45$ b. $\sum f = 112$ c. $\sum m^2 = 495$ d. $\sum mf = 975$ e. $\sum m^2f = 9855$ f. $\sum f^2 = 2994$

Chapter Two

2.1   Data in their original form are usually too large and unmanageable. By grouping data, we make them manageable. It is easier to make decisions and draw conclusions using grouped data than ungrouped data.

2.2   The relative frequency for a category is obtained by dividing the frequency of that category by the sum of the frequencies of all categories. The percentage for a category is obtained by multiplying the relative frequency of that category by 100. Exercise 2.3 can be considered as an example to show how relative frequencies and percentages are calculated.

2.3   a. & b.

Category	Frequency	Relative Frequency	Percentage
A	8	.267	26.7
B	8	.267	26.7
C	14	.467	46.7

c. 26.7 % of the elements in this sample belong to category B.

d. 26.7% + 46.7% = 73.4% of the elements in this sample belong to category A or C.

e.

2.4   a. & b.

Category	Frequency	Relative Frequency	Percentage
Y	23	.575	57.5
N	13	.325	32.5
D	4	.100	10.0

c. 57.5% of the elements belong to category Y.

d. 32.5 + 10 = 42.5% of the elements belong to categories N or D.

e.

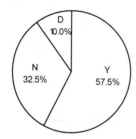

2.5   a. & b.

Category	Frequency	Relative Frequency	Percentage
F	12	.24	24
SO	12	.24	24
J	15	.30	30
SE	11	.22	22

c.   30 + 22 = 52% of the students are juniors or seniors.

d.

2.6   a. & b.

Category	Frequency	Relative Frequency	Percentage
T	4	.133	13.3
R	10	.333	33.3
A	7	.233	23.3
P	8	.267	26.7
M	1	.033	3.3

c.   33.3 + 23.3 = 56.6% of the adults ranked refrigerators or air conditioning as the convenience they would find the most difficult to do without.

d.

2.7   a. & b.

Category	Frequency	Relative Frequency	Percentage
C	9	9/20= .45	45
F	5	5/20= .25	25
T	6	6/20= .30	30

c. 45% of the employees would prefer a four-day work week.

d.

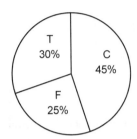

2.8   a. & b.

Category	Frequency	Relative Frequency	Percentage
C	4	.250	25.0
CK	5	.313	31.3
CC	4	.250	25.0
D	2	.125	12.5
O	1	.063	6.3

c.

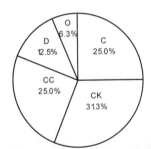

2.9   Let the seven categories listed in the table be denoted by S, HC, R, O, E, P, and U respectively.

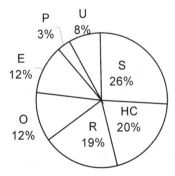

2.10   Let the four categories listed in the table be denoted by E, D, B, and O respectively.

2.11   1. The number of classes to be used to group the given data.

2. The width of each class.

3. The lower limit of the first class.

2.12   The relative frequency for a class is obtained by dividing the frequency of that class by the sum of frequencies of all classes. The percentage for a class is obtained by multiplying the relative frequency of that class by 100. Exercise 2.18 can be considered as an example to illustrate the calculation of relative frequencies and percentages.

2.13   A data set that does not contain fractional values is usually grouped by using classes with limits. Suppose we have data on ages of 100 managers, and ages are rounded to years. Then, the following table could be an example of grouped data that uses classes with limits.

Ages (years)	Frequency
21 to 30	12
31 to 40	27
41 to 50	31
51 to 60	22
61 to 70	8

A data set that contains fractional values is grouped by using the *less than* method. Suppose we have data on sales of 100 medium sized companies. The following table shows a frequency table for such data.

Sales(millions of dollars)	Frequency
0 to less than 10	27
10 to less than 20	31
20 to less than 30	19
30 to less than 40	14
40 to less than 50	9

Single valued classes are used to group a data set that contains only a few distinct values. As an example, suppose we have a data set on the number of children for 100 families. The following table is an example of a frequency table using single valued classes.

Number of Children	Frequency
0	13
1	26
2	38
3	18
4	5

2.14   a. & c.

Class Boundaries	Class Midpoint	Relative Frequency	Percentage
-0.5 to less than 99.5	49.5	.39	39
99.5 to less than 199.5	149.5	.21	21
199.5 to less than 299.5	249.5	.18	18
299.5 to less than 399.5	349.5	.15	15
399.5 to less than 499.5	449.5	.07	7

b. Yes, each class has a width of 100.

d. 18 + 15 + 7 = 40% of the students wrote 200 or more checks during 2002.

2.15   a. & c.

Class Boundaries	Class Midpoint	Relative Frequency	Percentage
17.5 to less than 30.5	24	.24	24
30.5 to less than 43.5	37	.38	38
43.5 to less than 56.5	50	.28	28
56.5 to less than 69.5	63	.10	10

b. Yes, each class has a width of 13.

d. 24 + 38 = 62% of the employees are 43 years old or younger.

2.16   a. & b.

Class Limits	Class Boundaries	Class Midpoints
1 to 200	.5 to less than 200.5	100.5
201 to 400	200.5 to less than 400.5	300.5
401 to 600	400.5 to less than 600.5	500.5
601 to 800	600.5 to less than 800.5	700.5
801 to 1000	800.5 to less than 1000.5	900.5
1001 to 1200	1000.5 to less than 1200.5	1100.5

2.17   a., b., & c.

Class Limits	Class Boundaries	Class Width	Class Midpoint
1 to 25	.5 to less than 25.5	25	13
26 to 50	25.5 to less than 50.5	25	38
51 to 75	50.5 to less than 75.5	25	63
76 to 100	75.5 to less than 100.5	25	88
101 to 125	100.5 to less than 125.5	25	113
126 to 150	125.5 to less than 150.5	25	138

2.18   a. & b.

Team Revenue	Frequency	Relative Frequency	Percentage
125 to 130	2	.133	13.3
131 to 136	5	.333	33.3
137 to 142	4	.267	26.7
143 to 148	2	.133	13.3
149 to 154	0	.000	0
155 to 160	2	.133	13.3

c.   The data are skewed to the right.

d.   26.7 + 13.3 + 0 +13.3 = 53.3% of these teams have revenue of $137 million or more.

2.19   a. & b.

Number of Computer Terminals Manufactured	Frequency	Relative Frequency	Percentage
21 - 23	7	.233	23.3
24 - 26	4	.133	13.3
27 - 29	9	.300	30.0
30 - 32	4	.133	13.3
33 - 35	6	.200	20.0

c.

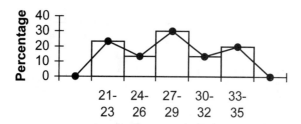

d.   For 30 % of the days, the number of computer terminals manufactured is in the interval 27 to 29.

2.20   a.

Keyboards Assembled	Frequency	Relative Frequency
41 – 44	5	.20
45 – 48	8	.32
49 – 52	8	.32
53 – 56	4	.16

b. & c.

2.21   a. & b.

Median Income (thousands of dollars)	Frequency	Relative Frequency	Percentage
30 − 35	4	.211	21.1
35 − 40	7	.368	36.8
40 − 45	2	.105	10.5
45 − 50	4	.211	21.1
50 − 55	2	.105	10.5

2.22   a., b., & c. The minimum rent paid in the sample is $540, and the maximum rent paid is $1306. One possible answer to this question is given in the following table. However, we can obtain a different frequency distribution if we choose different classes.

Apartment Rent (dollars)	Frequency	Relative Frequency	Percentage	Class Midpoint
540 − 694	11	.44	44	617
695 − 849	7	.28	28	772
850 − 1004	2	.08	8	927
1005 − 1159	2	.08	8	1082
1160 − 1314	3	.12	12	1237

2.23   a., b., & c.   The lowest phone bill in our data is $16.99 and the highest amount is $33.71. The following table provides one way to group these data into five classes of equal width.

Telephone Bill (dollars)	Frequency	Relative Frequency	Percentage	Class Midpoint
15 to less than 19	4	.16	16	17
19 to less than 23	11	.44	44	21
23 to less than 27	7	.28	28	25
27 to less than 31	2	.08	8	29
31 to less than 35	1	.04	4	33

2.24   a. & b.   The minimum price of gasoline in this data set is $1.299, and the maximum price is $1.643. One way to group these data into four classes is shown below.

Gasoline Prices (dollars)	Frequency	Relative Frequency	Percentage
1.299 − 1.398	12	.48	48
1.399 − 1.498	8	.32	32
1.499 − 1.598	3	.12	12
1.599 − 1.698	2	.08	8

c.

2.25   a. & b. The lowest cost of a visit to the doctor in our data is \$52.33, and the highest cost is \$88.67. One way to group these data using four classes is shown in the table here. However, your answer can be different depending on the number of classes used to group the data and the class width used.

Visit to Doctor (dollars)	Frequency	Relative Frequency	Percentage
50 to less than 60	9	.36	36
60 to less than 70	7	.28	28
70 to less than 80	7	.28	28
80 to less than 90	2	.08	8

c.

2.26   a. & b.

Price of Beer (dollars)	Frequency	Relative Frequency	Percentage
6.30 – 6.90	4	.16	16
6.90 – 7.50	13	.52	52
7.50 – 8.10	6	.24	24
8.10 – 8.70	2	.08	8

2.27   a. & b

ERA	Frequency	Relative Frequency	Percentage
3.00 to less than 3.50	1	.063	6.3
3.50 to less than 4.00	6	.375	37.5
4.00 to less than 4.50	6	.375	37.5
4.50 to less than 5.00	2	.133	13.3
5.00 to less than 5.50	1	.06	6

2.28   a. & b.

Turnovers	Frequency	Relative Frequency	Percentage
0	4	.167	16.7
1	5	.208	20.8
2	7	.292	29.2
3	5	.208	20.8
4	3	.125	12.5

c.   The total for games with two or more turnovers is 7+ 5 + 3 = 15.

d.

2.29   a. & b.

Number of Children less than 18 Years of Age	Frequency	Relative Frequency	Percentage
0	8	.267	26.7
1	10	.333	33.3
2	10	.333	33.3
3	2	.067	6.7

c. 10 + 2 = 12 families have 2 or 3 children under 18 years of age.

d.

2.30

The truncated graph exaggerates the difference in the number of students with different numbers of tickets.

2.31

The graph with the truncated frequency axis exaggerates the differences in the frequencies of the various classes.

2.32   The cumulative frequency distribution gives the total number of values that fall below the upper boundary of each class. The cumulative relative frequencies are obtained by dividing the cumulative frequencies by the total number of observations in the data. The cumulative percentages are obtained by multiplying the cumulative relative frequencies by 100.

2.33   An ogive is drawn for a cumulative frequency distribution, a cumulative relative frequency distribution, or a cumulative percentage distribution. An ogive can be used to find the approximate cumulative frequency (cumulative relative frequency or cumulative percentage) for any class interval.

2.34   a. & b.

Number of Checks	Cumulative Frequency	Cumulative Relative Frequency	Cumulative Percentage
0 to 99	39	.39	39
0 to 199	60	.60	60
0 to 299	78	.78	78
0 to 399	93	.93	93
0 to 499	100	1.00	100

c.   60% of the students wrote 199 or fewer checks in 2002.

d.

e.   About 69% of the students wrote 250 or fewer checks in 2002.

2.35   a. & b.

Age (years)	Cumulative Frequency	Cumulative Relative Frequency	Cumulative Percentage
18 to 30	12	.24	24
18 to 43	31	.62	62
18 to 56	45	.90	90
18 to 69	50	1.00	100

c. $100 - 62 = 38\%$ of the employees are 44 years of age or older.

d.

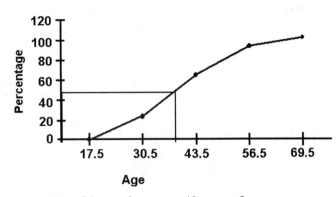

Age

e. About 52% of the employees are 40 years of age or younger.

2.36

Team Revenue	Cumulative Frequency	Cumulative Relative Frequency	Cumulative Percentage
125 to 130	2	.133	13.3
125 to 136	7	.467	46.7
125 to 142	11	.733	73.3
125 to 148	13	.867	86.7
125 to 154	13	.867	86.7
125 to 160	15	1.000	100.0

2.37

Number of Computer Terminals Manufactured	Cumulative Frequency	Cumulative Relative Frequency	Cumulative Percentage
21 – 23	7	.233	23.3
21 – 26	11	.367	36.7
21 – 29	20	.667	66.7
21 – 32	24	.800	80.0
21 – 35	30	1.00	100.0

2.38

Keyboards Assembled	Cumulative Frequency	Cumulative Relative Frequency	Cumulative Percentage
41 – 44	5	.20	20
41 – 48	13	.52	52
41 – 52	21	.84	84
41 – 56	25	1.00	100

2.39

Telephone Bill (dollars)	Cumulative Frequency	Cumulative Relative Frequency	Cumulative Percentage
15 to less than 19	4	.16	16
15 to less than 23	15	.60	60
15 to less than 27	22	.88	88
15 to less than 31	24	.96	96
15 to less than 35	25	1.00	100

2.40

Visit to Doctor (dollars)	Cumulative Frequency	Cumulative Relative Frequency	Cumulative Percentage
50 to less than 60	9	.36	36
50 to less than 70	16	.64	64
50 to less than 80	23	.92	92
50 to less than 90	25	1.00	100

2.41

Price of Beer (dollars)	Cumulative Frequency	Cumulative Relative Frequency	Cumulative Percentage
6.30 – 6.90	4	.16	16
6.30 – 7.50	17	.68	68
6.30 – 8.10	23	.92	92
6.30 – 8.70	25	1.00	100

2.42

Median Income (thousands of dollars)	Cumulative Frequency	Cumulative Relative Frequency	Cumulative Percentage
30 to less than 35	4	.211	21.1
30 to less than 40	11	.579	57.9
30 to less than 45	13	.684	68.4
30 to less than 50	17	.899	89.5
30 to less than 55	19	1.000	100.0

About 12 states in the table have median household income less than $43,000 for that year.

2.43

ERA	Cumulative Frequency	Cumulative Relative Frequency	Cumulative Percentage
3.00 to less than 3.50	1	.063	6.3
3.00 to less than 4.00	7	.438	43.8
3.00 to less than 4.50	13	.813	81.3
3.00 to less than 5.00	15	.938	93.8
3.00 to less than 5.50	16	1.000	100.0

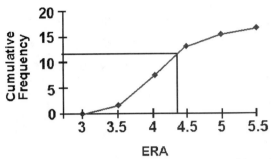

Approximately 11 of the teams had an ERA of less than 4.20.

2.44   To prepare a stem and leaf display for a data set, each value is divided into two parts – the first part is called the *stem* and the second part is called the *leaf*. The stems are written on the left side of a vertically drawn line and the leaves for each stem are written on the right side of the vertical line next to the corresponding stem. Exercise 2.46 is an example.

2.45   The advantage of a stem-and-leaf display over a frequency distribution is that by preparing a stem-and-leaf display we do not lose information on individual observations. From a stem-and-leaf display we can obtain the original data. However, we cannot obtain the original data from a frequency distribution table. Consider the following stem-and-leaf display.

```
3 | 3  5  6  6
4 | 2  4  4  7  9  9
5 | 0  1  1  3  4  5  5
```

The data that were used to make this stem-and-leaf display are:
33, 35, 36, 36, 42, 44, 44, 47, 49, 49, 50, 51, 51, 53, 54, 55, 55

2.46

```
43   46   50   51   54   55   63   64   66   67   67   67
68   69   72   72   73   75   76   76   79   80   87   88   89
```

2.47

```
218   245   256   329   367   383   397   404   427   433   471   523
537   551   563   581   592   622   636   647   655   678   689   810   841
```

2.48

```
0 | 8  5  6  5  3          0 | 3  5  5  6  8
1 | 5  7  0  4  7  6  9     1 | 0  4  5  6  7  7  9
2 | 3  1  2  5             2 | 1  2  3  5
3 | 1  4  0  1             3 | 0  1  1  4
```

2.49

```
 7 | 45  75              7 | 45  75
 8 | 48  00  57          8 | 00  48  57
 9 | 21  33  67  95      9 | 21  33  67  95
10 | 24  09             10 | 09  24
11 | 33  45             11 | 33  45
12 | 75                 12 | 75
```

2.50

```
2 | 4  7  3  9  5  3  8  1  6  2  7  7  3  8  9  2  6  8  3  7
3 | 2  3  3  1  3  1  5  4  5  1
```

```
2 | 1  2  2  3  3  3  3  4  5  6  6  7  7  7  7  8  8  8  9  9
3 | 1  1  1  2  3  3  3  4  5  5
```

2.51

```
4 | 5  8  1  6  4  2  8  8  6  3  7  4  7  9
5 | 2  6  3  1  3  1  2  0  4  0  2
```

```
4 | 1  2  3  4  4  5  6  6  7  7  8  8  8  9
5 | 0  0  1  1  2  2  2  3  3  4  6
```

2.52   Cost of a visit to the doctor rounded to the nearest dollar:

```
52    89    55    70    75    72    80    64    57    56    78    59    72
64    71    53    58    59    60    72    63    68    60    62    65
```

Stem-and-leaf display: Ranked stem-and-leaf display:

```
5 | 2  5  7  6  9  3  8  9            5 | 2  3  5  6  7  8  9  9
6 | 4  4  0  3  8  0  2  5            6 | 0  0  2  3  4  4  5  8
7 | 0  5  2  8  2  1  2              7 | 0  1  2  2  2  5  8
8 | 9  0                            8 | 0  9
```

2.53

```
0 | 5  7                    0 | 5  7
1 | 0  1  7  5  9            1 | 0  1  5  7  9
2 | 3  6  6  9  1  2         2 | 1  2  3  6  6  9
3 | 3  9  2                 3 | 2  3  9
4 | 8  3                    4 | 3  8
5 | 0                       5 | 0
6 | 5                       6 | 5
```

2.54   a.

```
0 | 6  5  8  9
1 | 0  2  5  3  8
2 | 0  5  4  0  4  6  2  8
3 | 7  0  3  6  6  2  8
4 | 0  5
7 | 2  0  2
8 | 0  4  6
9 | 6  2  0
```

b.

```
0-2 | 6  5  8  9  *  0  2  5  3  8  *  0  5  4  0  4  6  2  8
3-5 | 7  0  3  6  6  2  8  *  0  5  *  *
6-9 | *  *  2  0  2  *  0  4  6  *  6  2  0
```

2.55   a.

```
1 | 58
2 | 10  20  45  65  68  70
3 | 20  45  50  68  90
4 | 30  38  57  60  75  87  90
5 | 05  28  30  38  40  60  65  70
6 | 17  35  38
7 | 02  05  06  20  21
```

b.

1-3	58	*	10	20	45	65	68	70	*	20	45	50	68	90
4-5	30	38	57	60	75	87	90	*	05	28	30	38	40	60	65	70
6-7	17	35	38	*	02	05	06	20	21					

2.56   a. & b.

Political Party	Frequency	Relative Frequency	Percentage
D	9	.300	30.0
DR	4	.133	13.3
F	2	.067	6.7
R	11	.367	36.7
W	4	.133	13.3

c.

d.   13.3% of these presidents were Whigs.

2.57   a. & b.

Favorite Sport	Frequency	Relative Frequency	Percentage
CF	5	.139	13.9
CB	3	.083	8.3
MLB	9	.250	25.0
NFL	13	.361	36.1
NBA	6	.167	16.7

c.

d.   25% of these people prefer Major League Baseball.

2.58 a. & b.

TV sets owned	Frequency	Relative Frequency	Percentage
0	1	.025	2.5
1	14	.350	35.0
2	14	.350	35.0
3	8	.200	20.0
4	3	.075	7.5

c.

d. 62.5% of the households own 2 or more television sets.

2.59 a. & b.

Correct Names	Frequency	Relative Frequency	Percentage
0	1	.042	4.2
1	3	.125	12.5
2	4	.167	16.7
3	6	.250	25.0
4	4	.167	16.7
5	6	.250	25.0

c. 4.2 + 12.5 = 16.7% of the students name less than two representatives correctly.

d.

2.60 a. & b.

Amount Spent on Video Rentals (in dollars)	Frequency	Relative Frequency	Percentage
1 – 200	20	.667	66.7
201 – 400	4	.133	13.3
401 – 600	3	.100	10.0
601 – 800	2	.067	6.7
801 – 1000	1	.033	3.3

c. 10.0 + 6.7 + 3.3 = 20.0% of the households in this sample spent more than $400 on video rentals.

2.61   a. & b.

Number of Orders	Frequency	Relative Frequency	Percentage
23 – 29	4	.133	13.3
30 – 36	9	.300	30.0
37 – 43	6	.200	20.0
44 – 50	8	.267	26.7
51 – 57	3	.100	10.0

c. For 20.0 + 26.7 + 10.0 = 56.7% of the hours in this sample, the number of orders was more than 36.

2.62   a. & b.

Concession (dollars)	Frequency	Relative Frequency	Percentage
0 to less than 6	9	.300	30.0
6 to less than 12	10	.333	33.3
12 to less than 18	5	.167	16.7
18 to less than 24	4	.133	13.3
24 to less than 30	2	.067	6.7

c.

2.63   a. & b.

Car Repair Costs (dollars)	Frequency	Relative Frequency	Percentage
1 – 1400	11	.367	36.7
1401 – 2800	10	.333	33.3
2801 – 4200	3	.100	10.0
4201 – 5600	2	.067	6.7
5601 – 7000	4	.133	13.3

c.

d. The class boundaries of the fourth class are $4200.50 and 5600.50. The width of this class is $1400.

2.64   a. & b.

Amount Spent on Video Rentals (in dollars)	Cumulative Frequency	Cumulative Relative Frequency	Cumulative Percentage
1 – 200	20	.667	66.7
1 – 400	24	.800	80.0
1 – 600	27	.900	90.0
1 – 800	29	.967	96.7
1 – 1000	30	1.000	100.0

2.65

Number of Orders	Cumulative Frequency	Cumulative Relative Frequency	Cumulative Percentage
23 – 29	4	.133	13.3
23 – 36	13	.433	43.3
23 – 43	19	.633	63.3
23 – 50	27	.900	90.0
23 – 57	30	1.000	100.0

2.66

Concession(dollars)	Cumulative Frequency	Cumulative Relative Frequency	Cumulative Percentage
0 to less than 6	9	.300	30.0
0 to less than 12	19	.633	63.3
0 to less than 18	24	.800	80.0
0 to less than 24	28	.933	93.3
0 to less than 30	30	1.000	100.0

2.67

Car Repair Costs (dollars)	Cumulative Frequency	Cumulative Relative Frequency	Cumulative Percentage
1 – 1400	11	.367	36.7
1 – 2800	21	.700	70.0
1 – 4200	24	.800	80.0
1 – 5600	26	.867	86.7
1 – 7000	30	1.000	100.0

2.68

```
0 | 24  06  40  90  55  90  70  88  83  22  70  15
1 | 00  00  55  80  00  27  60  11
2 | 05  40
3 | 10  50
4 | 05  05
5 | 95
6 | 22
7 | 60
8 | 08
```

2.69

```
2 | 8  4  7  7
3 | 4  1  8  5  2  9  3  7  0  8  4  6  0
4 | 4  1  7  6  1  9  5  6  7
5 | 2  3  7  0
```

2.70   Let the six categories listed in the table be denoted by S, ML, M, O, V, and F respectively.

The truncated graph exaggerates the differences in the sale prices of houses in these towns.

2.71   Let the seven categories listed in the table be denoted by LA, TX, MI, OK, AL, GA, and SC respectively.

The truncated graph exaggerates the differences in the incarceration rates for the seven states.

2.72   a.

Age	Frequency	Relative Frequency
18 to less than 20	7	.060
20 to less than 25	12	.103
25 to less than 30	18	.154
30 to less than 40	14	.120
40 to less than 50	15	.128
50 to less than 60	16	.137
60 and over	35	.299

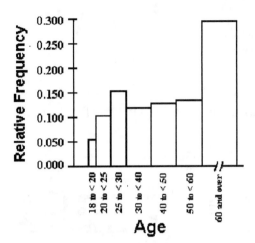

b. & c.   This histogram is misleading because the class widths differ. If you were to change the frequency distribution to reflect equal class widths, the resulting histogram would give a clearer picture.

2.73   The greater relative frequency of accidents in the older age group does not imply that they are more accident-prone than the younger group. The older group may drive more miles during a week than the younger group.

2.74   a.   Sturge's formula: $c = 1 + 3.3 \log n$ where c = number of classes and n = number of observations in data set. $c = 1 + 3.3 \log (135) \approx 8$

b.   Class width = (53-20)/8 = 4.125 which we round up to a class width of 5.

2.75   a.

```
1 | 7  8  8  8  8  9  9  9  9  9  9  9  9  9  9
2 | 0  0  0  0  0  0  0  1  1  1  2  2  3  6  8
3 | 0  8
4 | 3
5 | 1
6 | 2  4
```

b.

Enrolled Students	Frequency
15 to less than 25	27
25 to less than 35	3
35 to less than 45	2
45 to less than 55	1
45 to less than 65	2

The data is very skewed to the left. 77% shows up in the first class.

c.

Enrolled Students	Frequency
17 to less than 19	5
19 to less than 21	16
21 to less than 23	5
23 to less than 25	1
25 to less than 35	3
35 to less than 45	2
45 and over	3

2.76   a. Answers will vary.

b. i.
```
 9 | 9
10 | 2  8  8
11 | 0  4  5  5  6  9
12 | 3  3  3  5  8  8
13 | 2  3  8
14 | 6  7  7  8
15 | 5  9
16 | 1  2  4  8
17 | 4  4  5  9  9  9
18 | 0  2  3  9
19 | 3  3  5
20 | 2  4
```

ii. The display shows a bimodal distribution, due to the presence of both females and males in the sample. The males tend to be heavier, so their weights are concentrated in the larger values, while the females' weights are found primarily in the smaller values.

c.

Weight

Self-Review Test for Chapter Two

1. An ungrouped data set contains information on each member of a sample or population individually. The following data, which give the grade point averages of eight students, is an example of ungrouped data.

   3.45   2.98   3.81   2.04   3.67   3.01   1.88   2.59

Data presented in the form of a frequency table are called grouped data. The following table gives the frequency distribution table for the grade point averages of 100 students. This is an example of grouped data.

Grade Point Average	Frequency
1.5 to less than 2.0	5
2.0 to less than 2.5	13
2.5 to less than 3.0	34
3.0 to less than 3.5	31
3.5 to less than 4.0	17

2. a. 5   b. 7   c. 17   d. 6.5   e. 13   f. 90   g. .30

3. A histogram that is identical on both sides of its central point is called a symmetric histogram. A histogram that is skewed to the right has a longer tail on the right side, and a histogram that is skewed to the left has a longer tail on the left side. The following three histograms present these three cases.

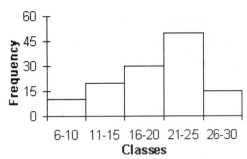

4. a. & b.

Category	Frequency	Relative Frequency	Percentage
B	8	.40	40
F	4	.20	20
M	7	.35	35
S	1	.05	5

c. 35% of the children live with their mothers only.

d.

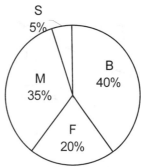

5. a. & b.

Number of Years	Frequency	Relative Frequency	Percentage
1 – 4	5	.208	20.8
5 – 8	6	.250	25.0
9 – 12	6	.250	25.0
13 – 16	5	.208	20.8
17 – 20	2	.083	8.3

c. 20.8 + 25.0 = 45.8% of the employees have been with their current employer for 8 or fewer years.

d.

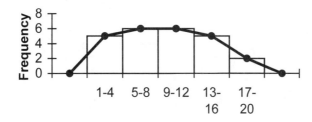

6. a. & b

Number of Years	Cumulative Frequency	Cumulative Relative Frequency	Cumulative Percentage
1 – 4	5	.208	20.8
1 – 8	11	.458	45.8
1 – 12	17	.708	70.8
1 – 16	22	.917	91.7
1 – 20	24	1.000	100.0

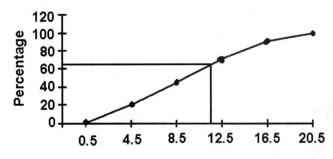

About 61 % of the employees have been with their current employer for 11 or fewer years.

7.

```
0 | 4   6   7   8
1 | 0   2   2   3   4   4   5   6   6   6   7   8   9
2 | 0   1   2   2   5   9
3 | 2
```

8.   30   33   37   42   44   46   47   49   51   53   53   56   60   67   67   71   79

Chapter Three

3.1   For a data set with an odd number of observations, first we rank the data set in increasing (or decreasing) order and then find the value of the middle term. This value is the median.

For a data set with an even number of observations, first we rank the data set in increasing (or decreasing) order and then find the average of the two middle terms. The average gives the median.

3.2   A few values that are either very small or very large relative to the majority of the values in a data set are called outliers or extreme values. Suppose the 2002 sales (in millions of dollars) of five companies are 10, 21, 14, 410, and 8. Then, $410 million is an outlier because this value is very large compared to the other values.

The median is a better measure of central tendency as compared to the mean for a data set that contains an outlier because the mean is affected more by outliers than the median.

3.3   Suppose the 2002 sales (in millions of dollars) of five companies are: 10, 21, 14, 410, and 8. The mean for the data set is:

$$\text{Mean} = (10 + 21 + 14 + 410 + 8) / 5 = \$92.60 \text{ million}$$

Now, if we drop the outlier (410), the mean is:

$$\text{Mean: } (10 + 21 + 14 + 8) / 4 = \$13.25 \text{ million}$$

This shows how an outlier can affect the value of the mean.

3.4   All three measures of central tendency (the mean, the median, and the mode) can be calculated for quantitative data. (Note that the mode may or may not exist for a data set.)

However, only the mode (if it exists) can be found for a qualitative data set. Examples given in Sections 3.1.1, 3.1.2, and 3.1.3 of the text show these cases.

3.5   The mode can assume more than one value for a data set. Examples 3–8 and 3–9 of the text present such cases.

3.6   A quantitative data set will definitely have a mean and a median but it may or may not have a mode. Example 3–7 of the text presents a data set that contains no mode.

3.7   For a symmetric histogram (with one peak), the values of the mean, median, and mode are all equal. Figure 3.2 of the text shows this case. For a histogram that is skewed to the right, the value of the mode is the smallest and the value of the mean is the largest. The median lies between the mode and the mean. Such a case is presented in Figure 3.3 of the text. For a histogram that is skewed to the left, the value of the mean is the smallest, the value of the mode is the largest, and the value of the median lies between the mean and the mode. Figure 3.4 of the text exhibits this case.

3.8   The median is the best measure to summarize this data set since it is not influenced by outliers.

3.9 $\sum x = 5 + (-7) + 2 + 0 + (-9) + 16 + 10 + 7 = 24$ $(N + 1) / 2 = (8 + 1) / 2 = 4.5$

$\mu = (\sum x) / N = 24 / 8 = 3$

Median = value of the 4.5th term in ranked data = $(2 + 5) / 2 = 3.50$

This data set has no mode.

3.10 $\sum x = 14 + 18 - 10 + 8 + 8 - 16 = 22$ $(n + 1) / 2 = (6 + 1) / 2 = 3.5$

$\bar{x} = (\sum x) / n = 22 / 6 = 3.67$

Median = value of the 3.5th term in ranked data = $(8 + 8) / 2 = 8$

Mode = 8

3.11 $\bar{x} = (\sum x) / n = 9907 / 12 = \$825.58$ $(n + 1) / 2 = (12 + 1) / 2 = 6.5$

Median = value of the 6.5th term in ranked data set = $(769 + 798) / 2 = \$783.50$

3.12 $\bar{x} = (\sum x) / n = 274.56 / 12 = \$22.88$ $(n + 1) / 2 = (12 + 1) / 2 = 6.5$

Median = $(21.16 + 22.39) / 2 = \$21.78$

3.13 $\bar{x} = (\sum x) / n = 16.682 / 12 = \$1.390$ $(n + 1) / 2 = (12 + 1) / 2 = 6.5$

Median = $(1.360 + 1.351) / 2 = \$1.356$

This data set has no mode.

3.14 $\bar{x} = (\sum x) / n = 770.08 / 12 = \$64.17$ $(n + 1) / 2 = (12 + 1) / 2 = 6.5$

Median = $(61.80 + 62.50) / 2 = \$62.15$

This data set has no mode because no value occurs more than once.

3.15 $\bar{x} = (\sum x) / n = 85.81 / 12 = \$7.15$ $(n + 1) / 2 = (12 + 1) / 2 = 6.5$

Median = (6.99 + 7.03) / 2 = $7.01

3.16 $\bar{x} = (\sum x) / n = 81 / 12 = 6.75$ cars $(n + 1) / 2 = (12 + 1) / 2 = 6.5$

Median = (6 + 7) / 2 = 6.5 cars

Mode = 3, 6, and 7 cars

3.17 $\mu = (\sum x) / N = 35{,}629 / 6 = \$5938.17$ thousand $(n + 1) / 2 = (6 + 1) / 2 = 3.5$

Median = (750 + 8500) / 2 = $4625 thousand

This data set has no mode because no value appears more than once.

3.18 $\mu = (\sum x) / N = 2464 / 14 = 176$ home runs $(n + 1) / 2 = (14 + 1) / 2 = 7.5$

Median = (167 + 177) / 2 = 172 home runs

The mode is 152 since this value occurs twice and no other value occurs more than once.

3.19 $\bar{x} = (\sum x) / n = 64 / 10 = 6.40$ hours $(n + 1) / 2 = (10 + 1) / 2 = 5.5$

Median = (7 + 7) / 2 = 7 hours

Mode = 0 and 7 hours

3.20 $\mu = (\sum x) / N = 1514 / 16 = 94.63$ stolen bases $(n + 1) / 2 = (16 + 1) / 2 = 8.5$

Median = (87 + 92) / 2 = 89.5 stolen bases

The data has two modes at 71 and 86 as the two values each occur twice and no other value occurs

more than once.

3.21 $\bar{x} = (\sum x) / n = 294 / 10 = 29.4$ computer terminals $(n + 1) / 2 = (10 + 1) / 2 = 5.5$

Median = (28 + 29) / 2 = 28.5 computer terminals

Mode = 23 computer terminals

3.22 $\bar{x} = (\sum x) / n = 107 / 12 = 8.92$ students $(n + 1) / 2 = (12 + 1) / 2 = 6.5$

Median = (9 + 9) / 2 = 9 students

Mode = 6 and 9 students

3.23   a. $\bar{x} = (\sum x) / n = 257 / 13 = 19.77$ newspapers $(n + 1) / 2 = (13 + 1) / 2 = 7$

Median = 12 newspapers

b.   Yes, 92 is an outlier. When we drop this value,

Mean = 165 / 12 = 13.75 newspapers $(n + 1) / 2 = (12 + 1) / 2 = 6.5$

Median = (11 + 12) / 2 = 11.5 newspapers

As we observe, the mean is affected more by the outlier.

c. The median is a better measure because it is not sensitive to outliers.

3.24 $\bar{x} = (\sum x) / n = 147{,}528.4 / 8 = \$18{,}441.05$ million $(n + 1) / 2 = (8 + 1) / 2 = 4.5$

Median $= (14002.7 + 18338.2) / 2 = \$16{,}170.45$ million.

This data set has no mode because no value appears more than once.

3.25 From the given information: $n_1 = 10$, $n_2 = 8$, $\bar{x}_1 = \$95$, $\bar{x}_2 = \$104$

$$\bar{x} = \frac{n_1 \bar{x}_1 + n_2 \bar{x}_2}{n_1 + n_2} = \frac{(10)(95) + (8)(104)}{10 + 8} = \frac{1782}{18} = \$99$$

3.26 From the given information: $n_1 = 18$, $n_2 = 20$, $\bar{x}_1 = \$144$, $\bar{x} = \$150$

$$\bar{x} = \frac{n_1 \bar{x}_1 + n_2 \bar{x}_2}{n_1 + n_2} = \frac{(18)(144) + (20)(\bar{x}_2)}{18 + 20} = 150 \quad \text{This becomes:}$$

$$\bar{x}_2 = \frac{(n_1 + n_2)\bar{x} - n_1 \bar{x}_1}{n_2} = \frac{(18 + 20)150 - (18)(144)}{20} = 155.4$$

3.27 Total money spent by 10 persons $= \sum x = n\bar{x} = 10(85.50) = \$855$

3.28 Total 2002 incomes of five families $= \sum x = n\bar{x} = 5(69{,}520) = \$347{,}600$

3.29 Sum of the ages of six persons $= 6 \times 46 = 276$ years, so the age of sixth person $= 276 - (57 + 39 + 44 + 51 + 37) = 48$ years.

3.30 Sum of the prices paid by the seven passengers $= 7 \times 361 = \$2527$

Total price paid by the couple $= 2527 - (402 + 210 + 333 + 695 + 485) = \$384$

Price paid by each of the couple $= 384 / 2 = \$192$

3.31 For Data Set I: Mean $= 123 / 5 = 24.60$ For Data Set II: Mean $= 158 / 5 = 31.60$

The mean of the second data set is greater than the mean of the first data set by 7.

3.32 For Data Set I: Mean $= 47 / 5 = 9.40$ For Data Set II: Mean $= 94 / 5 = 18.80$

The mean of the second data set is twice the mean of the first data set.

3.33 The ranked data are: 19   23   26   31   38   39   47   49   53   67

By dropping 19 and 67, we obtain: $\sum x = 23 + 26 + 31 + 38 + 39 + 47 + 49 + 53 = 306$

10% Trimmed Mean $= (\sum x) / n = 306 / 8 = 38.25$ years

3.34   The ranked data are: 165   179   184   187   195   195   238   245   259   271

278   297   307   309   323   369   390   390   410   457

To calculate the 20% trimmed mean, drop 20% of the smallest values and 20% of the largest values. This data set contains 20 values, and 20% of 20 is 4. Hence, drop the 4 smallest values and the 4 largest values. By dropping 165, 179, 184, 187, 390, 390, 410, and 457, we obtain:

$\sum x = 195 + 195 + 238 + 245 + 259 + 271 + 278 + 297 + 307 + 309 + 323 + 369 = 3286$

20% Trimmed Mean $= (\sum x) / n = 3286 / 12 = \$273.83$ thousand $= \$273,830$

3.35   From the given information: $x_1 = 73$, $x_2 = 67$, $x_3 = 85$, $w_1 = w_2 = 1$, $w_3 = 2$

Weighted mean $= \dfrac{\sum xw}{\sum w} = \dfrac{(1)(73) + (1)(67) + (2)(85)}{4} = \dfrac{310}{4} = 77.5$

3.36   From the given information: $n_1 = 10$, $n_2 = 8$, $\bar{x}_1 = \$95$, $\bar{x}_2 = \$104$

Geometric mean $= \sqrt[n]{x_1 \cdot x_2 \cdot x_3 \cdot \ldots \cdot x_n} = \sqrt[5]{1.04 \cdot 1.03 \cdot 1.05 \cdot 1.06 \cdot 1.08} = \sqrt[5]{1.287625248} = 1.05$

The inflation rate is $1 -$ Geometric mean $= 1.05 - 1 = .05 = 5\%$ inflation

3.37   Suppose the monthly income of five families are: \$1445   \$2310   \$967   \$3195   \$24,500

Then, Range = Largest value – Smallest value $= 24,500 - 967 = \$23,533$

Now, if we drop the outlier (\$24,500) and calculate the range, then:

Range = Largest value – Smallest value $= 3195 - 967 = \$2228$

Thus, when we drop the outlier (\$24,500), the range decreases from \$23,533 to \$2228. This exhibits the sensitivity of the range with respect to outliers.

3.38   No, the value of the standard deviation cannot be negative.

3.39   The value of the standard deviation is zero when all values in a data are the same. For example, suppose the scores of a sample of six students in an examination are: 87   87   87   87   87   87

As this data set has no variation, the value of the standard deviation is zero for these observations. This is shown below.

$\sum x = 522$ and $\sum x^2 = 45,414$

$$s = \sqrt{\frac{\sum x^2 - \frac{(\sum x)^2}{n}}{n-1}} = \sqrt{\frac{45{,}414 - \frac{(522)^2}{6}}{6-1}} = 0$$

3.40   A summary measure calculated for a population data set is called a population parameter. For example, suppose the mean age of all students enrolled in a class is 23.5 years. Then, this is an example of a population parameter.

A summary measure calculated for a sample data set is called a sample statistic. For example, suppose the mean age of a sample of 20 students selected from a university is 23 years. Then, this is an example of a sample statistic.

3.41   Range = Largest value – Smallest value = $16 - (-9) = 25$, $\sum x = 24$, $\sum x^2 = 564$ and $N = 8$

$$\sigma^2 = \frac{\sum x^2 - \frac{(\sum x)^2}{N}}{N} = \frac{564 - \frac{(24)^2}{8}}{8} = \frac{564 - 72}{8} = 61.5 \qquad \text{and} \qquad \sigma = \sqrt{61.5} = 7.84$$

3.42   Range = Largest value – Smallest value = $18 - (-16) = 34$, $\sum x = 22$, $\sum x^2 = 1004$, and $n = 6$

$$s^2 = \frac{\sum x^2 - \frac{(\sum x)^2}{n}}{n-1} = \frac{1004 - \frac{(22)^2}{6}}{6-1} = \frac{1004 - 80.6667}{5} = 184.6667 \text{ and } \quad s = \sqrt{184.6667} = 13.59$$

3.43   a. $\bar{x} = (\sum x)/n = 72/8 = 9$ shoplifters caught

Shoplifters caught	Deviations from the Mean
7	7 – 9 = –2
10	10 – 9 = 1
8	8 – 9 = –1
3	3 – 9 = –6
15	15 – 9 = 6
12	12 – 9 = 3
6	6 – 9 = –3
11	11 – 9 = 2
	Sum = 0

The sum of the deviations from the mean is zero.

b.   Range = Largest value – Smallest value = $15 - 3 = 12$, $\sum x = 72$, $\sum x^2 = 748$, and $n = 8$

$$s^2 = \frac{\sum x^2 - \frac{(\sum x)^2}{n}}{n-1} = \frac{748 - \frac{(72)^2}{8}}{8-1} = 14.2857 \qquad \text{and} \qquad s = \sqrt{14.2857} = 3.78$$

3.44 a. $\bar{x} = (\sum x) / n = 551/ 7 = \$78.71$

Prices	Deviations from the Mean
$89	89 – 78.71 = 10.29
$57	57 – 78.71 = –21.71
$104	104 – 78.71 = 25.29
$73	73 – 78.71 = –5.71
$26	26 – 78.71 = –52.71
$121	121 – 78.71 = 42.29
$81	81 – 78.71 = 2.29
	Sum = 0

  b. $\sum x = 551$, $\sum x^2 = 49{,}193$, and $n = 7$

   Range = 121 – 26 = \$95

$$s^2 = \frac{\sum x^2 - \frac{(\sum x)^2}{n}}{n-1} = \frac{49{,}193 - \frac{(551)^2}{7}}{7-1} = 970.2381 \qquad \text{and} \qquad s = \sqrt{970.2381} = \$31.15$$

3.45 $\sum x = 81$, $\sum x^2 = 699$, and $n = 12$

   Range = Largest value – Smallest value = 15 – 2 = 13 thefts

$$s^2 = \frac{\sum x^2 - \frac{(\sum x)^2}{n}}{n-1} = \frac{699 - \frac{(81)^2}{12}}{12-1} = 13.8409 \qquad \text{and} \qquad s = \sqrt{13.8409} = 3.72 \text{ thefts}$$

3.46 Range = Largest value – Smallest value = 36.7 – 3 = \$33.7 billion

$$s^2 = \frac{\sum x^2 - \frac{(\sum x)^2}{n}}{n-1} = \frac{3845.13 - \frac{(147.5)^2}{8}}{8-1} = 160.7998 \qquad \text{and} \qquad s = \sqrt{160.7998} = \$12.68 \text{ billion}$$

3.47 Range = Largest value – Smallest value = 41 – 14 = 27 pieces

$$s^2 = \frac{\sum x^2 - \frac{(\sum x)^2}{n}}{n-1} = \frac{9171 - \frac{(291)^2}{10}}{10-1} = 78.1 \qquad \text{and} \qquad s = \sqrt{78.1} = 8.84 \text{ pieces}$$

3.48 Range = Largest value – Smallest value = 10 – 2 = 8 collisions

$$s^2 = \frac{\sum x^2 - \frac{(\sum x)^2}{n}}{n-1} = \frac{397 - \frac{(55)^2}{8}}{8-1} = 7.6111 \qquad \text{and} \qquad s = \sqrt{7.6111} = 2.76 \text{ collisions}$$

3.49 Range = Largest value – Smallest value = 25 – 5 = 20 pounds

$$s^2 = \frac{\sum x^2 - \frac{(\sum x)^2}{n}}{n-1} = \frac{2480 - \frac{(174)^2}{15}}{15-1} = 32.9714 \qquad \text{and} \qquad s = \sqrt{32.9714} = 5.74 \text{ pounds}$$

3.50   Range = Largest value – Smallest value = 79 – 68 = 11 miles per hour

$$s^2 = \frac{\sum x^2 - \frac{(\sum x)^2}{n}}{n-1} = \frac{69,455 - \frac{(949)^2}{13}}{13-1} = 14.8333 \quad \text{and} \quad s = \sqrt{14.8333} = 3.85 \text{ miles per hour}$$

3.51   Range = Largest value – Smallest value = 23 – (–7) = 30° Fahrenheit

$$s^2 = \frac{\sum x^2 - \frac{(\sum x)^2}{n}}{n-1} = \frac{1552 - \frac{(80)^2}{8}}{8-1} = 107.4286 \quad \text{and} \quad s = \sqrt{107.4286} = 10.36° \text{ Fahrenheit}$$

3.52   Range = Largest value – Smallest value = 14 – 0 = 14 hours

$$s^2 = \frac{\sum x^2 - \frac{(\sum x)^2}{n}}{n-1} = \frac{580 - \frac{(64)^2}{10}}{10-1} = 18.9333 \quad \text{and} \quad s = \sqrt{18.9333} = 4.35 \text{ hours}$$

3.53   Range = Largest value – Smallest value = 83.4 – 31.2 = $52.2 sales in billions

$$s^2 = \frac{\sum x^2 - \frac{(\sum x)^2}{n}}{n-1} = \frac{23615.58 - \frac{(459.6)^2}{10}}{10-1} = 276.9293 \quad \text{and} \quad s = \sqrt{276.9293} = \$16.64 \text{ billion}$$

3.54   Range = Largest value – Smallest value = 5,841 – 660 = $5181 thousand

$$s^2 = \frac{\sum x^2 - \frac{(\sum x)^2}{n}}{n-1} = \frac{138,629,656 - \frac{(33412)^2}{11}}{11-1} = 3714222.4727$$

$$s = \sqrt{3714222.4727} = \$1,927.23 \text{ thousand}$$

3.55   From the given data: $\sum x = 96$, $\sum x^2 = 1152$, and $n = 8$

$$s = \sqrt{\frac{\sum x^2 - \frac{(\sum x)^2}{n}}{n-1}} = \sqrt{\frac{1152 - \frac{(96)^2}{8}}{8-1}} = \sqrt{\frac{1152 - 1152}{7}} = 0$$

The standard deviation is zero because all these data values are the same and there is no variation among them.

3.56   For the given data: $\sum x = 114$, $\sum x^2 = 2166$, and $n = 6$

$$s = \sqrt{\frac{\sum x^2 - \frac{(\sum x)^2}{n}}{n-1}} = \sqrt{\frac{2166 - \frac{(114)^2}{6}}{6-1}} = \sqrt{\frac{2166 - 2166}{5}} = 0$$

The standard deviation is zero because all these data values are the same and there is no variation among them.

3.57 For the yearly salaries of all employees: $CV = (\sigma/\mu) \times 100\% = (3{,}820/42{,}350) \times 100 = 9.02\%$

For the years of schooling of these employees: $CV = (\sigma/\mu) \times 100\% = (2/15) \times 100 = 13.33\%$

The relative variation in salaries is lower than that in years of schooling.

3.58 For the **SAT** scores of the 100 students: $CV = (s/\bar{x}) \times 100\% = (105/915) \times 100 = 11.48\%$

For the **GPA's** of these students: $CV = (s/\bar{x}) \times 100\% = (.22/3.06) \times 100 = 7.19\%$

The relative variation in **SAT** scores is higher than that in **GPA's.**

3.59 For Data Set I: $\Sigma x = 123$, $\Sigma x^2 = 3883$, and $n = 5$

$$s = \sqrt{\frac{\Sigma x^2 - \frac{(\Sigma x)^2}{n}}{n-1}} = \sqrt{\frac{3883 - \frac{(123)^2}{5}}{5-1}} = \sqrt{214.300} = 14.64$$

For Data Set II: $\Sigma x = 158$, $\Sigma x^2 = 5850$, and $n = 5$

$$s = \sqrt{\frac{\Sigma x^2 - \frac{(\Sigma x)^2}{n}}{n-1}} = \sqrt{\frac{5850 - \frac{(158)^2}{5}}{5-1}} = \sqrt{214.300} = 14.64$$

The standard deviations of the two data sets are equal.

3.60 For Data Set I: $\Sigma x = 47$, $\Sigma x^2 = 507$, and $n = 5$

$$s = \sqrt{\frac{\Sigma x^2 - \frac{(\Sigma x)^2}{N}}{N}} = \sqrt{\frac{507 - \frac{(47)^2}{5}}{5}} = \sqrt{13.04} = 3.61$$

For Data Set II: $\Sigma x = 94$, $\Sigma x^2 = 2028$, and $n = 5$

$$s = \sqrt{\frac{\Sigma x^2 - \frac{(\Sigma x)^2}{N}}{N}} = \sqrt{\frac{2028 - \frac{(94)^2}{5}}{5}} = \sqrt{52.16} = 7.22$$

The standard deviation of the second data set is twice the standard deviation of the first data set.

3.61 The values of the mean and standard deviation for a grouped data set are the approximate values of the mean and standard deviation. The exact values of the mean and standard deviation are obtained only when ungrouped data are used.

3.62

Classes	f	m	mf	m^2f
2– 4	5	3	15	45
5– 7	9	6	54	324
8–10	14	9	126	1134
11–13	7	12	84	1008
14–16	5	15	75	1125
$N = \Sigma f = 40$			$\Sigma mf = 354$	$\Sigma m^2 f = 3636$

$$\mu = (\Sigma mf) / N = 354 / 40 = 8.85$$

$$\sigma^2 = \frac{\Sigma m^2 f - \dfrac{(\Sigma mf)^2}{N}}{N} = \frac{3636 - \dfrac{(354)^2}{40}}{40} = 12.5775 \qquad \text{and} \qquad \sigma = \sqrt{12.5775} = 3.55$$

3.63 For this given data: $n = 80$, $\Sigma mf = 752$, and $\Sigma m^2 f = 10{,}048$

$$\bar{x} = (\Sigma mf) / n = 752 / 80 = 9.40$$

$$s^2 = \frac{\Sigma m^2 f - \dfrac{(\Sigma mf)^2}{n}}{n-1} = \frac{10{,}048 - \dfrac{(752)^2}{80}}{80-1} = 37.7114 \qquad \text{and} \qquad s = \sqrt{37.7114} = 6.14$$

3.64 For the given data: $n = 50$, $\Sigma mf = 5420$, and $\Sigma m^2 f = 651{,}050$

$$\bar{x} = (\Sigma mf) / n = 5420 / 50 = \$108.40$$

$$s^2 = \frac{\Sigma m^2 f - \dfrac{(\Sigma mf)^2}{n}}{n-1} = \frac{651{,}050 - \dfrac{(5420)^2}{50}}{50-1} = 1296.3673 \qquad \text{and} \qquad s = \sqrt{1296.3673} = \$36.01$$

3.65 For the given data: $N = 30$, $\Sigma mf = 2640$, and $\Sigma m^2 f = 273{,}150$

$$\mu = (\Sigma mf) / N = 2640 / 30 = 88 \text{ hours}$$

$$\sigma^2 = \frac{\Sigma m^2 f - \dfrac{(\Sigma mf)^2}{N}}{N} = \frac{273{,}150 - \dfrac{(2640)^2}{30}}{30} = 1361 \qquad \text{and} \qquad \sigma = \sqrt{1361} = 36.89 \text{ hours}$$

3.66 For the given data: $N = 100$, $\Sigma mf = 780$, and $\Sigma m^2 f = 6440$

$$\mu = (\Sigma mf) / N = 780 / 100 = 7.80 \text{ pounds}$$

$$\sigma^2 = \frac{\Sigma m^2 f - \dfrac{(\Sigma mf)^2}{N}}{N} = \frac{6440 - \dfrac{(780)^2}{100}}{100} = 3.5600 \qquad \text{and} \qquad \sigma = \sqrt{3.5600} = 1.89 \text{ pounds}$$

3.67 For the given data: $n = 300$, $\Sigma mf = 5900$, and $\Sigma m^2 f = 136{,}275$

$$\bar{x} = (\Sigma mf) / n = 5900 / 300 = 19.67 \text{ or } 19{,}670 \text{ miles}$$

$$s^2 = \frac{\sum m^2 f - \frac{(\sum mf)^2}{n}}{n-1} = \frac{136{,}275 - \frac{(5900)^2}{300}}{300-1} = 67.6979 \quad \text{and} \quad s = \sqrt{67.6979} = 8.23 \text{ or } 8230 \text{ miles}$$

Each value in the column labeled *mf* gives the approximate total mileage for the car owners in the corresponding class. For example, the value of *mf* =17.5 for the first class indicates that the seven car owners in this class drove a total of approximately 17,500 miles. The value $\sum mf$ = 5900 that the total mileage for all 300 car owners was approximately 5,900,000 miles.

3.68   For the given data: $n = 50$, $\sum mf = 2500$, and $\sum m^2 f = 156{,}200$

$\bar{x} = (\sum mf) / n = 2500 / 50 = \$50$

$$s^2 = \frac{\sum m^2 f - \frac{(\sum mf)^2}{n}}{n-1} = \frac{156{,}200 - \frac{(2500)^2}{50}}{50-1} = 636.7347 \quad \text{and} \quad s = \sqrt{636.7347} = \$25.23$$

The values in the column labeled *mf* give the approximate amounts of electric bills for the families belonging to corresponding classes. For example, the five families belonging to the first class paid a total of $50 for electricity in August 2002. The value $\sum mf$ = $2500 is the approximate total amount of the electric bills for all 50 families included in the sample.

3.69   For the given data: $n = 50$, $\sum mf = 1840$, and $\sum m^2 f = 97{,}000$

$\bar{x} = (\sum mf) / n = 1840 / 50 = 36.80$ minutes

$$s^2 = \frac{\sum m^2 f - \frac{(\sum mf)^2}{n}}{n-1} = \frac{97{,}000 - \frac{(1840)^2}{50}}{50-1} = 597.7143 \quad \text{and} \quad s = \sqrt{597.7143} = 24.45 \text{ minutes}$$

3.70   For the given data: $n = 45$, $\sum mf = 70$, and $\sum m^2 f = 186$

$\bar{x} = (\sum mf) / n = 70 / 45 = 1.56$ errors

$$s^2 = \frac{\sum m^2 f - \frac{(\sum mf)^2}{n}}{n-1} = \frac{186 - \frac{(70)^2}{45}}{45-1} = 1.7525 \quad \text{and} \quad s = \sqrt{1.7525} = 1.32 \text{ errors}$$

3.71   For the given data, $n = 15$,

a. $\bar{x} = \sum x / n = 2142.5 / 15 = 142.83$ cents per gallon

b.

Price of Gas (in pennies)	Frequency
137 to less than 140	5
140 to less than 143	2
143 to less than 146	4
146 to less than 149	3
149 to less than 152	1

c. For the given data: $n = 15$, $\sum mf = 2146.5$,

$\bar{x} = (\sum mf) / n = 2146.5 / 15 = 143.1$ cents per gallon.

d. The two means are not equal because the second method uses approximations (mid points of the range) and the first one does not. This leads to slightly different results.

3.72   Chebyshev's theorem is applied to find the area under a distribution curve between two points that are on opposite sides of the mean and at the same distance from the mean. According to this theorem, for any number k greater than 1, at least $(1 - (1/k^2))$ of the data values lie within k standard deviations of the mean.

3.73   The empirical rule is applied to a bell–shaped distribution. According to this rule, approximately
(1) 68% of the observations lie within one standard deviation of the mean.
(2) 95% of the observations lie within two standard deviations of the mean.
(3) 99.7% of the observations lie within three standard deviations of the mean.

3.74   For the interval $\bar{x} \pm 2s$: $k = 2$, and $1 - \dfrac{1}{k^2} = 1 - \dfrac{1}{(2)^2} = 1 - .25 = .75$ or 75%. Thus, at least 75% of the observations fall in the interval $\bar{x} \pm 2s$.

For the interval $\bar{x} \pm 2.5s$: $k = 2.5$, and $1 - \dfrac{1}{k^2} = 1 - \dfrac{1}{(2.5)^2} = 1 - .16 = .84$ or 84%. Thus, at least 84% of the observations fall in the interval $\bar{x} \pm 2.5s$.

For the interval $\bar{x} \pm 3s$: $k = 3$, and $1 - \dfrac{1}{k^2} = 1 - \dfrac{1}{(3)^2} = 1 - .11 = .89$ or 89%. Thus, at least 89% of the observations fall in the interval $\bar{x} \pm 3s$.

3.75   For the interval $\mu \pm 2\sigma$: $k = 2$, and $1 - \dfrac{1}{k^2} = 1 - \dfrac{1}{(2)^2} = 1 - .25 = .75$ or 75%. Thus, at least 75% of the observations fall in the interval $\mu \pm 2\sigma$.

For the interval $\mu \pm 2.5\sigma$: $k = 2.5$, and $1 - \dfrac{1}{k^2} = 1 - \dfrac{1}{(2.5)^2} = 1 - .16 = .84$ or 84%. Thus, at least 84% of the observations fall in the interval $\mu \pm 2.5\sigma$.

For the interval $\mu \pm 3\sigma$: $k = 3$, and $1 - \dfrac{1}{k^2} = 1 - \dfrac{1}{(3)^2} = 1 - .11 = .89$ or 89%. Thus, at least 89% of the observations fall in the interval $\mu \pm 3\sigma$.

3.76	Approximately 68% of the observations fall in the interval $\mu \pm 1\sigma$, approximately 95% fall in the interval $\mu \pm 2\sigma$, and about 99.7% fall in the interval $\mu \pm 3\sigma$.

3.77	Approximately 68% of the observations fall in the interval $\mu \pm 1s$, approximately 95% fall in the interval $\mu \pm 2s$, and about 99.7% fall in the interval $\mu \pm 3s$.

3.78	a.	Each of the two values is 40 minutes from $\mu = 220$. Hence, $k = 40 / 20 = 2$

$$1 - \frac{1}{k^2} = 1 - \frac{1}{(2)^2} = 1 - .25 = .75 \text{ or } 75\%.$$

Thus, at least 75% of the runners ran the race in 180 to 260 minutes.

b.	Each of the two values is 60 minutes from $\mu = 220$. Hence,

$$k = 60 / 20 = 3 \qquad \text{and} \qquad 1 - \frac{1}{k^2} = 1 - \frac{1}{(3)^2} = 1 - .11 = .89 \text{ or } 89\%.$$

Thus, at least 89% of the runners ran the race in 160 to 280 minutes.

c.	Each of the two values is 50 minutes from $\mu = 220$. Hence,

$$k = 50 / 20 = 2.5 \qquad \text{and} \qquad 1 - \frac{1}{k^2} = 1 - \frac{1}{(2.5)^2} = 1 - .16 = .84 \text{ or } 84\%.$$

Thus, at least 84% of the runners ran this race in 170 to 270 minutes.

3.79	a.	Each of the two values is $1.2 million from $\mu = \$2.3$ million. Hence,

$$k = 1.2 / .6 = 2 \qquad \text{and} \qquad 1 - \frac{1}{k^2} = 1 - \frac{1}{(2)^2} = 1 - .25 = .75 \text{ or } 75\%.$$

Thus, at least 75% of all firms had 1999 gross sales of $1.1 to $3.5 million.

b.	Each of the two values is $1.5 million from $\mu = \$2.3$ million. Hence,

$$k = 1.5 / .6 = 2.5 \qquad \text{and} \qquad 1 - \frac{1}{k^2} = 1 - \frac{1}{(2.5)^2} = 1 - .16 = .84 \text{ or } 84\%.$$

Thus, at least 84% of all firms had 1999 gross sales of $.8 to $3.8 million.

c. Each of the two values is $1.8 million from μ = $2.3 million. Hence,

$$k = 1.8 / .6 = 3 \qquad \text{and} \qquad 1 - \frac{1}{k^2} = 1 - \frac{1}{(3)^2} = 1 - .11 = .89 \text{ or } 89\%.$$

Thus, at least 89% of all firms had 1999 gross sales of $.5 to $ 4.1 million.

3.80 μ = $8367 and σ = $2400

a. i. Each of the two values is $4800 from μ = $8367. Hence,

$$k = 4800 / 2400 = 2 \text{and} \qquad 1 - \frac{1}{k^2} = 1 - \frac{1}{(2)^2} = 1 - .25 = .75 \text{ or } 75\%.$$

Thus, at least 75% of all households have credit card debt between $3567 and $13,167.

ii. Each of the two values is $6000 from μ = $8367. Hence,

$$k = 6000 / 2400 = 2.5 \qquad \text{and} \qquad 1 - \frac{1}{k^2} = 1 - \frac{1}{(2.5)^2} = 1 - .16 = .84 \text{ or } 84\%.$$

Thus, at least 84% of all households have credit card debt between $2,367 and $14, 367.

b. $1 - \frac{1}{k^2} = .89\,\text{gives}\,\frac{1}{k^2} = 1 - .89 = .11 \text{ or } k^2 = \frac{1}{.11}$, so $k = 3$.

$$\mu - 3\sigma = 8367 - 3(2400) = \$1167 \qquad \text{and} \qquad \mu + 3\sigma = 8367 + 3(2400) = \$15,567$$

Thus, the required interval is $1167 to $15,567.

3.81 a. i. Each of the two values is $480 from μ = $1365. Hence,

$$k = 480 / 240 = 2 \qquad \text{and} \qquad 1 - \frac{1}{k^2} = 1 - \frac{1}{(2)^2} = 1 - .25 = .75 \text{ or } 75\%.$$

Thus, at least 75% of all homeowners pay a monthly mortgage of $885 to $1845.

ii. Each of the two values is $720 from μ = $1365. Hence,

$$k = 720 / 240 = 3 \quad \text{and} \qquad 1 - \frac{1}{k^2} = 1 - \frac{1}{(3)^2} = 1 - .11 = .89 \text{ or } 89\%.$$

Thus, at least 89% of all homeowners pay a monthly mortgage of $645 to $2085.

b. $1 - \frac{1}{k^2} = .84\,\text{gives}\,\frac{1}{k^2} = 1 - .84 = .16 \text{ or } k^2 = 1/16 \text{ so } k = 2.5$

$$\mu - 2.5\sigma = 1365 - 2.5(240) = \$765 \qquad \text{and} \qquad \mu + 2.5\sigma = 1365 + 2.5(240) = \$1965$$

Thus, the required interval is $765 to $1965.

3.82 $\mu = 44$ months and $\sigma = 3$ months.

   a.   The interval 41 to 47 months is $\mu - \sigma$ to $\mu + \sigma$. Hence, approximately 68% of the batteries have a life of 41 to 47 months.

   b.   The interval 38 to 50 months is $\mu - 2\sigma$ to $\mu + 2\sigma$. Hence, approximately 95% of the batteries have a life of 38 to 50 months.

   c.   The interval 35 to 53 months is $\mu - 3\sigma$ to $\mu + 3\sigma$. Hence, approximately 99.7% of the batteries have a life of 35 to 53 months.

3.83 $\mu = \$24,317$ and $\sigma = \$2,000$

   a.   The interval \$20,317 to \$28,317 is $\mu - 2\sigma$ to $\mu + 2\sigma$. Hence, approximately 95% of teacher assistants in Connecticut have annual salaries between \$20,317 and \$28,317.

   b.   The interval \$18,317 to \$30,317 is $\mu - 3\sigma$ to $\mu + 3\sigma$. Hence, approximately 99.7% of teacher assistants in Connecticut have annual salaries between \$18,317 and \$30,317.

   c.   The interval \$22,317 to \$26,317 is $\mu - \sigma$ to $\mu + \sigma$. Hence, approximately 68% of teacher assistants in Connecticut have annual salaries between \$22,317 and \$26,317.

3.84 $\mu = 16$ hours of housework per week and $\sigma = 3.5$ hours

   a.   i.   The interval 12.5 to 19.5 hours is $\mu - \sigma$ to $\mu + \sigma$. Hence, approximately 68% of all men in the U.S. spent 12.5 to 19.5 hours per week on housework in 2002.

       ii.   The interval 9 to 23 hours is $\mu - 2\sigma$ to $\mu + 2\sigma$. Hence, approximately 95% of all men in the U.S. spent 9 to 23 hours per week on housework in 2002.

   b.   The interval that contains 99.7% of U.S. men's housework hours is $\mu - 3\sigma$ to $\mu + 3\sigma$. Hence, this interval is $16 - 3(3.5)$ to $16 + 3(3.5)$ or 5.5 to 26.5 hours of housework per week.

3.85 $\mu = 72$ MPH and $\sigma = 3$ MPH

a. i. The interval 63 to 81 MPH $\mu - 3\sigma$ to $\mu + 3\sigma$. Hence, about 99.7% of speeds of all vehicles are between 63 and 81 MPH.

    ii. The interval 69 to 75 MPH is $\mu - \sigma$ to $\mu + \sigma$. Hence, about 68% of the speeds of all vehicles are between 69 and 75 MPH.

b. The interval that contains the speeds of 95% of the vehicles is $\mu - 2\sigma$ to $\mu + 2\sigma$. Hence, this interval is $72 - 2(3)$ to $72 + 2(3)$ or 66 to 78 MPH.

3.86   To find the three quartiles:

1. Rank the given data set in increasing order.
2. Find the median by the procedure in Section 3.1.2. The median is the second quartile, Q_2.
3. The first quartile, Q_1, is the value of the middle term among the (ranked) observations that are less than Q_2.
4. The third quartile, Q_3, is the value of the middle term among the (ranked) observations that are greater that Q_2.

    Example 3–20 and 3–21 of the text exhibit how to calculate the three quartiles for data sets with an even and odd number of observations, respectively.

3.87   The interquartile range (IQR) is given by $Q_3 - Q_1$, where Q_1 and Q_3 are the first and third quartiles, respectively. Examples 3–20 and 3–21 of the text show how to find the IQR for a data set.

3.88   Given a data set of n values, to find the kth percentile (P_k):

1. Rank the given data in increasing order.
2. Calculate $kn/100$. Then, P_k is the term that is approximately ($kn/100$) in the ranking. If $kn/100$ falls between two consecutive integers a and b, it may be necessary to average the ath and b^{th} values in the ranking to obtain P_k.

3.89   If x_i is a particular observation in the data set, the percentile rank of x_i is the percentage of the values in the data set that are less than x_i. Thus,

$$\text{Percentile rank of } x_i = \frac{\text{Number of values less than } x_i}{\text{Total number of values in the data set}} \times 100$$

3.90   The ranked data are: 5 5 7 8 8 9 10 10 11 11 12 14 18 21 25

a. The three quartiles are: $Q_1 = 8$, $Q_2 = 10$, and $Q_3 = 14$

    $IQR = Q_3 - Q_1 = 14 - 8 = 6$

b. $kn/100 = 82(15)/100 = 12.30 \approx 12$

Thus, the 82nd percentile can be approximated by the value of the 12th term in the ranked data, which is 14. Therefore, $P_{82} = 14$.

   c. Six values in the given data are smaller than 10. Hence, percentile rank of $10 = (6/15) \times 100 = 40\%$

3.91   The ranked data are: 68 68 69 69 71 72 73 74 75 76 77 78 79

   a. The three quartiles are: $Q_1 = (69 + 69) / 2 = 69$, $Q_2 = 73$, and $Q_3 = (76 + 77) / 2 = 76.5$
     $IQR = Q_3 - Q_1 = 76.5 - 69 = 7.5$

   b. $kn/ 100 = 35(13) / 100 = 4.55 \approx 5$
     Thus, the 35th percentile can be approximated by the value of the 5th term in the ranked data, which is 71. Therefore, $P_{35} = 71$.

   c. Four values in the given data set are smaller than 71. Hence, the percentile rank of
     $71 = (4/13) \times 100 = 30.77\%$.

3.92   The ranked data are: 41 42 43 44 44 45 46 46 47 47 48 48
                                      48 49 50 50 51 51 52 52 52 53 53 54 56

   a. The three quartiles are: $Q_1 = (45 + 46) / 2 = 45.5$, $Q_2 = 48$, and $Q_3 = (52 + 52) / 2 = 52$
     $IQR = Q_3 - Q_1 = 52 - 45.5 = 6.5$

   b. $kn/100 = 53(25)/100 = 13.25 \approx 13$
     Thus, the 53rd percentile can be approximated by the value of the thirteenth term in the ranked data, which is 48. Therefore, $P_{53} = 48$.

   c. Fourteen values in the given data are less than 50. Therefore,
     Percentile rank of $50 = (14/25) \times 100 = 56\%$

3.93   The ranked data are: 16 21 23 27 31 34 34 35 36 36 38 39 39 40 40
                                      40 40 40 40 41 42 42 43 45 47 48 48 51 51 53

   a. The quartiles are : $Q_1 = 35$, $Q_2 = (40 + 40)/ 2 = 40$, and $Q_3 = 43$
     $IQR = Q_3 - Q_1 = 43 - 35 = 8$

   b. $kn/100 = 79(30) / 100 = 23.70 \approx 24$

Thus, the 79th percentile can be approximated by the value of the 24th term in the ranked data, which is 45. Therefore, $P_{79} = 45$.

    c. Eleven values in the given data are smaller than 39. Hence, percentile rank of $39 = (11/30) \times 100 = 36.67\%$.

3.94   The ranked data are: 3 5 6 6 7 9 9 10 11 12 14 15

    a. The three quartiles are:

        $Q_1 = (6+6)/2 = 6$, $Q_2 = (9 +9)/ 2 = 9$, and $Q_3 = (11 +12) /2 = 11.5$

        $IQR = Q_3 - Q_1 = 11.5 - 6 = 5.5$

        The value 10 lies between Q_2 and Q_3, which means it lies in the third 25% group from the bottom in the ranked data set.

    b. $kn/100 = 55(12)/100 = 6.6$

        Thus, the 55th percentile can be approximated by the average of the six and seventh terms in the ranked data. Therefore, $P_{55} = (9 + 9) /2 = 9$.

    c. Four values in the given data set are less than 7. Hence, percentile rank of 7 is $(4/12) \times 100 = 33.33\%$.

3.95   The ranked data are: 20 22 23 23 23 23 24 25 26 26 27 27 27 28 28

                                                29 29 31 31 31 32 33 33 33 34 35 35 36 37 43

    a. The three quartiles are: $Q_1 = 25$, $Q_2 = (28+ 29)/ 2 = 28.5$,   and $Q_3 = 33$

        $IQR = Q_3 - Q_1 = 33 - 25 = 8$

        The value 31 lies between Q_2 and Q_3, which means that it is in the third 25% group from the bottom in the (ranked) data set.

    b. $kn/ 100 = 65(30)/100 \approx 19.5$

        Thus, the 65th percentile may be approximated by the average of the nineteenth and twentieth terms in the ranked data. Therefore, $P_{65} = (31 = 31) /2 = 31$.

        Thus, we can state that the number of computer terminals produced by Nixon Corporation is less than 31 for approximately 65% of the days in this sample.

    c. Twenty values in the given data are less than 32. Hence, percentile rank 32 = (20/30) x100 = 66.67%. Thus, on 66.67% of the days in this sample, fewer than 32 terminals were produced. Hence, for 100 – 66.67 = 33.33% of the days, the company produced 32 or more terminals.

3.96   The ranked data are: 3 3 4 5 5 6 7 7 8 8 8 9 9 10 10 11 11 12 12 16

    a. The three quartiles are: $Q_1 = (5 + 6)/2 = 5.5$, $Q_2 = (8 + 8)/2 = 8$, and $Q_3 = (10 + 11)/2 = 10.5$

        $IQR = Q_3 - Q_1 = 10.5 - 5.5 = 5$

        The value 4 lies below Q_1, which indicates that it is in the bottom 25% of the value in the (ranked) data set.

    b. *kn*/ 100 = 25(20) / 100 = 5

        Thus, the 25th percentile may be approximated by the value of the fifth term in the ranked data, which is 5. Therefore, $P_{25} = 5$. Thus, the number of new cars sold at this dealership is less than 5 for approximately 25% of the days in this sample.

    c. Thirteen values in the given data are less than 10. Hence, percentile rank of 10 = (13/20) × 100 = 65%. Thus, on 65% of the days in the sample, this dealership sold fewer than 10 cars.

3.97   The ranked data in thousands of dollars are:

        9.4   13.1   14.9   15.3   17.2   18.1   21.3   21.6   21.8   22

        23.9   24   27.4   33   35.1   37.8   42.1   44.9   50.3

    a. The three quartiles are: $Q_1 = 17.2$, $Q_2 = 22$,   and $Q_3 = 35.1$

        $IQR = Q_3 - Q_1 = 35.1 - 17.2 = 17.9$

        The value 23.9 lies between Q_2 and Q_3, which indicates that it is in the second 25% group from the bottom in the (ranked) data set between 22 and 35.1.

    b. *kn*/100 = 89(19)/100 = 16.91 ≈ 17

        Thus, the 89th percentile may be approximated by the value of the 17th term in the ranked data, which is $42,100. Therefore, $P_{89} = \$42,100$. Thus, approximately 89% of the wedding costs are below $42,100.

    c. Eleven values in the given data are smaller that $24,000. Hence, percentile rank of $24,000 = (11/19) × 100 = 57.89%. Thus, 57.89% of the weddings costs were below $24,000.

3.98   A box–and–whisker plot is based on five summary measures: the median, the first quartile, the third quartile, and the smallest and largest value in the data set between the lower and upper inner fences.

3.99   The ranked data are: 22 24 25 28 31 32 34 35 36 41 42 43

47 49 52 55 58 59 61 61 63 65 73 98

For these data,

Median = (43 +47) /2 = 45, Q_1 = (32 + 34) / 2 = 33, and Q_3 = (59+61) /2= 60,

IQR = $Q_3 - Q_1$ = 60 – 33 = 27, 1.5 x IQR = 1.5 x 27 = 40.5,

Lower inner fence = Q_1 – 40.5 = 33 – 40.5 = – 7.5,

Upper inner fence = Q_3 + 40.5 = 60 + 40.5 = 100.5

The smallest and largest values within the two inner fences are 22 and 98, respectively. The data set has no outliers. The box–and–whisker plot is shown below.

3.100   The ranked data are: 3 6 7 8 11 13 14 15 16 18 19 23 26 29 30 31 33 42 62 75

For the data,

Median = (18+19) /2 = 18.5, Q_1 = (11 +13) / 2 = 12, and Q_3 = (30+31) /2= 30.5,

IQR = $Q_3 - Q_1$ = 30.5 – 12 = 18.5, 1.5 x IQR = 1.5 x 18.5 = 27.75,

Lower inner fence = Q_1 – 27.75 = 12 – 27.75= – 15.75,

Upper inner fence = Q_3 + 27.75= 30.5 + 27.75= 58.25

The smallest and the largest values within the two inner fences are 3 and 42, respectively.

There are two outliers, 62 and 75. To classify them, we compute:

3.0 x IQR = 3.0 x 18.5 = 55.5. Hence, the upper outer fence is: Q_3 + 55.5 = 86

Since 62 and 75 are both less than 86, they are within the upper outer fence and are called mild outliers.

3.101   The ranked data are: 3 5 5 6 8 10 14 15 16 17 17 19 21 22 23 25 30 31 31 34

Median = 17, Q_1 = 9, Q_3 = 24, IQR = $Q_3 - Q_1$ = 24 – 9 = 15, 1.5 x IQR = 1.5 x 15 = 22.5,

Lower inner fence = Q_1 – 22.5 = 9 – 22.5 = –13.5,

Upper inner fence = Q_3 + 22.5 = 24 + 22.5 = 46.5

The smallest and the largest values within the two inner fences are 3 and 34, respectively. The data set contains no outliers.

The data are skewed slightly to the right.

3.102   Median = 22,000, Q_1 = 17200, Q_3 = 35100, IQR = $Q_3 - Q_1$ =35,100 – 17,200 = 17,900,

1.5 x IQR = 1.5 x 17,200 = 26,850,

Lower inner fence = Q_1 – 26,850 = 17,200 – 26,850 = –9650,

Upper inner fence = Q_3 + 26,850 = 35,100 + 26,850 = 61,950

The smallest and the largest values within the two inner fences are 9400 and 50,300, respectively. The

data set contains no outliers.

The data are skewed to the right.

3.103   Median = 10, Q_1 = 8, Q_3 = 14, IQR = $Q_3 - Q_1$ = 14 – 8 = 6, 1.5 x IQR = 1.5 x 6 = 9,

Lower inner fence = Q_1 – 9 = 8 – 9 = –1,

Upper inner fence = Q_3 + 9 = 14 + 9 = 23

The smallest and largest values within the two inner fences are 5 and 21, respectively. The value 25 is

an outlier.

The data are skewed to the right.

3.104   Median = 48, Q_1 = 45.5, Q_3 = 52, IQR = $Q_3 - Q_1$ = 52 –45.5 = 6.5, 1.5 x IQR = 1.5 x 6.5 = 9.75,

Lower inner fence = Q_1 – 9.75 = 45.5 – 9.75 = 35.75,

Upper inner fence = Q_3 + 9.75 = 52 + 9.75 = 61.75

The smallest and largest values within the two inner fences are 41 and 56, respectively. There are no

outliers.

The data are skewed slightly to the right.

3.105   Median = 40, Q_1 = 35, Q_3 = 43, IQR = $Q_3 - Q_1$ = 43 – 35 = 8, 1.5 x IQR = 1.5 x 8 = 12,

Lower inner fence = Q_1 – 12 = 35 – 12 = 23,

Upper inner fence = Q_3 + 12 = 43 + 12 = 55

The smallest and largest values within the two inner fences are 23 and 53, respectively. The values 16 and 21 are the outliers.

The data are skewed to the left.

3.106   Median = 9, $Q_1 = 6$, $Q_3 = 11.5$, IQR = $Q_3 - Q_1 = 11.5 - 6 = 5.5$, 1.5 x IQR = 1.5 x 5.5 = 8.25,

Lower inner fence = $Q_1 - 8.25 = 6 - 8.25 = -2.75$,

Upper inner fence = $Q_3 + 8.25 = 11.5 + 8.25 = 19.75$

The smallest and largest values within the two inner fences are 3 and 15, respectively. There are no outliers.

The data are nearly symmetric.

3.107   Median = 27.5, $Q_1 = 24$, $Q_3 = 31$, IQR = $Q_3 - Q_1 = 31 - 24 = 7$, 1.5 x IQR = 1.5 x 7 = 10.5,

Lower inner fence = $Q_1 - 10.5 = 24 - 10.5 = 13.5$,

Upper inner fence = $Q_3 + 10.5 = 31 + 10.5 = 41.5$

The smallest and largest values within the two inner fences are 20 and 35, respectively. There are no outliers.

The data are nearly symmetric.

3.108   Median = 7.5, $Q_1 = 5$, $Q_3 = 9.5$, IQR = $Q_3 - Q_1 = 9.5 - 5 = 4.5$, 1.5 x IQR = 1.5 x 4.5 = 6.75,

Lower inner fence = $Q_1 - 6.75 = -1.75$ and   Upper inner fence = $Q_3 + 6.75 = 16.25$

The smallest and largest values within the two inner fences are 3 and 12, respectively. There are no outliers.

The data are nearly symmetric.

3.109   a. $\bar{x} = (\Sigma x)/n = 431 / 10 = \$43.1$ thousand $(n + 1)/2 = (10+1)/2 = 5.5$

Median = $(39 + 40)/2 = \$39.5$ thousand

b.   Yes, 84 is an outlier. After dropping this value,

$\bar{x} = (\Sigma x)/ n = 347 / 9 = \$38.56$ thousand

Median = \$39 thousand

The value of the mean changes by a larger amount.

c.   The median is a better summary measure for these data since it is influenced less by outliers.

3.110   a. $\bar{x} = (\Sigma x)/ n = 1{,}471{,}311 / 10 = \$147{,}131.10$

Median = \$147,195 and Mode = \$125,000

b.   Range = Largest value – Smallest value = 170,000 – 125,000 = \$45,000

$$s^2 = \frac{\Sigma x^2 - \dfrac{(\Sigma x)^2}{n}}{n-1} = \frac{218{,}975{,}790{,}881 - \dfrac{(1{,}471{,}311)^2}{10}}{10-1} = 277{,}798{,}334.3222$$

$$s = \sqrt{277{,}798{,}334.3222} = \$16{,}667.28$$

3.111   a. $\mu = (\Sigma x)/ n = 56{,}987 /15 = \$3799.13$ thousand

Median = \$2833 thousand

This data set has no mode because no value appears more than once.

b.   Range = Largest value – Smallest value = 13,350 –215 = \$13,135 thousand

$$\sigma^2 = \frac{\Sigma x^2 - \dfrac{(\Sigma x)^2}{N}}{N} = \frac{432{,}163{,}861 - \dfrac{(56{,}987)^2}{15}}{15} = 14{,}377{,}510$$

$$\sigma = \sqrt{14{,}377{,}510} = \$3{,}791.77 \text{ thousand}$$

3.112   a. $\bar{x} = (\Sigma x)/ n = 88/12 = 7.33$ citations $(n + 1) /2 = (12 + 1) /2 = 6.5$

Median = $(7 + 8) /2 = 7.5$ citations

Mode = 4, 7, and 8 citations

b.   Range = Largest value – Smallest value = 14 – 0 = 14 citations

$$s^2 = \frac{\Sigma x^2 - \dfrac{(\Sigma x)^2}{n}}{n-1} = \frac{834 - \dfrac{(88)^2}{12}}{12-1} = 17.1515 \qquad \text{and} \qquad s = \sqrt{17.1515} = 4.14 \text{ citations}$$

c.   The values of the summary measures in parts a and b are sample statistics because the data are based on a sample of 12 drivers.

3.113 $n = 50$, $\sum mf = 254$, $\sum m^2 f = 1626$,   and $\overline{x} = (\sum mf)/n = 254 / 50 = 5.08$ inches

$$s^2 = \frac{\sum m^2 f - \frac{(\sum mf)^2}{n}}{n-1} = \frac{1626 - \frac{(254)^2}{50}}{50-1} = 6.8506 \quad \text{and} \quad s = \sqrt{6.8506} = 2.62 \text{ inches}$$

The values of these summary measures are sample statistics since they are based on a sample of 50 cities.

3.114 $n = 50$, $\sum mf = 620$, $\sum m^2 f = 8680$ and $\overline{x} = (\sum mf)/n = 620 / 50 = 12.4$ minutes

$$s^2 = \frac{\sum m^2 f - \frac{(\sum mf)^2}{n}}{n-1} = \frac{8680 - \frac{(620)^2}{50}}{50-1} = 20.2449 \quad \text{and} \quad s = \sqrt{20.2449} = 4.50 \text{ minutes}$$

The values of these summary measures are sample statistics, since they are based on a sample of 50 students.

3.115   a   i.   Each of the two values is 40 minutes from $\mu = 200$. Hence,

$$k = 40 / 20 = 2 \quad \text{and} \quad 1 - \frac{1}{k^2} = 1 - \frac{1}{(2)^2} = 1 - .25 = .75 \text{ or } 75\%.$$

Thus, at least 75% of the students will learn the basics in 160 to 240 minutes.

ii. Each of the two values is 60 minutes from $\mu = 200$. Hence,

$$k = 60 / 20 = 3 \quad \text{and} \quad 1 - \frac{1}{k^2} = 1 - \frac{1}{(3)^2} = 1 - .11 = .89 \text{ or } 89\%.$$

Thus, at least 89% of the students will learn the basics in 140 to 260 minutes.

b. $1 - \dfrac{1}{k^2} = .75$ gives $\dfrac{1}{k^2} = 1 - .75 = .25$ or $k^2 = \dfrac{1}{.25}$, so $k = 2$

The required interval is: $\mu - k\sigma$ to $\mu + k\sigma = 200 - 2(20)$ to $200 + 2(20) = 160$ to 240 minutes.

3.116   a.   i.   Each of the two values is \$900 from $\mu = \$1100$. Hence,

$$k = 900/300 = 3 \quad \text{and} \quad 1 - \frac{1}{k^2} = 1 - \frac{1}{3^2} = 1 - .11 = .89 \text{ or } 89\%.$$

Thus, at least 89% of households will have holiday expenditures between \$200 and \$2000.

ii. Each of the two values is \$600 from $\mu = \$1100$. Hence,

$$k = 600/300 = 2 \quad \text{and} \quad 1 - \frac{1}{k^2} = 1 - \frac{1}{(2)^2} = 1 - .25 = .75 \text{ or } 75\%.$$

Thus, at least 75% of households will have holiday expenditures between \$500 and \$1700.

b. $1 - \dfrac{1}{k^2} = .84$ gives $\dfrac{1}{k^2} = 1 - .84 = .16$ or $k^2 = \dfrac{1}{.16}$, so $k = 2.5$

The required interval is:

$\mu - k\sigma$ to $\mu + k\sigma = \{1100 - 2.5(300)\}$ to $\{1100 + 2.5(300)\} = \$350$ to $\$1850$

3.117 $\mu = 200$ minutes and $\sigma = 20$ minutes

a. i. The interval 180 to 220 minutes is $\mu - \sigma$ to $\mu + \sigma$. Thus, approximately 68% of the students will learn the basics in 180 to 220 minutes.

   ii. The interval 160 to 240 minutes is $\mu - 2\sigma$ to $\mu + 2\sigma$. Hence, approximately 95% of the students will learn the basics in 160 to 240 minutes.

b. The interval that contains the learning time of 99.7% of the students is $\mu - 3\sigma$ to $\mu + 3\sigma$. Hence, this interval is: $\mu - 3\sigma$ to $\mu + 3\sigma = \{200 - 3(20)\}$ to $\{200 + 3(20)\} = 140$ to 260 minutes.

3.118 $\mu = \$134,000$ and $\sigma = \$12,000$

a. i. The interval \$98,000 to \$170,000 is $\mu - 3\sigma$ to $\mu + 3\sigma$. Thus, approximately 99.7% of CPAs have salaries between \$98,000 and \$170,000.

   ii. The interval \$110,000 to \$158,000 is $\mu - 2\sigma$ to $\mu + 2\sigma$. Thus, approximately 95% of CPAs have salaries between \$110,000 and \$158,000.

b. The interval that contains salaries of 68% of all such CPAs is $\mu - \sigma$ to $\mu + \sigma$. Hence, this interval is: $(134,000 - 12,000)$ to $(134,000 + 12,000) = \$122,000$ to $\$146,000$.

3.119 The ranked data are: 27 34 36 38 39 40 41 44 48 84

a. The three quartiles are: $Q_1 = 36$, $Q_2 = (39 + 40 / 2 = 39.5$, and $Q_3 = 44$,

$IQR = Q_3 - Q_1 = 8$

The value 40 lies below the third quartile (and above the second).

b. $kn / 100 = 70(10)/100 = 7$. Thus, 70th percentile occurs at the seventh term in the ranked data which is 41, so $P_{70} = 41$. This means that about 70% of the values in the data set are smaller than $P_{70} = 41$.

c. Five values in the given data are smaller than 40. Hence, percentile rank of 40 = (5/10) x100 = 50%. This means approximately 50% of the values in the data set are less than 40.

3.120 The ranked data are: 0 3 4 4 7 7 8 8 9 11 13 14

a. The three quartiles are: $Q_1 = (4 + 4)/2 = 4$, $Q_2 = (7 + 8)/2 = 7.5$, $Q_3 = (9 + 11)/2 = 10$

$IQR = Q_3 - Q_1 = 10 - 4 = 6$

The value 4 is equal to Q_1, which indicates that approximately 25% of the values in the data set are less than this value.

b. $kn/100 = 70(12)/100 = 8.4$.   Thus, the 70th percentile may be approximated by the mean of the eighth and ninth terms in the ranked data. Therefore, $P_{70} = (8 + 9)/2 = 8.5$.   Thus, approximately 70% of these drivers had fewer than 8.5 citations.

c. One value in the given data is less than 3. Hence, the percentile rank of 3 = $(1/12) \times 100 = 8.33\%$. Thus, 8.33% of these drivers had fewer than 3 citations.

3.121 Median = 39, $Q_1 = 27$, $Q_3 = 47$, $IQR = Q_3 - Q_1 = 47 - 27 = 20$,

1.5 x IQR = 1.5 x 20 = 30,

Lower inner fence = $Q_1 - 30 = 27 - 30 = -3$ and   Upper inner fence = $Q_3 + 30 = 47 + 30 = 77$

The smallest and largest values in the data set within the two inner fences are 19 and 65, respectively.

The data set does not contain any outliers.

The data are skewed slightly to the right.

3.122 Median = (11 + 12)/2 = 11.5, $Q_1 = 8$, $Q_3 = 16$, $IQR = Q_3 - Q_1 = 16 - 8 = 8$

1.5 x IQR = 1.5 x 8 = 12

Lower inner fence = $Q_1 - 12 = 8 - 12 = -4$

Upper inner fence = $Q_3 + 12 = 16 + 12 = 28$

The smallest and largest values in the data set within the two inner fences are 2 and 24, respectively.

The data are skewed to the right.

The values 33 and 42 are outliers. The upper outer fence is $Q_3 + 3(IQR) = 16 + 3(8) = 40$. Since 42 is greater than 40, it is an extreme outlier.

3.123 Let y = Mellisa's score on the final exam. Then her grade is $\dfrac{75+69+87+y}{5}$. To get a B, she needs this to be at least 80. So we solve,

$$80 = \frac{75+69+87+y}{5}$$

$$5(80) = 75 + 69 + 87 + y$$

$$400 = 231 + y$$

$$y = 169$$

Thus, the minimum score that Mellisa needs on the final exam in order to get a B grade is 169 out of 200 points.

3.124 a. Let y = amount that Jeffery suggests. Then, to insure the outcome Jeffery wants, we need

$$\frac{y+12{,}000(5)}{6} = 20{,}000$$

$$y + 12{,}000(5) = 6(20{,}000)$$

$$y + 60{,}000 = 120{,}000$$

$$y = 60{,}000$$

So Jeffery would have to suggest $60,000 be awarded to the plaintiff.

b. To prevent a juror like Jeffery from having an undue influence on the amount of damage to be awarded to the plaintiff, the jury could revise its procedure by throwing out any amounts that are outliers and than recalculate the mean, or by using the median, or by using a trimmed mean.

3.125 a. Since $\bar{x} = \dfrac{\sum x}{n}$, we have $76 = \dfrac{\sum x}{5}$ so $\sum x = 5(76) = 380$ inches. If we replace the tallest player by a substitute who is two inches taller, the sum of the new heights is $380 + 2 = 382$ inches. Thus, the new mean is $\bar{x} = \dfrac{382}{5} = 76.4$ inches. Since, Range = Largest value – Smallest value and the largest value has increased by two while the smallest value is unchanged, the range has increased by two. While the smallest value is unchanged, the range has increased by two. Thus, the new range is $11 + 2 = 13$ inches. The median is the height of the third player (if their heights are ranked) and this doesn't change. So the median remains 78 inches.

b. If we replace the tallest player by a substitute who is four inches shorter, then by reasoning similar to that in part a, we have a new mean of $\bar{x} = \dfrac{(380-4)}{5} = \dfrac{376}{5} = 75.2$ inches. You cannot determine the new median or range with only the information given.

3.126   a. To calculate how much time the trip requires, divide miles driven by miles per hour for each 100 mile segment. Time = 100 / 52+ 100 / 65 +100/ 58= 1.92 + 1.54 + 1.72 = 5.18 hours.

b. Linda's average speed for the 300 – mile trip is not equal to (52 + 65 + 58) / 3 = 58.33 MPH. This would assume that she spent an equal amount of time on each 100–mile segment, which is not true, because her average speed is different on each segment. Linda's average speed for the entire 300–mile trip is given by (miles driven) / (elapsed time) = 300 / 5.18 = 57.92 MPH.

3.127   The mean price per barrel of oil purchased in that week is

$$\left[\underbrace{1000(31)}_{\substack{\text{oil purchased from}\\ \text{Mexico}}} + \underbrace{200(34)}_{\substack{\text{oil purchased}\\ \text{from Kuwait}}} + \underbrace{100(44)}_{\substack{\text{oil purchased}\\ \text{from Spot Market}}} \right] / 1300 = \frac{42,200}{1300} \approx \$32.46 \text{ per barrel}$$

3.128   The method of calculating the mean is wrong in this case because it does not take into account the fact that the homeowner bought different amounts of heating oil in the four deliveries. The correct value of the mean is: $\bar{x} = (\sum mf) / n = [1.10(198) + 1.25(173) + 1.28(130) +1.33(124)] / (198 +173+130+124) = \$1.22$ per gallon.

3.129   a. Mean = $\dfrac{9.4+9.5+9.5+9.5+9.6}{5} = 9.5$

b. The percentage of trimmed mean is $2/7 \times 100 \approx 28.6 /2 = 14.3$ % since we dropped two of the seven values.

c. Suppose gymnast B has the following scores: 9.4, 9.4, 9.5, 9.5, 9.5, 9.5, 9.9. Then the mean for the gymnast B is: (9.4 + 9.4 + 9.5 + 9.5 + 9.5 + 9.5 + 9.9) / 7 = 9.5286, and the mean for gymnast A is (9.4 + 9.7 + 9.5 + 9.5 + 9.4 + 9.6 + 9.5) / 7 = 9.5143. So, Gymnast B would win if all seven scores were counted. The trimmed mean of B is (9.4 + 9.5 + 9.5 + 9.5 + 9.5) / 5 = 9.4800. This is less than the trimmed mean for A (9.500), so gymnast A would win using the trimmed mean.

3.130   a. Total amount spent per month by the 2000 shoppers = $(14 \times 8 \times 1100) + (18 \times 11 \times 900) = \$301,400$

b. Total number of trips per month by the 2000 shoppers = $(8 \times 1100) + (11 \times 900) = 18,700$
Mean number of trips per month per shopper = 18,700/ 2000 = 9.35 trips

c. Total amount spent per month by the 2000 shoppers = $301,400, from part a

Mean amount spent per person per month = 301,400 / 2000 = $150.70

3.131 For people age 30 and under, we have the following death rates from heart attack:

Country A: # of patients = $\dfrac{\text{\# of deaths}}{}$ = $(1 / 40) \times 1000 = 25$

Country B: # of patients = $\dfrac{\text{\# of deaths}}{}$ = $(.5 / 25) \times 1000 = 20$

So the death rate for people 30 and under is lower in Country B.

b. For people age 31 and older, the death rates from heart attack are as follows:

Country A: $(2/20) \times 1000 = 100$

Country B: $(3/35) \times 1000 = 85.7$

Thus, the death rate for Country A is greater than that for Country B for people age 31 and older.

c. The overall death rates are as follows:

Country A: $(3 / 60) \times 1000 = 50$

Country B: $(3.5 / 60) \times 1000 = 58.3$

Thus, overall the death rate for country A is *lower* than the death rate for Country B.

d. In both countries people age 30 and under have a lower percentage of death due to heart attack than people age 31 and over. Country A has 2/3 of its population age 30 and under while more than 1/2 of the people in Country B are age 31 and over. Thus, more people in Country B than in A fall into the higher risk group which drives up Country B's overall death rate from heart attacks..

3.132 Total distance for the first 100 students = 100 x 8.73 = 873 miles

Total distance for all 103 students = 873 + 11.5 + 7.6 + 10.0 = 902.1 miles

Mean distance for all 103 students = 902.1 / 103 = 8.76 miles

3.133 $\mu = 70$ minutes and $\sigma = 10$

a. Using Chebyshev's theorem, we need to find k so that

$$1 - \frac{1}{k^2} = .50$$

$$.50 = \frac{1}{k^2}$$

$$k^2 = \frac{1}{.50} = 2$$

So $k = \sqrt{2} \approx 1.4$

Thus, at least 50% of the scores are within 1.4 standard deviations of the mean.

b. Using Chebyshev's theorem, we first find k so that at least $1 - .20 = .80$ of the scores are within k standard deviations of the mean.

$$1 - \frac{1}{k^2} = .80$$

$$.20 = \frac{1}{k^2}$$

$$k^2 = \frac{1}{.20} = 5$$

So $k = \sqrt{5} \approx 2.2$

Thus, at least 80% of the scores are within 2.2 standard deviations of the mean, but this means that at most 10% of the scores are greater than 2.2 standard deviations above the mean.

3.134 a. Since we are dealing with a normal distribution and we know that 16% of all students scored above 85, which is $\mu + 15$, we must also have that 16% of all students scored below $\mu - 15 = 55$. Therefore, the remaining 68% of students scored between 55 and 85. By the empirical rule, we know that 68% of the scores fall in the interval $(\mu - \sigma)$ to $(\mu + \sigma)$, so we have $\mu - \sigma = 70 - \sigma = 55$ and $\mu + \sigma = 70 + \sigma = 85$. Thus, $\sigma = 15$.

b. We know that 95% of the scores are between 60 and 80 and that $\mu = 70$. By the empirical rule, 95% of the scores fall in the interval $(\mu - 2\sigma)$ to $(\mu + 2\sigma)$. So $60 = \mu - 2\sigma = 70 - 2\sigma$ and $80 = \mu + 2\sigma = 70 + 2\sigma$. Therefore $10 = 2\sigma$ and so $\sigma = 5$.

3.135 a.

$\bar{x} = \$13,872$	$Q_1 = \$50$	$s = \$64,112.47$
Median = $500	$Q_3 = \$1400$	Lowest = $0
Mode = $0	IQR = $1350	Highest = $321,500

Below are the box–and–whisker plot and the histogram for the given data.

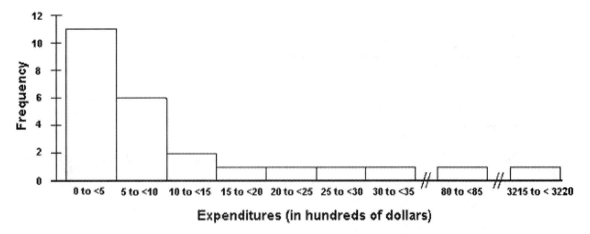

Expenditures (in hundreds of dollars)

The vacation expenditures are strongly skewed to the right. Most of the expenditures are relatively small ($3400 or less) but there are two extreme outliers ($8200 and $321,500).

b. Neither the mode ($0) nor the mean ($13,872) are typical of these expenditures. Thus, the median, $500 is the best indicator of the average family's vacation expenditures.

3.136 a. Mean = $600.35, Mean = $90, and Mode = $0, s = $2347.33,

Most of the losses are zero or near zero; however there is a relatively infrequent occurrence of very large losses.

b. The mean is the largest.

c. Q_1 = $0, Q_3 = $272.50, IQR = $272.50, 1.5 x IQR = $408.75

Lower Inner fence is Q_1 – 408.75 = 0 – 408.75 = –408.75

Upper Inner Fence is Q_3 + 408.75 = 272.50 +408.75 = 681.25

The largest and smallest values within the two inner fences are 0 and 501 respectively. There are three outliers at 1127, 3709 and 14,589.

Below are the box–and–whisker plot and the histogram for the given data.

The data are skewed to the right.

d. Because the data are skewed to the right, the insurance company should use the mean when considering the center of the data as it is more affected by the extreme values. The insurance company would want to use a measure that takes into consideration the possibility of extremely large losses. The standard deviation should be used as a measure of variation.

3.137 a.

The box–and–whisker plots show that the men's scores tend to be lower and more varied than the women's scores. The men's scores are skewed to the right, while the women's are more nearly symmetric.

b. **Men**   **Women**

    $\bar{x}$ = 82 $\bar{x}$ = 97.53

    s = 12.08 s = 8.44

    Median = 79   Median = 98

    Modes = 75, 79, and 92   Modes = 94 and 100

    Q_1 = 73.5 Q_1 = 94

    Q_3 = 89.5 Q_3 = 101

    IQR = 16   IQR = 7

These numerical measures confirm the observations based on the box–and–whisker plots.

3.138 a. Since, $\bar{x} = \dfrac{\sum x}{n}$ we have: $n = \dfrac{\sum x}{\bar{x}}$ = 12,372 / 51.55 = 240 pieces of luggage.

b. Since, $\bar{x} = \dfrac{\sum x}{n}$, $\sum x = n\bar{x}$. Thus, the total score for the seven students is 7 x 81 = 567. Let x = seventh student's score. Then $x + 81 + 75 + 93 + 88 + 82 + 85 = 567$. Hence, $x + 504 = 567$, so x = 567 – 504 = 63.

3.139 a. The total enrollment in the 25 freshman engineering classes is   24 x 25 + 150 = 750

Thus, the mean size of these 25 classes is $\dfrac{750}{25}$ = 30.

b. Each student attends five classes with total enrollment of   25 + 25 + 25 + 25 + 150 = 250

Thus, the mean size of the class is $\dfrac{250}{5}$ = 50.

The means in parts a and b are not equal because:

From the college's point of view, the large class of 150 is just one of 25 classes, so its influence on the mean is strongly offset by the 24 small classes. This leads to a relatively small mean of 30 students per class.

From the point of view of each student, the larger class is one of just five, so it has a stronger influence on the mean. This results in a larger mean of 50.

3.140 For all students:

$n = 44$, $\sum x = 6597$, $\sum x^2 = 1,030,639$ and   Median = 147.50 pounds

$$\bar{x} = \frac{\sum x}{n} = \frac{6597}{44} = 149.93 \text{ pounds}$$

$$s = \sqrt{\frac{\sum x^2 - \frac{(\sum x)^2}{n}}{n-1}} = \sqrt{\frac{1,030,639 - \frac{(6597)^2}{44}}{44-1}} = 31.0808 \text{ pounds}$$

For men only:

$n = 22$, $\sum x = 3848$, $\sum x^2 = 680,724$ and

Median = 179 pounds

$$\bar{x} = \frac{\sum x}{n} = \frac{3848}{22} = 174.91 \text{ pounds}$$

$$s = \sqrt{\frac{\sum x^2 - \frac{(\sum x)^2}{n}}{n-1}} = \sqrt{\frac{680,724 - \frac{(3848)^2}{22}}{22-1}} = 19.1160 \text{ pounds}$$

For women:

$n = 22$, $\sum x = 2749$, $\sum x^2 = 349,915$ and   Median = 123 pounds

$$\bar{x} = \frac{\sum x}{n} = \frac{2749}{22} = 124.95 \text{ pounds}$$

$$s = \sqrt{\frac{\sum x^2 - \frac{(\sum x)^2}{n}}{n-1}} = \sqrt{\frac{349,915 - \frac{(2749)^2}{22}}{22-1}} = 17.4778 \text{ pounds}$$

In this case, the median may be more informative than the mean, since it is less influenced by extremely high or low weights.

As one might expect, the mean and median weights for men are higher than those of women. For the entire group, the mean and median weights are about midway between the corresponding values for men and women.

The standard deviations are roughly the same for men and women. The standard deviation for the whole group is much larger than for men or women only, due to the fact that it includes the lower weights of women and the heavier weights of men.

3.141 $\mu = 6$ inches and $\sigma = 2$ inches

    a. $3 = 6 - 3 = \mu - 1.5\sigma$, and $9 = 6 + 3 = \mu + 1.5\sigma$

       Using Chebyshev's theorem with $k = 1.5$: $1 - \dfrac{1}{k^2} = 1 - \dfrac{1}{(1.5)^2} = .556$ or 55.6%

       Thus, at least 55.6% of the fish are between 3 and 9 inches in length.

    b. $1 - \dfrac{1}{k^2} = .84$ gives $.16 = \dfrac{1}{k^2}$; $k^2 = \dfrac{1}{.16}$ or $k = \sqrt{\dfrac{1}{.16}} = 2.5$

       The required interval is:

       $\mu - k\sigma$ to $\mu + k\sigma = \mu - 2.5\sigma$ to $\mu + 2.5\sigma = \{6 - 2.5(2)\}$ to $\{6 + 2.5(2)\} = 1$ to 11 inches

    c. $100 - 36 = 64\%$ of the fish have lengths *inside* the required interval .

       Thus, $1 - \dfrac{1}{k^2} = .64$, so $.36 = \dfrac{1}{k^2}$, and $k^2 = \dfrac{1}{.36}$

       Hence, $k = \sqrt{\dfrac{1}{.36}} = 1.67$

       Thus, the required interval is:

       $\mu - k\sigma$ to $\mu + k\sigma = \mu - 1.67\sigma$ to $\mu + 1.67\sigma = \{6 - 1.67(2)\}$ to $\{6 + 1.67(2)\} = 2.66$ to 9.34 inches.

3.142   The given data are: 3   6   9   12   18   15   11   10   15   25   21   26   38   41   62

    Ranked data are: 3   6   9   10   11   12   15   15   18   21   25   26   38   41   62

    a. $\bar{x} = 20.80$ thousand miles, Median = 15 thousand miles, Mode = 15 thousand miles

    b.   Range = 59 thousand miles, $s^2 = 249.03$, $s = 15.78$ thousand miles

    c. $Q_1 = 10$ thousand miles and $Q_3 = 26$ thousand miles

    d.   IQR $= Q_3 - Q_1 = 26 - 10 = 16$ thousand miles

       Since the interquartile range is based on the middle 50% of the observations it is not affected by outliers. The standard deviation, however, is strongly affected by outliers. Thus, the inter–quartile range is preferable in applications in which a measure of variation is required that is unaffected by extreme values.

Self–Review Test for Chapter Three

1. b   **2.** a and d   **3.** c   **4.** c   **5.** b   **6.** b   **7.** a

8. a   **9.** b   **10.** a   **11.** b   **12.** c   **13.** a   **14.** a

15.   For the given data: $n = 10, \sum x = 109, \sum x^2 = 1775$

$\bar{x} = (\sum x)/ n = 109/10 = 10.9$ $(n +1) /2 = (10 +1) /2 = 5.5$

Median $= (7 +9) /2 = 8$

Mode $= 6$

Range = Largest value – Smallest value $= 28 – 2 = 26$

$$s^2 = \frac{\sum x^2 - \frac{(\sum x)^2}{n}}{n-1} = \frac{1775 - \frac{(109)^2}{10}}{10-1} = 65.2111$$

$$s = \sqrt{65.2111} = 8.08$$

16.   Suppose the 2002 gross sales (in millions of dollars) of six companies are:

1.2   1.9   .5   2.1   3.4   110.5

Then, $\sum x = 1.2 + 1.9 + .5 + 2.1 + 3.4 + 110.5 = 119.6$

Mean $= (\sum x) /n = 119.6 / 6 = \$19.93$ million

The value of \$110.5 million is an outlier. When we drop it:

$\sum x = 1.2 + 1.9 + .5 + 2.1 + 3.4 = 9.1$

Mean $= (\sum x) /n = 9.1 / 5 = \$1.82$ million

Thus, when we drop the value of 110.5 million, which is an outlier, the value of the mean decreases from \$19.93 million to \$1.82 million.

17.   Reconsider the data on the 2002 gross sales (in millions of dollars) of six companies given in Problem 16, which are reproduced below.

1.2   1.9   .5   2.1   3.4   110.5

Then, Range = Largest value – Smallest value $= 110.5 – .5 = \$110$ million

When we drop the value of \$110.5 million, which is an outlier: Range $= 3.4 – .5 = \$2.9$ million

Thus, when we drop the value of \$110.5 million, which is an outlier, the value of the range decreases from \$110 million to \$2.9 million.

18.   The value of the standard deviation is zero when all the values in a data set are the same. For example, suppose the heights (in inches) of five women are: 67   67   67   67   67

This data set has no variation. As shown below the value of the standard deviation is zero for this data set. For these data: $n = 5$, $\sum x = 335$, and $\sum x^2 = 22{,}445$.

$$s = \sqrt{\frac{\sum x^2 - \frac{(\sum x)^2}{n}}{n-1}} = \sqrt{\frac{22{,}445 - \frac{(335)^2}{5}}{5-1}} = \sqrt{\frac{22{,}445 - 22{,}445}{4}} = 0$$

19.   a. The frequency column gives the number of weeks for which the number of computers sold were in the corresponding class.

b. For the given data: $n = 25$, $\sum mf = 486.50$, and $\sum m^2 f = 10{,}524.25$

$$\bar{x} = (\sum mf) / n = 486.50 / 25 = 19.46 \text{ computers}$$

$$s^2 = \frac{\sum m^2 f - \frac{(\sum mf)^2}{n}}{n-1} = \frac{10{,}524.25 - \frac{(486.50)^2}{25}}{25-1} = 44.0400$$

$$s = \sqrt{44.0400} = 6.64 \text{ computers}$$

20.   a. i. Each of the two values is 5.5 years from $\mu = 7.3$ years. Hence,

$$k = 5.5 / 2.2 = 2.5 \qquad \text{and} \qquad 1 - \frac{1}{k^2} = 1 - \frac{1}{(2.5)^2} = 1 - .16 = .84 \text{ or } 84\%$$

Thus, at least 84% of the cars are 1.8 to 12.8 years old.

ii. Each of the two values is 6.6 years from $\mu = 7.3$ years. Hence

$$k = 6.6 / 2.2 = 3 \qquad \text{and} \qquad 1 - \frac{1}{k^2} = 1 - \frac{1}{(3)^2} = 1 - .11 = .89 \text{ or } 89\%$$

Thus, at least 89% of the cars are .7 to 13.9 years old.

b. $1 - \frac{1}{k^2} = .75$ gives $\frac{1}{k^2} = 1 - .75 = .25$ or $k^2 = \frac{1}{2.5}$ or $k = 2$

Thus, the required interval is $\mu - k\sigma$ to $\mu + k\sigma = \{7.3 - 2(2.2)\}$ to $\{7.3 + 2(2.2)\} = 2.9$ to 11.7 years.

21.   a. $\mu = 7.3$ years and $\sigma = 2.2$ years

i. The intervals 5.1 to 9.5 years is $\mu - \sigma$ to $\mu + \sigma$. Hence, approximately 68% of the cars are 5.1 to 9.5 years old.

ii. The interval .7 to 13.9 years is $\mu - 3\sigma$ to $\mu + 3\sigma$. Hence, approximately 99.7% of the cars are .7 to 13.9 years.

b. The interval that contains ages of 95% of the cars will be $\mu - 2\sigma$ to $\mu + 2\sigma$. Hence, this interval is: μ $- 2\sigma$ to $\mu + 2\sigma = \{7.3 - 2(2.2)\}$ to $\{7.3 + 2(2.2)\} = 2.9$ to 11.7 years. Thus, approximately 95% of the cars are 2.9 to 11.7 years old.

22.   The ranked data are: 0 1 2 3 4 5 7 8 10 11 12 13 14 15 20

a. The three quartiles are: $Q_1 = 3$, $Q_2 = 8$, and $Q_3 = 13$. IQR $= Q_3 - Q_1 = 13 - 3 = 10$. The value 4 lies between Q_1 and Q_2, which indicates that this value is in the second from the bottom 25% group in the ranked data.

b. $kn/100 = 60(15)/100 = 9$. Thus, the 60th percentile may be represented by the value of the ninth term in the ranked data, which is 10. Therefore, $P_{60} = 10$. Thus, approximately 60% of the half hour time periods had fewer than 10 passengers set off the metal detectors during this day.

c. Ten values in the given data are less than 12. Hence, percentile rank of 12 $= (10/15) \times 100 =$ 66.67%. Thus, 66.67% of the half hour time periods had fewer than 12 passengers set off the metal detectors during this day.

23.   The ranked data are: 0 1 2 3 4 5 7 8 10 11 12 13 14 15 20

$Q_1 = 3$, $Q_2 = 8$, and $Q_3 = 13$. IQR $= Q_3 - Q_1 = 13 - 3 = 10$, $1.5 \times$ IQR$= 1.5 \times 10 = 15$
Lower inner fence $= Q_1 - 15 = 3 - 15 = -12$
Upper inter fence $= Q_3 + 15 = 13 + 15 = 28$
The smallest and largest values in the data set within the two inner fences are 0 and 20, respectively.
The data does not contain any outliers.

The data are skewed slightly to the right.

24.   From the given information: $n_1 = 15$, $n_2 = 20$, $\bar{x}_1 = \$435$, $\bar{x}_2 = \$490$

$$\bar{x} = \frac{n_1\bar{x}_1 + n_1\bar{x}_2}{n_1 + n_2} = \frac{(15)(435) + (20)(490)}{15 + 20} = \frac{16{,}325}{35} = \$466.43$$

25.   Sum of the GPAs of five students $= 5 \times 3.21 = 16.05$
Sum of the GPAs of four students $= 3.85 + 2.67 + 3.45 + 2.91 = 12.88$
GPA of the fifth student $= 16.05 - 12.88 = 3.17$

26.   The ranked data are: 58   149   163   166   179   193   207   238   287   2534

Thus, to find the 10% trimmed mean, we drop the smallest value and the largest value (10% of 10 is 1) and find the mean of the remaining 8 values. For these 8 values,

$\Sigma x = 149 + 163 + 166 + 179 + 193 + 207 + 238 + 287 = 1582$

10% trimmed mean = $(\Sigma x) / 8 = 1582 / 8 = \$197.75$ thousand = $\$197,750$. The 10% trimmed mean is a better summary measure for these data than the mean of all 10 values because it eliminates the effect of the outliers, 58 and 2534.

27.   a. For Data Set I: $\bar{x} = (\Sigma x) / n = 79 / 4 = 19.75$

For Data Set II: $\bar{x} = (\Sigma x) / n = 67 / 4 = 16.75$

The mean of Data Set II is smaller than the mean of Data Set I by 3.

b. For Data Set I: $\Sigma x = 79$, $\Sigma x^2 = 1945$, and $n = 4$

$$s = \sqrt{\frac{\Sigma x^2 - \frac{(\Sigma x)^2}{n}}{n-1}} = \sqrt{\frac{1995 - \frac{(79)^2}{4}}{4-1}} = 11.32$$

c. For Data Set II: $\Sigma x = 67$, $\Sigma x^2 = 1507$, and $n = 4$

$$s = \sqrt{\frac{\Sigma x^2 - \frac{(\Sigma x)^2}{n}}{n-1}} = \sqrt{\frac{1507 - \frac{(67)^2}{4}}{4-1}} = 11.32$$

The standard deviations of the two data sets are equal.

Chapter Four

4.1   **Experiment:** When a process that results in one and only one of many observations is performed, it is called an experiment.

       Outcome: The result of the performance of an experiment is called an outcome.

       Sample space: The collection of all outcomes for an experiment is called a sample space.

       Simple event: A simple event is an event that includes one and only one of the final outcomes of an experiment.

       Compound event: A compound event is an event that includes more than one of the final outcomes of an experiment.

4.2   a. $S = \{1,2,3,4,5,6\}$

       b. $S = \{HHH,\ HHT,\ HTH,\ HTT,\ THH,\ THT,\ TTH,\ TTT\}$

       c. $S = \{H1,\ H2,\ H3,\ H4,\ H5,\ H6,\ T1,\ T2,\ T3,\ T4,\ T5,\ T6\}$

4.3   The experiment of selecting two items from the box without replacement has the following six possible outcomes: *AB, AC, BA, BC, CA, CB*. Hence, the sample space is written as
$S = \{AB,\ AC,\ BA,\ BC,\ CA,\ CB\}$

4.4   Let: $Y =$ student selected suffers from math anxiety
             $N =$ student selected does not suffer from math anxiety

    The experiment of selecting two students has four outcomes: *YY, YN, NY,* and *NN.*

Venn Diagram

Tree Diagram

4.5   Let: L = person is computer literate I = person is computer illiterate

The experiment has four outcomes: *LL, LI, IL,* and *II.*

Tree Diagram

4.6   Let: C = the answer selected is correct W = the answer selected is wrong

This experiment has four outcomes: *CC, CW, WC,* and *WW*

Venn Diagram

Tree Diagram

4.7   Let: G = the selected part is good D = the selected part is defective

The four outcomes for this experiment are: *GG, GD, DG,* and *DD*

4.8   Let: F = person selected is in favor of tax increase and A = person selected is against tax increase.

The experiment of selecting three persons has eight outcomes: *FFF, FFA, FAF, FAA, AFF, AFA, AAF,* and *AAA.* Hence the sample space is written as

S = {*FFF, FFA, FAF, FAA, AFF, AFA, AAF, AAA*}.

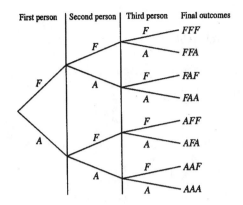

4.9   Let: *H* = a toss results in a head   and   *T* = a toss results in a tail

Thus the sample space is written as *S* = {*HHH, HHT, HTH, HTT, THH, THT, TTH, TTT*}

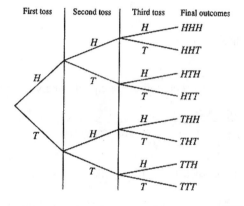

4.10   a.   {*YY*}; a simple event   c.   {*NY*}; a simple event

           b.   {*YN, NY*}; a compound event   d.   {*NN*}; a simple event

4.11   a.   {*LI, IL*}; a compound event   c.   {*II, IL, LI*}; a compound event

           b.   {*LL, LI, IL*}; a compound event   d.   {*LI*}; a simple event

4.12   a.   {*CC*}; a simple event   c.   {*CW*}; a simple event

           b.   {*CC, CW, WC*}; a compound event   d.   {*CW, WC*}; a compound event

4.13   a.   {*DG, GD, GG*}; a compound event   c.   {*GD*}; a simple event

           b.   {*DG, GD*}; a compound event   d.   {*DD, DG, GD*}; a compound event

4.14   a.   {*FFF, FFA, FAF, AFF* }; a compound event

           b.   {*FFA, FAF, AFF* }; a compound event

           c.   {*FFA, FAF, FAA, AFF, AFA, AAF, AAA*}; a compound event

    d. {*FAA, AFA, AAF, AAA*}; a compound event

4.15   The following are the two properties of probability.

    1. The probability of an event always lies in the range zero to 1, that is:

$$0 \le P(E_i) \le 1 \text{ and } 0 \le P(A) \le 1$$

    2. The sum of the probabilities of all simple events for an experiment is always 1, that is:

$$\sum P(E_i) = P(E_1) + P(E_2) + P(E_3) + \cdots = 1$$

4.16   **Impossible event:** An event that cannot occur is called an impossible event. The probability that such an event will occur is zero.

    Sure event: An event that is certain to occur is called a sure event. The probability that this event will occur is 1.

4.17   The following are three approaches to probability.

    1. **Classical probability approach:** When all outcomes are equally likely, the probability of an event A is given by: $P(A) = \dfrac{\text{Number of outcomes in } A}{\text{Total Number of outcomes ub the experiment}}$

    For example, the probability of observing a 1 when a fair die is tossed once is 1/6.

    2. **Relative frequency approach:** If an event A occurs f times in n repetitions of an experiment, then $P(A)$ is approximately f/n. As the experiment is repeated more and more times, f/n approaches $P(A)$. For example, if 510 of the last 1000 babies born in a city are male, the probability of the next baby being male is approximately 510/1000 = .510

    3. **Subjective probability approach:** Probabilities are derived from subjective judgment, based on experience, information and belief. For example, a banker might estimate the probability of a new donut shop surviving for two years to be 1/3 based on prior experience with similar businesses.

4.18   The classical approach is used when all the outcomes of an experiment are equally likely. The relative frequency approach is used when all the outcomes are not equally likely, but the experiment can be performed repeatedly to generate data.

4.19   The following cannot be the probabilities of events: −.55,   1.56,   5/3,   and   −2/7

    This is because the probability of an event can never be less than zero or greater than one.

4.20   The following cannot be the probabilities of events: −.09,   1.42,   9/4,   and   −1/4

This is because the probability of an event can never be less than zero or greater than one.

4.21   These two outcomes would not be equally likely unless exactly half of the passengers entering the metal detectors set it off, which is unlikely. We would have to obtain a random sample of passengers going through New York's JFK airport, collect information on whether they set off the metal detector or not, and use the relative frequency approach to find the probabilities.

4.22   We would use the classical approach, since each of the 32 applicants is equally likely to be selected. Thus, the probability of selecting an experienced candidate is 7/32 and that of selecting an inexperienced candidate is 25/32.

4.23   This is a case of subjective probability because the given probability is based on the president's judgment.

4.24   This is a case of subjective probability because the given probability is based on the coach's judgment.

4.25   a. *P*(marble selected is red) = 18/40 = .450

   b. *P*(marble selected is green) = 22/40 = .550

4.26   a. *P*(a number less than 5 is obtained) = 4/6 = .6667

   b. *P*(a number 3 to 6 is obtained) = 4/6 = .6667

4.27   *P*(adult selected has shopped on the internet) = 860/2000 = .430

4.28   *P*(student selected has volunteered before) = 28 /42 = .6667

4.29   *P*(executive selected has a type A personality) = 29/50 = .580

4.30   Number of families who paid income tax last year = 3000-600 = 2400

   P(family selected paid income tax last year) = 2400/3000 = .800

4.31   a. *P*(her answer is correct) = 1/5 = .200

   b. *P*(her answer is wrong) = 4/5 = .800

   Yes, these probabilities add up to 1.0 because this experiment has two and only two outcomes, and according to the second property of probability, the sum of their probabilities must be equal to 1.0.

4.32   a. *P*(professor selected is female) = 105/320 = .3281

   b. Number of male professors = 320 − 105 = 215

   P(professor selected is male) = 215/320 = .6719

   Yes, these probabilities add up to 1.0 because this experiment has two and only two outcomes, and according to the second property of probability, the sum of their probabilities must be equal to 1.0.

4.33   *P*(person selected is a woman) = 4/6 = .6667

P(person selected is a man) $= 2/6 = .3333$

Yes, the sum of these probabilities is 1.0 because of the second property of probability.

4.34 P(company selected offers free psychiatric help) $= 120/500 = .240$

Number of companies that do not offer free psychiatric help $= 500-120 = 380$

P(company selected does not offer free psychiatric help) $= 380/500 = .760$

Yes, these probabilities add up to 1.0 because of the second property of probability.

4.35 P(company selected offers free health fitness center) $= 130/400 = .325$

Number of companies that do not offer free health fitness center $= 400-130 = 270$

P(company selected does not offer free health fitness center) $= 270/400 = .675$

Yes, the sum of the probabilities is 1.0 because of the second property of probability.

4.36   a. P(company closed down or moved) $= 7400 / 15,000 = .4933$

       b. P(insufficient work) $= 4600/15,000 = .3067$

       c. P(position abolished) $= \dfrac{15,000-(7400+4600)}{15,000} = \dfrac{3000}{15,000} = .2000$

Yes, the sum of these three probabilities is 1.0 because of the second property of probability.

4.37

Credit Cards	Frequency	Relative Frequency
0	80	.098
1	116	.141
2	94	.115
3	77	.094
4	43	.052
5 or more	410	.500

a. P(person selected has three credit cards) $= .0939$

b. P(person selected has five or more cards) $= .5000$

4.38

Income	Frequency	Relative Frequency
Less than $30,000	90	.180
$20,000 to $70,000	270	.540
More than $70,000	140	.280

a. P(income is less than $30,000) $= .180$

b. P(income is more than $70,000) $= .280$

4.39   Take a random sample of families from Los Angeles and determine how many of them earn more than $75,000 per year. Then use the relative frequency approach.

4.40   Roll the die repeatedly for a large number of times, recording how many times each of the six outcomes occurs. Then apply the relative frequency approach.

4.41   The marginal probability of an event is determined without reference to any other event; the conditional probability of an event depends upon whether or not another event has occurred. Example: When a single die is rolled, the marginal probability of a 2-spot occurring is 1/6; the conditional probability of a 2-spot given that an even number has occurred is 1/3.

4.42   Events that cannot occur together are called mutually exclusive events. Example: A die is rolled once. The events "odd number" and "even number" are mutually exclusive events. The events "a number less than 5" and "a number less than 4" are mutually nonexclusive events.

4.43   Two events are independent if the occurrence of one does not affect the probability of the other. Two events are dependent if the occurrence of one affects the probability of the other. If two events A and B satisfy the condition $P(A|B) = P(A)$, or $P(B|A) = P(B)$, they are independent; otherwise they are dependent.

4.44   The complement of event A consists of all the outcomes for an experiment that are not included in A. The sum of the probabilities of two complementary events is 1.

4.45   Total outcomes for four rolls of a die $= 6 \times 6 \times 6 \times 6 = 1296$

4.46   Total outcomes for 10 tosses of a coin $= 2 \times 2 \times 2 \times 2 \times 2 \times 2 \times 2 \times 2 \times 2 \times 2 = 1024$

4.47   a. Events A and B are not mutually exclusive since they have the element "2" in common.

b. $P(A) = 3/8$ and $P(A|B) = 1/3$

Since these probabilities are not equal, A and B are dependent.

c. $\overline{A} = \{1,3,4,6,8\};$ $P(\overline{A}) = 5/8 = .625$

$\overline{B} = \{1,3,5,6,7\};$ $P(\overline{B}) = 5/8 = .625$

4.48   a. Events A and B are mutually exclusive since they have no common element.

b. $P(A) = 4/10$ and $P(A|B) = 0$

Since these two probabilities are not equal, A and B are dependent events.

c. $\overline{A} = \{1,2,5,7,8,10\};$ $P(\overline{A}) = 6/10 = .600$

$\overline{B} = \{3,4,6,7,8,9,10\};$ $P(\overline{B}) = 7/10 = .700$

4.49   Total selections $= 10 \times 5 = 50$

4.50   Total selections $= 4 \times 8 \times 12 = 384$

4.51   Total outcomes = $4 \times 8 \times 5 \times 6 = 960$

4.52   Total outcomes = $8 \times 6 \times 5 = 240$

4.53   a.   i. P(selected adult has never shopped on the internet) = $1500/2000 = .750$

       ii. P(selected adult is a male) = $1200/2000 = .600$

       iii. P(selected adult has shopped on the internet given that this adult is a female) = $200/800 = .250$

       iv. P(selected adult is a male given that this adult has never shopped on the internet) = $900/1500 = .600$

     b.   The events "male" and "female" are mutually exclusive because they cannot occur together. The events "have shopped" and "male" are not mutually exclusive because they can occur together.

     c. P(female) = $800/2000 = .400$

        P(female | have shopped) = $200/500 = .400$

        Since these probabilities are equal, the events "female" and "have shopped" are independent.

4.54   a.   i. P(selected person is unemployed) = $\dfrac{1615+1435}{1615+1435+11{,}232+10{,}353} = \dfrac{3050}{24{,}635} = .1238$

       ii. P(selected person is female) = $\dfrac{10{,}353+1435}{24{,}635} = \dfrac{11{,}788}{24{,}635} = .4785$

       iii. P(selected person is employed | male) = $\dfrac{11{,}232}{11{,}232+1615} = \dfrac{11{,}232}{12{,}847} = .8743$

       iv. P(selected person is female | unemployed) = $\dfrac{1435}{1435+1615} = \dfrac{1435}{3050} = .4705$

     b.   The events "employed" and "unemployed" are mutually exclusive because they cannot occur together. The events "unemployed" and "male" are not mutually exclusive because they can occur together.

     c. P(female) = $\dfrac{11{,}788}{24{,}635} = .4785$ and P(female | employed) = $\dfrac{10{,}353}{21{,}585} = .4796$

        Since these two probabilities are not equal, the events "female" and "employed" are not independent.

4.55   a.   i. P(in favor) = $695/2000 = .3475$

       ii. P(against) = $1085/2000 = .5425$

       iii. P(in favor | female) = $300/1100 = .2727$

       iv. P(male | no opinion) = $100/220 = .4545$

b. The events "male" and "in favor" are not mutually exclusive because they can occur together. The events "in favor" and "against" are mutually exclusive because they cannot occur together.

c. P(female) = 1100/2000 = .5500

P(female | no opinion) = 120/220 = .5455

Since these two probabilities are not equal, the events "female" and "no opinion" are not independent.

4.56   a.   i. P(woman) = 200/500 = .400

ii. P(has retirement benefits) = 375/500 = .750

iii. P(has retirement benefits | man) = 225/300 = .750

iv. P(woman | does not have retirement benefits) = 50/125 = .400

b. The events "man" and "yes" are not mutually exclusive because they can occur together. The Events "yes" and "no" are mutually exclusive because they cannot occur together.

c. P(woman) = 200/500 = .400   and P(woman | yes) = 150/375 = .400

Since these two probabilities are equal, the events "woman" and "yes" are independent.

4.57   a.   i. P(more than one hour late) = (92 + 80) /1700 = .1012

ii. P(less than 30 minutes late) = (429 + 393) / 1700 = .4835

iii. P(Airline A's flight | 30 minutes to one hour late) = 390 / (390 + 316) = .5524

iv. P(more than one hour late | Airline B's flight) = 80/ (393 + 316 + 80) = .1014

b. The events "Airline A" and "more than one hour late" are not mutually exclusive because they can occur together. The events "less than 30 minutes late" and "more than one hour late" are mutually exclusive because they cannot occur together.

c. P(Airline B) = (393 + 316 + 80) /1700 = .4641

P(Airline B | 30 minutes to one hour late) = 316/ (390 + 316) = .4476

Since these two probabilities are not equal, the events "Airline B" and "30 minutes to one hour late" are not independent.

4.58   a.   i. P(better off) = 1010/2000 = .5050

ii. P(better off | less than high school) = 140/400 = .3500

iii. P(worse off | high school) = 300/1000 = .3000

iv. P(the same | more than high school) = 110/600 = .1833

b. The events "better off" and "high school education" are not mutually exclusive because they can occur together. The events "less than high school" and "more than high school" are mutually exclusive because they cannot occur together.

c. $P(\text{worse off}) = 570/2000 = .2850$

$P(\text{worse off} \mid \text{more than high school}) = 70/600 = .1167$

Since these two probabilities are not equal, the events "worse off" and "more than high school" are not independent.

4.59 $P(\text{pediatrician}) = 25/160 = .1563$ $P(\text{pediatrician} \mid \text{female}) = 20/75 = .2667$

Since these two probabilities are not equal, the events "female" and "pediatrician" are not independent. The events are not mutually exclusive because they can occur together.

4.60 Let D = the diskette selected is defective F = the diskette selected is made on Machine I

Then, $P(D) = 20/100 = .2000$, and $P(D|F) = 10/60 = .1667$

Since these two probabilities are not equal, the events "machine type" and "defective diskettes" are not independent.

4.61 $P(\text{business major}) = 11/30 = .3667$. $P(\text{business major} \mid \text{female}) = 9/16 = .5625$

Since these two probabilities are not equal, the events "female" and "business major" are not independent. The events are not mutually exclusive because they can occur together.

4.62 The experiment involving two tosses of a coin has four outcomes: $HH, HT, TH,$ and TT where H denotes the event that a head is obtained and T that a tail is obtained on any toss. Events A and B contain the following outcomes: $A = \{HH, HT, TH\}$ and $B = \{TT\}$

a. Since events A and B do not contain any common outcome, they are mutually exclusive events.

$P(A) = \frac{3}{4} = .750$ and $P(A|B) = 0$

Since these two probabilities are not equal, A and B are dependent events.

b. Events A and B are complementary events because they do not contain any common outcome and taken together they contain all the outcomes for this experiment.

$P(B) = 1/4 = .250$ and $P(A) = 1 - P(B) = 1 - .250 = .750$

4.63 Event A will occur if either a 1-spot or a 2-spot is obtained on the die. Thus, $P(A) = 2/6 = .3333$. The complementary event of A is that either a 3-spot, or a 4-spot, or a 5-spot, or a 6-spot is obtained on the die. Hence, $P(A) = 1 - .3333 = .6667$.

4.64 The two complementary events are that the person selected has health insurance and that the person selected has no health insurance.

$$P(\text{has health insurance}) = \frac{231{,}533}{274{,}087} = .8447 \quad P(\text{has no health insurance}) = \frac{42{,}554}{274{,}087} = .1553$$

4.65   The complementary event is that the college student attended no MLB games last year. The probability of this complementary event is $1 - .12 = .88$

4.66   The intersection of two events is the collection of outcomes they have in common. For example, if $A = \{1,2,3\}$ and $B = \{1,3,5\}$, then the intersection of A and B is the event $\{1,3\}$.

4.67   The joint probability of two or more events is the probability that all those events occur simultaneously. For example, suppose a die is rolled once. Let:
A = an even number occurs = $\{2, 4, 6\}$ B = a number less than 3 occurs = $\{1, 2\}$
Then the probability $P(A \text{ and } B) = P(2)$ is the joint probability of A and B.

4.68   Unlike the rule for independent events, the rule for dependent events requires a conditional probability. Thus, if A and B are dependent, then $P(A \text{ and } B) = P(A)P(B|A)$. If A and B are independent events, then $P(A \text{ and } B) = P(A)P(B)$.

4.69   The joint probability of two mutually exclusive events is zero. For example, consider one roll of a die. Let: A = an even number occurs   and B = an odd number occurs
Thus, A and B are mutually exclusive events. Hence, event (A and B) is impossible, and consequently $P(A \text{ and } B) = 0$.

4.70   a. $P(A \text{ and } B) = P(A)P(B|A) = (.40)(.25) = .100$
b. $P(A \text{ and } B) = P(B \text{ and } A) = P(B)P(A|B) = (.65)(.36) = .234$

4.71   a. $P(A \text{ and } B) = P(B \text{ and } A) = P(B)P(A|B) = (.59)(.77) = .4543$
b. $P(A \text{ and } B) = P(A)P(B|A) = (.28)(.35) = .0980$

4.72   a. $P(A \text{ and } B) = P(A)P(B) = (.61)(.27) = .1647$
b. $P(A \text{ and } B) = P(A)P(B) = (.39)(.63) = .2457$

4.73   a. $P(A \text{ and } B) = P(A)P(B) = (.20)(.76) = .1520$
b. $P(A \text{ and } B) = P(A)P(B) = (.57)(.32) = .1824$

4.74   a. $P(A \text{ and } B \text{ and } C) = P(A)P(B)P(C) = (.20)(.46)(.25) = .0230$
b. $P(A \text{ and } B \text{ and } C) = P(A)P(B)P(C) = (.44)(.27)(.43) = .0511$

4.75   a. $P(A \text{ and } B \text{ and } C) = P(A)P(B)P(C) = (.49)(.67)(.75) = .2462$
b. $P(A \text{ and } B \text{ and } C) = P(A)P(B)P(C) = (.71)(.34)(.45) = .1086$

4.76 $P(B|A) = P(A \text{ and } B) / P(A) = .24 / .30 = .800$

4.77 $P(A|B) = P(A \text{ and } B) / P(B) = .45 / .65 = .6923$

4.78 $P(B) = P(A \text{ and } B) / P(A|B) = .36 / .40 = .900$

4.79 $P(A) = P(A \text{ and } B) / P(B|A) = .58 / .80 = .725$

4.80 a. i. $P(V \text{ and } C) = P(V)P(C|V) = \left(\dfrac{312}{1800}\right)\left(\dfrac{61}{312}\right) = .0339$

        ii. $P(N \text{ and } A) = P(N)P(A|N) = \left(\dfrac{1488}{1800}\right)\left(\dfrac{698}{1488}\right) = .3878$

    b. This probability is zero since B and C are mutually exclusive events.

4.81 Let: M = male, F = female, G = graduated, N = did not graduate

    a. i. $P(F \text{ and } G) = P(F)P(G|F) = \left(\dfrac{165}{346}\right)\left(\dfrac{133}{165}\right) = .3844$

        ii. $P(M \text{ and } N) = P(M)P(N|M) = \left(\dfrac{181}{346}\right)\left(\dfrac{55}{181}\right) = .1590$

    b. This probability is zero since G and N are mutually exclusive events.

4.82 Let: M = man, W = woman, R = has retirement benefits,

        and N = does not have retirement benefits

    a. i. $P(W \text{ and } R) = P(W)P(R|W) = \left(\dfrac{200}{500}\right)\left(\dfrac{150}{200}\right) = .300$

        ii. $P(N \text{ and } M) = P(N)P(M|N) = \left(\dfrac{125}{500}\right)\left(\dfrac{75}{125}\right) = .150$

    b.

M $R|M$ — $P(M \text{ and } R) = (300/500)(225/300) = .450$

  $N|M$ — $P(M \text{ and } N) = (300/500)(75/300) = .150$

W $R|W$ — $P(W \text{ and } R) = (200/500)(150/200) = .300$

  $N|W$ — $P(W \text{ and } N) = (200/500)(50/200) = .100$

4.83 Let: Y = this adult has shopped at least once on the internet, M = male,

        N = this adult has never shopped on the internet, and F = female.

    a. i. $P(N \text{ and } M) = P(N)P(M|N) = \left(\dfrac{1500}{2000}\right)\left(\dfrac{900}{1500}\right) = .450$

ii. $P(Y \text{ and } F) = P(Y)P(F|Y) = \left(\dfrac{500}{2000}\right)\left(\dfrac{200}{500}\right) = .100$

b.

P(Y and M) = (500/2000)(300/500) = .150
P(Y and F) = (500/2000)(200/500) = .100
P(N and M) = (1500/2000)(900/1500) = .450
P(N and F) = (1500/2000)(600/1500) = .300

4.84 a.   i. *P*(more than one hour late *and* on Airline A)

= *P*(more than one hour late) *P*(Airline A | more than one hour late)

= (172 / 1700)(92 / 172) = .0541

ii. *P*(on Airline B and less than 30 minutes late)

= *P*(Airline B) *P*(less than 30 minutes late | Airline B)

= (789 / 1700)(393 / 789) = .2312

b. This probability is zero since "30 minutes to one hour late" and "more than one hour late" are mutually exclusive events.

4.85 a.   i. *P*(better off and high school) = *P*(better off) *P*(high school | better off)

= (1010 / 2000)(450 / 1010) = .225

ii. *P*(more than high school and worse off)

= *P*(more than high school) *P*(worse off | more than high school)

= (600 / 2000)(70 / 600) = .035

b. *P*(worse off and better off) = *P*(worse off) *P*(better off | worse off) = (570/2000)(0) = 0

This probability is zero because "worse off" and "better off" are mutually exclusive events.

4.86 Let: *A* = first student selected has a volunteered before,

B = first student selected has not volunteered before,

C = second student selected has volunteered before,

D = second student selected has not volunteered before.

The probability that both students have volunteered before is:

$$P(A \text{ and } C) = P(A)P(C \mid A) = (28 / 42)(27 / 41) = .4390$$

4.87 Let: A = first selected student favors abolishing the Electoral College

B = first selected student favors keeping the Electoral College

C = second selected student favors abolishing the Electoral College

D = second selected student favors keeping the Electoral College

The following is a tree diagram for the experiment of selecting two students.

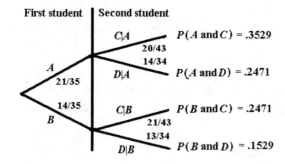

The probability that the both student favors abolishing the Electoral College is:

$$P(A \text{ and } C) = P(A)P(C \mid A) = (21 / 35)(20 / 34) = .3529$$

4.88 Let: A = first candidate selected is a woman

B = first candidate selected is a man

C = second candidate selected is a woman

D = second candidate selected is a man

The probability that both candidates selected are women is:

$$P(A \text{ and } C) = P(A)P(C \mid A) = (5 / 8)(4 / 7) = .3571$$

4.89   Let: C = first selected person has a type A personality

D = first selected person has a type B personality

E = second selected person has a type A personality

F = second selected person has a type B personality

The following is the tree diagram for the experiment of selecting two persons.

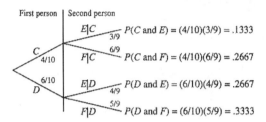

The probability that the first person has a type A personality and the second has a type B personality is:

$$P(C \text{ and } F) = P(C)P(F|C) = \left(\frac{4}{10}\right)\left(\frac{6}{9}\right) = .2667$$

4.90   Let: A = first selected senior has spent Spring Break in Florida

B = first selected senior has never spent Spring Break in Florida

C = second selected senior has spent Spring Break in Florida

D = second selected senior has never spent Spring Break in Florida

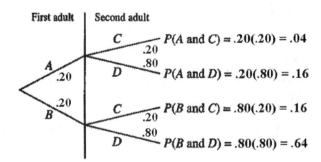

$P(B \text{ and } C) = P(B)P(C) = (.80)(.20) = .160$

Note that all seniors are independent in this exercise.

4.91   Let: N_1 = first adult selected has spent less than \$100 on lottery tickets last year

N_2 = second adult selected spent less than \$100 on lottery tickets last year

Because all adults are independent:

$P(N_1) = 1 - .35 = .65$, and $P(N_2) = 1 - .35 = .65$

$P(N_1 \text{ and } N_2) = P(N_1)P(N_2) = (.65)(.65) = .4225$

4.92 Let: A = wins first contract, C = wins second contract,

B = does not win first contract, and D = does not win second contract

   a. $P(A \text{ and } C) = P(A)P(C) = (.25)(.25) = .0625$

   b. $P(B \text{ and } D) = P(B)P(D) = (.75)(.75) = .5625$

4.93 Let: A = first item is returned C = second item is returned

B = first item is not returned and D = second item is not returned

   a. $P(A \text{ and } C) = P(A)P(C) = (.05)(.05) = .0025$

   b. $P(B \text{ and } D) = P(B)P(D) = (.95)(.95) = .9025$

4.94 Let: N_1 = first person selected is not allergic to the drug

N_2 = second person selected is not allergic to the drug

N_3 = third person selected is not allergic to the drug

Then, $P(N_1 \text{ and } N_2 \text{ and } N_3) = P(N_1)P(N_2)\,P(N_3) = (.97)(.97)(.97) = .9127$

4.95 Let: D_1 = first farmer selected is in debt

D_2 = second farmer selected is in debt

D_3 = third farmer selected is in debt

Then, $P(D_1 \text{ and } D_2 \text{ and } D_3) = P(D_1)P(D_2)\,P(D_3) = (.80)(.80)(.80) = .5120$

4.96 Let: L = spent more than \$100 on lottery tickets last year and M = person is married.

It is given that $P(L) = .35$ and $P(L \text{ and } M) = .16$

Then, $P(M\,|\,L) = P(L \text{ and } M)\,/\,P(L) = .16\,/\,.35 = .4571$

4.97 Let: F = employee selected is a female and M = employee selected is married

It is given that $P(F) = .36$ and $P(F$ and $M) = .19$

Hence, $P(M \mid F) = P(M$ and $F) / P(F) = .19 / .36 = .5278$

4.98   Let: F = person selected visiting doctor is a female   and Y = person selected is 15 to 24 years old

From the data, we can compute $P(Y$ and $F) = \dfrac{44{,}313}{823{,}542} = .0538$ and $P(F) = \dfrac{488{,}199}{823{,}542} = .5928$,

hence, $P(T|F) = P(T$ and $F) / P(F) = .0538 / .5928 = .0908$

4.99   Let: A = adult in small town lives alone,   and

   P = adult in small town has at least one pet.

It is given that $P(A) = .20$ and $P(A$ and $P) = .08$, hence $P(P|A) = P(A$ and $P) / P(A) = .08 / .20 = .400$

4.100   The union of two events A and B is the event "A or B", which is made up of all outcomes belonging to either A or B or both.

   Example: Let $A = \{1,2,3\}$ and $B = \{1,3,5\}$

   Then $(A$ or $B) = \{1,2,3,5\}$

4.101   When two events are mutually exclusive, their joint probability is zero and is dropped from the formula.

   Let A and B be two events. If A and B are mutually nonexclusive events, then:
$$P(A \text{ or } B) = P(A) + P(B) - P(A \text{ and } B)$$
   However, if A and B are mutually exclusive events, then $P(A \text{ or } B) = P(A) + P(B)$

4.102   When events A and B are not mutually exclusive, $P(A$ and $B) > 0$, so it is subtracted from $P(A) + P(B)$ to avoid double counting. For example, consider one roll of a die. Let:

   A = an even number is obtained   and B = a number less than 3 is obtained

   Then, $P(A \text{ or } B) = P(A) + P(B) - P(A \text{ and } B) = \dfrac{3}{6} + \dfrac{2}{6} - \dfrac{1}{6} = \dfrac{4}{6} = .6667$

4.103   The formula $P(A$ or $B) = P(A) + P(B)$ is used when A and B are mutually exclusive events.

   For example, consider one roll of a die. Let:

   A = outcome is a 1-spot   and B = outcome is an even number. Then

$$P(A \text{ and } B) = P(A) + P(B) = \frac{1}{6} + \frac{3}{6} = \frac{4}{6} = .6667$$

4.104   a. $P(A$ or $B) = P(A) + P(B) - P(A$ and $B) = .58 + .66 - .57 = .67$

   b. $P(A$ or $B) = P(A) + P(B) - P(A$ and $B) = .72 + .42 - .39 = .75$

4.105   a. $P(A \text{ or } B) = P(A) + P(B) - P(A \text{ and } B) = .18 + .49 - .11 = .56$

   b. $P(A \text{ or } B) = P(A) + P(B) - P(A \text{ and } B) = .73 + .71 - .68 = .76$

4.106   a. $P(A \text{ or } B) = P(A) + P(B) = .47 + .32 = .79$

   b. $P(A \text{ or } B) = P(A) + P(B) = .16 + .59 = .75$

4.107   a. $P(A \text{ or } B) = P(A) + P(B) = .25 + .27 = .52$

   b. $P(A \text{ or } B) = P(A) + P(B) = .58 + .09 = .67$

4.108   a. $P(V \text{ or } B) \quad = P(V) + P(B) - P(V \text{ or } B)$

$$= P(V) + P(B) - P(V)P(B|V) = \frac{312}{1800} + \frac{592}{1800} - \left(\frac{312}{1800}\right)\left(\frac{145}{312}\right) = .4217$$

   b. $P(N \text{ or } C) \quad = P(N) + P(C) - P(N \text{ or } C)$

$$= P(N) + P(C) - P(N)P(C|N) = \frac{1488}{1800} + \frac{404}{1800} - \left(\frac{1488}{1800}\right)\left(\frac{343}{1488}\right) = .8606$$

4.109   Let: M = basketball player selected is a male F = basketball player selected is a female

   G = player selected has graduated N = player selected has not graduated

   a. $P(F \text{ or } N) \quad = P(F) + P(N) - P(F \text{ and } N)$

$$= P(F) + P(N) - P(F)P(N|F) = \frac{165}{346} + \frac{87}{346} - \left(\frac{165}{346}\right)\left(\frac{32}{165}\right) = .6358$$

   b. $P(G \text{ or } M) \quad = P(G) + P(M) - P(G \text{ and } M)$

$$= P(G) + P(M) - P(G)P(M|G) = \frac{259}{346} + \frac{181}{346} - \left(\frac{259}{346}\right)\left(\frac{126}{259}\right) = .9075$$

4.110   Let: M = man, W = woman,

   Y = has retirement benefits, and N = does not have retirement benefits

   a. $P(W \text{ or } Y) = P(W) + P(Y) - P(W \text{ and } Y) = \dfrac{200}{500} + \dfrac{375}{500} - \dfrac{150}{500} = .850$

   b. $P(N \text{ or } M) = P(N) + P(M) - P(N \text{ and } M) = \dfrac{125}{500} + \dfrac{300}{500} - \dfrac{75}{500} = .700$

4.111   Let: Y = this adult has shopped on the internet M = male

   N = this adult has never shopped on the internet   and F = female

   a. $P(N \text{ or } F) = P(N) + P(F) - P(N \text{ and } F) = \dfrac{1500}{2000} + \dfrac{800}{2000} - \dfrac{600}{2000} = .850$

   b. $P(M \text{ or } Y) = P(M) + P(Y) - P(M \text{ and } Y) = \dfrac{1200}{2000} + \dfrac{500}{2000} - \dfrac{300}{2000} = .700$

c. Since Y and N are mutually exclusive events,

$$P(Y \text{ or } N) = P(Y) + P(N) = \frac{500}{2000} + \frac{1500}{2000} = 1.0$$

4.112 Let: E = less than 30 minutes late, $\qquad\qquad A$ = Airline A, and

$\qquad\qquad\quad F$ = 30 minutes to one hour late, $\qquad\quad B$ = Airline B

$\qquad\qquad\quad G$ = more than one hour late

a. $P(G \text{ or } A) = P(G) + P(A) - P(G \text{ and } A) = \dfrac{172}{1700} + \dfrac{911}{1700} - \dfrac{92}{1700} = .5829$

b. $P(B \text{ or } E) = P(B) + P(E) - P(B \text{ and } E) = \dfrac{789}{1700} + \dfrac{822}{1700} - \dfrac{393}{1700} = .7165$

c. Since A and B are mutually exclusive events, $P(A \text{ or } B) = P(A) + P(B) = \dfrac{789}{1700} + \dfrac{911}{1700} = 1.0$

4.113 B = better off, $\qquad\qquad\qquad S$ = same, $\qquad\qquad\qquad W$ = worse of

$\qquad\quad L$ = less than high school, $\qquad\quad H$ = high school, and $\qquad M$ = more than high school

a. $P(B \text{ or } H) = P(B) + P(H) - P(B \text{ and } H) = \dfrac{1010}{2000} + \dfrac{1000}{2000} - \dfrac{450}{2000} = .780$

b. $P(M \text{ or } W) = P(M) + P(W) - P(M \text{ and } W) = \dfrac{600}{2000} + \dfrac{570}{2000} - \dfrac{70}{2000} = .550$

c. Since B and W are mutually exclusive events, $P(B \text{ or } W) = P(B) + P(W) = \dfrac{1010}{2000} + \dfrac{570}{2000} = .790$

4.114 Let: V = vehicle selected vandalized $\qquad$ and $\qquad$ T = vehicle selected ticketed

Then, $P(T \text{ or } V) = P(T) + P(V) - P(T \text{ and } V) = .35 + .15 - .10 = .40$

4.115 Let: W = family selected owns a washing machine $\qquad V$ = family selected owns a VCR

Then, $P(W \text{ or } V) = P(W) + P(V) - P(W \text{ and } V) = .68 + .81 - .58 = .91$

4.116 Let: B = wedding day has bad weather $\qquad\qquad D$ = wedding day has a disruptive incident

Then, $P(B \text{ or } D) = P(B) + P(D) - P(B \text{ and } D) = .25 + .15 - .08 = .32$

4.117 Let: F = teacher selected is a female $\qquad\qquad S$ = teacher selected holds a second job

Then, $P(F \text{ or } S) = P(F) + P(S) - P(F \text{ and } S) = .68 + .38 - .29 = .77$

4.118 Let: H = death by heart attack, and $\qquad\qquad C$ = death by cancer

$$P(H \text{ or } C) = P(H) + P(C) = \frac{710{,}760}{2{,}403{,}351} + \frac{553{,}091}{2{,}403{,}351} = .5259$$

This probability is not equal to 1.0 because heart attacks and cancer are not the only causes of death in the U.S.

4.119 Let: A = response to poll was "player", and B = response to poll was "announcer/reporter"

$$P(A \text{ or } B) = P(A) + P(B) = \frac{11,715}{30,270} + \frac{9982}{30,270} = .7168$$

This probability is not equal to 1.0 because some respondents have other answers, like "cheerleader".

4.120 Let: A = student selected will get an A in economics

B = student selected will get a B in economics

Since A and B are mutually exclusive events, $P(A \text{ or } B) = P(A) + P(B) = .24 + .28 = .52$

This probability is not equal to 1.0 because not all students will get an A or B grade; some may get a C, D, or F grade.

4.121 $P(\text{against or indifferent}) = P(\text{against}) + P(\text{indifferent}) .63 + .17 = .80$

This probability is not equal to 1.0 because some voters favor the entry of a major discount store in their neighborhood.

4.122 Let: A = first selected corporation makes charitable contributions

B = first selected corporation does not make charitable contributions

C = second selected corporation makes charitable contributions

D = second selected corporation does not make charitable contributions

a.

b. $P(\text{at most one corporation makes charitable contributions})$

$= P(A \text{ and } D) + P(B \text{ and } C) + P(B \text{ and } D)$

$= (.72)(.28) + (.28)(.72) + (.28)(.28) = .2016 + .2016 + .0784 = .4816$

4.123 Let: A = first open-heart operation is successful

B = first open-heart operation is not successful

C = second open-heart operation is successful

D = second open-heart operation is not successful

$P(\text{at least one open-heart operation is successful}) = P(A \text{ and } C) + P(A \text{ and } D) + P(B \text{ and } C)$

$$= (.84)(.84) + (.84)(.16) + (.16)(.84) = .7056 + .1344 + .134 = .9744$$

4.124 a. *P*(selected car has a GPS navigation system) = 12 / 44 = .2727

   b. Number of cars which do not have a GPS navigation system = 44 – 12 = 32

   P(selected car does not have a GPS navigation system) = 32 / 44 = .7273

4.125 a. *P*(student selected is a junior) = 9 / 35 = .2571

   b. *P*(student selected is a freshman) = 5 / 35 = .1429

4.126 Let: *P* = psychology major, *C* = communications major,

   H = happy with major, and *U* = unhappy with major

   a. i. $P(H) = 195 / 250 = .7800$

   ii. $P(P) = 100 / 250 = .4000$

   iii. $P(C|H) = 115 / 195 = .5897$

   iv. $P(U|P) = 20/ 100 = .2000$

   v. $P(P \text{ and } H) = P(P)P(H/P) = \left(\dfrac{100}{250}\right)\left(\dfrac{80}{100}\right) = .3200$

   vi. $P(C \text{ or } U) = P(C) + P(U) - P(C \text{ and } U) = \dfrac{150}{250} + \dfrac{55}{120} - \left(\dfrac{150}{250}\right)\left(\dfrac{35}{150}\right) = .6800$

   b. $P(P) = 100/250 = .4000$ and $P(P|H) = 80/195 = .4103$

   Since these two probabilities are not equal, the events "*P*" and "*H*" are not independent.

   The events "*P*" and "*H*" are not mutually exclusive because they can occur together.

4.127 Let: *M* = adult selected is a male   *A* = adult selected prefers watching sports

   F = adult selected is a female   *B* = adult selected prefers watching opera

   a. i. $P(B) = 109 /250 = .4360$

   ii. $P(M) = 120 / 250 = .4800$

   iii. $P(A|F) = 45 / 130 = .3462$

   iv. $P(M|A) = 96 / 141 = .6809$

v. $P(F \text{ and } B) = P(F)P(B|F) = \left(\dfrac{130}{250}\right)\left(\dfrac{85}{130}\right) = .3400$

vi. $P(A \text{ or } M) = P(A) + P(M) - P(A \text{ and } M) = \dfrac{141}{250} + \dfrac{120}{250} - \left(\dfrac{141}{250}\right)\left(\dfrac{96}{141}\right) = .6600$

b. $P(F) = 130 / 250 = .5200$ $P(F|A) = 45 / 141 = .3191$

Since these two probabilities are not equal, the events "female" and "prefers watching sports" are not independent.

Events "female" and "prefers watching sports" are not mutually exclusive because they can occur together. There are a total of 45 adults who are female and prefer watching sports. If the selected adult is one of these 45, then both of these events occur at the same time.

4.128 Let: M = lawyer selected is a male A = lawyer selected favors capital punishment

            F = lawyer selected is a female B = lawyer selected opposes capital punishment

a.   i. $P(A) = 45 / 80 = .5625$

     ii. $P(F) = 24 / 80 = .3000$

     iii. $P(B|F) = 11 / 24 = .4583$

     iv. $P(M|A) = 32 / 45 = .7111$

     v. $P(F \text{ and } A) = P(F)P(A/F) = \left(\dfrac{24}{80}\right)\left(\dfrac{13}{24}\right) = .1625$

     vi. $P(B \text{ or } M) = P(B) + P(M) - P(B \text{ and } M) = \dfrac{35}{80} + \dfrac{56}{80} - \left(\dfrac{35}{80}\right)\left(\dfrac{24}{35}\right) = .8375$

b. $P(F) = 24 / 80 = .3000$ $P(F|B) = 11 / 35 = .3143$

Since these two probabilities are not equal, the events "female" and "opposes capital punishment" are not independent.

Events "female" and "opposes capital punishment" are not mutually exclusive because they can occur together. There are a total of 11 lawyers who are female and oppose capital punishment. If the selected lawyer is one of these 11, then both of these events occur at the same time.

4.129 Let: A = student selected is an athlete F = student selected favors paying college athletes

            B = student selected is a nonathlete N = student selected is against paying college athletes

a.   i. $P(F) = 300 / 400 = .750$

     ii. $P(F|B) = 210 / 300 = .700$

     iii. $P(A \text{ and } F) = P(A)P(F|A) = \left(\dfrac{100}{400}\right)\left(\dfrac{90}{100}\right) = .225$

iv. $P(B \text{ or } N) = P(B) + P(N) - P(B \text{ and } N) = \dfrac{300}{400} + \dfrac{100}{400} - \left(\dfrac{300}{400}\right)\left(\dfrac{90}{300}\right) = .775$

b. $P(A) = 100 / 400 = .250$ and $P(A|F) = 90 / 300 = .300$

Since these two probabilities are not equal, the events "athlete" and "should be paid" are not independent.

Events "athlete" and "should be paid" are not mutually exclusive because they can occur together. There are a total of 90 students who are athletes and favor paying athletes. If the selected student is one of these 90, then both of these events occur at the same time.

4.130 Let: A = first service call due to customer error C = second service call due to customer error

B = first service call due to broken appliance D = second service call due to broken appliance

a. $P(A \text{ and } C) = P(A)P(C) = (.05)(.05) = .0025$

b. $P(\text{at least one service call not due to customer error})$

$= P(A \text{ and } D) + P(B \text{ and } C) + P(B \text{ and } D) = .0475 + .0475 + .9025 = .9975$

4.131 Let: A = first person selected said they would trust public defender,

B = first person selected said they would not trust public defender,

C = second person selected said they would trust public defender, and

D = second person selected said they would not trust public defender.

a. $P(B \text{ and } D) = (211/290)(211/290) = .5294$

b. $P(\text{at most one would trust public defender})$

$= P(A \text{ and } D) + P(B \text{ and } C) + P(B \text{ and } D) = .1982 + .1982 + .5294 = .9258$

4.132   Let: G_1 = first car selected has a GPS system.

                         G_2 = second car selected has a GPS system.

$$P(G_1 \text{ and } G_2) = P(G_1)P(G_2|G_1) = \left(\frac{12}{44}\right)\left(\frac{11}{43}\right) = .0698$$

4.133   Let: J_1 = first student selected is a junior S_2 = second student selected is a sophomore

$$P(J_1 \text{ and } S_2) = P(J_1)P(S_2|J_1) = \left(\frac{9}{35}\right)\left(\frac{8}{34}\right) = .0605$$

4.134 P(company loses both sources of power) = P(power fails)P(generator stops working)= $(.30)(.09)= .027$

4.135 P(both machines are not working properly)

              = P(first machine is not working properly)P(second machine is not working properly)

              = $(.08)(.06) = .0048$

4.136   a. P(player wins for first time on tenth bet)

                     = P(player loses first nine bets and wins the tenth) = $\left(\frac{37}{38}\right)^9\left(\frac{1}{38}\right) = .0207$

              b. P(player loses first 50 bets) = $\left(\frac{37}{38}\right)^{50} = .2636$

              c. P(he wins at least once in 38 plays)

                     = $1 - P$(he loses on all 38 plays) = $1 - \left(\frac{37}{38}\right)^{38}$ = $1 - .3630 = .6370$

4.137   a.   There are 26 possibilities for each letter and 10 possibilities for each digit, hence there are

                     $26^3 \cdot 10^3 = 17,576,000$ possible different license places.

              b.   There are 2 possibilities for the second letter, 26 possibilities for the third letter and 100

                     possibilities for the two missing numbers, hence $2 \times 26 \times 100 = 5200$ license plates fit the

                     description.

4.138   Since the median life of these batteries is 100 hours, the probability that a given battery lasts longer

              than 100 hours is ½.

              P(exactly two will last longer than 100 hours)

               = P(first and second last longer but third does not)+ P(first and third last longer but second does not)

               + P(second and third last longer but first does not)

               = $(.5)(.5)(.5)+ (.5)(.5)(.5)+(.5)(.5)(.5) = .375$

4.139   Note: This exercise requires the use of combinations, which students may have studied in algebra. Combinations are covered in Section 5.5.2 in the text.

a. Let: A = the player's first five numbers match the numbers on the five white balls drawn by the lottery organization

B = the player's powerball number matches the powerball number drawn by the lottery organization, and

C = the player's powerball number does not match the powerball number drawn by the lottery organization

There are $_{53}C_5$ ways for the lottery organization to draw five different white balls from a set of 53 balls. Thus the sample space for this phase of the drawing consists of $(_{53}C_5)$ equally likely outcomes. Hence, $P(A) = 1 / {_{53}C_5}$.

Since the sample space for the drawing of the powerball number consists of 42 equally likely outcomes, $P(B) = 1 / 42$

Because the powerball number is drawn independently of the five white balls, A and B are independent events. Therefore, $P(\text{player wins jackpot}) = P(A \text{ and } B)$

$$= P(A)P(B) = \frac{1}{_{53}C_5} \times \frac{1}{42} = \frac{1}{120{,}526{,}770} = .0000000083$$

b. To win the \$100,000 prize, events A and C must occur. In the sample space of 42 equally likely outcomes for the drawing of the powerball numbers, there are 41 outcomes which do not match the powerball number, and thus result in event C. Hence, $P(C) = 41 / 42$. Therefore,

$P(\text{player wins the \$100,000 prize}) \ = \ P(A \text{ and } C) = \ P(A)\,P(C)$

$$= \frac{1}{(_{53}C_5)} \times \frac{41}{42} = \frac{41}{120{,}526{,}770} = .00000034$$

4.140   Let: L = left wing engine fails R = right wing engine fails
C = central engine fails

$P(\text{crash}) = P(C \text{ and } L, \text{ but not } R) + P(C \text{ and } R, \text{ but not } L) + P(C \text{ and } R \text{ and } L)$
$= (.005)(.008)(.992)+(.005)(.008)(.992)+(.005)(.008)(.008) = .00008$

4.141   a. $P(\text{sixth marble is red}) = 10 / 20 = .5000$

b. $P(\text{sixth marble is red}) = 5 / 15 = .3333$

c. The probability of obtaining a head on the sixth toss is .5, since each toss is independent of the previous outcomes.

Tossing a coin is mathematically equivalent to the situation in part a. Each drawing in part a is independent of previous drawings and the probability of drawing a red marble is .5 each time.

4.142   Let: C_1 = first card picked is a club D_1 = first card picked is a diamond

C_2 = second card picked is a club D_2 = second card picked is a diamond

$$P(\text{you win}) = P(C_1) \cdot P(C_2 \mid C_1) + P(D_1) \cdot P(D_2 \mid D_1) = \frac{2}{4} \cdot \frac{1}{3} + \frac{2}{4} \cdot \frac{1}{3} = \frac{4}{12} = .3333$$

Do not accept this proposition, since the chance of winning $10 is only .3333.

4.143   a.   The thief has three attempts to guess the correct PIN. Since there are 100 possible numbers in the beginning, the probability that he finds the number on the first attempt is 1/100. Assuming that the first guess is wrong, there are 99 numbers left, etc. Hence,

$P(\text{thief succeeds}) = 1 - P(\text{thief fails}) = 1 - P(\text{thief guessed incorrectly on all three attempts})$

$$= 1 - \left(\frac{99}{100} \cdot \frac{98}{99} \cdot \frac{97}{98} \right) = 1 - \frac{97}{100} = \frac{3}{100} = .030$$

b.   Since the first two digits of the four-digit PIN must be 3 and 5, respectively, and the third digit must be 1 or 7, the possible PINs are 3510 to 3519 and 3570 to 3579, a total of 20 possible PINs.

Hence, $P(\text{thief succeeds}) = 1 - P(\text{thief fails}) = 1 - \left(\frac{19}{20} \cdot \frac{18}{19} \cdot \frac{17}{18} \right) = 1 - \frac{17}{20} = \frac{3}{20} = .150$

4.144   a. $P(\text{you must pay gambler}) = P(\text{gambler rolls at least one 6 in four tries})$
         $= 1 - P(\text{he rolls no 6 in four tries}) = 1 - (5/6)^4 = 1 - .4823 = .5177$

b.   Let: E = you obtain at least one double six in 24 rolls,

A_1 = you obtain a double six on the first roll,

A_2 = you obtain a double six on the second roll,

.
.
.

A_{24} = you obtain a double six on the 24th roll

Then, $P(E) = P(A_1 \text{ or } A_2 \text{ or } \cdots \text{ or } A_{24})$

However, $A_1, A_2, \ldots, A_{24}$ are not mutually exclusive, since it is possible to obtain a double 6 on more than one roll. Thus, we cannot find $P(A_1 \text{ or } A_2 \text{ or } \ldots \text{ or } A_{24})$ by simply adding $P(A_1) + P(A_2) + \ldots + P(A_{24})$ as the gambler does to obtain 24 (1/36).

To find $P(E)$ we may use complementary events, as in part a.

The probability of failing to roll a double six on any one attempt is $1 - 1/36 = 35/36$.

Hence, $P(E) = 1 - P(\text{you roll no double six in 24 tries}) = 1 - \left(\frac{35}{36} \right)^{24} = 1 - .5086 = .4914$

Since your chance of winning is less than 50%, the gambler has the advantage and you should not accept his proposition.

4.145   a.   Answers may vary.

b.   Let: B = no two of the 25 students share the same birthday

Note that if B is to occur, the second student cannot have the same birthday as the first student, so the second may have any of the remaining 364 birthdays.

Thus, P(first two students do not share a birthday) = 364 / 365

Furthermore, if B is to occur, the third student cannot have the same birthday as either of the first two. Thus, the third may have any of the remaining 363 birthdays.

Therefore, P(first three students do not share a birthday) = $\dfrac{364}{365} \cdot \dfrac{363}{365}$. By extending this logic we can show that P(no two of the 25 students share a birthday) = $\dfrac{364}{365} \cdot \dfrac{363}{365} \cdot \dfrac{362}{365} \cdots \dfrac{341}{365}$ = .4313

Thus, since A and B are complementary events, $P(A) = 1 - P(B) = 1 - .4313 = .5687$

Therefore, $P(A)$ is greater than 50%.

4.146   Let: S = Smith family is selected, J = Jones family is selected

G = a girl is selected,   and B = a boy is selected

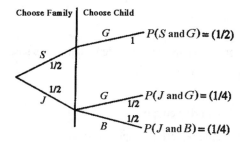

Note that "the other child is a girl" is equivalent to P(the Smith family was selected).

Thus, we must find $P(S|G)$, which is given by $\dfrac{P(S \text{ and } G)}{P(G)}$

From the tree diagram: $P(S \text{ and } G)$ = ½, and $P(G) = P(S \text{ and } G) + P(J \text{ and } G) = \dfrac{1}{2} + \dfrac{1}{4} = \dfrac{3}{4}$

Thus, $P(S|G) = \dfrac{1/2}{3/4} = 2/3 = .667$

Therefore, the probability that the other child is a girl is .6667.

4.147   a.   Let E = neither topping is anchovies

A = customer's first selection is not anchovies

B = customer's second selection is not anchovies

For A to occur, the customer may choose any of 11 toppings from the 12 available.

Thus $P(A) = \dfrac{11}{12}$.

For B to occur, given that A has occurred, the customer may choose any of 10 toppings from the remaining 11. Thus $P(B|A) = \dfrac{10}{11}$.

Therefore, $P(E) = P(A \text{ and } B) = P(A \text{ and } B) = P(A)P(B|A) = \left(\dfrac{11}{12}\right)\left(\dfrac{10}{11}\right) = .8333$

b. Let   C = pepperoni is one of the toppings

Then $\overline{C}$ = neither topping is pepperoni

By the same form of argument used to find $P(E)$ in part a, we obtain

$$P(\overline{C}) \ = \ \left(\dfrac{11}{12}\right)\left(\dfrac{10}{11}\right) = .8333$$

Thus, $P(C) = 1 - P(\overline{C}) \ = 1 - .8333 = .1667$

4.148   Note: This exercise requires the use of combinations, which students may have studied in algebra. Combinations are covered in Section 5.5.2 of the text.

Let: A = none of the six boxes sold contains a prize coupon,

   B = at least one of the six boxes sold contain a prize coupon, and

   C = two of the six boxes sold contain prize coupons

a. The sample space for picking six boxes at random from a set of nine boxes has $\left({}_9C_6\right)$ equally likely outcomes. For A to occur, any six of the seven boxes containing no coupons may be selected. Thus, there are ${}_7C_6$ ways for A to occur. Therefore, $P(A) = \dfrac{{}_7C_6}{{}_9C_6} = \dfrac{7}{84} = .0833$

b. A and B are complementary events, so $P(B) = 1 - P(A) = 1 - .0833 = .9167$

c. The probability that two of the six boxes sold contain prize coupons is the same as the probability that none of the three remaining boxes on the shelf contains a prize coupon. Using the same

reasoning as in part a, $P(C) = \dfrac{{}_7C_3}{{}_9C_3} = \dfrac{35}{84} = .4167$

4.149   Let: C = auto policy holders with collision coverage

U = auto policy holders with uninsured motorist coverage

a. $P(C \text{ or } U) = P(C) + P(U) - P(C \text{ and } U)$.

Hence, $P(C \text{ and } U) = P(C) + P(U) - P(C \text{ or } U) = .80 + .60 - .93 = .47$

Thus 47% of the policy holders have both collision and uninsured motorist coverage.

b. The event "$\overline{C}$ and $\overline{U}$" is the complement of the event "C or U".

Hence, $P(\overline{C} \text{ and } \overline{U}) = 1 - P(C \text{ or } U) = 1 - .93 = .07$

Thus, 7% of the policy holders have neither collision nor uninsured motorist coverage.

c. The group of policy holders who have collision but not uninsured motorist coverage may be formed by considering the policy holders that have collision, then removing those who have both collision and uninsured motorist coverage. Hence,

$P(C \text{ and } \overline{U}) = P(C) - P(C \text{ and } U) = .80 - .47 = .33$

Thus 33% of policy holders have collision but not uninsured motorist coverage.

4.150   Let: S = an apple is satisfactory Y = Andy judges an apple as satisfactory

U = an apple is unsatisfactory N = Andy judges an apple as unsatisfactory

The following is the tree diagram for the experiment of picking an apple and judging it.

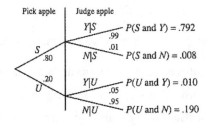

a. $P(S \text{ and } Y) = P(S)P(Y|S) = .792$ or 79.2%

b. $P(\text{marketed}) = P(Y) = P(S \text{ and } Y) + P(U \text{ and } Y) = .792 + .010 = .802$ or 80.2%

c. Using the solutions to parts *a* and *b*:

$P(\text{satisfactory given marketed}) = P(S|Y) = P(S \text{ and } Y) / P(Y) = .792 / .802 = .988$ or 98.8%

4.151   Let: p = the proportion of the students who used drugs during the past week

D = a student used drugs during the past week

N = a student did not use drugs during the past week

T = a student answers the question truthfully

U = a student answers the question untruthfully

The following is the tree diagram for this problem:

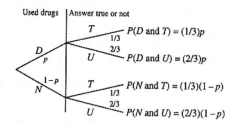

The proportion of students who said that they used drugs during the past week is given by the sum of the probabilities of the events "D and T" and "N and U". Hence,

P(a student says he or she used drugs during the past week)

$$= P(D \text{ and } T) + P(N \text{ and } U) = \frac{1}{3}p + \frac{2}{3}(1-p) = .60, \text{ which gives } p = .20$$

Thus, we estimate that 20% of the students used drugs during the past week.

Self-Review Test for Chapter Four

1. a   **2.** b   **3.** c   **4.** a   **5.** a   **6.** b

7. c   **8.** b   **9.** b   **10.** c   **11.** b

12. Total outcomes = 4 × 3 × 5 × 2 = 120

13. a. *P*(job offer selected is from the insurance company) = 1/ 3 = .3333

     b. *P*(job offer selected is not from the accounting firm) = 2 / 3 = .6667

14. a. *P*(out of state) = 125 / 200 = .6250   *P*(out of state | female) = 70 / 110 = .6364

Since these two probabilities are not equal, the two events are not independent. Events "female" and "out of state" are not mutually exclusive because they can occur together. There are a total of 70 students who are female and out of state. If any of these students is selected then both of these events occur at the same time.

     b. i. In 200 students, there are 90 males. Hence, *P*(a male is selected) = 90 / 200= .4500

       ii. There are a total of 110 female students and 70 of them are out of state students.

          Hence, *P*(out of state | female) = 70 / 110 = .6364

15. $P(\text{out of state or female}) = P(\text{out of state}) + P(\text{female}) - P(\text{out of state and female})$

$= P(\text{out of state}) + P(\text{female}) - P(\text{out of state}) P(\text{female }|\text{out of state}) = \dfrac{125}{200} + \dfrac{110}{200} - \left(\dfrac{110}{200}\right)\left(\dfrac{70}{110}\right) = .825$

16. Let: $(S_1) = $ first student selected is from out of state $(S_2) = $ second student selected is from out of state

The probability that both students selected are from out of state is calculated as follows:

$$P(S_1 \text{ and } S_2) = P(S_1)P(S_2|S_1) = \left(\dfrac{125}{200}\right)\left(\dfrac{124}{199}\right) = .3894$$

17. Let: $(F_1) = $ first adult selected has experienced a migraine headache

$(N_1) = $ first adult selected has never experienced a migraine headache

$(F_2) = $ second adult selected has experienced a migraine headache

$(N_2) = $ second adult selected has never experienced a migraine headache

The probability that neither of the two adults have ever experienced a migraine headache is calculated as follows. Note that the two adults are independent. From the given information: $P(F_1) = .35$, and $P(F_2) = .35$. Hence, $P(N_1) = 1 - .35 = .65$ and $P(N_2) = 1 - .35 = .65$. $P(N_1 \text{ and } N_2) = P(N_1)P(N_2) = (.65)(.65) = .4225$

18. $P(A) = 8 / 20 = .400$ The complementary event of A is that the selected marble is not red, that is, the selected marble is either green or blue. Hence, $P(A) = 1 - .400 = .600$.

19. Let: $M = $ male, $F = $ female, $W = $ works at least 10 hours, and $N = $ does not work 10 hours
  a. $P(M \text{ and } W) = P(M)P(W) = (.45)(.62) = .279$
  b. $P(F \text{ or } W) = P(F) + P(W) - P(F \text{ and } W) = .55 + .62 - (.55)(.62) = .829$

20. a. i. $P(Y) = (77 + 104) / 506 = .3577$

   ii. $P(Y|W) = 104 / (104 + 119 + 34) = .4047$

   iii. $P(W \text{ and } N) = P(W)P(N/W) = \left(\dfrac{257}{506}\right)\left(\dfrac{119}{257}\right) = .2352$

   iv. $P(N \text{ or } M) = P(N) + P(M) - P(N \text{ and } M) = \dfrac{66}{506} + \dfrac{249}{506} - \left(\dfrac{66}{506}\right)\left(\dfrac{32}{66}\right) = .5593$

  b. $P(W) = 257/506 = .5079$ $\qquad\qquad$ $P(W|Y) = 104 / 181 = .5746$

   Since these two probabilities are not equal, the events "woman" and "yes" are not independent. Events "woman" and "yes" are not mutually exclusive because they can occur together. There are a total of 104 adults who are a woman and belong to the "yes" category. If the selected adult is one of these 104, then both of these events occur at the same time.

Chapter Five

5.1 **Random variable:** A variable whose value is determined by the outcome of a random experiment is called a random variable. An example of this is the income of a randomly selected family.

Discrete random variable: A random variable whose values are countable is called a discrete random variable. An example of this is the number of cars in a parking lot at any particular time.

Continuous random variable: A random variable that can assume any value in one or more intervals is called a continuous random variable. An example of this is the time taken by a person to travel by car from New York City to Boston.

5.2 a. a continuous random variable   b. a discrete random variable

      c. a discrete random variable   d. a continuous random variable

      e. a discrete random variable   f. a continuous random variable

5.3 a. a discrete random variable   b. a continuous random variable

      c. a continuous random variable   d. a discrete random variable

      e. a discrete random variable   f. a continuous random variable

5.4 The number of households x watching ABC is a discrete random variable because the values of x are countable: 0, 1, 2, 3, 4 and 5.

5.5 The number of cars x that stop at the Texaco station is a discrete random variable because the values of x are countable: 0, 1, 2, 3, 4, 5 and 6.

5.6 The probability distribution of a discrete random variable lists all the possible values that the random variable can assume and their corresponding probabilities. As an example, the following table lists the probability distribution of x where x is the number of cars owned by a family living in a city.

x	$P(x)$
0	0.04
1	0.32
2	0.35
3	0.18
4	0.11

The probability distribution of a discrete random variable can be presented in the form of a mathematical formula, a table, or graph.

5.7   The two characteristics of the probability distribution of a discrete random variable x are:

1. The probability that x assumes any single value lies in the range 0 to 1, that is, $0 \leq P(x) \leq 1$.

2. The sum of the probabilities of all values of x for an experiment is equal to 1, that is: $\sum P(x) = 1$.

5.8   a. This table does represent a valid probability distribution of x because it satisfies both conditions required for a valid probability distribution.

b. This table does not represent a valid probability distribution of x because the sum of the probabilities of all outcomes listed in the table is not 1, which violates the second condition of a probability distribution.

c. This table does not represent a valid probability distribution of x because the probability of $x = 7$ is negative, which violates the first condition of a probability distribution.

5.9   a. This table does not satisfy the first condition of a probability distribution because the probability of $x = 5$ is negative. Hence, it does not represent a valid probability distribution of x.

b. This table represents a valid probability distribution of x because it satisfies both conditions required for a valid probability distribution.

c. This table does not represent a valid probability distribution of x because the sum of the probabilities of all outcomes listed in the table is not 1, which violates the second condition of a probability distribution.

5.10   a. $P(x = 3) = .15$

b. $P(x \leq 2) = P(0) + P(1) + P(2) = .11 + .19 + .28 = .58$

c. $P(x \geq 4) = P(4) + P(5) + P(6) = .12 + .09 + .06 = .27$

d. $P(1 \leq x \leq 4) = P(1) + P(2) + P(3) + P(4) = .19 + .28 + .15 + .12 = .74$

e. $P(x < 4) = P(0) + P(1) + P(2) + P(3) = .11 + .19 + .28 + .15 = .73$

f. $P(x > 2) = P(3) + P(4) + P(5) + P(6) = .15 + .12 + .09 + .06 = .42$

g. $P(2 \leq x \leq 5) = P(2) + P(3) + P(4) + P(5) = .28 + .15 + .12 + .09 = .64$

5.11   a. $P(x = 1) = .17$

b. $P(x \leq 1) = P(0) + P(1) = .03 + .17 + = .20$

c. $P(x \geq 3) = P(3) + P(4) + P(5) = .31 + .15 + .12 = .58$

d. $P(0 \leq x \leq 2) = P(0) + P(1) + P(2) = .03 + .17 + .22 = .42$

e. $P(x < 3) = P(0) + P(1) + P(2) = .03 + .17 + .22 = .42$

f. $P(x > 3) = P(4) + P(5) = .15 + .12 = .27$

g. $P(2 \leq x \leq 4) = P(2) + P(3) + P(4) = .22 + .31 + .15 = .68$

5.12   Let x = number of exercise machines sold at Elmo's on a given day.

a.

b.   i. $P(\text{exactly } 6) = P(6) = .14$

     ii. $P(\text{more than } 8) = P(x > 8) = P(9) + P(10) = .16 + .12 = .28$

     iii. $P(5 \text{ to } 8) = P(5 \leq x \leq 8) = P(5) + P(6) + P(7) + P(8) = .11 + .14 + .19 + .20 = .64$

     iv. $P(\text{at most } 6) = P(x \leq 6) = P(4) + P(5) + P(6) = .08 + .11 + .14 = .33$

5.13   a.

b.   i. $P(\text{exactly } 3) = P(3) = .25$

     ii. $P(\text{at least } 4) = P(x \geq 4) = P(4) + P(5) + P(6) = .14 + .07 + .03 = .24$

     iii. $P(\text{less than } 3) = P(x < 3) = P(0) + P(1) + P(2) = .10 + .18 + .23 = .51$

     iv. $P(2 \text{ to } 5) = P(2) + P(3) + P(4) + P(5) = .23 + .25 + .14 + .07 = .69$

5.14   a. Let x denote the number of TV sets owned by a family. The following table gives the probability distribution of x.

x	$P(x)$
0	120 / 2500 = .048
1	970 / 2500 = .388
2	730 / 2500 = .292
3	410 / 2500 = .164
4	270 / 2500 = .108

b. The probabilities listed in the table of part a are exact because they are based on data from the entire population.

c. i. $P(x = 1) = .388$

ii. $P(x > 2) = + P(3) + P(4) = .164 + .108 = .272$

iii. $P(x \leq 1) = P(0) + P(1) = .048 + .338 = .436$

iv. $P(1 \leq x \leq 3) = P(1) + P(2) + P(3) = .388 + .292 + .164 = .844$

5.15 a.

x	$P(x)$
1	8 / 80 = .10
2	20 / 80 = .25
3	24 / 80 = .30
4	16 / 80 = .20
5	12 / 80 = .15

b. The probabilities listed in the table of part a are approximate because they are obtained from a sample of 80 days.

c. i. $P(x = 3) = .30$

ii. $P(x \geq 3) = P(3) + P(4) + P(5) = .30 + .20 + .15 = .65$

iii. $P(2 \leq x \leq 4) = P(2) + P(3) + P(4) = .25 + .30 + .20 = .75$

iv. $P(x \leq 4) = P(2) + P(3) + P(4) = .10 + .25 + .30 = .65$

5.16 Let L = car selected is a lemon   and G = car selected is good

Then, $P(L) = .05$ and $P(G) = 1 - .05 = .95$

Let x be the number of lemons in two selected cars. The following table lists the probability distribution of x. Note that $x = 0$ if both cars are good, $x = 1$ if one car is good and the other car is a lemon, and $x = 2$ if both cars are lemons. The probabilities are written in the table using the above tree diagram. The probability of $x = 1$ is obtained by adding the probabilities of *LG* and *GL*.

Outcomes	x	$P(x)$
GG	0	.9025
LG or GL	1	.0950
LL	2	.0025

5.17 Let Y = owns a cell phone and N = does not own a cell phone.

Then $P(Y) = .64$ and $P(N) = 1 - .64 = .36$

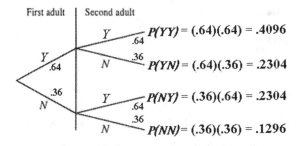

Let x be the number of adults in a sample of two who own a cell phone. The following table lists the probability distribution of x. Note that $x = 0$ if neither adult owns a cell phone, $x = 1$ if one adult owns a cell phone and the other does not, and $x = 2$ if both adults own a cell phone. The probabilities are written in the table using the tree diagram above. The probability that $x = 1$ is obtained by adding the probabilities of *YN* and *NY*

Outcomes	x	$P(x)$
NN	0	.1296
YN or NY	1	.4608
YY	2	.4096

5.18 Let: A = adult selected is against using animals for research

N = adult selected is not against using animals for research

Then, $P(A) = .30$ and $P(N) = 1 - .30 = .70$

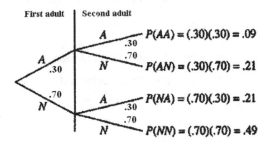

Let x be the number of adults in a sample of two adults who are against using animals for research. The following table lists the probability distribution of x.

Outcomes	x	$P(x)$
NN	0	.49
AN or NA	1	.42
AA	2	.09

5.19   Let: Y = teen said teachers were "totally clueless" about using the internet for teaching and learning.

N = teen said teachers were not "totally clueless" about using the internet for teaching and learning.

Then $P(Y) = .78$ and $P(N) = 1 - .78 = .22$

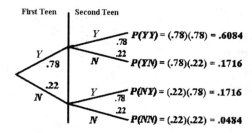

Let x denote the number of teens in a sample of two teens who believe that teachers are "totally clueless" about using the internet for teaching and learning. The following table lists the probability distribution of x.

Outcomes	x	$P(x)$
NN	0	.0484
AN or NA	1	.3432
AA	2	.6084

5.20   Let: A = first person selected is left-handed C = second person selected is left-handed

B = first person selected is right-handed D = second person selected is right-handed

First person | Second person

C|A $P(AC) = (3/12)(2/11) = .0455$
2/11
A   D|A $P(AD) = (3/12)(9/11) = .2045$
3/12   9/11
9/12   C|B $P(BC) = (9/12)(3/11) = .2045$
B   3/11
D|B $P(BD) = (9/12)(8/11) = .5455$
8/11

Outcomes	x	$P(x)$
BD	0	.5455
AD or BC	1	.4090
AC	2	.0455

5.21   Let A = first athlete selected used drugs C = second athlete selected used drugs

B = first athlete selected did not use drugs D = second athlete selected did not use drugs

First athlete | Second athlete

$P(AC) = (6/20)(5/19) = .0789$

$P(AD) = (6/20)(14/19) = .2211$

$P(BC) = (14/20)(6/19) = .2211$

$P(BD) = (14/20)(13/19) = .4789$

Let x be the number of athletes who used illegal drugs in a sample of two athletes. The following table lists the probability distribution of x.

Outcomes	x	$P(x)$
BD	0	.4789
AD or BC	1	.4422
AC	2	.0789

5.22 The mean of discrete random variable x is the value that is expected to occur per repetition, on average, if an experiment is repeated a large number of times. It is denoted by μ and calculated as

$$\mu = \sum xP(x)$$

The standard deviation of a discrete random variable x measures the spread of its probability distribution. It is denoted by σ and is calculated as $\sigma = \sqrt{\sum x^2 P(x) - \mu^2}$

5.23 a.

x	$P(x)$	$xP(x)$	$x^2P(x)$
0	.16	0	0
1	.27	.27	.27
2	.39	.78	1.56
3	.18	.54	1.62
		$\sum xP(x) = 1.59$	$\sum x^2 P(x) = 3.45$

$$\mu = \sum xP(x) = 1.590$$
$$\sigma = \sqrt{\sum x^2 P(x) - \mu^2} = \sqrt{3.45 - (1.59)^2} = .960$$

b.

x	$P(x)$	$xP(x)$	$x^2P(x)$
6	.40	2.40	14.40
7	.26	1.82	12.74
8	.21	1.68	13.44
9	.13	1.17	10.53
		$\sum xP(x) = 7.07$	$\sum x^2 P(x) = 51.11$

$$\mu = \sum xP(x) = 7.070$$
$$\sigma = \sqrt{\sum x^2 P(x) - \mu^2} = \sqrt{51.11 - (7.07)^2} = 1.061$$

5.24   a.

x	$P(x)$	$xP(x)$	$x^2P(x)$
3	.09	.27	.81
4	.21	.84	3.36
5	.34	1.70	8.50
6	.23	1.38	8.28
7	.13	.91	6.37
		$\sum xP(x) = 5.10$	$\sum x^2P(x) = 27.32$

$$\mu = \sum xP(x) = 5.10$$
$$\sigma = \sqrt{\sum x^2 P(x) - \mu^2} = \sqrt{27.32 - (5.10)^2} = 1.145$$

b.

x	$P(x)$	$xP(x)$	$x^2P(x)$
0	.43	0	0
1	.31	.31	.31
2	.17	.34	.36
3	.09	.27	.81
		$\sum xP(x) = .92$	$\sum x^2P(x) = 1.80$

$$\mu = \sum xP(x) = .92$$
$$\sigma = \sqrt{\sum x^2 P(x) - \mu^2} = \sqrt{1.80 - (.92)^2} = .977$$

5.25

x	$P(x)$	$xP(x)$	$x^2 P(x)$
0	.73	0	.00
1	.16	.16	.16
2	.06	.12	.24
3	.04	.12	.36
4	.01	.04	.16
		$\sum xP(x) = .44$	$\sum x^2P(x) = .92$

$$\mu = \sum xP(x) = .440 \text{ error}$$
$$\sigma = \sqrt{\sum x^2 P(x) - \mu^2} = \sqrt{.92 - (.44)^2} = .852 \text{ error}$$

5.26

x	$P(x)$	$xP(x)$	$x^2 P(x)$
4	.17	.68	2.27
5	.26	1.30	6.50
6	.19	1.14	6.84
7	.13	.91	6.37
8	.11	.88	7.04
9	.10	.90	8.10
10	.04	.40	4.00
		$\sum xP(x) = 6.21$	$\sum x^2P(x) = 41.57$

$$\mu = \sum xP(x) = 6.21 \text{ pigs}$$
$$\sigma = \sqrt{\sum x^2 P(x) - \mu^2} = \sqrt{41.57 - (6.21)^2} = 1.734 \text{ pigs}$$

5.27

x	$P(x)$	$xP(x)$	$x^2 P(x)$
0	.05	0	0
1	.12	.12	.12
2	.19	.38	.76
3	.30	.90	2.70
4	.20	.80	3.20
5	.10	.50	2.50
6	.04	.24	1.44
		$\sum xP(x) = 2.94$	$\sum x^2 P(x) = 10.72$

$\mu = \sum xP(x) = 2.94$ camcorders

$\sigma = \sqrt{\sum x^2 P(x) - \mu^2} = \sqrt{10.72 - (2.94)^2} = 1.441$ camcorders

On average, 2.94 camcorders are sold per day at this store.

5.28 Let x be the number of exercise machines sold on a given day at Elmo's.

x	$P(x)$	$xP(x)$	$x^2 P(x)$
4	.08	.32	1.28
5	.11	.55	2.75
6	.14	.84	5.04
7	.19	1.33	9.31
8	.20	1.60	12.80
9	.16	1.44	12.96
10	.12	1.20	12.00
		$\sum xP(x) = 7.28$	$\sum x^2 P(x) = 56.14$

$\mu = \sum xP(x) = 7.28$ machines

$\sigma = \sqrt{\sum x^2 P(x) - \mu^2} = \sqrt{56.14 - (7.28)^2} = 1.772$ machines.

The value of the mean, $\mu = 7.28$, indicates that Elmo's sells an average of 7.28 exercise machines per day.

5.29

x	$P(x)$	$xP(x)$	$x^2 P(x)$
0	.25	0	0
1	.50	.50	.50
2	.25	.50	1.00
		$\sum xP(x) = 1.00$	$\sum x^2 P(x) = 1.50$

$\mu = \sum xP(x) = 1.000$ head

$\sigma = \sqrt{\sum x^2 P(x) - \mu^2} = \sqrt{1.50 - (1.00)^2} = .707$ heads

The value of the mean, $\mu = 1.00$, indicates that, on average, we will expect to obtain 1 head in every two tosses of the coin.

5.30

x	$P(x)$	$xP(x)$	$x^2 P(x)$
0	.14	.00	.00
1	.28	.28	.28
2	.22	.44	.88
3	.18	.54	1.62
4	.12	.48	1.92
5	.06	.30	1.50
		$\sum xP(x) = 2.04$	$\sum x^2 P(x) = 6.20$

$\mu = \sum xP(x) = 2.04$ potential weapons

$\sigma = \sqrt{\sum x^2 P(x) - \mu^2} = \sqrt{6.2 - (2.04)^2} = 1.428$ potential weapons

The value of the mean, $\mu = 2.04$, indicates, on average, 2.04 potential weapons are found per day.

5.31

x	$P(x)$	$xP(x)$	$x^2 P(x)$
0	.048	0	0
1	.388	.388	.388
2	.292	.584	1.168
3	.164	.492	1.476
4	.108	.432	1.728
		$\sum xP(x) = 1.896$	$\sum x^2 P(x) = 4.760$

$\mu = \sum xP(x) = 1.896 \approx 1.9$ TV sets

$\sigma = \sqrt{\sum x^2 P(x) - \mu^2} = \sqrt{4.76 - (1.896)^2} = 1.079$ TV sets

Thus, there is an average of 1.90 TV sets per family in this town, with a standard deviation of 1.079 sets.

5.32

x	$P(x)$	$xP(x)$	$x^2 P(x)$
1	.10	.10	.00
2	.25	.50	1.00
3	.30	.90	2.70
4	.20	.80	3.20
5	.15	.75	3.75
		$\sum xP(x) = 3.05$	$\sum x^2 P(x) = 10.75$

$\mu = \sum xP(x) = 3.05$ installations

$\sigma = \sqrt{\sum x^2 P(x) - \mu^2} = \sqrt{10.75 - (3.05)^2} = 1.203$ installations

Thus, the average number of installations is 3.05 per day with a standard deviation of 1.203 installations.

5.33

x	P(x)	xP(x)	$x^2 P(x)$
0	.9025	0	0
1	.0950	.0950	.0950
2	.0025	.0050	.0100
		$\sum xP(x) = .1000$	$\sum x^2 P(x) = .1050$

$\mu = \sum xP(x) = .10$ car

$\sigma = \sqrt{\sum x^2 P(x) - \mu^2} = \sqrt{.1050 - (.10)^2} = .308$ car

5.34

x	P(x)	xP(x)	$x^2 P(x)$
0	.1296	0	0
1	.4608	.4608	.4608
2	.4096	.8192	1.6384
		$\sum xP(x) = 1.280$	$\sum x^2 P(x) = 2.0992$

$\mu = \sum xP(x) = 1.280$ adults

$\sigma = \sqrt{\sum x^2 P(x) - \mu^2} = \sqrt{20992 - (1.280)^2} = .679$ adult

5.35

x	P(x)	xP(x)	$x^2 P(x)$
10	.15	1.50	15.00
5	.30	1.50	7.50
2	.45	0.90	1.80
0	.10	0	0
		$\sum xP(x) = 3.9$	$\sum x^2 P(x) = 24.30$

$\mu = \sum xP(x) = \$3.9$ million

$\sigma = \sqrt{\sum x^2 P(x) - \mu^2} = \sqrt{24.3 - (3.9)^2} = \$3.015$ million.

Thus, the contractor is expected to make $3.9 million profit with a standard deviation of $3.015 million.

5.36   Note that the price of the ticket ($2) must be deducted from the amount won. For example, the $5 prize results in a net gain of $5 – $2 = $3.

x	P(x)	xP(x)	$x^2 P(x)$
3	1000/10,000=.10	.30	0.9
8	100/10,000=.01	.08	0.64
998	5/10,000=.0005	.499	498.002
4998	1/10,000=.0001	.4998	2498
-2	8894/10,000=.8894	-1.7788	3.5576
		$\sum xP(x) = -.40$	$\sum x^2 P(x) = 3001.10$

$\mu = \sum xP(x) = \$-.40$

$\sigma = \sqrt{\sum x^2 P(x) - \mu^2} = \sqrt{3001.10 - (-.40)^2} = \$54.78$

Thus, on average, the players who play this game are expected to lose \$.40 per ticket with a standard deviation of \$54.78.

5.37

x	$P(x)$	$xP(x)$	$x^2 P(x)$
0	.5455	0	0
1	.4090	.4090	.4090
2	.0455	.0910	.1820
		$\sum xP(x) = .5000$	$\sum x^2 P(x) = .5910$

$\mu = \sum xP(x) = .500 \text{ person}$

$\sigma = \sqrt{\sum x^2 P(x) - \mu^2} = \sqrt{.5910 - (.50)^2} = .584 \text{ person}$

5.38

x	$P(x)$	$xP(x)$	$x^2 P(x)$
0	.4789	.0000	.0000
1	.4422	.4422	.4422
2	.0789	.1578	.3156
		$\sum xP(x) = .600$	$\sum x^2 P(x) = .7578$

$\mu = \sum xP(x) = .600 \text{ athlete}$

$\sigma = \sqrt{\sum x^2 P(x) - \mu^2} = \sqrt{.7578 - (.600)^2} = .631 \text{ athlete}$

5.39 $3! = 3 \times 2 \times 1 = 6$

$(9 - 3)! = 6! = 6 \times 5 \times 4 \times 3 \times 2 \times 1 = 720$

$9! = 9 \times 8 \times 7 \times 6 \times 5 \times 4 \times 3 \times 2 \times 1 = 362,880$

$(14 - 12)! = 2! = 2 \times 1 = 2$

$_5C_3 = \dfrac{5!}{(5-3)! \times 3!} = \dfrac{5!}{2! \times 3!} = \dfrac{5 \times 4 \times 3 \times 2 \times 1}{2 \times 1 \times 3 \times 2 \times 1} = 10$

$_7C_4 = \dfrac{7!}{(7-4)! \times 4!} = \dfrac{7!}{3! \times 4!} = \dfrac{7 \times 6 \times 5 \times 4 \times 3 \times 2 \times 1}{3 \times 2 \times 1 \times 4 \times 3 \times 2 \times 1} = 35$

$_9C_3 = \dfrac{9!}{(9-3)! \times 3!} = \dfrac{9!}{6! \times 3!} = \dfrac{9 \times 8 \times 7 \times 6 \times 5 \times 4 \times 3 \times 2 \times 1}{6 \times 5 \times 4 \times 3 \times 2 \times 1 \times 3 \times 2 \times 1} = 84$

$_4C_0 = \dfrac{4!}{(4-0)! \times 0!} = \dfrac{4!}{4! \times 0!} = \dfrac{4 \times 3 \times 2 \times 1}{4 \times 3 \times 2 \times 1 \times 1} = 1$

$_3C_3 = \dfrac{3!}{(3-3)! \times 3!} = \dfrac{3!}{0! \times 3!} = \dfrac{3 \times 2 \times 1}{1 \times 3 \times 2 \times 1} = 1$

5.40 $6! = 6 \times 5 \times 4 \times 3 \times 2 \times 1 = 720$

$11! = 11 \times 10 \times 9 \times 8 \times 7 \times 6 \times 5 \times 4 \times 3 \times 2 \times 1 = 39,916,800$

$$(7-2)! = 5! = 5 \times 4 \times 3 \times 2 \times 1 = 120$$

$$(15-5)! = 10! = 10 \times 9 \times 8 \times 7 \times 6 \times 5 \times 4 \times 3 \times 2 \times 1 = 3,628,800$$

$$_8C_2 = \frac{8!}{(8-2)! \times 2!} = \frac{8!}{6! \times 2!} = \frac{8 \times 7 \times 6 \times 5 \times 4 \times 3 \times 2 \times 1}{6 \times 5 \times 4 \times 3 \times 2 \times 1 \times 2 \times 1} = 28$$

$$_5C_0 = \frac{5!}{(5-0)! \times 0!} = \frac{5!}{5! \times 0!} = \frac{5 \times 4 \times 3 \times 2 \times 1}{5 \times 4 \times 3 \times 2 \times 1 \times 1} = 1$$

$$_5C_5 = \frac{5!}{(5-5)! \times 5!} = \frac{5!}{0! \times 5!} = \frac{5 \times 4 \times 3 \times 2 \times 1}{1 \times 5 \times 4 \times 3 \times 2 \times 1} = 1$$

$$_6C_4 = \frac{6!}{(6-4)! \times 4!} = \frac{6!}{2! \times 4!} = \frac{6 \times 5 \times 4 \times 3 \times 2 \times 1}{2 \times 1 \times 4 \times 3 \times 2 \times 1} = 15$$

$$_{11}C_7 = \frac{11!}{(11-7)! \times 7!} = \frac{11!}{4! \times 7!} = \frac{11 \times 10 \times 9 \times 8 \times 7 \times 6 \times 5 \times 4 \times 3 \times 2 \times 1}{4 \times 3 \times 2 \times 1 \times 7 \times 6 \times 5 \times 4 \times 3 \times 2 \times 1} = 330$$

5.41 The total number of ways to select two faculty members from 16 is:

$$_{16}C_2 = \frac{16!}{(16-2)! \times 2!} = \frac{16 \times 15 \times 14 \times 13 \times 12 \times 11 \times 10 \times 9 \times 8 \times 7 \times 6 \times 5 \times 4 \times 3 \times 2 \times 1}{14 \times 13 \times 12 \times 11 \times 10 \times 9 \times 8 \times 7 \times 6 \times 5 \times 4 \times 3 \times 2 \times 1 \times 2 \times 1} = 120$$

5.42 Total number of ways to select two different flavors from 25 is:

$$_{25}C_2 = \frac{25!}{(25-2)! \times 2!} =$$

$$\frac{25 \times 24 \times 23 \times 22 \times 21 \times 20 \times 19 \times 18 \times 17 \times 16 \times 15 \times 14 \times 13 \times 12 \times 11 \times 10 \times 9 \times 8 \times 7 \times 6 \times 5 \times 4 \times 3 \times \times 2 \times 1}{23 \times 22 \times 21 \times 20 \times 19 \times 18 \times 17 \times 16 \times 15 \times 14 \times 13 \times 12 \times 11 \times 10 \times 9 \times 8 \times 7 \times 6 \times 5 \times 4 \times 3 \times 2 \times 1 \times 2 \times 1} = 300$$

5.43 Total possible selections for selecting three horses from 12 are:

$$_{12}C_3 = \frac{12!}{(12-3)! \times 3!} = \frac{12 \times 11 \times 10 \times 9 \times 8 \times 7 \times 6 \times 5 \times 4 \times 3 \times 2 \times 1}{9 \times 8 \times 7 \times 6 \times 5 \times 4 \times 3 \times 2 \times 1 \times 3 \times 2 \times 1} = 220$$

5.44 Total number of ways of selecting four houses from a block containing 25 houses are:

$$_{25}C_4 = \frac{25!}{(25-4)! \times 4!} = 12,650$$

5.45 Total number of ways of selecting six stocks from 20 are: $_{20}C_6 = \frac{20!}{(20-6)! \times 6!} = 38,760$

5.46 Total number of ways of selecting two workers from 16 are: $_{16}C_2 = \frac{16!}{(16-2)! \times 2!} = 120$

5.47 Total number of ways of selecting 9 items from a population of 20 are: $_{20}C_9 = \dfrac{20!}{(20-9)! \times 9!} = 167{,}960$

5.48 Total number of ways selecting 5 items from a population of 15 are: $_{15}C_5 = \dfrac{15!}{(15-5)! \times 5!} = 3.003$

5.49 a. An experiment that satisfies the following four conditions is called a binomial experiment:

    i. There are n identical trials. In other words, the given experiment is repeated n times. All these repetitions are performed under similar conditions.

    ii. Each trial has two and only two outcomes. These outcomes are usually called a *success* and a *failure*.

    iii. The probability of success is denoted by p and that of failure by q, and $p + q = 1$. The probability of p and q remain constant for each trial.

    iv. The trials are independent. In other words, the outcome of one trial does not affect the outcome of another trial.

  b. Each repetition of a binomial experiment is called a trial.

  c. A binomial random variable x represents the number of successes in n independent trials of a binomial experiment.

5.50 The parameters of the binomial distribution are n and p, the total number of trials and the probability of success. If we know the values of n and p for the binomial distribution, we can find the probability of any number of successes in n trials by using either the binomial formula or the table of binomial probabilities.

5.51 a. This is not a binomial experiment because there are more than two outcomes for each repetition.

  b. This is an example of a binomial experiment because it satisfies all four conditions of a binomial experiment:

    i. There are many identical rolls of the die.

    ii. Each trial has two outcomes: an even number and an odd number.

    iii. The probability of obtaining an even number is ½ and that of an odd number is ½. These probabilities add up to 1, and they remain constant for all trials.

    iv. All rolls of the die are independent.

  c. This is an example of a binomial experiment because it satisfies all four conditions of a binomial experiment:

    i. There are many identical trials (selection of voters).

     ii.   Each trial has two outcomes: a voter favors the proposition and a voter does not favor the proposition.

     iii. The probability of the two outcomes are .54 and .46 respectively. These probabilities add up to 1. These two probabilities remain the same for all selections.

     iv. All voters are independent.

5.52   a. This is an example of a binomial experiment because it satisfies all four conditions of a binomial experiment.

     i.   There are three identical trials (selections)

     ii. Each trial has two outcomes: a red ball is drawn and a blue ball is drawn.

     iii. The probability of drawing a red ball is 6/10 and that of a blue ball is 4/10. These probabilities add up to 1. The two probabilities remain constant for all draws because the draws are made with replacement.

     iv. All draws are independent.

    b. This is not a binomial experiment because the draws are not independent since the selections are made without replacement and, hence, the probabilities of drawing a red and a blue ball change with every selection.

    c. This is an example of a binomial experiment because it satisfies all four conditions of a binomial experiment:

     i   There are many identical trials (selection of households).

     ii. Each trial has two outcomes: a household holds stocks and a household does not hold stocks.

     iii. The probabilities of these two outcomes are .28 and .72 respectively. These probabilities add up to 1. The two probabilities remain the same for all selections.

     iv. All households are independent.

5.53   a. $n = 8$, $x = 5$, $n - x = 8 - 5 = 3$, $p = .70$, and $q = 1 - p = 1 - .70 = .30$

$$P(x = 5) = {}_nC_x p^x q^{n-x} = {}_8C_5 \,(.70)^5 (.30)^3 = (56)\,(.16807)\,(.027) = .2541$$

    b. $n = 4$, $x = 3$, $n - x = 4 - 3 = 1$, $p = .40$, and $q = 1 - p = 1 - .40 = .60$

$$P(x = 3) = {}_nC_x p^x q^{n-x} = {}_4C_3 \,(.40)^3 (.60)^1 = (4)\,(.064)\,(.60) = .1536$$

    c. $n = 6$, $x = 2$, $n - x = 6 - 2 = 4$, $p = .30$, and $q = 1 - p = 1 - .30 = .70$

$$P(x = 2) = {}_nC_x p^x q^{n-x} = {}_6C_2 \,(.30)^2 (.70)^4 = (15)\,(.09)\,(.2401) = .3241$$

5.54   a. $n = 5$, $x = 0$, $n - x = 5 - 0 = 5$, $p = .05$, and $q = 1 - p = 1 - .05 = .95$

$$P(x = 0) = {}_nC_x\, p^x\, q^{n-x} = {}_5C_0\ (.05)^0\ (.95)^5 = (1)\,(1)\,(.77378) = .7738$$

b. $n = 7$, $x = 4$, $n - x = 7 - 4 = 3$, $p = .90$, and $q = 1 - p = 1 - .90 = .10$

$$P(x=4) = {}_nC_x\, p^x\, q^{n-x} = {}_7C_4\ (.90)^4(.10)^3 = (35)\,(.6561)\,(.001) = .0230$$

c. $n = 10$, $x = 7$, $n - x = 10 - 7 = 3$, $p = .60$, and $q = 1 - p = 1 - .60 = .40$

$$P(x = 7) = {}_nC_x\, p^x\, q^{n-x} = {}_{10}C_7\ (.60)^7(.40)^3 = (120)\,(.0279936)\,(.064) = .2150$$

5.55   a.

x	$P(x)$
0	.0824
1	.2471
2	.3177
3	.2269
4	.0972
5	.0250
6	.0036
7	.0002

b. $\mu = np = (7)(.30) = 2.100$

$\sigma = \sqrt{npq} = \sqrt{(7)(.30)(.70)} = 1.212$

5.56   a.

x	$P(x)$
0	.0003
1	.0064
2	.0512
3	.2048
4	.4096
5	.3277

b. $\mu = np = (5)(.80) = 4.000$

$\sigma = \sqrt{npq} = \sqrt{(5)(.80)(.20)} = .894$

5.57   i.   Let $n = 5$ and $p = .50$. The probability distribution and probability graph for this case are shown below. As we can observe, the probability distribution is symmetric in this case.

x	$P(x)$
0	.0312
1	.1562
2	.3125
3	.3125
4	.1562
5	.0312

ii. Let $n = 5$ and $p = .20$. The probability distribution and probability graph for this case are shown below. As we can observe, the probability distribution is skewed to the right in this case.

x	P(x)
0	.3277
1	.4096
2	.2048
3	.0512
4	.0064
5	.0003

iii. Let $n = 5$ and $p = .70$. The probability distribution and probability graph for this case are shown below. As we can observe, the probability distribution is skewed to the left in this case.

x	P(x)
0	.0024
1	.0284
2	.1323
3	.3087
4	.3601
5	.1681

5.58 a. Here, $n = 12$ and $p = .85$

The random variable x can assume any of the values 0, 1, 2, 3, 4, 5, 6, 7, 8, 9, 10, 11, or 12.

b. $P(x = 10) = {}_nC_x p^x q^{n-x} = {}_{12}C_{10} (.85)^{10} (.15)^2 = (66) (.19687440) (.0225) = .2924$

5.59 a. Here, $n = 10$ and $p = .24$

The random variable x can assume any of the values 0, 1, 2, 3, 4, 5, 6, 7, 8, 9, or 10.

b. $P(x = 4) = {}_nC_x p^x q^{n-x} = {}_{10}C_4 (.24)^4 (.76)^6 = (210) (.00331776) (.192699928) = .1343$

5.60 Here, $n = 16$ and $p = .40$

Let x denote the number of 12 to 18 year old females in a random sample of 16 who expect the United States to have a female president within 10 years.

a. $P(\text{at least } 9) = P(x \geq 9) = P(9) + P(10) + P(11) + P(12) + P(13) + P(14) + P(15) + P(16)$

$= .0840 + .0392 + .0142 + .0040 + .0008 + .0001 + .0000 + .0000 = .1423$

b. $P(\text{at most } 5) = P(x \leq 5) = P(0) + P(1) + P(2) + P(3) + P(4) + P(5)$

$= .0003 + .0030 + .0150 + .0468 + .1014 + .1623 = .3288$

   c. $P(6 \text{ to } 9)$ $= P(6 \le x \le 9) = P(6) + P(7) + P(8) + P(9) = .1983 + .1889 + .1417 + .0840 = .6129$

5.61   Here, $n = 15$ and $p = .80$

Let x denote the number of adults in a random sample of 15 who feel stress "frequently" or "sometimes" in their daily lives.

   a. $P(\text{at most } 9) = P(x \le 9) = P(0) + P(1) + P(2) + P(3) + P(4) + P(5) + P(6) + P(7) + P(8) + P(9) =$

          $= .0000 + .0000 + .0000 = .0000 + .0001 + .0007 + .0035 + .0138 + .0430 = .0611$

   b. $P(\text{at least } 11) = P(x \ge 11) = P(11) + P(12) + P(13) + P(14) + P(15)$

          $= .1876 + .2501 + .2309 + .1319 + .0352 = .8357$

   c. $P(10 \le x \le 12) = + P(10) + P(11) + P(12) = .1032 + .1876 + .2501 = .5409$

5.62   Here, $n = 5$ and $p = .20$

   a. $P(\text{exactly } 2) = P(2) = {}_nC_x p^x q^{n-x} = {}_5C_2 (.20)^2 (.80)^3 = (10)(.04)(.512) = .2048$

   b. $P(\text{none}) = P(0) = {}_nC_x p^x q^{n-x} = {}_5C_0 (.20)^0 (.80)^5 = (1)(1)(.32768) = .3277$

   c. $P(\text{exactly } 4) = P(4) = {}_nC_x p^x q^{n-x} = {}_5C_4 (.20)^4 (.80)^1 = (5)(.0016)(.8) = .0064$

5.63   Here, $n = 10$ and $p = .349$

   a. $P(\text{exactly } 4) = P(4) = {}_nC_x p^x q^{n-x} = {}_{10}C_4 (.349)^4 (.651)^6 = (210)(.014835483)(.076117748) =$

      $.2371$

   b. $P(\text{none}) = P(0) = {}_nC_x p^x q^{n-x} = {}_{10}C_0 (.349)^0 (.651)^{10} = (1)(1)(.013671302) = .0137$

   c. $P(\text{exactly } 8) = P(8) = {}_nC_x p^x q^{n-x} = {}_{10}C_8 (.349)^8 (.651)^2 = (45)(.00022009)(.423801) = .0042$

5.64   Here, $n = 10$, $x = 0$, $n - x = 10 - 0 = 10$, $p = .25$,   and $q = .75$

    $P(x = 0) = {}_nC_x p^x q^{n-x} = {}_{10}C_0 (.25)^0 (.75)^{10} = (1)(1)(.0563135) = .0563$

5.65   Here, $n = 8$ and $p = .85$

   a. $P(\text{exactly } 8) = P(8) = {}_nC_x p^x q^{n-x} = {}_8C_8 (.85)^8 (.15)^0 = (1)(.27249053)(1) = .2725$

   b. $P(\text{exactly } 5) = P(5) = {}_nC_x p^x q^{n-x} = {}_8C_5 (.85)^5 (.15)^3 = (56)(.4437053)(.003375) = .0839$

5.66   Here, $n = 15$ and $p = .10$

a. i. $P(\text{exactly } 4) = P(4) = {_n}C_x\, p^x q^{n-x} = {_{15}}C_4 (.10)^4 (.90)^{11} = (1365)\,(.0001)\,(.313810596) =$
        $.0428$

    ii. $P(\text{none}) = P(0) = {_n}C_x\, p^x q^{n-x} = {_{15}}C_0 (.10)^0 (.90)^{15} = (1)\,(1)\,(.205891132) = .2059$

b. i. $P(\text{at least } 4) = P(x \geq 4) = P(4) + P(5) + P(6) + P(7) + P(8) + P(9) + P(10)\,P(11) + P(12) +$
        $P(13) + P(14) + P(15) = .0428 + .0105 + .0019 + .0003 + .0000 + .0000 + .0000 + .0000 +$
        $.0000 + .0000 + .0000 + .0000 = .0555$

    ii. $P(\text{at most } 2) = P(x \leq 2) = P(0) + P(1) + P(2) = .2059 + .3432 + .2669 = .8160$

    iii. $P(1 \text{ to } 4) = P(1 \leq x \leq 4) = P(1) + P(2) + P(3) + P(4) = .3432 + .2669 + .1285 + .0428 = .7814$

5.67   Here, $n = 7$ and $p = .80$

a.

x	$P(x)$
0	.0000
1	.0004
2	.0043
3	.0287
4	.1147
5	.2753
6	.3670
7	.2097

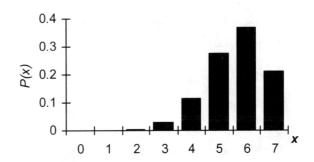

The mean and standard deviation of x are:

$$\mu = np = (7)(.80) = 5.6 \text{ customers}$$

$$\sigma = \sqrt{npq} = \sqrt{(7)(.80)(.20)} = 1.058 \text{ customers}$$

b. $P(\text{exactly } 4) = P(4) = .1147$

5.68   a. Here, $n = 10$ and $p = .05$

x	$P(x)$
0	.5987
1	.3151
2	.0746
3	.0105
4	.0010
5	.0001
6	.0000
7	.0000
8	.0000
9	.0000
10	.0000

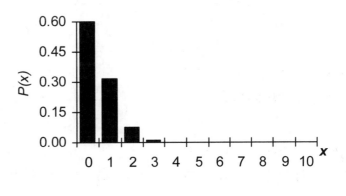

The mean and standard deviation of x are:

$$\mu = np = (10)(.05) = .50 \text{ calculator}$$

$$\sigma = \sqrt{npq} = \sqrt{(10)(.05)(.95)} = .689 \text{ calculator}$$

b. $P(\text{exactly } 2) = P(2) = .0746$

5.69   a.   Here, $n = 8$ and $p = .70$

x	$P(x)$
0	.0001
1	.0012
2	.0100
3	.0467
4	.1361
5	.2541
6	.2965
7	.1977
8	.0576

The mean and standard deviation of x are:

$$\mu = np = (8)(.70) = 5.600 \text{ customers}$$

$$\sigma = \sqrt{npq} = \sqrt{(8)(.70)(.30)} = 1.296 \text{ customers}$$

b. $P(\text{exactly 3 customers like the hamburger}) = P(3) = .0467$

5.70   The hypergeometric probability distribution gives probabilities for the number of successes in a fixed number of trials. It is used for sampling without replacement from a small population, because the trials are not independent, which rules out the use of the binomial probability distribution.

Example: A customs inspector is confronted with 10 suitcases, 2 of which (unknown to him) contain contraband. If he only has time to inspect 3 of them, we would use the hypergeometric distribution to find the probability that none of the 3 contain contraband.

5.71   a. $P(x=2) = \dfrac{{}_rC_x \; {}_{N-r}C_{n-x}}{{}_NC_n} = \dfrac{{}_3C_2 \; {}_{8-3}C_{4-2}}{{}_8C_4} = (3)(10)/70 = .4286$

b. $P(x=0) = \dfrac{{}_rC_x \; {}_{N-r}C_{n-x}}{{}_NC_n} = \dfrac{{}_3C_0 \; {}_{8-3}C_{4-0}}{{}_8C_4} = (1)(5)/70 = .0714$

c. $P(x \le 1) = P(0) + P(1) = .0714 + \dfrac{{}_3C_1 \; {}_{8-3}C_{4-1}}{{}_8C_4} = .0714 + (3)(10)/70 = .0714 + .4286 = .5000$

5.72   a. $P(x=4) = \dfrac{{}_rC_x \; {}_{N-r}C_{n-x}}{{}_NC_n} = \dfrac{{}_6C_4 \; {}_{14-6}C_{5-4}}{{}_{14}C_5} = (15)(8)/2002 = .0599$

b. $P(x=5) = \dfrac{{}_rC_x \; {}_{N-r}C_{n-x}}{{}_NC_n} = \dfrac{{}_6C_5 \; {}_{14-6}C_{5-5}}{{}_{14}C_5} = (6)(1)/2002 = .0030$

c. $P(x \le 1) = P(0) + P(1) = \dfrac{{}_6C_0 \; {}_{14-6}C_{5-0}}{{}_{14}C_5} + \dfrac{{}_6C_1 \; {}_{14-6}C_{5-1}}{{}_{14}C_5} = (1)(56)/2002 + (6)(70)/2002$

$= .0280 + .2098 = .2378$

5.73 a. $P(x = 2) = \dfrac{{}_rC_x \; {}_{N-r}C_{n-x}}{{}_NC_n} = \dfrac{{}_4C_2 \; {}_{11-4}C_{4-2}}{{}_{11}C_4} = (6)(21)/330 = .3818$

b. $P(x = 4) = \dfrac{{}_rC_x \; {}_{N-r}C_{n-x}}{{}_NC_n} = \dfrac{{}_4C_4 \; {}_{11-4}C_{4-4}}{{}_{11}C_4} = (1)(1)/330 = .0030$

c. $P(x \le 1) = P(0) + P(1) = \dfrac{{}_4C_0 \; {}_{11-4}C_{4-0}}{{}_{11}C_4} + \dfrac{{}_4C_1 \; {}_{11-4}C_{4-1}}{{}_{11}C_4} = (1)(35)/330 + (4)(35)/330$

$= .1061 + .4242 = .5303$

5.74 a. $P(x = 5) = \dfrac{{}_rC_x \; {}_{N-r}C_{n-x}}{{}_NC_n} = \dfrac{{}_{10}C_5 \; {}_{16-10}C_{5-5}}{{}_{16}C_5} = (252)(1)/4368 = .0577$

b. $P(x = 0) = \dfrac{{}_rC_x \; {}_{N-r}C_{n-x}}{{}_NC_n} = \dfrac{{}_{10}C_0 \; {}_{16-10}C_{5-0}}{{}_{16}C_5} = (1)(6)/4368 = .0014$

c. $P(x \le 1) = P(0) + P(1) = .0014 + \dfrac{{}_{10}C_1 \; {}_{16-10}C_{5-1}}{{}_{16}C_5} = .0014 + (10)(15)/4368 = .0014 + .0343 = .0357$

5.75 Let x be the number of corporations that incurred losses in a random sample of 3 corporations, and r be the number of corporations in 15 that incurred losses. Then, $N = 15, r = 9, N - r = 6,$ and $n = 3$.

a. $P(\text{exactly } 2) = P(x = 2) = \dfrac{{}_9C_2 \; {}_6C_1}{{}_{15}C_3} = (36)(6)/455 = .4747$

b. $P(\text{none}) = P(x = 0) = \dfrac{{}_9C_0 \; {}_6C_3}{{}_{15}C_3} = (1)(20)/455 = .0440$

c. $P(\text{at most } 1) = P(x \le 1) = P(0) + P(1) = \dfrac{{}_9C_0 \; {}_6C_3}{{}_{15}C_3} + \dfrac{{}_9C_1 \; {}_6C_2}{{}_{15}C_3} = (1)(20)/455 + (9)(15)/455$

$= .0440 + .2967 = .3407$

5.76 a. $P(x = 1) = \dfrac{{}_rC_x \; {}_{N-r}C_{n-x}}{{}_NC_n} = \dfrac{{}_4C_1 \; {}_{20-4}C_{6-1}}{{}_{20}C_6} = \dfrac{4(4368)}{38,760} = .4508$

b. $P(x = 0) = \dfrac{{}_rC_x \; {}_{N-r}C_{n-x}}{{}_NC_n} = \dfrac{{}_4C_0 \; {}_{20-4}C_{6-0}}{{}_{20}C_6} = \dfrac{1(8008)}{38,760} = .2066$

c. $P(x \le 2) = P(0) + P(1) + P(2) = \dfrac{{}_4C_0 \; {}_{20-4}C_{6-0}}{{}_{20}C_6} + \dfrac{{}_4C_1 \; {}_{20-4}C_{6-1}}{{}_{20}C_6} + \dfrac{{}_4C_2 \; {}_{20-4}C_{6-2}}{{}_{20}C_6}$

$$= \frac{1(8008)}{38,760} + \frac{4(4368)}{38,760} + \frac{6(1820)}{38,760} + .2066 + .4508 + .2817 = .9391$$

5.77   Let x be the number of extremely violent games in a random sample of three, and r be the number of extremely violent games in eleven. Then, $N = 11$, $r = 4$, and $N - r = 7$, and $n = 3$.

   a. $P(x = 2) = \dfrac{{_r}C_x \; {_{N-r}}C_{n-x}}{{_N}C_n} = \dfrac{{_4}C_2 \; {_{11-4}}C_{3-2}}{{_{11}}C_3} = \dfrac{6(7)}{165} = .2545$

   b. $P(x > 1) = P(2) + P(3) = .2545 + \dfrac{{_4}C_3 \; {_{11-4}}C_{3-3}}{{_{11}}C_3} = .2545 + \dfrac{4(1)}{165} = .2545 + .0242 = .2787$

   c. $P(x = 0) = \dfrac{{_r}C_x \; {_{N-r}}C_{n-x}}{{_N}C_n} = \dfrac{{_4}C_0 \; {_{11-4}}C_{3-0}}{{_{11}}C_3} = \dfrac{1(35)}{165} = .2121$

5.78   Let x be the number of defective keyboards in a sample of 5 from the box selected, and r be the number of defective keyboards in 20. Then, $N = 20$, $r = 6$, $N - r = 14$, and $n = 5$.

   a. $P(\text{shipment accepted}) = P(x \le 1) = P(0) + P(1) = \dfrac{{_6}C_0 \; {_{14}}C_5}{{_{20}}C_5} + \dfrac{{_6}C_1 \; {_{14}}C_4}{{_{20}}C_5}$

$$= (1)(2002)/15,504 + (6)(1001)/15,504 = .1291 + .3874 = .5165$$

   b. $P(\text{shipment not accepted}) = 1 - P(\text{shipment accepted}) = 1 - .5165 = .4835$

5.79   The following three conditions must be satisfied to apply the Poisson probability distribution:

   i. x is a discrete random variable.

   ii.   The occurrences are random, that is, they do not follow any pattern.

   iii.   The occurrences are independent.

5.80   The parameter of the Poisson probability distribution is λ, which represents the mean number of occurrences in an interval. If we know λ, we can find the probability of any given x.

5.81   a. $P(x \le 1) = P(x = 0) + P(x = 1) = \dfrac{(5)^0 e^{-5}}{0!} + \dfrac{(5)^1 e^{-5}}{1!} = \dfrac{(1)(.00673795)}{1} + \dfrac{(5)(.00673795)}{1}$

$$= .0067 + .0337 = .0404$$

    Note that the value of e^{-5} is obtained from Table V of Appendix C of the text.

   b. $P(x = 2) = \dfrac{(2.5)^2 e^{-2.5}}{2!} = \dfrac{(6.25)(.08208500)}{2} = .2565$

5.82   a. $P(x < 2) = P(x = 0) + P(x = 1) = \dfrac{(3)^0 e^{-3}}{0!} + \dfrac{(3)^1 e^{-3}}{1!} = \dfrac{(1)(.04978707)}{1} + \dfrac{(3)(.04978707)}{1}$

$$= .0498 + .1494 = .1992$$

Note that the value of e^{-3} is obtained form Table V of Appendix C of the text.

b. $P(x = 8) = \dfrac{(5.5)^8 e^{-5.5}}{8!} = \dfrac{(837,339.3789)(.00408677)}{40,320} = .0849$

5.83   a. Probability distribution of x for $\lambda = 1.3$

x	$P(x)$
0	.2725
1	.3543
2	.2303
3	.0998
4	.0324
5	.0084
6	.0018
7	.0003
8	.0001

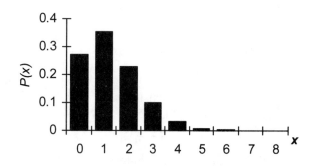

The mean, variance, and standard deviation are:

$\mu = \lambda = 1.3, \qquad \sigma^2 = \lambda = 1.3, \text{ and } \sigma = \sqrt{\lambda} = \sqrt{1.3} = 1.140$

b. Probability distribution of x for $\lambda = 2.1$

x	$P(x)$
0	.1225
1	.2572
2	.2700
3	.1890
4	.0992
5	.0417
6	.0146
7	.0044
8	.0011
9	.0003
10	.0001

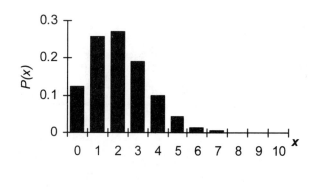

The mean, variance, and standard deviation are:

$\mu = \lambda = 2.1, \qquad \sigma^2 = \lambda = 2.1, \text{ and } \sigma = \sqrt{\lambda} = \sqrt{2.1} = 1.449$

5.84   a. Probability distribution of x for $\lambda = .6$

x	$P(x)$
0	.5488
1	.3293
2	.0988
3	.0198
4	.0030
5	.0004

The mean, variance and standard deviation are:

$\mu = \lambda = .6,\qquad \sigma^2 = \lambda = .6,\ \text{and}\ \sigma = \sqrt{\lambda} = \sqrt{.6} = .775$

b. Probability distribution of x for $\lambda = 1.8$

x	$P(x)$
0	.1653
1	.2975
2	.2678
3	.1607
4	.0723
5	.0260
6	.0078
7	.0020
8	.0005
9	.0001

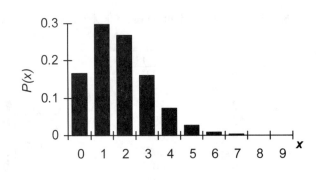

The mean, variance, and standard deviation are:

$\mu = \lambda = 1.8,\qquad \sigma^2 = \lambda = 1.8,\ \text{and}\ \sigma = \sqrt{\lambda} = \sqrt{1.8} = 1.342$

5.85 $\lambda = 1.7$ pieces of junk mail per day and $x = 3$

$$P(x = 3) = \frac{\lambda^x e^{-\lambda}}{x!} = \frac{(1.7)^3 e^{-1.7}}{3!} = \frac{(4.913)(18268352)}{6} = .1496$$

5.86 $\lambda = 9.7$ complaints per day and $x = 6$

$$P(x = 6) = \frac{\lambda^x e^{-\lambda}}{x!} = \frac{(9.7)^6 e^{-9.7}}{6!} = \frac{(832{,}972.0049)(.00006128)}{720} = .0709$$

5.87 $\lambda = 5.4$ shoplifting incidents per day and $x = 3$

$$P(x = 3) = \frac{\lambda^x e^{-\lambda}}{x!} = \frac{(5.4)^3 e^{-5.4}}{3!} = \frac{(157.464)(.00451658)}{6} = .1185$$

5.88 $\lambda = 12.5$ rooms per day and $x = 3$

$$P(x = 3) = \frac{\lambda^x e^{-\lambda}}{x!} = \frac{(12.5)^3 e^{-12.5}}{3!} = \frac{(1953.125)(.00000373)}{6} = .0012$$

5.89 $\lambda = 3.7$ reports of lost students' ID cards per day

a. $P(\text{at most } 1) = P(0) + P(1) = \dfrac{(3.7)^0 e^{-3.7}}{0!} + \dfrac{(3.7)^1 e^{-3.7}}{1!} = \dfrac{(1)(.02472353)}{1} + \dfrac{(3.7)(.02472353)}{1}$

$\qquad\qquad\qquad\qquad = .0247 + .0915 = .1162$

b.   i. $P(1 \text{ to } 4) = P(1) + P(2) + P(3) + P(4) = .0915 + .1692 + .2087 + .1931 = .6625$

ii. $P(\text{at least } 6) = P(6) + P(7) + P(8) + P(9) + P(10) + P(11) + P(12) + P(13)$

$\qquad = .0881 + .0466 + .0215 + .0089 + .0033 + .0011 + .0003 + .0001 = .1699$

iii. $P(\text{at most } 3) = P(0) + P(1) + P(2) + P(3) = .0247 + .0915 + .1692 + .2087 = .4941$

5.90 Let x be the number of businesses that file for bankruptcy on a given day in this city. Then λ = mean number of filings per day = 1.6.

a. $P(3 \text{ filings}) = P(x = 3) = \dfrac{\lambda^x e^{-\lambda}}{x!} = \dfrac{(1.6)^3 e^{-1.6}}{3!} = \dfrac{(4.096)(.20189652)}{6} = .1378$

b. i. $P(2 \text{ to } 3) = P(2) + P(3) = .2584 + .1378 = .3962$

ii. $P(\text{more than } 3) = P(4) + P(5) + P(6) + P(7) + P(8) + P(9)$

$\qquad = .0551 + .0176 + .0047 + .0011 + .0002 + .0000 = .0787$

iii. $P(\text{less than } 3) = P(0) + P(1) + P(2) = .2019 + .3230 + .2584 = .7833$

5.91 λ = .5 defect per 500 yards

a. $P(x = 1) = \dfrac{\lambda^x e^{-\lambda}}{x!} = \dfrac{(.5)^1 e^{-.5}}{1!} = \dfrac{(.5)(.60653066)}{1} = .3033$

b. i. $P(2 \text{ to } 4) = P(2) + P(3) + P(4) = .0758 + .0126 + .0016 = .0900$

ii. $P(\text{more than } 3) = P(4) + P(5) + P(6) + P(7) = .0016 + .0002 + .0000 + .0000 = .0018$

iii. $P(\text{less than } 2) = P(0) + P(1) = .6065 + .3033 = .9098$

5.92 Let x be the number of drive offs this week. Since the average number of drive offs is 1.6, $\lambda = 1.6$.

a. $P(\text{exactly } 2) = P(x = 2) = \dfrac{\lambda^x e^{-\lambda}}{x!} = \dfrac{(1.6)^2 e^{-1.6}}{2!} = \dfrac{(2.56)(.20189652)}{2} = .2584$

b. i. $P(\text{less than } 3) = P(x < 3) = P(0) + P(1) + P(2) = .2019 + .3230 + .2584 = .7833$

ii. $P(\text{more than } 5) = P(x > 5) = P(6) + P(7) + P(8) + P(9) + \ldots = .0047 + .0011 + .0002 + .0000 = .0060$

iii. $P(2 \text{ to } 5) = P(2 \leq x \leq 5) = P(2) + P(3) + P(4) + P(5) = .2584 + .1378 + .0551 + .0176 = .4689$

5.93 Let x be the number of customers that come to this savings and loan during a given hour. Since the average number of customers per half hour is 4.8, the average number per hour is $2 \times 4.8 = 9.6$. Thus, $\lambda = 9.6$.

a. $P(\text{exactly } 2) = P(x = 2) = \dfrac{\lambda^x e^{-\lambda}}{x!} = \dfrac{(9.6)^2 e^{-9.6}}{2!} = \dfrac{(92.16)(.00006773)}{2} = .0031$

b. i. $P(2 \text{ or less}) = P(x \leq 2) = P(0) + P(1) + P(2) = .0001 + .0007 + .0031 = .0039$

ii. $P(10 \text{ or more}) = P(x \geq 10) = P(10) + P(11) + P(12) + \ldots P(24) = .1241 + .1083 + .0866 + .0640 +$

$\qquad .0439 + .0281 + .0168 + .0095 + .0051 + .0026 + .0012 + .0006 + .0002 + .0001 + .0000 = .4911$

5.94   Let x be the number of typographical errors on a randomly selected page of this newspaper. Since it contains an average of 1.1 typographical errors per page, $\lambda = 1.1$.

a. $P(\text{exactly } 4) = P(x = 4) = \dfrac{\lambda^x e^{-\lambda}}{x!} = \dfrac{(1.1)^4 e^{-1.1}}{4!} = \dfrac{(1.4641)(.33287108)}{24} = .0203$

b.   i. $P(\text{more than } 3) = P(x \geq 3) = P(4) + P(5) + P(6) + P(7) = .0203 + .0045 + .0008 + .0001 = .0257$

   ii. $P(\text{less than } 4) = P(x < 4) = P(0) + P(1) + P(2) + P(3) = .3329 + .3662 + .2014 + .0738 = .9743$

5.95   Let x be the number of policies sold by this salesperson on a given day. Since the salesperson sells an average of 1.4 policies per day, $\lambda = 1.4$.

a. $P(\text{none}) = P(x = 0) = \dfrac{\lambda^x e^{-\lambda}}{x!} = \dfrac{(1.4)^0 e^{-1.4}}{0!} = \dfrac{(1)(.24659696)}{1} = .2466$

b.

x	$P(x)$
0	.2466
1	.3452
2	.2417
3	.1128
4	.0395
5	.0111
6	.0026
7	.0005
8	.0001

c.   The mean, variance, and standard deviation are:

$\mu = \lambda = 1.4, \qquad \sigma^2 = \lambda = 1.4, \text{ and } \sigma = \sqrt{\lambda} = \sqrt{1.4} = 1.183$

5.96   Let x denote the number of accidents on a given day. Since the average number of accidents per day is .8, $\lambda = .8$.

a. $P(\text{no accidents}) = P(x = 0) = \dfrac{\lambda^x e^{-\lambda}}{x!} = \dfrac{(.8)^0 e^{-.8}}{0!} = \dfrac{(1)(.44932896)}{1} = .4493$

b.

x	$P(x)$
0	.4493
1	.3595
2	.1438
3	.0383
4	.0077
5	.0012
6	.0002

c.   The mean, variance, and standard deviation are:

$\mu = \lambda = .8, \qquad \sigma^2 = \lambda = .8, \text{ and } \sigma = \sqrt{\lambda} = \sqrt{.8} = .894$

5.97   Let x denote the number of households in a random sample of 50 who own answering machines. Since, on average, 20 households in 50 own answering machines, $\lambda = 20$.

   a. $P(\text{exactly 25 own answering machines}) = P(x = 25) = \dfrac{\lambda^x e^{-\lambda}}{x!} = \dfrac{(20)^{25} e^{-20}}{25!} = .0446$

   b.   i. $P(\text{at most 12 own answering machines}) = P(x \le 12) = P(0) + P(1) + P(2) + P(3) + P(4) + P(5)$
   $+ P(6) + P(7) + P(8) + P(9) + P(10) + P(11) + P(12) = .0000 + .0000 + .0000 + .0000 + .0000$
   $+ .0001 + .0002 + .0005 + .0013 + .0029 + .0058 + .0106 + .0176 = .0390$

   ii. $P(13 \text{ to } 17) = P(13 \le x \le 17) = P(13) + P(14) + P(15) + P(16) + P(17)$
   $= .0271 + .0387 + .0516 + .0646 + .0760 = .2580$

   iii. $P(\text{at least 30 own answering machines}) = P(x \ge 30)\quad = P(30) + P(31) + P(32) + \ldots + P(39)$
   $= .0083 + .0054 + .0034 + .0020 + .0012 + .0007 + .0004 + .0002 + .0001 + .0001 = .0218$

5.98   Let x denote the number of cars passing through a school zone exceeding the speed limit. The average number of cars speeding by at least ten miles per hour is 20 percent Thus, $\lambda = .20 \, x \, 100 = 20$.

   a. $P(\text{exactly 25 speed}) = P(x = 25) = \dfrac{\lambda^x e^{-\lambda}}{x!} = \dfrac{(20)^{25} e^{-20}}{25!} = .0446$

   b.   i. $P(\text{at most 8 speed}) = P(x \le 8) = P(0) + P(1) + P(2) + P(3) + P(4) + P(5) + P(6) + P(7) + P(8)$
   $= .0000 + .0000 + .0000 + .0000 + .0000 + .0001 + .0002 + .0005 + .0013 = .0021$

   ii. $P(15 \text{ to } 20 \text{ speed})\ = P(15 \le x \le 20) = P(15) + P(16) + P(17) + P(18) + P(19) + P(20)$
   $= .0516 + .0646 + .0760 + .0844 + .0888 + .0888 = .4542$

   iii. $P(\text{at least 30}) = P(x \ge 30)\quad = P(30) + P(31) + P(32) + P(33) + P(34) + P(35) + P(36) + P(37)$
   $+ P(38) + P(39) = .0083 + .0054 + .0034 + .0020 + .0012 + .0007 + .0004 + .0002 + .0001 +$
   $.0001 = .0218$

5.99

x	$P(x)$	$xP(x)$	$x^2P(x)$
2	.05	.10	.20
3	.22	.66	1.98
4	.40	1.60	6.40
5	.23	1.15	5.75
6	.10	.60	3.60
		$\sum xP(x) = 4.11$	$\sum x^2 P(x) = 17.93$

$\mu = \sum xP(x) = 4.11$ cars

$\sigma = \sqrt{\sum x^2 P(x) - \mu^2} = \sqrt{17.93 - (4.11)^2} = 1.019$ cars

This mechanic repairs, on average, 4.11 cars per day.

5.100

x	$P(x)$	$xP(x)$	$x^2 P(x)$
0	.13	.00	.00
1	.28	.28	.28
2	.30	.60	1.20
3	.17	.51	1.53
4	.08	.32	1.28
5	.04	.20	1.00
		$\sum xP(x)=1.91$	$\sum x^2 P(x)=5.29$

$\mu = \sum xP(x) = 1.91$ root canal surgeries

$\sigma = \sqrt{\sum x^2 P(x) - \mu^2} = \sqrt{5.29 - (1.91)^2} = 1.281$ root canal surgeries

The average (mean) number of root canals preformed on a Monday by Dr. Sharp is 1.91.

5.101   a. & b.

x	$P(x)$	$xP(x)$	$X^2 P(x)$
-1.2	.17	-.204	.2448
-.7	.21	-.147	.1029
.9	.37	.333	.2997
2.3	.25	.575	1.3225
		$\sum xP(x)=.557$	$\sum x^2 P(x)=1.9699$

$\mu = \sum xP(x) = \$.557$ million $= \$557,000$

$\sigma = \sqrt{\sum x^2 P(x) - \mu^2} = \sqrt{1.9699 - (.557)^2} = \$1.288274$ million $= \$1,288,274$

The value of $\mu = \$557,000$ indicates that the company has an expected profit of $557,000 for next year.

5.102   a. & b.

Note that if the policyholder dies next year, the company's net loss of $100,000 is offset by the $350 premium. Thus, in this case, $x = 350 - 10,000 = -99,650$.

x	$P(x)$	$xP(x)$	$X^2 P(x)$
-99,650	.002	-199.30	19,860,245
350	.998	-349.30	122,255
		$\sum xP(x)=150.00$	$\sum x^2 P(x)=19,982,500$

$\mu = \sum xP(x) = \$150.00$

$\sigma = \sqrt{\sum x^2 P(x) - \mu^2} = \sqrt{19,982,500 - (150.00)^2} = \$4467.66$

The value of $\mu = \$150.00$ indicates the company's expected gain for the next year on this policy is $150.00.

5.103 Let x denote the number of machines that are broken down at a given time. Assuming machines are independent, x is a binomial random variable with $n = 8$ and $p = .04$.

a. $P(\text{all 8 are broken down}) = P(x = 8) = {}_{n}C_{x}p^{x}q^{n-x} = {}_{8}C_{8}(0.04)^{8}(.96)^{0}$

$$= (1)(.000000000007)(1) \approx .0000$$

b. $P(\text{exactly 2 are broken down}) = P(x = 2) = {}_{n}C_{x}p^{x}q^{n-x} = {}_{8}C_{2}(0.04)^{2}(.96)^{6}$

$$= (28)(.0016)(.78275779) \approx .0351$$

c. $P(\text{none is broken down}) = P(x = 0) = {}_{n}C_{x}p^{x}q^{n-x} = {}_{8}C_{0}(0.04)^{0}(.96)^{8} = (1)(1)(.72138958) \approx .7214$

5.104 Let x denote the number of the 12 new credit card holders who will eventually default. Then x is a binomial random variable with $n = 12$ and $p = .08$.

a. $P(\text{exactly 3 will default}) = P(x = 3) = {}_{n}C_{x}p^{x}q^{n-x} = {}_{12}C_{3}(.08)^{3}(.92)^{9}$

$$= (220)(.000512)(.47216136) \approx .0532$$

b. $P(\text{exactly 1 will default}) = P(x = 1) = {}_{n}C_{x}p^{x}q^{n-x} = {}_{12}C_{1}(.08)^{1}(.92)^{11} = (12)(.08)(.39963738) \approx .3837$

c. $P(\text{none will default}) = P(x = 0) = {}_{n}C_{x}p^{x}q^{n-x} = {}_{12}C_{0}(0.8)^{0}(.92)^{12} = (1)(1)(.36766639) \approx .3677$

5.105 Let x denote the number of defective motors in a random sample of 20. Then x is a binomial random variable with $n = 20$ and $p = .05$.

a. $P(\text{shipment accepted}) = P(x \leq 2) = P(0) + P(1) + P(2) = .3585 + .3774 + .1887 = .9246$

b. $P(\text{shipment rejected}) = 1 - P(\text{shipment accepted}) = 1 - .9246 = .0754$

5.106 a. Here, $n = 15$ and $p = .10$

x	$P(x)$
0	.2059
1	.3432
2	.2669
3	.1285
4	.0428
5	.0105
6	.0019
7	.0003

Note that the probabilities of $x = 8$ to $x = 15$ are all equal to zero from Table IV of Appendix C.

$$\mu = np = (15)(.10) = 1.50$$

$$\sigma = \sqrt{npq} = \sqrt{(15)(.10)(.90)} = 1.162$$

b. $P(\text{exactly } 5) = P(x = 5) = .0105$

5.107 Let x denote the number of households who own homes in the random sample of 4 households. Then x is a hypergeometric random variable with $N = 15$, $r = 9$, and $n = 4$.

a. $P(\text{exactly } 3) = P(x = 3) = \dfrac{_rC_x \ _{N-r}C_{n-x}}{_NC_n} = \dfrac{_9C_3 \ _6C_1}{_{15}C_4} = \dfrac{(84)(6)}{1365} = .3692$

b. $P(\text{at most } 1) = P(x \le 1) = P(0) + P(1) = \dfrac{_9C_0 \ _6C_4}{_{15}C_4} + \dfrac{_9C_1 \ _6C_3}{_{15}C_4} = \dfrac{(1)(15)}{1365} + \dfrac{(9)(20)}{1365}$

$$= .0110 + .1319 = .1429$$

c. $P(\text{exactly } 4) = P(x = 4) = \dfrac{_rC_x \ _{N-r}C_{n-x}}{_NC_n} = \dfrac{_9C_4 \ _6C_0}{_{15}C_4} = \dfrac{(126)(1)}{1365} = .0923$

5.108 Let x denote the number of corporations in the random sample of 5 that provide retirement benefits. Then x is a hypergeometric random variable with $N = 20$, $r = 14$, and $n = 5$.

a. $P(\text{exactly } 2) = P(x = 2) = \dfrac{_rC_x \ _{N-r}C_{n-x}}{_NC_n} = \dfrac{_{14}C_2 \ _6C_3}{_{20}C_5} = \dfrac{(91)(20)}{15,504} = .1174$

b. $P(\text{none}) = P(x = 0) = \dfrac{_rC_x \ _{N-r}C_{n-x}}{_NC_n} = \dfrac{_{14}C_0 \ _6C_5}{_{20}C_5} = \dfrac{(1)(6)}{15,504} = .0004$

c. $P(\text{at most one}) = P(x \le 1) = P(0) + P(1) = .0004 + \dfrac{_{14}C_1 \ _6C_4}{_{20}C_5} = .0004 + \dfrac{(14)(15)}{15,504}$

$$= .0004 + .0135 = .0139$$

5.109 Let x denote the number of defective parts in a random sample of 4. Then x is a hypergeometric random variable with $N = 16$, $r = 3$, and $n = 4$.

a. $P(\text{shipment accepted}) = P(x \le 1) = P(0) + P(1) = \dfrac{_3C_0 \ _{13}C_4}{_{16}C_4} + \dfrac{_3C_1 \ _{13}C_3}{_{16}C_4}$

$$= \dfrac{(1)(715)}{1820} + \dfrac{(3)(286)}{1820} = .3929 + .4714 = .8643$$

b. $P(\text{shipment not accepted}) = 1 - P(\text{shipment accepted}) = 1 - .8643 = .1357$

5.110 Let x denote the number of tax returns in a random sample of 3 that contain errors. Then x is a hypergeometric random variable with $N = 12$, $r = 2$, and $n = 3$.

a. $P(\text{exactly 1 contains errors}) = P(x = 1) = \dfrac{_rC_x \ _{N-r}C_{n-x}}{_NC_n} = \dfrac{_2C_1 \ _{10}C_2}{_{12}C_3} = \dfrac{(2)(45)}{220} = .4091$

b. $P(\text{none contains errors}) = P(x = 0) = \dfrac{_rC_x \ _{N-r}C_{n-x}}{_NC_n} = \dfrac{_2C_0 \ _{10}C_3}{_{12}C_3} = \dfrac{(1)(120)}{220} = .5455$

c. $P(\text{exactly 2 contain errors}) = P(x = 2) = \dfrac{{}_rC_x \; {}_{N-r}C_{n-x}}{{}_NC_n} = \dfrac{{}_2C_2 \; {}_{10}C_1}{{}_{12}C_3} = \dfrac{(1)(10)}{220} = .0455$

5.111 Here, $\lambda = 7$ cases per day

a. $P(x = 4) = \dfrac{\lambda^x e^{-\lambda}}{x!} = \dfrac{(7)^4 e^{-7}}{4!} = \dfrac{(2401)(.00091188)}{24} = .0912$

b. i. $P(\text{at least 7}) = P(7) + P(8) + \cdots + P(18) = .1490 + .1304 + .1014 + .0710 + .0452 + .0263 +$
   $.0142 + .0071 + .0033 + .0014 + .0006 + .0002 + .0001 = .5502$

   ii. $P(\text{at most 3}) = P(0) + P(1) + P(2) + P(3) = .0009 + .0064 + .0223 + .0521 = .0817$

   iii. $P(2 \text{ to } 5) = P(2) + P(3) + P(4) + P(5) = .0223 + .0521 + .0912 + .1277 = .2933$

5.112 Here, $\lambda = 6.3$ robberies per day

a. $P(x = 3) = \dfrac{\lambda^x e^{-\lambda}}{x!} = \dfrac{(6.3)^3 e^{-6.3}}{3!} = \dfrac{(250.047)(.00183630)}{6} = .0765$

b. i. $P(\text{at least 12}) = P(12) + P(13) + \cdots + P(19)$
   $= .0150 + .0073 + .0033 + .0014 + .0005 + .0002 + .0001 + .0000 = .0278$

   ii. $P(\text{at most 3}) = P(0) + P(1) + P(2) + P(3) = .0018 + .0116 + .0364 + .0765 = .1263$

   iii. $P(2 \text{ to } 6) = P(2) + P(3) + P(4) + P(5) + P(6) = .0364 + .0765 + .1205 + .1519 + .1595 = .5448$

5.113 Here, $\lambda = 1.4$ airplanes per hour

a. $P(x = 0) = \dfrac{\lambda^x e^{-\lambda}}{x!} = \dfrac{(1.4)^0 e^{-1.4}}{0!} = \dfrac{(1)(.24659696)}{1} = .2466$

b.

x	$P(x)$
0	.2466
1	.3452
2	.2417
3	.1128
4	.0395
5	.0111
6	.0026
7	.0005
8	.0001

5.114 Here, $\lambda = 1.2$ technical fouls per game

a. $P(x = 3) = \dfrac{\lambda^x e^{-\lambda}}{x!} = \dfrac{(1.2)^3 e^{-1.2}}{3!} = \dfrac{(1.728)(.30119421)}{6} = .0867$

b.

x	P(x)
0	.3012
1	.3614
2	.2169
3	.0867
4	.0260
5	.0062
6	.0012
7	.0002

5.115 Let x be a random variable that denotes the gain you have from this game. The probability for each number is not the same, however. There are 36 different outcomes for two dice: (1,1), (1,2), (1,3), (1,4), (1,5), (1.6), (2,1), (2,2),..., (6,6).

$P(\text{sum} = 2) = P(\text{sum} = 12) = \dfrac{1}{36}$

$P(\text{sum} = 3) = P(\text{sum} = 11) = \dfrac{2}{36}$

$P(\text{sum} = 4) = P(\text{sum} = 10) = \dfrac{3}{36}$

$P(\text{sum} = 9) = \dfrac{4}{36}$

$P(x = 20) = P(\text{you win}) = P(2) + P(3) + P(4) + P(9) + P(10) + P(11) + P(12)$

$= \dfrac{1}{36} + \dfrac{2}{36} + \dfrac{3}{36} + \dfrac{4}{36} + \dfrac{3}{36} + \dfrac{2}{36} + \dfrac{1}{36} = \dfrac{16}{36}$

$P(x = -20) = P(\text{you lose}) = 1 - P(\text{you win}) = 1 - \dfrac{16}{36} = \dfrac{20}{36}$

x	P(x)	xP(x)
20	$\dfrac{16}{36} = .4444$	8.89
-20	$\dfrac{20}{36} = .5556$	-11.11
		$\sum xP(x) = -2.22$

The value of $\sum xP(x) = -2.22$ indicates that your expected "gain" is -\$2.22, so you should not accept this offer. This game is not fair to you since you are expected to lose \$2.22.

5.116 Let x be a random variable that denotes the net profit of the venture, and p denotes the probability for a success.

x	$P(x)$	$xP(x)$
10,000,000	p	$10{,}000{,}000 \cdot p$
-4,000,000	$1 - p$	$-4{,}000{,}000 \cdot (1 - p)$
		$\sum xP(x) = 10{,}000{,}000\,p - 4{,}000{,}000(1 - p)$

$\mu = \sum xP(x) = 10{,}000{,}000p - 4{,}000{,}000(1 - p) = 14{,}000{,}000p - 4{,}000{,}000$

a. If $p = .40$, $\mu = 14{,}000{,}000p - 4{,}000{,}000 = 14{,}000{,}000(.40) - 4{,}000{,}000 = 1{,}600{,}000$.

Yes, the owner will be willing to take the risk as the expected net profit is above 500,000.

b. As long as $\mu \geq 500{,}000$ the owner will be willing to take the risk. This means as long as

$\mu = 14{,}000{,}000p - 4{,}000{,}000 \geq 500{,}000$, the owner will do it. This inequality holds for $p \geq .3214$.

Thus the owner will go ahead with the project if p is at least .3214.

5.117 a. Team A needs to win four games in a row, each with probability .5, so P(team A wins the series in

four games) $= .5^4 = .0625$

In order to win in five games, Team A needs to win 3 of the first four games as well as the fifth

game, so P(team A wins the series in five games) $= {}_4C_3(.5)^3(.5)(.5) = .125$.

b. If seven games are required for a team to win the series, then each team needs to win three of the

first six games, so P(seven games are required to win the series) $= {}_6C_3(.5)^3(.5)^3 = .3125$.

5.118 Let x denote the number of bearings in a random sample of 15 that do not meet the required

specifications. Then x is a binomial random variable with $n = 15$ and $p = .10$.

a. P(production suspended) $= P(x > 2) = 1 - P(x \leq 2) = 1 - [P(0) + P(1) + P(2)]$

$= 1 - (.2059 + .3432 + .2669) = .1840$

b. The 15 bearings are sampled without replacement, which normally requires the use of the

hypergeometric distribution. We are assuming that the population from which the sample is drawn

is so large that each time a bearing is selected, the probability of being defective is .10. Thus, the

sampling of 15 bearings constitutes 15 independent trials, so that the distribution of x is

approximately binomial.

5.119 a. Let x denote the number of drug deals on this street on a given night. Note that x is discrete. This

text has covered two discrete distributions, the binomial and the Poisson. The binomial distribution

does not apply here, since there is no fixed number of "trials". However, the Poisson distribution

might be appropriate.

b. To use the Poisson distribution we would have to assume that the drug deals occur randomly and independently.

c. The mean number of drug deals per night is three, so for the Poisson distribution for one night, $\lambda = 3$. If the residents tape for two nights, then $\lambda = 2 \times 3 = 6$.

Thus, $P(\text{film at least 5 drug deals}) = P(x \geq 5) = 1 - P(x < 5)$

$= 1 - [P(0) + P(1) + P(2) + P(3) + P(4)] = 1 - (.0025 + .0149 + .0446 + .0892 + .1339) = .7149$

d. Part c. shows that two nights of taping are insufficient, since $P(x \geq 5) = .7149 < .90$. Try taping for three nights. Then $\lambda = 3 \times 3 = 9$. $P(x \geq 5) = 1 - (.0001 + .0011 + .0050 + .0150 + .0337) = .9451$. This exceeds the required probability of .90, so the camera should be rented for three nights.

5.120   a. It's easier to do part b first.

b. Let x be a random variable that denotes the score after guessing on one multiple choice question.

x	$P(x)$	$xP(x)$
1	.25	.250
$-\dfrac{1}{2}$	.75	-.375
		$\sum xP(x) = -.125$

So a student decreases his expected score by guessing on a question, if he has no idea what the correct answer is.

Back to part a.: Guessing on 12 questions lowers the expected score by $12 \cdot (-.125) = -1.5$, so the expected score for 38 correct answers and 12 random guesses is $38 \cdot 1 - 12 \cdot (-.125) = 36.5$.

c. If the student can eliminate one of the wrong answers, then we get the following:

x	$P(x)$	$xP(x)$
1	$\dfrac{1}{3}$	$\dfrac{1}{3}$
$-\dfrac{1}{2}$	$\dfrac{2}{3}$	$-\dfrac{1}{3}$
		$\sum xP(x) = 0$

So in this case guessing does not affect the expected score.

5.121   Let λ be the mean number of cheesecakes sold per day. Here $\lambda = 5$. Let x be the number of sales per day. We want to find k such that $P(x > k) < .1$. Using the Poisson probability distribution we find that $P(x > 7) = 1 - P(x \leq 7) = 1 - .867 = .133$ and $P(x > 8) = 1 - P(x \leq 8) = 1 - .932 = .068$. So, if the baker wants the probability of losing a sale to be less than .1, he needs to make 8 cheesecakes.

5.122   a. For the $1 outcome:

P(gambler wins) = 22/50 = .44; net gain = $1

P(gambler loses) = 1 – 22/50 = .56; net gain = -$1

Therefore, the expected net payoff for a $1 bet on the $1 outcome is $1 \times .44 – 1 \times .56 = -\$.12 =$ -12¢. Thus, the gambler has an average net loss of 12¢ per $1 bet on the $1 outcome.

b. When the gambler bets $1 and loses, his loss is $1 regardless of the outcome bet on. The expected values for a $1 bet on each of the other six outcomes are shown in the following table.

Outcome	P(win)	P(lose)	Expected net payoff
$2	14/50 = .28	.72	$2 \times .28 – 1 \times .72 = -.16$
5	7/50 = .14	.86	$5 \times .14 – 1 \times .86 = -.16$
10	3/50 = .06	.94	$10 \times .06 – 1 \times .94 = -.34$
20	2/50 = .04	.96	$20 \times .04 – 1 \times .96 = -.16$
flag	1/50 = .02	.98	$40 \times .02 – 1 \times .98 = -.18$
joker	1/50 = .02	.98	$40 \times .02 – 1 \times .98 = -.18$

c. In terms of expected net payoff, the $1 outcome is best (-12¢), while the $10 outcome is worst (-34¢).

5.123 a. There are $_7C_4 = 35$ ways to choose four questions from the set of seven.

b. The teacher must choose both questions that the student did not study ($_2C_2$ ways to do this), and any two of the remaining five questions ($_5C_2$ ways to do this).

Thus, there are $_2C_2\,_5C_2 = (1)(10) = 10$ ways to choose four questions that include the two that the student did not study.

c. From the answers to parts a and b, P(the four questions on the test include both questions that the student did not study) = 10/35 = .2857.

5.124 Let y = number of dice showing the number the gambler bet on. Then y is a binomial random variable. Since there are three dice, $n = 3$.

For each die, the probability of showing the number the gambler bet on is 1/6, so $p = 1/6$ and $q = 1 - p$ = 5/6. For each $1 bet, let x = amount gambler wins.

Now, $P(x = -1) = P(y = 0) = {}_3C_0(1/6)^0(5/6)^3 = 125/216,$

$P(x = 1) = P(y = 1) = {}_3C_1(1/6)^1(5/6)^2 = 75/216,$

$P(x = 2) = P(y = 2) = {}_3C_2(1/6)^2(5/6)^1 = 15/216,$

$P(x = 3) = P(y = 3) = {}_3C_3(1/6)^3(5/6)^0 = 125/216$

x	P(x)	xP(x)
-1	125/216	-125/216
1	75/516	75/216
2	15/216	30/216
3	1/216	3/216
		$\sum xP(x) = -17/216 = \$.08$

Thus, the gambler should expect to lose about 8¢ per $1 bet, which agrees with the columnist's assertion.

5.125   For each game, let x = amount you win

Game I:

Outcome	x	P(x)	xP(x)	$X^2 P(x)$
Head	3	.50	1.50	4.50
Tail	-1	.50	-.50	.50
			$\sum xP(x) = 1.00$	$\sum x^2 P(x) = 5.00$

$$\mu = \sum xP(x) = \$1.00$$

$$\sigma = \sqrt{\sum x^2 P(x) - \mu^2} = \sqrt{5 - (1)^2} = \$2.00$$

Game II:

Outcome	x	P(x)	xP(x)	$X^2 P(x)$
First ticket	300	1/500	.60	180
Second ticket	150	1/500	.30	45
Neither	0	498/500	.00	0
			$\sum xP(x) = .90$	$\sum x^2 P(x) = 225$

$$\mu = \sum xP(x) = \$.90$$

$$\sigma = \sqrt{\sum x^2 P(x) - \mu^2} = \sqrt{225 - (.90)^2} = \$14.97$$

Game III:

Outcome	x	P(x)	xP(x)	$X^2 P(x)$
Head	1,000,002	.50	500,001	5×10^{11}
Tail	-1,000,000	.50	-500,000	5×10^{11}
			$\sum xP(x) = 1.00$	$\sum x^2 P(x) = 10^{12}$

$$\mu = \sum xP(x) = \$1.00$$

$$\sigma = \sqrt{\sum x^2 P(x) - \mu^2} = \sqrt{10^{12} - (1)^2} = \$1,000,000$$

Game I is preferable to Game II because the mean for Game I is greater than the mean for Game II. Although the mean for Game III is the same as Game I, the standard deviation for Game III is extremely high, making it very unattractive to a risk-adverse person. Thus, for most people, Game I is the best and, probably, Game III is the worst (due to its very high standard deviation).

5.126 The probability distribution for x is obtained from the given formula:

$$P(0) = \frac{0^2 - 0 + 2}{8} = \frac{2}{8} = .25$$

$$P(1) = \frac{1^2 - 1 + 2}{8} = \frac{2}{8} = .25$$

$$P(2) = \frac{2^2 - 2 + 2}{8} = \frac{4}{8} = .50$$

x	$P(x)$	$xP(x)$	$x^2 P(x)$
0	.25	.00	.00
1	.25	.25	.25
2	.50	1.00	2.00
		$\sum xP(x) = 1.25$	$\sum x^2 P(x) = 2.25$

a. P(at least one exam)$= P(1) + P(2) = .25 + .50 = .75$

b. $\mu = \sum xP(x) = 1.25$ Thus, on average, the expected number of exams is 1.25.

c. $\sigma = \sqrt{\sum x^2 P(x) - \mu^2} = \sqrt{2.25 - (1.25)^2} = .829$

5.127 Let: x_1 = the number of contacts on the first day

x_2 = the number of contacts on the second day

The following table, which may be constructed with the help of a tree diagram, lists the various combinations of contacts during the two days and their probabilities. Note that the probability of each combination is obtained by multiplying the probabilities of the two events included in that combination.

x_1, x_2	Probability	y
(1, 1)	.0144	2
(1, 2)	.0300	3
(1, 3)	.0672	4
(1, 4)	.0084	5
(2, 1)	.0300	3
(2, 2)	.0625	4
(2, 3)	.1400	5
(2. 4)	.0175	6
(3, 1)	.0672	4
(3, 2)	.1400	5
(3, 3)	.3136	6
(3, 4)	.0392	7
(4, 1)	.0084	5
(4, 2)	.0175	6
(4, 3)	.0392	7
(4, 4)	.0049	8

The following table gives the probability distribution of y. This table is prepared from the previous table.

y	P(y)
2	.0144
3	.0600
4	.1969
5	.2968
6	.3486
7	.0784
8	.0049

5.128 a. $\lambda = 5$ calls per 15-minute period

Using the table of Poisson probabilities: $P(x \geq 9) = P(9) + P(10) + P(11) + \cdots = .0680$

b. $\lambda = 7.5$ calls per 15-minute period

$P(x \geq 9) = P(9) + P(10) + P(11) + \cdots = .3380$

c. There is a 6.8% chance of observing 9 (or more) calls in the 15-minute period if the rate is actually 20 per hour. This low probability suggests that the true rate may exceed 20 calls per hour.

d. In view of this evidence, it may be wise to hire a second operator.

Self-Review Test for Chapter Five

1. **Random variable:** A variable whose value is determined by the outcome of a random experiment is called a random variable.

   Discrete random variable: A random variable whose values are countable is called a discrete random variable. An example of this is the number of students in a class.

   Continuous random variable: A random variable that can assume any value in one or more intervals is called a continuous random variable. An example of this is the height of a person.

2. The probability distribution table.   **3.** a   **4.** b   **5.** a

6. Following are the four conditions of a binomial experiment.

   i. There are n identical trials. In other words, the given experiment is repeated n times. All these repetitions are performed under similar conditions.

   ii. Each trial has two and only two outcomes. These outcomes are usually called a *success* and a *failure*.

   iii. The probability of success is denoted by p and that of failure by q, and $p + q = 1$. The probabilities p and q remain constant for each trial.

   iv. The trials are independent.

   Example 5-16 in the text can be considered as an example of a binomial experiment.

7. b   **8.** a   **9.** b   **10.** a   **11.** c   **13.** a

12. A hypergeometric probability distribution is used to find probabilities for the number of successes in a fixed number of trials, when the trials are not independent (such as sampling without replacement from a small population.) Example: Select 2 balls without replacement from an urn that contains 3 red balls and 5 black balls. The number of red balls in the sample is a hypergeometric random variable.

14. Following are the three conditions that must be satisfied to apply the Poisson probability distribution.

   i. x is a discrete random variable.

   ii. The occurrences are random, that is, they do not follow any pattern.

   iii. The occurrences are independent, that is, the occurrence (or nonoccurrence) of an event does not influence the successive occurrences (or nonoccurrences) of that event.

15.

x	$P(x)$	$xP(x)$	$x^2 P(x)$
0	.15	.00	.00
1	.24	.24	.24
2	.29	.58	1.16
3	.14	.42	1.26
4	.10	.40	1.60
5	.08	.40	2.00
		$\sum xP(x) = 2.04$	$\sum x^2 P(x) = 6.26$

$\mu = \sum xP(x) = 2.04$ homes

$\sigma = \sqrt{\sum x^2 P(x) - \mu^2} = \sqrt{6.26 - (2.04)^2} = 1.449$ homes

The four real estate agents sell an average of 2.04 homes per week.

16.   Here, $n = 12$ and $p = .60$

   a. i. $P(\text{exactly } 4) = P(4) = {}_nC_x p^x q^{n-x} = {}_{12}C_8 (.60)^8 (.40)^4 = (495) (.01679616) (.0256) = .2128$

   ii. $P(\text{at least } 6) = P(x \geq 6) = P(6) + P(7) + P(8) + P(9) + P(10) + P(11) + P(12)$

   $= .1766 + .2270 + .2128 + .1419 + .0639 + .0174 + .0022 = .8418$

   iii. $P(\text{less than } 4) = P(x < 4) = P(0) + P(1) + P(2) + P(3) = .0000 + .0003 + .0025 + .0125 = .0153$

   b. $\mu = np = 12(.60) = 7.2$ adults   and $\sigma = \sqrt{npq} = \sqrt{12(.60)(.40)} = 1.697$ adults

17. Let x denote the number of females in a sample of 4 volunteers from the 12 nominees. Then x is a hypergeometric random variable with: $N = 12$, $r = 8$, $N - r = 4$ and $n = 4$.

   a. $P(x = 3) = \dfrac{{}_rC_x \, {}_{N-r}C_{n-x}}{{}_NC_n} = \dfrac{{}_8C_3 \, {}_{12-8}C_{4-3}}{{}_{12}C_4} = \dfrac{56(4)}{495} = .4525$

b. $P(x=1) = \dfrac{{}_rC_x \; {}_{N-r}C_{n-x}}{{}_NC_n} = \dfrac{{}_8C_1 \; {}_{12-8}C_{4-1}}{{}_{12}C_4} = \dfrac{8(4)}{495} = .0646$

c. $P(x \le 1) = P(0) + P(1) = \dfrac{{}_8C_0 \; {}_{12-8}C_{4-0}}{{}_{12}C_4} + .0646 = \dfrac{1(1)}{495} + .0646 = .0020 + .0646 = .0666$

18. Here, $\lambda = 10$ red light runners are caught per day.

Let $x =$ number of drivers caught during rush hour on a given weekday.

a. i. $P(x = 14) = \dfrac{\lambda^x e^{-\lambda}}{x!} = \dfrac{(10)^{14} e^{-10}}{14!} = \dfrac{(1000000000000000)(.0000453999)}{87,178,291,200} = .0521$

ii. Using Table VI of Appendix C of the text, we obtain:

$P(\text{at most } 7) = P(0) + P(1) + P(2) + P(3) + P(4) + P(5) + P(6) + P(7) = .0000 + .0005 + .0023 + .0076 + .0189 + .0378 + .0631 + .0901 = .2203$

iii. $P(13 \text{ to } 18) = P(13) + P(14) + P(15) + P(16) + P(17) + P(18) = .0729 + .0521 + .0347 + .0217 + .0128 + .0071 = .2013$

b.

x	$P(x)$
0	.0000
1	.0005
2	.0023
3	.0076
4	.0189
5	.0378
6	.0631
7	.0901
8	.1126
9	.1251
10	.1251
11	.1137
12	.0948
13	.0729
14	.0521
15	.0347
16	.0217
17	.0128
18	.0071
19	.0037
20	.0019
21	.0009
22	.0004
23	.0002
24	.0001

19. See solution to Exercise 5.57.

Chapter Six

6.1 The probability distribution of a discrete random variable assigns probabilities to points while that of a continuous random variable assigns probabilities to intervals.

6.2 The probability that a continuous random variable x assumes a single value is zero, that is, $P(x = a) = 0.$

6.3 Since $P(a) = 0$ and $P(b) = 0$ for a continuous random variable, $P(a \leq x \leq b) = P(a < x < b)$.

6.4 The following are the three main characteristics of a normal distribution.
1. The total are under a normal curve is 1.0 or 100%.
2. A normal curve is symmetric about the mean. For this reason ½ of the total area under a normal curve lies on the left side of the mean and ½ lies on the right side of the mean.
3. The tails of a normal distribution curve extend indefinitely in both directions without touching or crossing the horizontal axis. Although a normal curve never meets the horizontal axis, beyond the points represented by $\mu - 3\sigma$ and $\mu + 3\sigma$ it becomes so close to this axis that the area under the curve beyond these points in both directions is very close to zero.

These three characteristics are described with graphs on pages 262 and 263 of the text.

6.5 The standard normal distribution is a special case of the normal distribution. For the standard normal distribution, the value of the mean is equal to zero and the value of the standard deviation is 1. In other words, the units of the standard normal distribution curve are denoted by z and are called the z values or z scores. The z values on the right side of the mean (which is zero) are positive and those on the left side are negative. The z value for a point on the horizontal axis gives the distance between the mean and that point in terms of the standard deviation.

6.6 The parameters of the normal distribution are the mean μ and the standard deviation σ.

6.7 As its standard deviation decreases, the width of a normal distribution curve decreases and its height increases.

6.8   The width and height of a normal distribution do not change when its standard deviation remains the same but its mean increases.

6.9   For a standard normal distribution, z gives the distance between the mean and the point represented by z in terms of the standard deviation.

6.10 Area between $\mu - \sigma$ and $\mu + \sigma$ is the area from $z = -1$ to $z = 1$, which is: $P = (-1 \le z \le 1) =$
$P(-1 \le z \le 0) + P(0 \le z \le 1) = .3413 + .3413 = .6826$

6.11 Area between $\mu - 1.5\sigma$ and $\mu + 1.5\sigma$ is the area from $z = -1.50$ to $z = 1.50$, which is:
$P(-1.5 \le z \le 1.5) = P(-1.5 \le z \le 0) + P(0 \le z \le 1.5) = .4332 + .4332 = .8664$

6.12 Area within two standard deviations of the mean is:
$P(-2 \le z \le 2) = P(-2 \le z \le 0) + P(0 \le z \le 2) = .4772 + .4772 = .9544$

6.13 Area within 2.5 standard deviations of the mean is:
$P(-2.5 \le z \le 2.5) = P(-2.5 \le z \le 0) + P(0 \le z \le 2.5) = .4938 + .4938 = .9876$

6.14 Area within three standard deviations of the mean is:
$P(-3 \le z \le 3) = P(-3 \le z \le 0) + P(0 \le z \le 3) = .4987 + .4987 = .9974$

6.15 a. $P(0 < z < 1.95) = .4744$

b. $P(-1.85 < z < 0) = .4678$

c. $P(1.15 < z < 2.37) = P(0 \le z \le 2.37) - P(0 \le z \le 1.15) = .4911 - .3749 = .1162$

d. $P(-2.88 \le z \le -1.53) = P(-2.88 \le z \le 0) - P(-1.53 \le z \le 0) = .4980 - .4370 = .0610$

e. $P(-1.67 \le z \le 2.44) = P(-1.67 \le z \le 0) + P(0 \le z \le 2.44) = .4525 + .4927 = .9452$

6.16 a. $P(0 \le z \le 2.34) = .4904$

b. $P(-2.78 < z < 0) = .4973$

c. $P(.84 \le z \le 1.95) = P(0 \le z \le 1.95) - P(0 \le z \le .84) = .4744 - .2995 = .1749$

d. $P(-2.39 < z < -.57) = P(-2.39 < z < 0) - P(-.57 < z < 0) = .4916 - .2157 = .2759$

e. $P(-2.15 < z < 1.67) = P(-2.15 < z < 0) + P(0 \le z < 1.67) = .4842 + .4525 = .9367$

6.17   a. $P(z > 1.56) = .5 - P(0 \leq z \leq 1.56) = .5 - .4406 = .0594$

        b. $P(z < -1.97) = .5 - P(-1.97 \leq z \leq 0) = .5 - .4756 = .0244$

        c. $P(z > -2.05) = P(-2.05 < z < 0) + .5 = .4798 + .5 = .9798$

        d. $P(z < 1.86) = .5 + P(0 < z < 1.86) = .5 + .4686 = .9686$

6.18   a. $P(z > 1.73) = .5 - P(0 \leq z \leq 1.73) = .5 - .4582 = .0418$

        b. $P(z < -1.55) = .5 - P(-1.55 \leq z \leq 0) = .5 - .4394 = .0606$

        c. $P(z > -.65) = P(-.65 < z < 0) + .5 = .2422 + .5 = .7422$

        d. $P(z < .89) = .5 + P(0 < z < .89) = .5 + .3133 = .8133$

6.19   a. $P(0 < z < 4.28) = .5$ approximately

        b. $P(-3.75 \leq z \leq 0) = .5$ approximately

        c. $P(z > 7.43) = .5 - P(0 \leq z \leq 7.43) = .5 - .5 = .0000$ approximately

        d. $P(z < -4.49) = .5 - P(-4.49 \leq z \leq 0).5 - .5 = .0000$ approximately

6.20   a. $P(0 \leq z \leq 3.94) = .5$ approximately

        b. $P(-5.16 < z < 0) = .5$ approximately

        c. $P(z > 5.42) = .5 - P(0 \leq z \leq 5.42) = .5 - .5 = .0000$ approximately

        d. $P(z < -3.68) = .5 - P(-3.68 \leq z \leq 0) = .5 - .5 = .0000$ approximately

6.21   a. $P(-1.83 \leq z \leq 2.57) = P(-1.83 \leq z \leq 0) + (0 \leq z \leq 2.57) = .4664 + .4949 = .9613$

        b. $P(0 \leq z \leq 2.02) = .4783$

        c. $P(-1.99 \leq z \leq 0) = .4767$

        d. $P(z \geq 1.48) = .5 - P(0 < z < 1.48).5 - .4036 = .0694$

6.22   a. $P(-2.46 \leq z \leq 1.68) = P(-2.46 \leq z \leq 0) + P(0 \leq z \leq 1.68) = .4931 + .4535 = .9466$

        b. $P(0 \leq z \leq 1.86) = .4686$

        c. $P(-2.58 \leq z \leq 0) = .4951$

        d. $P(z \geq .83) = .5 - P(0 < z < .83) = .5 - .2967 = .2033$

6.23   a. $P(z < -2.14) = .5 - P(-2.14 \leq z \leq 0) = .5 - .4838 = .0162$

        b. $P(.67 \leq z \leq 2.49) = P(0 \leq z \leq 2.49) - P(0 \leq z \leq .67) = .4936 - .2486 = .2450$

c. $P(-2.07 \le z \le -.93) = P(-2.07 \le z \le 0) - P(-.93 \le z \le 0) = .4808 - .3238 = .1570$

d. $P(z < 1.78) = .5 + P(0 < z < 1.78) = .5 + .4625 = .9625$

6.24 a. $P(z < -1.31) = .5 - P(-1.31 \le z \le 0) = .5 - .4049 = .0951$

b. $P(1.03 \le z \le 2.89) = P(0 \le z \le 2.89) - P(0 \le z \le 1.03) = .4981 - .3485 = .1496$

c. $P(-2.24 \le z \le -1.09) = P(-2.24 \le z \le 0) - P(-1.09 < z \le 0) = .4875 - .3621 = .1254$

d. $P(z < 2.12) = .5 + P(0 < z < 2.12) = .5 + .4830 = .9830$

6.25 a. $P(z > -.98) = P(-.98 < z < 0) + .5 = .3365 + .5 = .8365$

b. $P(-2.47 \le z \le 1.19) = P(-2.47 \le z \le 0) + P(0 \le z \le 1.19) = .4932 + .3830 = .8762$

c. $P(0 \le z \le 4.25) = .5$ approximately

d. $P(-5.36 \le z \le 0) = .5$ approximately

e. $P(z > 6.07) = .5 - P(0 \le z \le 6.07) = .5 - .5 = .0000$ approximately

f. $P(z < -5.27) = .5 - P(-5.27 \le z \le 0) = .5 - .5 = .0000$ approximately

6.26 a. $P(z > -1.06) = P(-1.06 < z < 0) + .5 = .3554 + .5 = .8554$

b. $P(-.68 \le z \le 1.84) = P(-.68 \le z \le 0) + P(0 \le z \le 1.84) = .2517 + .4671 = .7188$

c. $P(0 \le z \le 3.85) = .5$ approximately

d. $P(-4.34 \le z \le 0) = .5$ approximately

e. $P(z > 4.82) = .5 - P(0 \le z \le 4.82) = .5 - .5 = .0000$ approximately

f. $P(z < -6.12) = .5 - .5 = .0000$ approximately

6.27 Here, $\mu = 30$ and $\sigma = 5$

a. $z = (x - \mu)/\sigma = (39 - 30)/5 = 9/5 = 1.80$

b. $z = (x - \mu)/\sigma = (17 - 30)/5 = -13/5 = -2.60$

c. $z = (x - \mu)/\sigma = (22 - 30)/5 = -8/5 = -1.60$

d. $z = (x - \mu)/\sigma = (42 - 30)/5 = 12/5 = 2.40$

6.28 Here, $\mu = 16$ and $\sigma = 3$

a. $z = (x - \mu)/\sigma = (12 - 16)/3 = -4/3 = -1.33$

b. $z = (x - \mu)/\sigma = (21 - 16)/3 = 5/3 = 1.67$

c. $z = (x - \mu)/\sigma = (19 - 16)/3 = 3/3 = 1.00$

d. $z = (x - \mu)/\sigma = (13 - 16)/3 = -3/3 = -1.00$

6.29 Here, $\mu = 20$ and $\sigma = 3$

a. For $x = 20: z = (x - \mu)/\sigma = (20 - 20)/4 = 0$

For $x = 27: z = (x - \mu)/\sigma = (27 - 20)/4 = 7/4 = 1.75$

$P(20 < x < 27) = P(0 < z < 1.75) = .4599$

b. For $x = 23: z = (x - \mu)/\sigma = (23 - 20)/4 = 3/4 = .75$

For $x = 25: z = (x - \mu)/\sigma = (25 - 20)/4 = 5/4 = 1.25$

$P(23 \le x \le 25) = P(.75 \le z \le 1.25) = P(0 \le z \le 1.25) - P(0 \le z \le .75) = .3944 - .2734 = .1210$

c. For $x = 9.5: z = (x - \mu)/\sigma = (9.5 - 20)/4 = -10.5/4 = -2.63$

For $x = 17: z = (x - \mu)/\sigma = (17 - 20)/4 = -3/4 = -.75$

$P(9.5 < x < 17) = P(-2.63 < z < -.75) = P(-2.63 < z < 0) - P(-.75 \le z \le 0)$

$= .4957 - .2734 = .2223$

6.30 Here, $\mu = 12$ and $\sigma = 2$

a. For $x = 7.76: z = (x - \mu)/\sigma = (7.76 - 12)/2 = -4.24/2 = -2.12$

For $x = 12: z = (x - \mu)/\sigma = (12 - 12)/2 = 0$

$P(7.76 < x < 12) = P(-2.12 < z < 0) = .4830$

b. For $x = 14.48: z = (x - \mu)/\sigma = (14.48 - 12)/2 = 2.48/2 = 1.24$

For $x = 16.54: z = (x - \mu)/\sigma = (16.54 - 12)/2 = 4.54/2 = 2.27$

$P(14.48 < x < 16.54) = P(1.24 < z < 2.27) = P(0 \le z < 2.27) - P(0 \le z \le 1.24)$

$= .4844 - .3925 = .0959$

c. For $x = 8.22: z = (x - \mu)/\sigma = (8.22 - 12)/2 = -3.78/2 = -1.89$

For $x = 10.06: z = (x - \mu)/\sigma = (10.06 - 12)/2 = -1.94/2 = -.97$

$P(8.22 \le x \le 10.06) = P(-1.89 \le z \le -.97) = P(-1.89 \le z \le 0) - P(-.97 \le z \le 0)$

$= .4706 - .3340 = .1366$

6.31 Here, $\mu = 55$ and $\sigma = 7$

a. For $x = 58: z = (58 - 55)/7 = 3/7 = .43$

$$P(x > 58) = P(z > .43) = .5 - P(0 \leq z \leq .43) = .5 - .1664 = .3336$$

b. For $x = 43 : z = (43 - 55)/7 = -12/7 = -1.71$

$$P(x > 43) = P(z > -1.71) = P(-1.71 < z < 0) + .5 = .4564 + .5 = .9564$$

c. For $x = 67 : z = (67 - 55)/7 = 12/7 = 1.71$

$$P(x < 67) = P(z < 1.71) = .5 + p(0 < z < 1.71) = .5 + .4564 = .9564$$

d. For $x = 24 : z = (24 - 55)/7 = -31/7 = -4.43$

$$P(x < 24) = P(z < -4.43) = .5 - P(-4.43 \leq z \leq 0) = .5 - .5 = .00 \quad \text{approximately}$$

6.32   Here, $\mu = 37$ and $\sigma = 3$

a. For $x = 30 : z = (30 - 37)/3 = -7/3 = -2.33$

$$P(x < 30) = P(z < -2.33) = .5 - P(-2.33 \leq z \leq 0) = .5 - .4901 = .0099$$

b. For $x = 52 : z = (52 - 37)/3 = 15/3 = 5.00$

$$P(x > 52) = P(z > 5.00) = .5 - P(0 < z < 5.00) = .5 - .5 = .0000 \quad \text{approximately}$$

c. For $x = 44 : z = (44 - 37)/3 = 7/3 = 2.33$

$$p(x < 44) = P(z < 2.33) = .5 + P(0 < z < 2.33) = .5 + .4901 = .9901$$

d. For $x = 32 : z = (32 - 37)/3 = -5/3 = -1.67$

$$P(x > 32) = P(z > -1.67) = P(-1.67 < z < 0) + .5 = .4525 + .5 = .9525$$

6.33   Here, $\mu = 25$ and $\sigma = 6$

a. For $x = 29 : z = (29 - 25)/6 = 4/6 = .67$

For $x = 36 : z = (36 - 25)/6 = 11/6 = 1.83$

$$P(29 < x < 36) = P(.67 < z < 1.83) = P(0 < z < 1.83) - P(0 < z < .67) = .4664 - .2486 = .2178$$

b. For $x = 22 : z = (22 - 25)/6 = -3/6 = -.50$

For $x = 33 : z = (33 - 25)/6 = 8/6 = 1.33$

$$P(22 < x < 33) = P(-.50 < z < 1.33) = P(-.50 < z < 0) + P(0 < z < 1.33) = .1915 + .4082 = .5997$$

6.34   Here, $\mu = 40$ and $\sigma = 4$

a.   For $x = 29 : z = (29 - 40)/4 = -11/4 = -2.75$

For $x = 35 : z = (35 - 40)/4 = -5/4 = -1.25$

$P(29 < x < 35) = P(-2.75 < z < -1.25) = P(-2.75 < z < 0) - P(-1.25 < z < 0)$

$= .4970 - .3944 = .1026$

b.   For $x = 34 : z = (34 - 40)/4 = -6/4 = -1.50$

For $x = 51 : z = (51 - 40)/4 = 11/4 = 2.75$

$P(34 \leq x \leq 51) = P(-1.50 \leq z \leq 2.75) = P(-1.50 \leq z \leq 0) + P(0 < z \leq 2.75)$

$= .4332 + .4970 = .9302$

6.35   Here, $\mu = 80$ and $\sigma = 12$

a.   For $x = 69 : z = (69 - 80)/12 = -11/12 = -.92$

$P(x > 69) = P(z > -.92) = P(-.92 < z < 0) + .5 = .3212 + .5 = .8212$

b.   For $x = 74 : z = (74 - 80)/12 = -6/12 = -.50$

$P(x < 74) = P(z < -.50) = .5 - P(-.50 \leq z \leq 0) = .5 - .1915 = .3085$

c.   For $x = 101 : z = (101 - 80)/12 = 21/12 = 1.75$

$P(x > 101) = P(z > 1.75) = .5 - P(0 \leq z \leq 1.75) = .5 - .4599 = .0401$

d.   For $x = 88 : z = (88 - 80)/12 = 8/12 = .67$

$P(x < 88) = P(z < .67) = .5 + P(0 < z < .67) = .5 + .2486 = .7486$

6.36   Here, $\mu = 65$ and $\sigma = 15$

a.   For $x = 43 : z = (43 - 65)/15 = -22/15 = -1.47$

$P(x < 43) = P(z < -1.47) = .5 - P(-1.47 \leq z \leq 0) = .5 - .4292 = .0708$

b.   For $x = 74 : z = (74 - 65)/15 = 9/15 = .60$

$P(x > 74) = P(z > .60) = .5 - P(0 \leq z \leq .60) = .5 - .2257 = .2743$

c.   For $x = 56 : z = (56 - 65)/15 = -9/15 = -.60$

$P(x > 56) = P(z > -.60) = P(-.60 < z < 0) + .5 = .2257 + .5 = .7257$

d. For $x = 71 : z = (71 - 65)/15 = 6/15 = .40$

$P(x < 71) = P(z < .40) = .5 + P(0 < z < .40) = .5 + .1554 = .6554$

6.37 Here, $\mu = 190$ minutes and $\sigma = 21$ minutes

a. For $x = 150 : z = (150 - 190)/21 = -40/21 = -1.90$

$P(x < 150) = P(z < -1.90) = .5 - P(-1.90 \leq z \leq 0) = .5 - .4713 = .0287$

b. For $x = 205 : z = (205 - 190)/21 = 15/21 = .71$

For $x = 245 : z = (245 - 190)/21 = 2.62$

$P(205 \leq x \leq 245) = P(.71 \leq z \leq 2.62) = P(0 \leq z \leq 2.62) - P(0 < z < .71)$

$= .4956 - .2611 = .2345$

6.38 Here, $\mu = \$84.58$ and $\sigma = \$12$

a. For $x = 75 : z = (75 - 84.58)/12 = -.80$

For $x = 92 : z = (92 - 84.58)/12 = .62$

$P(75 < x < 92) = P(-.80 < z < .62) = P(-.80 < z < 0) + P(0 < z < .62)$

$= .2881 + .2324 = .5205 = 52.05\%$

b. For $x = 93 : z = (93 - 84.58)/12 = .70$

For $x = 100 : z = (100 - 84.58)/12 = 1.29$

$P(93 < x < 100) = P(.70 < z < 1.29) = P(0 < z < 1.29) - P(0 < z < .70)$

$= .4015 - .2580 = .1435 = 14.35\%$

6.39 Here, $\mu = \$10$ and $\sigma = \$1.10$

a. For $x = 12.00 : z = (12.00 - 10.00)/1.10 = 1.82$

$P(x > 12.00) = P(z > 1.82) = .5 - P(0 \leq z \leq 1.82) = .5 - .4656 = .0344$

b. For $x = 8.50 : z = (8.50 - 10.00)/1.10 = -1.36$

For $x = 10.80 : z = (10.80 - 10.00)/1.10 = .73$

$P(8.50 < x < 10.80) = P(-1.36 < z < .73) = P(-1.36 < z < 0) + P(0 < z < .73)$

$= .4131 + .2673 = .6804$

6.40   Here, $\mu = 36$ seconds and $\sigma = 2.5$ seconds

     a. For $x = 40 : z = (40 - 36) / 2.5 = 4 / 2.5 = 1.60$

       $P(x > 40) = P(z > 1.60) = .5 - P(0 < z < 1.60) = .5 - .4452 = .0548$

     b. For $x = 30 : z = (30 - 36) / 2.5 = -2.40$

       For $x = 34 : z = (34 - 36) / 2.5 = -.80$

       $P(30 < x < 34) = P(-2.40 < z < -.80) = P(-2.40 < z < 0) - P(-.80 < z < 0)$

          $= .4918 - .2881 = .2037$

6.41   Here, $\mu = 40$ miles per hour and $\sigma = 4$ miles per hour

     a. For $x = 46 : z = (40 - 46) / 4 = -6 / 4 = -1.50$

       $P(x > 0) = P(z > -1.50) = .5 + P(-1.50 \le z \le 0) = .5 + .4332 = .9332$

       Thus, about 93.32% of the cars traveling in the construction zone are exceeding the 40 miles per hour posted speed limit.

     b. For $x = 50 : z = (50 - 46) / 4 = 4 / 4 = 1.00$

       For $x = 55 : z = (55 - 46) / 4 = 9 / 4 = 2.25$

       $P(50 \le x \le 55) = P(1.00 \le z \le 2.25) = P(0 \le z \le 2.25) - P(0 \le z \le 1.00) = .4878 - .3413 = .1465$

       Thus, about 14.65% of the cars traveling in the construction zone have a speed of 50 to 55 miles per hour.

6.42   Here, $\mu = \$845$ and $\sigma = \$270$

     a. For $x = 1000 : z = (1000 - 845) / 270 = .57$

       For $x = 1400 : z = (1400 - 845) / 270 = 2.06$

       $P(1000 < x < 1400) = P(.57 < z < 2.06) = P(0 < z < 2.06) - P(0 < z < .57) = .4803 - .2157 = .2646$

     b. For $x = 750 : z = (750 - 845) / 270 = -.35$

       $P(x \ge 750) = P(z \ge -.35) = P(-.35 \le z \le 0) + .5 = .1368 + .5 = .6368$ or 63.68%.

6.43   Here, $\mu = \$2084$ and $\sigma = \$300$

     a. For $x = 2500 : z = (2500 - 2084) / 300 = 1.39$

       $P(x > 2500) = P(z > 1.39) = .5 - P(0 \le z \le 1.39) = .5 - .4177 = .0823$

b. For $x = 1800 : z = (1800 - 2084)/300 = -.95$

For $x = 2400 : z = (2400 - 2084)/300 = 1.05$

$P(1800 < x < 2400) = P(-.95 < z < 1.05) = P(-.95 < z < 0) + P(0 \le z \le 1.05)$

$$= .3289 + .3531 = .6820 \text{ or } 68.20\%$$

6.44 Here, $\mu = 72{,}000$ miles and $\sigma = 12{,}000$ miles

a. For $x = 40{,}000 : z = (40{,}000 - 72{,}000)/12{,}000 = -2.67$

$P(x \le 40{,}000) = P(z \le -2.67) = .5 - P(-2.67 < z < 0) = .5 - .4962 = .0038 \text{ or } .38\%$

b. For $x = 100{,}000 : z = (100{,}000 - 72{,}000)/12{,}000 = 2.33$

$P(x > 100{,}000) = P(z > 2.33) = .5 - P(0 \le z \le 2.33) = .5 - .4901 = .0099 \text{ or } .99\%$

6.45 Here, $\mu = 1650$ kwh and $\sigma = 320$ kwh

a. For $x = 1850 : z = (1850 - 1650)/320 = .63$

$P(x < 1850) = P(z < .63) = P(z \le 0) + P(0 < z < .63) = .5 + .2357 = .7357$

b. For $x = 900 : z = (900 - 1650)/320 = -2.34$

For $x = 1340 : z = (1340 - 1650)/320 = -.97$

$P(900 \le x \le 1340) = P(-2.34 \le z \le -.97) = P(-2.34 \le z \le 0) - P(.97 < z < 0)$

$$= .4904 - .3340 = .1564 \text{ or } 15.64\%$$

6.46 Here, $\mu = \$95$ and $\sigma = \$21$

For $x = 130 : z = (130 - 95)/21 = 1.67$

$P(x > 130) = P(z > 1.67) = .5 - P(0 \le z \le 1.67) = .5 - .4525 = .0475$

Thus, about 4.75% of the customers are expected to get free gifts.

6.47 Here, $\mu = \$43{,}070$ and $\sigma = \$3000$

a. For $x = 40{,}000 : z = (40{,}000 - 43{,}070)/3000 = -1.02$

$P(x < 40{,}000) = P(z < -1.02) = .5 - P(-1.02 \le z \le 0) = .5 - .3461 = .1539 \text{ or } 15.39\%$

b. For $x = 45{,}000 : z = (45{,}000 - 43{,}070)/3000 = .64$

$P(x > 45{,}000) = P(z > .64) = .5 - P(0 < z < .64) = .5 - .2389 = .2611 \text{ or } 26.11\%$

6.48   Here, $\mu = 7.59$ and $\sigma = .73$

a.   For $x = 6: z = (6 - 7.59)/.73 = -2.18$

$P(x < 6) = P(z < -2.18) = .5 - P(-2.18 \leq z \leq 0) = .5 - .4854 = .0146$

b.   For $x = 7: z = (7 - 7.59)/.73 = -.81$

For $x = 8: z = (8 - 7.59)/.73 = .56$

$P(7 < x < 8) = P(-.81 < z < .56) = P(-.81 < z < 0) + P(0 \leq z \leq .56) = .2910 + .2123 = .5033$

c.   For $x = 9: z = (9 - 7.59)/.73 = 1.93$

$P(x \geq 9) = P(z \geq 1.93) = .5 - P(0 < z < 1.93) = .5 - .4732 = .0268$ or 2.68%

6.49   Here, $\mu = 516$ and $\sigma = 90$

a.   For $x = 600: z = (600 - 516)/90 = .93$

$P(x > 600) = P(z > .93) = .5 - P(0 \leq z \leq .93) = .5 - .3238 = .1762$ or 17.62%

b.   For $x = 450: z = (450 - 516)/90 = -.73$

$P(x < 450) = P(z < -.73) = .5 - P(-.73 \leq z \leq 0) = .5 - .2673 = .2327$ or 23.27%

c.   For $x = 700: z = (700 - 516)/90 = 2.04$

$P(x \geq 700) = P(z \geq 2.04) = .5 - P(0 \leq z \leq 2.04) = .5 - .4793 = .0207$ or 2.07%

6.50   Here, $\mu = 15$ minutes and $\sigma = 2.4$ minutes

a.   For $x = 20: z = (20 - 15)/2.4 = 2.08$

$P(x > 20) = P(z > 2.08) = .5 - P(0 \leq z \leq 2.08) = .5 - .4812 = .0188$ or 1.88%

b.   For $x = 25: z = (25 - 15)/2.4 = 4.17$

$P(x > 25) = P(z > 4.17) = .5 - P(0 \leq z \leq 4.17) = .5 - .5 = .0000$ approximately

Although it is possible that a given car may take more than 25 minutes for oil and lube service, the probability is almost zero.

6.51   Here, $\mu = 3.0$ inches and $\sigma = .009$ inch

For $x = 2.98: z = (2.98 - 3.00)/.009 = -2.22$

For $x = 3.02: z = (3.02 - 3.00)/.009 = 2.22$

$P(x < 2.98) + P(x > 3.02) = P(z < -2.22) + P(z > 2.22)$

$= \{.5 - P(-2.22 \le z \le 0)\} + \{.5 - P(0 \le z \le 2.22)\} = \{.5 - .4868\} + \{.5 - .4868\} = .0132 + .0132 = .0264$

Thus, about 2.64% of the nails produced by this machine are unusable.

6.52   Here, $\mu = 5.75$ ounces and $\sigma = .11$ ounce

For $x = 5.5 : z = (5.5 - 5.75)/.11 = -2.27$

For $x = 6.0 : z = (6.0 - 5.75)/.11 = 2.27$

$P(x < 5.5) + P(x > 6.0) = 1 - P(5.5 < x < 6.0) = 1 - P(-2.27 < z < 2.27) =$

$= 1 - [P(-2.27 < z < 0) + P(0 \le z \le 2.27)] = 1 - [.4884 + .4884] = 1 - .9768 = .0232$ or 2.32%

6.53   a. $z = 2.00$ b. $z = -2.02$ approximately

c. $z = -.37$ approximately   d. $z = 1.02$ approximately

6.54   a. $z = .51$ approximately   b. $z = -.75$ approximately

c. $z = -.82$ approximately   d. $z = 1.25$ approximately

6.55   a. $z = 1.65$ approximately   b. $z = -1.96$

c. $z = -2.33$ approximately   d. $z = 2.58$ approximately

6.56   a. $z = 1.96$ b. $z = -1.65$ approximately

c. $z = -3.08$ approximately   d. $z = 2.33$ approximately

6.57   Here, $\mu = 200$ and $\sigma = 25$

a.   Area to left of $x = .6330$

Area between μ and $x = .6330 - .5 = .1330$

z for area of .1330 = .34 approximately

$x = \mu + z\sigma = 200 + (.34)(25) = 208.50$ approximately

b.   Area to the right of $x = .05$

Area between μ and $x = .5 - .05 = .4500$

z for area of .4500 = 1.65 approximately

$x = \mu + z\sigma = 200 + (1.65)(25) = 241.25$ approximately

c. $x = \mu + z\sigma = 200 + (-.86)(25) = 178.50$

d. $x = \mu + z\sigma = 200 + (-2.17)(25) = 145.75$

e. $x = \mu + z\sigma = 200 + (-1.67)(25) = 158.25$

f. $x = \mu + z\sigma = 200 + (2.05)(25) = 251.25$ approximately

6.58 Here, $\mu = 550$ and $\sigma = 75$

a. $x = \mu + z\sigma = 550 + (-1.96)(75) = 403$

b. $x = \mu + z\sigma = 550 + (-1.51)(75) = 436.75$

c. $x = \mu + z\sigma = 550 + (1.92)(75) = 694$ approximately

d. $x = \mu + z\sigma = 550 + (1.75)(75) = 681.25$ approximately

e. $x = \mu + z\sigma = 550 + (-1.88)(75) = 409$ approximately

f. $x = \mu + z\sigma = 550 + (1.34)(75) = 650.5$ approximately

6.59 Here, $\mu = 15$ minutes and $\sigma = 2.4$ minutes

Let x denote the time to service a randomly chosen car. Then, we are to find x so that the area in the right tail of the normal distribution curve is 5%. Thus, $x = \mu + z\sigma = 15 + (1.65)(2.4) = 18.96 \approx 19$ minutes.

6.60 Here, $\mu = \$95$ and $\sigma = \$21$

Let x denote the amount spent by a randomly chosen customer on a visit to this store. Then, we are to find x such that the area in the right tail of the normal distribution curve is 10%. Thus,

$$x = \mu + z\sigma = 95 + (1.28)(21) = \$121.88 \text{ approximately}$$

So a required minimum purchase of $122 would meet the condition.

6.61 Here, $\mu = 1650$ kwh and $\sigma = 320$ kwh

Bill Johnson's monthly electric consumption is: $x = \mu + z\sigma = 1650 + (1.28)(320) = 2059.6 \approx 2060$ kwh

6.62 Here, $\mu = 70$ months and $\sigma = 8$ months

Let x denote the warranty period.

a. We are to find x such that the area in the left tail of the normal curve is 1%. Thus,

$$x = \mu + z\sigma = 70 + (-2.33)(8) = 51.36 \approx 51 \text{ months}$$

b. We are to find x such that the area in the left tail of the normal curve is 5%. Thus,

$$x = \mu + z\sigma = 70 + (-1.65)(8) = 56.8 \approx 57 \text{ months}$$

6.63 We are given that: $\sigma = \$9.50$ and $P(x \geq 90) = .20$

Hence, the area between μ and $x = 90$ is $.5 - .20 = .3000$. The z value for .3000 is .84 approximately.

Since $x = \mu + z\sigma$, we obtain: $\mu = x - z\sigma = 90 - (.84)(9.50) = \$82.02$ approximately

6.64   We are given that: $\sigma = .35$ and $P(x \geq 64) = .95$

Hence, the area between μ and $x = 64$ is .4500 and μ is greater than 64 ounces. Then,

$$\mu = x - z\sigma = 64 - (-1.65)(.35) = 64.58 \text{ ounces}$$

6.65   The normal distribution may be used as an approximation to a binomial distribution when both np and nq are greater than 5.

6.66   a. From Table IV of Appendix C, for $n = 20$ and $p = .60$, $P(x = 12) = .1797$

b. $\mu = np = 20(.60) = 12$ and $\sigma = \sqrt{npq} = \sqrt{20(.60)(.40)} = 2.19089023$

The binomial probability $P(x = 12)$ is approximated by the area under the normal distribution curve from $x = 11.5$ to $x = 12.5$.

For $x = 11.5$: $z = (11.5 - 12) / 2.19089023 = -.23$

For $x = 12.5$: $z = (12.5 - 12) / 2.19089023 = .23$

$P(11.5 \leq x \leq 12.5) = P(-.23 \leq z \leq .23) = P(-.23 \leq z \leq 0) + P(0 \leq z \leq .23) = .0910 + .0910 = .1820$

The difference between this approximation and the exact probability is: $.1820 - .1797 = .0023$

6.67   a. From Table IV of Appendix C, for $n = 25$ and $p = .40$,

$$P(8 \leq x \leq 13) = P(8) + P(9) + P(10) + P(11) + P(12) + P(13)$$

$$= .1200 + .1511 + .1612 + .1465 + .1140 + .0760 = .7688$$

b. $\mu = np = 25(.40) = 10$ and $\sigma = \sqrt{npq} = \sqrt{25(.40)(.60)} = 2.44948974$

The binomial probability $P(8 \leq x \leq 13)$ is approximated by the area under the normal distribution curve from $x = 7.5$ to 13.5.

For $x = 7.5$: $z = (7.5 - 10) / 2.44948974 = -1.02$

For $x = 13.5$: $z = (13.5 - 10) / 2.44948974 = 1.43$

$P(7.5 \leq x \leq 13.5) = P(-1.02 \leq z \leq 1.43) = P(-1.02 \leq z \leq 0) + P(0 \leq z \leq 1.43)$

$= .3461 + .4236 = .7697$

The difference between this approximation and the exact probability is: $.7697 - .7688 = .0009$

6.68   a. $\mu = np = 80(.50) = 40$ and $\sigma = \sqrt{npq} = \sqrt{80(.50)(.50)} = 4.47213596$

b. The binomial probability $P(x \geq 40)$ is approximated by the area under the normal distribution curve to the right of $x = 39.5$.

For $x = 39.5$: $z = (39.5 - 40) / 4.47213596 = -.11$

$P(x \geq 39.5) = P(z \geq -.11) = P(-.11 \leq z \leq 0) + .5 = .0438 + .5 = .5438$

c. The binomial probability $P(41 \leq x \leq 48)$ is approximated by the area under the normal distribution curve between $x = 40.5$ and 48.5.

For $x = 40.5$: $z = (40.5 - 40) / 4.47213596 = .11$

For $x = 48.5$: $z = (48.5 - 40) / 4.47213596 = 1.90$

$P(40.5 \leq x \leq 48.5) = P(.11 \leq z \leq 1.90) = P(0 \leq z \leq 1.90) - P(0 \leq z \leq .11) = .4713 - .0438 = .4275$

6.69 a. $\mu = np = 120(.60) = 72$ and $\sigma = \sqrt{npq} = \sqrt{120(.60)(.40)} = 5.36656315$

b. The binomial probability $P(x \leq 72)$ is approximated by the area under the normal distribution curve to the left of $x = 72.5$.

For $x = 72.5$: $z = (72.5 - 72) / 5.36656315 = .09$

$P(x \leq 72.5) = P(z \leq .09) = .5 + P(0 \leq z \leq .09) = .5 + .0359 = .5359$

c. The binomial probability $P(67 \leq x \leq 73)$ is approximated by the area under the normal distribution curve between $x = 66.5$ and 73.5.

For $x = 66.5$: $z = (66.5 - 72) / 5.36656315 = -1.02$

For $x = 73.5$: $z = (73.5 - 72) / 5.36656315 = .28$

$P(66.5 \leq x \leq 73.5) = P(-1.02 \leq z \leq .28) = P(-1.02 \leq z \leq 0) + P(0 \leq z \leq .28)$

$= .3461 + .1103 = .4564$

6.70 a. $\mu = np = 140(.45) = 63$ and $\sigma = \sqrt{npq} = \sqrt{140(.45)(.55)} = 5.88642506$

For $x = 66.5$: $z = (66.5 - 63) / 5.88642506 = 3.5 / 5.88642506 = .59$

For $x = 67.5$: $z = (67.5 - 63) / 5.88642506 = 4.5 / 5.88642506 = .76$

$P(66.5 \leq x \leq 67.5) = P(.59 \leq z \leq .76) = P(0 \leq z \leq .76) - P(0 \leq z \leq .59)$

$= .2764 - .2224 = .0540$

b. $\mu = np = 100(.55) = 55$ and $\sigma = \sqrt{npq} = \sqrt{100(.55)(.45)} = 4.97493719$

For $x = 51.5$: $z = (51.5 - 55) / 4.97493719 = -3.5 / 4.97493719 = -.70$

For $x = 60.5$: $z = (60.5 - 55) / 4.97493719 = 5.5 / 4.97493719 = 1.11$

$P(51.5 \le x \le 60.5) = P(-.70 \le z \le 1.11) = P(-.70 \le z \le 0) + P(0 \le z \le 1.11) = .2580 + .3665 = .6245$

c. $\mu = np = 90(.42) = 37.80$ and $\sigma = \sqrt{npq} = \sqrt{90(.42)(.58)} = 4.68230712$

For $x = 39.5$: $z = (39.5 - 37.80) / 4.68230712 = 1.7 / 4.68230712 = .36$

$P(x \ge 39.5) = P(z \ge .36) = 5 - P(0 < z < .36) = .5 - .1406 = .3594$

d. $\mu = np = 104(.75) = 78$ and $\sigma = \sqrt{npq} = \sqrt{104(.75)(.25)} = 4.41588043$

For $x = 72.5$: $z = (72.5 - 78) / 4.41588043 = -5.5 / 4.41588043 = -1.25$

$P(x \le 72.5) = P(z \le -1.25) = 5 - P(-1.25 < z < 0) = .5 - .3944 = .1056$

6.71 a. $\mu = np = 70(.30) = 21$ and $\sigma = \sqrt{npq} = \sqrt{70(.30)(.70)} = 3.38405790$

For $x = 17.5$: $z = (17.5 - 21) / 3.38405790 = -3.5 / 3.38405790 = -.91$

For $x = 18.5$: $z = (18.5 - 21) / 3.38405790 = -2.5 / 3.38405790 = -.65$

$P(17.5 \le x \le 18.5) = P(-.91 \le z \le -.65) = P(-.91 \le z \le 0) - P(-.65 \le z \le 0) = .3186 - .2422 = .0764$

b. $\mu = np = 200(.70) = 140$ and $\sigma = \sqrt{npq} = \sqrt{200(.70)(.30)} = 6.48074070$

For $x = 132.5$: $z = (132.5-140) / 6.48074070 = -7.5 / 6.48074070 = -1.16$

For $x = 145.5$: $z = (145.5 - 140) / 6.48074070 = 5.5 / 6.48074070 = .85$

$P(132.5 \le x \le 145.5) = P(-1.16 \le z \le .85) = P(-1.16 \le z \le 0) + P(0 \le z \le .85)$

$= .3370 + .3023 = .6793$

c. $\mu = np = 85(.40) = 34$ and $\sigma = \sqrt{npq} = \sqrt{85(.40)(.60)} = 4.51663592$

For $x = 29.5$: $z = (29.5 - 34) / 4.51663592 = -4.5 / 4.51663592 = -1.00$

$P(x \ge 29.5) = P(z \ge -1.00) = P(-1.00 \le z \le 0) + .5 = .3413 + .5 = .8413$

d. $\mu = np = 150(.38) = 57$ and $\sigma = \sqrt{npq} = \sqrt{150(.38)(.62)} = 5.94474558$

For $x = 62.5$: $z = (62.5 - 57) / 5.94474558 = 5.5 / 5.94474558 = .93$

$P(x \le 62.5) = P(z \le .93) = 5 + P(0 \le z \le .93) = .5 + .3238 = .8238$

6.72 $\mu = np = 500(.30) = 150$ and $\sigma = \sqrt{npq} = \sqrt{500(.30)(.70)} = 10.24695077$

For $x = 141.5$: $z = (141.5 - 150) / 10.24695077 = -.83$

For $x = 163.5$: $z = (163.5 - 150) / 10.24695077 = 1.32$

$P(141.5 \le x \le 163.5) = P(-.83 \le z \le 1.32) = P(-.83 \le z \le 0) + P(0 \le z \le 1.32) = .2467 + .4066 = .7033$

In a random sample of 500 young men, the probability that 142 to 163 have tattoos is .7033 or 70.33%.

6.73 $n = 250, p = .453, q = 1 - p = .547$

$\mu = 250(.453) = 113.25;\ \sigma = \sqrt{npq} = \sqrt{250(.453)(.547)} = 7.87068930$

For $x = 103.5$: $z = (103.5 - 113.25) / 7.87068930 = -1.24$

For $x = 120.5$: $z = (120.5 - 113.25) / 7.87068930 = .92$

$P(103.5 \leq x \leq 120.5) = P(-1.24 \leq z \leq .92) = P(-1.24 \leq z \leq 0) + P(0 \leq z \leq .92) = .3925 + .3212 = .7137$

The probability that in a random sample of 250 ninth-graders, 104 to 120 watch three or more hours of television on a typical school day is approximately .7137.

6.74   a. $n = 400, p = .64, q = 1 - p = .36$

$\mu = 400(.64) = 256;\ \sigma = \sqrt{npq} = \sqrt{400(.64)(.36)} = 9.6$

For $x = 250.5$: $z = (250.5 - 256) / 9.6 = -.57$

For $x = 249.5$: $z = (249.5 - 256) / 9.6 = -.68$

$P(249.5 \leq x \leq 250.5) = P(-.68 \leq z \leq -.57) = P(-.68 \leq z \leq 0) - P(-.57 \leq z \leq 0)$

$$= .2517 - .2157 = .0360$$

The probability that in a random sample of 400 Americans, the number who will say that mistakes have taught them the most about money management is exactly 250 is .0360.

   b. For $x = 259.5$: $z = (259.5 - 256) / 9.6 = .36$

For $x = 272.5$: $z = (272.5 - 256) / 9.6 = 1.72$

$P(259.5 \leq x \leq 272.5) = P(.36 \leq z \leq 1.72) = P(0 \leq z \leq 1.72) - P(0 \leq z \leq .36) = .4573 - .1406 = .3167$

The probability that in a random sample of 400 Americans, the number who will say that mistakes have taught them the most about money management is between 260 and 272 is .3167 or 31.67%.

   c. For $x = 244.5$: $z = (244.5 - 256) / 9.6 = -1.20$

$P(x \leq 244.5) = P(z \leq -1.20) = .5 - P(-1.20 \leq z \leq 0) = .5 - .3849 = .1151$

The probability that in a random sample of 400 Americans, the number who will say that mistakes have taught them the most about money management is at most 244 is .0968 or 9.687%.

6.75 $n = 720, p = .2, q = 1 - p = .8, \mu = np = 720(.2) = 144,$ and $\sigma = \sqrt{npq} = \sqrt{720(.2)(.8)} = 10.73312629$

   a. For $x = 139.5$: $z = (139.5 - 144) / 10.73312629 = -.42$

For $x = 140.5$: $z = (140.5 - 144) / 10.73312629 = -.33$

$P(139.5 \leq x \leq 140.5) = P(-.42 \leq z \leq -.33) = P(-.42 \leq z \leq 0) - P(-.33 \leq z \leq 0)$

$$= .1628 - .1293 = .0335$$

    b. For $x = 150.5$: $z = (150.5 - 144) / 10.73312629 = .61$

        $P(x \geq 150.5) = P(z \geq .61) = .5 - P(0 \leq z \leq .61) = .5 - .2291 = .2709$

    c. For $x = 129.5$: $z = (129.5 - 144) / 10.73312629 = -1.35$

        For $x = 152.5$: $z = (152.5 - 144) / 10.73312629 = .79$

        $P(129.5 \leq x \leq 152.5) = P(-1.35 \leq z \leq .79) = P(-1.35 \leq z \leq 0) + P(0 \leq z \leq .79)$

                            $= .4115 + .2852 = .6967$

6.76 $n = 300, p = .27, q = 1 - p = .73, \mu = np = 300(.27) = 81$, and $\sigma = \sqrt{npq} = \sqrt{300(.27)(.73)} = 7.68960337$

    a. For $x = 78.5$: $z = (78.5 - 81) / 7.68960337 = -.33$

        For $x = 79.5$: $z = (79.5 - 81) / 7.68960337 = -.20$

        $P(78.5 \leq x \leq 79.5) = P(-.33 \leq z \leq -.20) = P(-.33 \leq z \leq 0) - P(-.20 \leq z \leq 0) = .1293 - .0793 = .0500$

    b. For $x = 74.5$: $z = (74.5 - 81) / 7.68960337 = -.85$

        $P(x \leq 74.5) = P(z \leq -.85) = .5 - P(-.85 \leq z \leq 0) = .5 - .3023 = .1977$

    c. For $x = 74.5$: $z = (74.5 - 81) / 7.68960337 = -.85$

        For $x = 89.5$: $z = (89.5 - 81) / 7.68960337 = 1.11$

        $P(74.5 \leq x \leq 89.5) = P(-.85 \leq z \leq 1.11) = P(-.85 \leq z \leq 0) + P(0 \leq z \leq 1.11) = .3023 + .3665 = .6688$

6.77 $\mu = np = 100(.80) = 80$, and $\sigma = \sqrt{npq} = \sqrt{100(.80)(.20)} = 4$

    a. For $x = 74.5$: $z = (74.5 - 80) / 4 = -1.38$

        For $x = 75.5$: $z = (75.5 - 80) / 4 = -1.13$

        $P(74.5 \leq x \leq 75.5) = P(-1.38 \leq z \leq -1.13) = P(-1.38 \leq z \leq 0) - P(-1.13 \leq z \leq 0)$

                            $= .4126 - .3708 = .0454$

    b. For $x = 73.5$: $z = (73.5 - 80) / 4 = -1.63$

        $P(x \leq 73.5) = P(z \leq -1.63) = .5 - P(-1.63 \leq z \leq 0) = .5 - .4484 = .0516$

    c. For $x = 73.5$: $z = (73.5 - 80) / 4 = -1.63$

        For $x = 85.5$: $z = (85.5 - 80) / 4 = 1.38$

        $P(73.5 \leq x \leq 85.5) = P(-1.63 \leq z \leq 1.38) = P(-1.63 \leq z \leq 0) + P(0 \leq z \leq 1.38) = .4484 + .4162 = .8646$

6.78 $\mu = np = 500(.05) = 25$, and $\sigma = \sqrt{npq} = \sqrt{500(.05)(.95)} = 4.87339717$

a. For $x = 28.5$: $z = (28.5 - 25) / 4.87339717 = .72$

For $x = 29.5$: $z = (29.5 - 25) / 4.87339717 = .92$

$P(28.5 \leq x \leq 29.5) = P(.72 \leq z \leq .92) = P(0 \leq z \leq .92) - P(0 \leq z \leq .72) = .3212 - .2642 = .0570$

b. For $x = 26.5$: $z = (26.5 - 25) / 4.87339717 = .31$

$P(x \geq 26.5) = P(z \geq .31) = .5 - P(0 \leq z \leq .31) = .5 - .1217 = .3783$

c. For $x = 14.5$: $z = (14.5 - 25) / 4.87339717 = -2.15$

For $x = 22.5$: $z = (22.5 - 25) / 4.87339717 = -.51$

$P(14.5 \leq x \leq 22.5) = P(-2.15 \leq z \leq -.51) = P(-2.15 \leq z \leq 0) - P(-.51 \leq z \leq 0)$

$= .4842 - .1950 = .2892$

6.79 $\mu = np = 100(.05) = 5$, and $\sigma = \sqrt{npq} = \sqrt{100(.05)(.95)} = 2.17944947$

a. For $x = 6.5$: $z = (6.5 - 5) / 2.17944947 = .69$

$P(\text{shipment is accepted}) = P(x < 6.5) = P(z < .69) = .5 + P(0 < z < .69) = .5 + .2549 = .7549$

b. $P(\text{shipment is not accepted}) = 1 - P(\text{shipment is accepted}) = 1 - .7549 = .2451$

6.80 $\mu = 8$ minutes and $\sigma = 2$ minutes

a. For $x = 3$: $z = (3 - 8) / 2 = -2.50$

$P(x < 3) = P(z < -2.50) = .5 - P(-2.50 \leq z \leq 0) = .5 - .4938 = .0062$

b. For $x = 10$: $z = (10 - 8) / 2 = 1.00$

For $x = 13$: $z = (13 - 8) / 2 = 2.50$

$P(10 \leq x \leq 13) = P(1.00 \leq z \leq 2.50) = P(0 \leq z \leq 2.50) - P(0 \leq z \leq 1.00)$

$= .4938 - .3413 = .1525$ or 15.25%

c. For $x = 6$: $z = (6 - 8) / 2 = -1.00$

For $x = 12$: $z = (12 - 8) / 2 = 2.00$

$P(6 \leq x \leq 12) = P(-1.00 \leq z \leq 2.00) = P(-1.00 \leq z \leq 0) + P(0 \leq z \leq 2.00)$

$= .3413 + .4772 = .8185$ or 81.85%

d. For $x = 16$: $z = (16 - 8) / 2 = 4.00$

$P(x > 16) = P(z > 4.00) = .5 - P(0 \leq z \leq 4.00) = .5 - .5 = .00$ approximately

Although it is possible for a customer to wait for more than 16 minutes, the probability of this is very close to zero.

6.81 $\mu = \$87$ and $\sigma = \$22$

a.   For $x = 114$: $z = (114 - 87) / 22 = 1.23$

$P(x > 114) = P(z > 1.23) = .5 - P(0 \le z \le 1.23) = .5 - .3907 = .1093$

b.   For $x = 40$: $z = (40 - 87) / 22 = -2.14$

For $x = 60$: $z = (60 - 87) / 22 = -1.23$

$P(40 \le x \le 60) = P(-2.14 \le z \le -1.23) = P(-2.14 \le z \le 0) - P(-1.23 < z < 0)$

$= .4838 - .3907 = .0931$ or 9.31%

c.   For $x = 70$: $z = (70 - 87) / 22 = -.77$

For $x = 105$: $z = (105 - 87) / 22 = .82$

$P(70 \le x \le 105) = P(-.77 \le z \le .82) = P(-.77 \le z \le 0) + P(0 \le z \le .82) =$

$.2794 + .2939 = .5733$ or 57.33%

d.   For $x = 185$: $z = (185 - 87) / 22 = 4.45$

$P(x > 185) = P(z > 4.45) = .5 - P(0 \le z \le 4.45) = .5 - .5 = .00$ approximately

Although it is possible for a customer to write a check for more than \$185, its probability is very close to zero.

6.82 $\mu = 16$ ounces and $\sigma = .18$ ounce

a.   For $x = 16.20$: $z = (16.20 - 16) / .18 = 1.11$

For $x = 16.50$: $z = (16.50 - 16) / .18 = 2.78$

$P(16.20 \le x \le 16.50) = P(1.11 \le z \le 2.78) = P(0 \le z \le 2.78) - P(0 \le z \le 1.11)$

$= .4973 - .3665 = .1308$

b.   For $x = 15.70$: $z = (15.70 - 16) / .18 = -1.67$

$P(x < 15.70) = P(z < -1.67) = .5 - P(-1.67 \le z \le 0) = .5 - .4525 = .0475$ or 4.75%

c.   For $x = 15.20$: $z = (15.20 - 16) / .18 = -4.44$

$P(x < 15.20) = P(z < -4.44) = .5 - P(-4.44 \le z \le 0) = .5 - .5 = .00$ approximately

Although it is possible for a carton to contain less than 15.20 ounces, its probability is very close to zero.

6.83 $\mu = 50$ inches and $\sigma = .06$

For $x = 49.85$: $z = (49.85 - 50) / .06 = -2.50$

For $x = 50.15$: $z = (50.15 - 50) / .06 = 2.50$

$P(x < 49.85) + P(x > 50.15) = P(z < -2.50) + P(z > 2.50)$

$= [.5 - P(-2.50 \le z \le 0)] + [.5 - P(0 \le z \le 2.50)] = [.5 - .4938] + [.5 - .4938] = .0124$ or 1.24%

6.84 $\mu = 28$ minutes and $\sigma = 5$ minutes

Area in the right tail of the normal curve is given to be 1%, which gives $z = 2.33$ approximately.

$x = \mu + z\sigma = 28 + (2.33)(5) = 39.65$ minutes approximately. Thus Jenn must leave approximately 40 minutes before 8:30AM, so she must leave by 7:50 AM.

6.85 $\mu = 750$ hours and $\sigma = 50$ hours

a. Area in the right tail of the normal curve is given to be 2.5%, which gives $z = 1.96$.

     $x = \mu + z\sigma = 750 + (1.96)(50) = 848$ hours

b. Area to the left of x is 80%, which gives $z = .84$ approximately.

     $x = \mu + z\sigma = 750 + (.84)(50) = 792$ hours approximately.

6.86 $\mu = np = 100(.80) = 80$ and $\sigma = \sqrt{npq} = \sqrt{100(.80)(.20)} = 4$

a. For $x = 84.5$: $z = (84.5 - 80) / 4 = 1.13$

     For $x = 85.5$: $z = (85.5 - 80) / 4 = 1.38$

     $P(84.5 \le x \le 85.5) = P(1.13 \le z \le 1.38) = P(0 \le z \le 1.38) - P(0 \le z \le 1.13)$

                 $= .4162 - .3708 = .0454$

b. For $x = 74.5$: $z = (74.5 - 80) / 4 = -1.38$

     $P(x \le 74.5) = P(z \le -1.38) = .5 - P(-1.38 < z < 0) = .5 - .4162 = .0838$

c. For $x = 74.5$: $z = (74.5 - 80) / 4 = -1.38$

     For $x = 87.5$: $z = (87.5 - 80) / 4 = 1.88$

     $P(74.5 \le x \le 87.5) = P(-1.38 \le z \le 1.88) = P(-1.38 \le z \le 0) + P(0 \le z \le 1.88)$

                 $= .4162 + .4699 = .8861$

d. For $x = 71.5$: $z = (71.5 - 80) / 4 = -2.13$

     For $x = 77.5$: $z = (77.5 - 80) / 4 = -.63$

     $P(71.5 \le x \le 77.5) = P(-2.13 \le z \le -.63) = P(-2.13 \le z \le 0) - P(-.63 \le z \le 0)$

$$= .4834 - .2357 = .2477$$

6.87 $\mu = np = 200(.80) = 160$ and $\sigma = \sqrt{npq} = \sqrt{200(.80)(.20)} = 5.65685425$

a. For $x = 149.5$: $z = (149.5 - 160) / 5.65685425 = -1.86$

For $x = 150.5$: $z = (150.5 - 160) / 5.65685425 = -1.68$

$P(149.5 \le x \le 150.5) = P(-1.86 \le z \le -1.68) = P(-1.86 \le z \le 0) - P(-1.68 \le z \le 0)$

$$= .4686 - .4535 = .0151$$

b. For $x = 169.5$: $z = (169.5 - 160) / 5.65685425 = 1.68$

$P(x \ge 169.5) = P(z \ge 1.68) = .5 - P(0 < z < 1.68) = .5 - .4535 = .0465$

c. For $x = 165.5$: $z = (165.5 - 160) / 5.65685425 = .97$

$P(x \le 165.5) = P(z \le .97) = .5 + P(0 \le z \le .97) = .5 + .3340 = .8340$

d. For $x = 163.5$: $z = (163.5 - 160) / 5.65685425 = .62$

For $x = 172.5$: $z = (172.5 - 160) / 5.65685425 = 2.21$

$P(163.5 \le x \le 172.5) = P(.62 \le z \le 2.21) = P(0 \le z \le 2.21) - P(0 \le z \le .62)$

$$= .4864 - .2324 = .2540$$

6.88 We are given: $\sigma = \$350$ and $P(x > 2500) = .15$

Hence, the area between μ and $x = 2500$ is $.5 - .15 = .3500$, which gives $z = 1.04$ approximately.

Consequently, $\mu = x - z\sigma = 2500 - (1.04)(350) = \$2136$

6.89 We are given: $\sigma = .18$ ounce and $P(x > 16) = .90$

Hence, the area between $x = 16$ and μ is $.4000$, which gives $z = -1.28$ approximately. Note that x is less than μ. Consequently, $\mu = x - z\sigma = 16 - (-1.28)(.18) = 16.23$ ounces approximately

Thus, for at least 90% of the cartons to contain more than 16 ounces of ice cream, the mean should be 16.23 ounces or higher.

6.90 a. If $\$3500 = \$1000 + \dfrac{\$12}{ft} \cdot x$, where x is depth in feet, then $x = 208.33$, i.e., Company B charges

more for depths of more than 208.33 ft.

$\mu = 250$, $\sigma = 40$

For $x = 208.33$: $z = (208.33 - 250) / 40 = -1.04$

$P(x > 208.33) = P(z > -1.04) = .5 + .3508 = .8508$

Company B charges more in 85.08% of all wells.

b. $\mu = 250$, so the mean amount charged by Company B is $1000 + \dfrac{\$12}{ft} \cdot 250$ ft = $4000.

6.91 We are given $\mu = 290$ feet and $\sigma = 10$. The simplest solution to this exercise is obtained by using complementary events, i.e., P(at least one throw is 320 feet or longer) $= 1 - P$(all three throws are less than 320 feet). First we find P(any one throw is less than 320 feet).

For $x = 320$: $z = (320 - 290) / 10 = 3.00$. Thus, $P(x < 320) = P(z < 3.00) = .5 + .4987 = .9987$.

Hence, P(all three throws are less than 320 feet $) = (.9987)^3 = .9961$. Therefore, P(at least one throw is 320 feet or longer) $= 1 - .9961 = .0039$.

6.92 $\mu = 45,000$; $\sigma = 2000$

For $x = 46,000$: $z = (46,000 - 45,000) / 2000 = \dfrac{1}{2} = .50$

$P(x \geq 46,000 = P(z \geq .50) = .5 - P(0 \leq z < .50) = .5 - .1915 = .3085$

So the probability of one tire lasting at least 46,000 miles is .3085. Hence, the probability of all four lasting more than 46,000 miles is $(.3085)^4 = .0091$, i.e., less than 1%.

6.93 Plant A: $\mu = 20$, $\sigma = 2$

For $x = 18$: $z = (18 - 20) / 2 = -1.00$

For $x = 22$: $z = (22 - 20) / 2 = 1.00$

$P(18 \leq x \leq 22) = P(-1.00 \leq z \leq 1.00) = P(-1.00 \leq z \leq 0) + P(0 \leq z \leq 1.00) = .3413 + .3413 = .6826$

Plant B: $\mu = 19$, $\sigma = 1$

For $x = 18$: $z = (18 - 19) / 1 = -1.00$

For $x = 22$: $z = (22 - 19) / 1 = 3.00$

$P(18 \leq x \leq 22) = P(-1.00 \leq z \leq 3.00) = P(-1.00 \leq z \leq 0) + P(0 \leq z \leq 3.00) = .3413 + .4987 = .8400$

Plant B produces the greater proportion (84% vs. 68.26%) of pints that contain between 18% and 22% air.

6.94 $\mu = 0$, $\sigma = 2$ mph

a. Let x be the error of these estimates in mph.

For $x = 5$: $z = (5 - 0) / 2 = 2.50$

$P(x \geq 5) = P(z \geq 2.50) = .5 - P(0 \leq z \leq 2.50) = .5 - .4938 = .0062$

So the probability that an estimate is at least 5 mph higher than the actual speed is .0062 or .62%. Hence, the probability that a car traveling at 60 mph or less will be cited for speeding is at most .0062.

b. We are given that the area to the left of x is .99. So, the area between μ and x is .99-.5=.49, which gives $z \approx 2.33$.

$x = \mu + z\sigma = 0 + 2.33(2) = 4.66 \approx 5$ mph

So the minimum estimate of speed at which a car should be cited for speeding is $60 + 5 = 65$ mph.

6.95 Here, $\mu = 45$ min. and $\sigma = 3$ min.

Let x be the amount of time Ashley spends commuting to work.

Area to left of $x = .95$

Area between μ and $\sigma = .95 - .5 = .45$

z for area of .45 ≈ 1.65

$x = \mu + z\sigma = 45 + 1.65(3) \approx 49.95$ minutes

So Ashley should leave home at about 8:10 a.m. in order to arrive at work by 9 a.m. in 95% of all cases.

6.96 We are given: $\sigma = .07$ ounces and $P(x \geq 8) = .99$. Hence the area between μ and $x = 8$ is $.99 - .50 = .49$, which gives $z \approx -2.33$. Consequently, $\mu = x - z\sigma = 8 - (-2.33)(.07) = 8.16$

The mean should be set to 8.16 ounces.

6.97 We are given: $\mu = 500$ and $P(x < 430) = .20$

a. Hence, the area between μ and $x = 400$ is $.5 - .20 = .30$, which gives $z \approx -.84$.

Now $x = \mu + z\sigma$, so $\sigma = \dfrac{x - \mu}{z} = \dfrac{430 - 500}{-.84} = 83.33$

b. For $x = 520$: $z = (520 - 500) / 83.33 = .24$

$P(x > 520) = P(z > .24) = .5 - .0948 = .4052$

Hence, 40.52% of the students score higher than 520.

6.98 Company A: $\mu = 8$ mm and $\sigma = .15$ mm

Company A sells 10,000 rods for $400.

For $x = 7.8$: $z = (7.8 - 8) / .15 = -1.33$

For $x = 8.2$: $z = (8.2 - 8) / .15 = 1.33$

$$P(7.8 \le x \le 8.2) = P(-1.33 \le z \le 1.33) = P(-1.33 \le z \le 0) + P(0 \le z \le 1.33)$$

$$= .4082 + .4082 = .8164$$

Price per usable rod is: $400 / (.8164 \times 10,000) \approx \$0.048996$

Company B: $\mu = 8$ mm and $\sigma = .12$ mm

Company B sells 10,000 rods for \$460.

For $x = 7.8$: $z = (7.8 - 8) / .12 = -1.67$

For $x = 8.2$: $z = (8.2 - 8) / .12 = 1.67$

$$P(7.8 \le x \le 8.2) = P(-1.67 \le z \le 1.67) = P(-1.67 \le z \le 0) + P(0 \le z \le 1.67)$$

$$= .4525 + .4525 - .9050$$

Price per usable rod is: $460 / (.9050 \times 10,000) \approx \$0.050829$

Hence, the Alpha Corporation should choose Company A as a supplier.

6.99 a. Answers will vary.

b. Single-number bet: $n \cdot p = 25 \cdot \dfrac{1}{38} = .658 < 5$, so we cannot use the standard normal distribution to

find the probability. The gambler comes out ahead if his number comes up at least once:

$$P \text{ (at least one success)} = 1 - P \text{ (all losses)} = 1 - \left(1 - \frac{1}{38}\right)^{25} = .4866$$

Color bet: $n \cdot p = 25 \cdot \dfrac{18}{38} = 11.84$, so we can use the normal distribution to find the probability.

$$\mu = n \cdot p = 11.84, \sigma = \sqrt{npq} = \sqrt{25\left(\frac{18}{38}\right)\left(\frac{20}{38}\right)} = 2.49653500$$

The gambler will come out ahead if he wins 13 bets or more.

For $x = 12.5$: $z = (12.5 - 11.84) / 2.49653500 = .26$

$$P(x \ge 12.5) = P(z \ge .26) = .5 - .1026 = .3974$$

The gambler has a better chance of coming out ahead with the single-number bet.

6.100 a. Let $x =$ number of ticket holders who show up.

Then x is a binomial random variable with $n = 65$ and $p = 1 - .10 = .90$

$$\mu = np = 65(.90) = 58.5 \text{ and } \sigma = \sqrt{npq} = \sqrt{65(.90)(.10)} = 2.41867732$$

Using a normal approximation with correction for continuity:

For $x = 60.5$: $z = (60.5 - 58.5) / 2.41867732 = .83$

$$P(x \le 60.5) = P(z \le .83) = .5 + P(0 \le z \le .83) = .5 + .2967 = .7967$$

168

b. Let n = number of tickets sold

Since μ and σ both depend on n, it is not easy to solve for n directly. Instead, we will use trial and error.

In part a, $n = 65$ was too large, since $P(x \le 60.5) = .80 < .90$.

Try $n = 62$: $\mu = 62(.9) = 55.8$ and $\sigma = \sqrt{62(.9)(.1)} = 2.36220236$

For $x = 60.5$: $z = (60.5 - 55.8) / 2.36220236 = 1.99$

$P(x \le 60.5) = P(z \le 1.99) = .5 + .4767 = .9767 > .95$

Thus, $n = 62$ satisfies the requirement.

To see if n may be increased, try $n = 63$.

$\mu = 63(.9) = 56.7$ and $\sigma = \sqrt{63(.9)(.1)} = 2.38117618$

For $x = 60.5$: $z = (60.5 - 56.7) / 2.38117618 = 1.60$

$P(x \le 60.5) = P(z \le 1.60) = .5 + .4452 = .9452 < .95$

Thus, $n = 63$ is too large.

Therefore, the largest number of tickets the company can sell and be at least 95% sure that the bus can hold all ticket holders who show up is 62.

6.101 a. $\mu = 2$ minutes and $\sigma = .5$ minute

For $x = 1$: $z = (1 - 2) / .5 = -2.00$

$P(x < 1) = P(z < -2.00) = .5 - P(-2.00 \le z \le 0) = .5 - .4772 = .0228$

P(two customers take less than one minute each)

$= P$(first takes less than one minute)$\cdot P$ (second takes less than one minute)$= (.0228)(.0228) = .0005$

b. For $x = 2.25$: $z = (2.25 - 2) / .5 = .50$

$P(x > 2.25) = P(z > .50) = .5 - P(0 \le z \le .50) = .5 - .1915 = .3085$

Using complementary events:

$P(x \le 2.25) = 1 - P(x > 2.25) = 1 - .3085 = .6915$

P(at least one of the four needs 2.25 minutes)$= 1 - P$ (each of the four needs 2.25 minutes or less)

$= 1 - (.6915)^4 = .7714$

Self-Review Test for Chapter Six

1. a   **2.** a   **3.** d   **4.** b   **5.** a   **6.** c   **7.** b   **8.** b

9.   a. $P(.85 \le z \le 2.33) = P(0 \le z \le 2.33) - P(0 < z < .85) = .4901 - .3203 = .1878$

b. $P(-2.97 \le z \le 1.49) = P(-2.97 \le z \le 0) + P(0 < z \le 1.49) = .4985 + .4319 = .9304$

c. $P(z \le -1.29) = .5 - P(-1.29 < z < 0) = .5 - .4015 = .0985$

d. $P(z > -.74) = P(-.74 < z < 0) + .5 = .2704 + .5 = .7704$

10.   a.   -1.28 approximately   b.   .61

c.   1.65 approximately   d.   -1.07 approximately

11. $\mu = 16$ hours and $\sigma = 3.5$ hours

a.   For $x = 12$: $z = (12 - 16) / 3.5 = -1.14$

For $x = 22$: $z = (22 - 16) / 3.5 = 1.71$

$P(12 < x < 22) = P(-1.14 < z < 1.71) = P(-1.14 < z < 0) + P(0 < z < 1.71)$

$= .3729 + .4564 = .8293$

b.   For $x = 10$: $z = (10 - 16) / 3.5 = -1.71$

$P(x < 10) = P(z < -1.71) = .5 - P(-1.71 < z < 0) = .5 - .4564 = .0436$

c.   For $x = 20$: $z = (20 - 16) / 3.5 = 1.14$

$P(x > 20) = P(z > 1.14) = .5 - P(0 < z < 1.14) = .5 - .3729 = .1271$

d.   For $x = 20$: $z = (20 - 16) / 3.5 = 1.14$

For $x = 25$: $z = (25 - 16) / 3.5 = 2.57$

$P(20 < x < 25) = P(1.14 < z < 2.57) = P(0 < z < 2.57) - P(0 < z < 1.14) = .4949 - .3729 = .1220$

12. $\mu = 16$ hours and $\sigma = 3.5$ hours

a.   For .10 area in the left tail of the normal distribution curve, $z \approx -1.28$. Hence,

$x = \mu + z\sigma = 16 + (-1.28)(3.5) = 11.52$ hours

b.   For .05 area in the right tail of the normal distribution curve, $z \approx 1.65$. Hence,

$x = \mu + z\sigma = 16 + (1.65)(3.5) = 21.78$ hours

13. $n = 500$, $p = .19$, $q = 1 - p = .81$, $\mu = np = 500(.19) = 95$, and

$\sigma = \sqrt{npq} = \sqrt{500(.19)(.81)} = 8.77211491$

a.   i.   For $x = 99.5$: $z = (99.5 - 95) / 8.77211491 = .51$

For $x = 100.5$: $z = (100.5 - 95) / 8.77211491 = .63$

$P(99.5 \le x \le 100.5) = P(.51 \le z \le .63) = P(0 \le z \le .63) - P(0 \le z \le .51) = .2357 - .1950 = .0407$

ii. From 100 to 110 means finding the probability from 99.5 to 110.5.

For $x = 99.5$: $z = .51$ from part a

For $x = 110.5$: $z = (110.5 - 95) / 8.77211491 = 1.77$

$P(99.5 < x < 110.5) = P(.51 < z < 1.77) = P(0 < z \leq 1.77) - P(0 < z \leq .51)$

$= .4616 - .1950 = .2666$

iii. For $x = 90.5$: $z = (90.5 - 95) / 8.77211491 = -.51$

$P(x \leq 90.5) = P(z \leq -.51) = .5 - P(-.51 \leq z < 0) = .5 - .1950 = .3050$

iv. For $x = 97.5$: $z = (97.5 - 95) / 8.77211491 = .28$

$P(x \geq 97.5) = P(z \geq .28) = .5 - P(0 \leq z < .28) = .5 - .1103 = .3897$

v. Between 80 and 103 means finding the probability from 80.5 to 102.5.

For $x = 80.5$: $z = (80.5 - 95) / 8.77211491 = -1.65$

For $x = 102.5$: $z = (102.5 - 95) / 8.77211491 = .85$

$P(80.5 < x < 102.5) = P(-1.65 < z < .85) = P(-1.65 < z \leq 0) + P(0 \leq z < .85)$

$= .4505 + .3023 = .7528$

b. P(at most 397 of the 500 fear something else) = P(at least 103 fear losing their jobs)= $P(x \geq 102.5)$

For $x = 102.5$: $z = (102.5 - 95) / 8.77211491 = .85$

$P(x \geq 102.5) = P(z \geq .85) = .5 - P(0 \leq z \leq .85) = .5 - .3023 = .1977$

c. P(between 390 and 414 of the 500 fear something else) =

P(between 86 and 110 fear losing their job) = $P(86.5 \leq x \leq 109.5)$

For $x = 86.5$: $z = (86.5 - 95) / 8.77211491 = -.97$

For $x = 109.5$: $z = (109.5 - 95) / 8.77211491 = 1.65$

$P(86.5 \leq x \leq 109.5) = P(-.97 \leq z \leq 1.65) = P(-.97 \leq z \leq 0) + P(0 < z < 1.65) = .3340 + .4505 = .7845$

Chapter Seven

7.1 The probability distribution of the population data is called the population distribution. Tables 7.1 and 7.2 on page 309 of the text provide an example of such a distribution. The probability distribution of a sample statistic is called its sampling distribution. Tables 7.3 to 7.5 on page 311 of the text provide an example of the sampling distribution of the sample mean.

7.2 Sampling error is the difference between the value of the sample statistic and the value of the corresponding population parameter, assuming that the sample is random and no nonsampling error has been made. Example 7–1 on page 312 of the text exhibits sampling error. Sampling errors occur only in sample surveys.

7.3 Nonsampling errors are errors that may occur during collection, recording, and tabulation of data. The second part of Example 7–1 on pages 312 and 313 of the text exhibits nonsampling error. Nonsampling errors occur both in sample surveys and censuses.

7.4   a. $\mu = (15 + 13 + 8 + 17 + 9 + 12)/6 = 74/6 = 12.33$

      b. $\bar{x} = (13 + 8 + 9 + 12)/4 = 42/4 = 10.50$

          Sampling error $= \bar{x} - \mu = 10.50 - 12.33 = -1.83$

      c.   Liza's incorrect $\bar{x} = (13 + 8 + 6 + 12)/4 = 39/4 = 9.75$

          $\bar{x} - \mu = 9.75 - 12.33 = -2.58$

          Sampling error (from part b) $= -1.83$

          Nonsampling error $= -2.58 - (-1.83) = -.75$

d.

Sample	$\bar{x}$	$\bar{x} - \mu$
15, 13, 8, 17	13.25	.92
15, 13, 8, 9	11.25	−1.08
15, 13, 17, 9	13.50	1.17
15, 8, 17, 9	12.25	−.08
13, 8, 17, 9	11.75	−.58
15, 13, 8, 12	12.00	−.33
15, 13, 17, 12	14.25	1.92
15, 8, 17, 12	13.00	.67
13, 8, 17, 12	12.50	.17
15, 13, 9, 12	12.25	−.08
15, 8, 9, 12	11.00	−1.33
13, 8, 9, 12	10.50	−1.83
15, 17, 9, 12	13.25	.92
13, 17, 9, 12	12.75	.42
8, 17, 9, 12	11.50	−.83

7.5   a. $\mu = (20 + 25 + 13 + 19 + 9 + 15 + 11 + 7 + 17 + 30)/10 = 166/10 = 16.60$

b. $\bar{x} = (20 + 25 + 13 + 9 + 15 + 11 + 7 + 17 + 30)/9 = 147/9 = 16.33$

Sampling error = $\bar{x} - \mu = 16.33 - 16.60 = -.27$

c.   Rich's incorrect $\bar{x} = (20 + 25 + 13 + 9 + 15 + 11 + 17 + 17 + 30)/9 = 157/9 = 17.44$

$\bar{x} - \mu = 17.44 - 16.60 = .84$

Sampling error (from part b) = $-.27$

Nonsampling error = $.84 - (-.27) = 1.11$

d.

Sample	$\bar{x}$	$\bar{x} - \mu$
25, 13 19,9,15,11,7,17,30	16.22	−.38
20, 13, 19, 9, 15, 11, 7, 17, 30	15.67	−.93
20, 25 19, 9, 15, 11, 7, 17, 30	17.00	.40
20, 25, 13, 9, 15, 11, 7, 17, 30	16.33	−.27
20, 25, 13, 19, 15, 11, 7, 17, 30	17.44	.84
20, 25, 13, 19, 9, 11, 7, 17, 30	16.78	.18
20, 25, 13, 19, 9, 15, 7, 17, 30	17.22	.62
20, 25, 13, 19, 9, 15, 11, 17, 30	17.67	1.07
20, 25, 13, 19, 9, 15, 11, 7, 30	16.56	−.04
20, 25, 13, 19, 9, 15, 11, 7, 17	15.11	−1.49

7.6

x	$P(x)$	$x P(x)$	$x^2 P(x)$
70	.20	14.00	980.00
78	.20	15.60	1216.80
80	.40	32.00	2560.00
95	.20	19.00	1805.00
		$\sum x P(x) = 80.60$	$\sum x^2 P(x) = 6561.80$

$$\mu = \sum xP(x) = 80.60$$

$$\sigma = \sqrt{\sum x^2 P(x) - \mu^2} = \sqrt{6561.80 - (80.60)^2} = 8.09$$

7.7   a.

$\bar{x}$	$P(\bar{x})$
55	1/6=.167
53	1/6=.167
28	1/6=.167
25	1/6=.167
21	1/6=.167
15	1/6=.167

b.

Sample	$\bar{x}$
55, 53, 28, 25, 21	36.4
55, 53, 28, 25, 15	35.2
55, 53, 28, 21, 15	34.4
55, 53, 25, 21, 15	33.8
55, 28, 25, 21, 15	28.8
53, 28, 25, 21, 15	28.4

$\bar{x}$	$P(\bar{x})$
36.4	1/6=.167
35.2	1/6=.167
34.4	1/6=.167
33.8	1/6=.167
28.8	1/6=.167
28.4	1/6=.167

c. The mean for the population data is: $\mu = \dfrac{55 + 53 + 28 + 25 + 21 + 15}{6} = \dfrac{197}{6} = 32.83$

Suppose the random sample of five family members includes the observations: 55, 28, 25, 21, and

15. The mean for this sample is: $\bar{x} = \dfrac{55 + 28 + 25 + 21 + 15}{5} = \dfrac{144}{5} = 28.80$

Then the sampling error is: $\bar{x} - \mu = 28.80 - 32.83 = -4.03$

7.8   a.

$\bar{x}$	$P(\bar{x})$
14	1/5 = .20
8	2/5 = .40
7	1/5 = .20
20	1/5 = .20

b.

Sample	$\bar{x}$
14, 8, 7, 7	9.00
14, 8, 7, 20	12.25
14, 8, 7, 20	12.25
14, 7, 7, 20	12.00
8, 7, 7, 20	10.50

$\bar{x}$	$P(\bar{x})$
9.00	1/5 = .20
12.25	2/5 = .40
12.00	1/5 = .20
10.50	1/5 = .20

c. The mean for the population data is: $\mu = \dfrac{14+8+7+7+20}{5} = \dfrac{56}{5} = 11.20$

Suppose the random sample of four faculty members includes the observations: 14, 8, 7 and 20.

The mean for this sample is: $\mu = \dfrac{14+8+7+20}{4} = \dfrac{49}{4} = 12.25$

Then the sampling error is: $\bar{x} - \mu = 12.25 - 11.20 = 1.05$

7.9   a. Mean of $\bar{x} = \mu_{\bar{x}} = \mu$

b. Standard deviation of $\bar{x} = \sigma_{\bar{x}} = \sigma/\sqrt{n}$ where $\sigma =$ population standard deviation and $n =$ sample size.

7.10   A sample statistic used to estimate a population parameter is called an estimator. An estimator is unbiased when its mean is equal to the corresponding population parameter. The sample mean $\bar{x}$ is an unbiased estimator of μ, because the mean of $\bar{x}$ is equal to μ.

7.11   An estimator is consistent when its standard deviation decreases as the sample size is increased. The sample mean $\bar{x}$ is a consistent estimator of μ because its standard deviation decreases as the sample size increases. This is obvious from the formula of $\sigma_{\bar{x}} = \sigma/\sqrt{n}$.

7.12   Since $\sigma_{\bar{x}} = \sigma/\sqrt{n}$, increasing n decreases $\sigma_{\bar{x}}$. Hence, as n increases, $\sigma_{\bar{x}}$ decreases.

7.13 $\mu = 60$ and $\sigma = 10$

a. $\mu_{\bar{x}} = \mu = 60$ and $\sigma_{\bar{x}} = \sigma/\sqrt{n} = 10/\sqrt{18} = 2.357$

    b. $\mu_{\bar{x}} = \mu = 60$ and $\sigma_{\bar{x}} = \sigma/\sqrt{n} = 10/\sqrt{90} = 1.054$

7.14 $\mu = 90$ and $\sigma = 18$

    a. $\mu_{\bar{x}} = \mu = 90$ and $\sigma_{\bar{x}} = \sigma/\sqrt{n} = 18/\sqrt{10} = 5.692$

    b. $\mu_{\bar{x}} = \mu = 90$ and $\sigma_{\bar{x}} = \sigma/\sqrt{n} = 18/\sqrt{35} = 3.043$

7.15   a. $n/N = 300/5000 = .06 > .05$

$$\sigma_{\bar{x}} = \frac{\sigma}{\sqrt{n}} \sqrt{\frac{N-n}{N-1}} = \frac{25}{\sqrt{300}} \sqrt{\frac{5000-300}{5000-1}} = 1.400$$

    b. $n/N = 100/5000 = .02 < .05,\ \sigma_{\bar{x}} = \sigma/\sqrt{n} = 25/\sqrt{100} = 2.500$

7.16   a. $n/N = 2500/100{,}000 = .025 < .05,\ \sigma_{\bar{x}} = \sigma/\sqrt{n} = 40/\sqrt{2500} = .800$

    b. $n/N = 7000/100{,}000 = .07 > .05$

$$\sigma_{\bar{x}} = \frac{\sigma}{\sqrt{n}} \sqrt{\frac{N-n}{N-1}} = \frac{40}{\sqrt{7{,}000}} \sqrt{\frac{100{,}000-7{,}000}{100{,}000-1}} = .461$$

7.17 $\mu = 125$ and $\sigma = 36$

    a. $\sigma_{\bar{x}} = \sigma/\sqrt{n} = 3.6$ so $n = (\sigma/\sigma_{\bar{x}})^2 = (36/3.6)^2 = 100$

    b. $n = (\sigma/\sigma_{\bar{x}})^2 = (36/2.25)^2 = 256$

7.18 $\mu = 46$ and $\sigma = 10$

    a. $\sigma_{\bar{x}} = \sigma/\sqrt{n} = 2.0$ so $n = (\sigma/\sigma_{\bar{x}})^2 = (10/2.0)^2 = 25$

    b. $n = (\sigma/\sigma_{\bar{x}})^2 = (10/1.6)^2 = 39$ approximately

7.19 $\mu = \$961{,}000;\ \sigma = \$180{,}000$, and $n = 80$

$\mu_{\bar{x}} = \mu = \$961,000$ and $\sigma_{\bar{x}} = \sigma/\sqrt{n} = 180,000/\sqrt{80} = \$20,124.61$

7.20 $\mu = 2,300$ square feet; $\sigma = 450$ square feet; and $n = 25$

$\mu_{\bar{x}} = \mu = 2,300$ square feet and $\sigma_{\bar{x}} = \sigma/\sqrt{n} = 450/\sqrt{25} = 90$ square feet

7.21 $\mu = \$233$, $\sigma = \$72$, and $n = 25$, $\mu_{\bar{x}} = \mu = \$233$ and $\sigma_{\bar{x}} = \sigma/\sqrt{n} = 72/\sqrt{25} = \$14.40$

7.22 $\mu = 594$ minutes, $\sigma = 160$ minutes, and $n = 400$

$\mu_{\bar{x}} = \mu = 594$ minutes and $\sigma_{\bar{x}} = \sigma/\sqrt{n} = 160/\sqrt{400} = 8$ minutes

7.23 $\sigma = \$405$ and $\sigma_{\bar{x}} = \$45$

$\sigma_{\bar{x}} = \sigma/\sqrt{n}$, so $n = (\sigma/\sigma_{\bar{x}})^2 = (405/45)^2 = 81$ players

7.24 $\sigma = \$139.50$ million and $\sigma_{\bar{x}} = \$15.50$ million

$\sigma_{\bar{x}} = \sigma/\sqrt{n}$ so $n = (\sigma/\sigma_{\bar{x}})^2 = (139.50/15.50)^2 = 81$

7.25   a.

$\bar{x}$	$P(\bar{x})$	$\bar{x}\,P(\bar{x})$	$\bar{x}^2\,P(\bar{x})$
76.00	.20	15.200	1155.200
76.67	.10	7.667	587.829
79.33	.10	7.933	629.325
81.00	.10	8.100	656.100
81.67	.20	16.334	1333.998
84.33	.20	16.866	1422.310
85.00	.10	8.500	722.500
		$\sum \bar{x}\,P(\bar{x})=80.600$	$\sum \bar{x}^2\,P(\bar{x}) = 6507.262$

$\sum \bar{x}\,P(\bar{x})=80.600 =$ same value found in Exercise 7.6 for μ.

b. $\sigma_{\bar{x}} = \sqrt{\sum \bar{x}^2 P(\bar{x}) - (\mu_{\bar{x}})^2} = \sqrt{6507.262 - (80.60)^2} = 3.302$

c. $\sigma/\sqrt{n} = 8.09/\sqrt{3} = 4.67$ is not equal to $\sigma_{\bar{x}} = 3.30$ in this case because $n/N = 3/5 = .60$, which is greater than .05.

d. $\sigma_{\bar{x}} = \dfrac{\sigma}{\sqrt{n}}\sqrt{\dfrac{N-n}{N-1}} = \dfrac{8.09}{\sqrt{3}}\sqrt{\dfrac{5-3}{5-1}} = 3.302$

7.26   The population from which the sample is drawn must be normally distributed.

7.27   The central limit theorem states that for a large sample, the sampling distribution of the sample mean is approximately normal, irrespective of the shape of the population distribution. Furthermore, $\mu_{\bar{x}} = \mu$ and $\sigma_{\bar{x}} = \sigma/\sqrt{n}$, where μ and σ are the population mean and standard deviation, respectively. A sample size of 30 or more is considered large enough to apply the central limit theorem to $\bar{x}$.

7.28   The central limit theorem will apply in cases a and c since $n \geq 30$. It will not apply in case b because $n < 30$.

7.29   a. Slightly skewed to the right

b   Approximately normal because $n > 30$ and the central limit theorem applies

c. Close to normal with perhaps a slight skew to the right

7.30   In both cases the sampling distribution of $\bar{x}$ would be normal because the population distribution is normal.

7.31   In both cases the sampling distribution of $\bar{x}$ would be normal because the population distribution is normal.

7.32 $\mu = 6.7$ minutes, σ = 2.1 minutes, and $n = 16$

$\mu_{\bar{x}} = \mu = 6.7$ minutes and $\sigma_{\bar{x}} = \sigma/\sqrt{n} = 2.1/\sqrt{16} = .525$ minute

The sampling distribution of $\bar{x}$ is normal because the population is normally distributed.

7.33 $\mu = 46$ miles per hour, $\sigma = 3$ miles per hour, and $n = 20$

$\mu_{\bar{x}} = \mu = 46$ miles per hour and $\sigma_{\bar{x}} = \sigma/\sqrt{n} = 3/\sqrt{20} = .671$ mile per hour

The sampling distribution of $\bar{x}$ is normal because the population is normally distributed.

7.34 $\mu = \$80$, $\sigma = \$15$, and $n = 25$

$\mu_{\bar{x}} = \mu = \$80$ and $\sigma_{\bar{x}} = \sigma/\sqrt{n} = 15/\sqrt{25} = \$3$

The sampling distribution of $\bar{x}$ is approximately normal because the population is approximately normally distributed.

7.35 $\mu = 3.02$, $\sigma = .29$, $N = 5540$ and $n = 48$

$\mu_{\bar{x}} = \mu = 3.02$

Since $n/N = 48/5540 = .009$ which is less than .05,

$\sigma_{\bar{x}} = \sigma/\sqrt{n} = .29/\sqrt{48} = .042$

The sampling distribution of $\bar{x}$ is approximately normal because the population is approximately normally distributed.

7.36 $\mu = 133$ pounds, $\sigma = 24$ pounds, and $n = 45$

$\mu_{\bar{x}} = \mu = 133$ pounds and $\sigma_{\bar{x}} = \sigma/\sqrt{n} = 24/\sqrt{45} = 3.578$ pounds

The sampling distribution of $\bar{x}$ is approximately normal because the sample size is large ($n > 30$).

7.37 $\mu = \$96$, $\sigma = \$27$, and $n = 90$

$\mu_{\bar{x}} = \mu = \$96$ and $\sigma_{\bar{x}} = \sigma/\sqrt{n} = 27/\sqrt{90} = \$2.846$

The sampling distribution of $\bar{x}$ is approximately normal because the sample size is large ($n > 30$).

7.38 $\mu = \$5800$; $\sigma = \$8300$, and $n = 50$

$\mu_{\bar{x}} = \mu = \$5800$ and $\sigma_{\bar{x}} = \sigma/\sqrt{n} = 8300/\sqrt{50} = \$1173.80$

The sampling distribution of $\bar{x}$ is normal because the population distribution is normal.

7.39 $\mu = \$8367$, $\sigma = \$2400$, and $n = 625$

$\mu_{\bar{x}} = \mu = \$8367$ and $\sigma_{\bar{x}} = \sigma/\sqrt{n} = 2400/\sqrt{625} = \$96.00$

The sampling distribution of $\bar{x}$ is approximately normal because the sample size is large ($n > 30$).

7.40 $P(\mu - 2.50\sigma_{\bar{x}} \le \bar{x} \le \mu + 2.50\sigma_{\bar{x}}) = P(-2.50 \le z \le 2.50) = .4938 + .4938 = .9876$ or 98.76%.

7.41 $P(\mu - 1.50\sigma_{\bar{x}} \le \bar{x} \le \mu + 1.50\sigma_{\bar{x}}) = P(-1.50 \le z \le 1.50) = .4332 + .4332 = .8664$ or 86.64%.

7.42 $\mu = 124$, $\sigma = 18$, and $n = 36$

$$\sigma_{\bar{x}} = \sigma/\sqrt{n} = 18/\sqrt{36} = 3.00$$

a. $z = (\bar{x} - \mu)/\sigma_{\bar{x}} = (128.60 - 124)/3.00 = 1.53$

b. $z = (\bar{x} - \mu)/\sigma_{\bar{x}} = (119.30 - 124)/3.00 = -1.57$

c. $z = (\bar{x} - \mu)/\sigma_{\bar{x}} = (116.88 - 124)/3.00 = -2.37$

d. $z = (\bar{x} - \mu)/\sigma_{\bar{x}} = (132.50 - 124)/3.00 = 2.68$

7.43 $\mu = 66$, $\sigma = 7$, and $n = 49$

$$\sigma_{\bar{x}} = \sigma/\sqrt{n} = 7/\sqrt{49} = 1$$

a. $z = (\bar{x} - \mu)/\sigma_{\bar{x}} = (68.44 - 66)/1.00 = 2.44$

b. $z = (\bar{x} - \mu)/\sigma_{\bar{x}} = (58.75 - 66)/1.00 = -7.25$

c. $z = (\bar{x} - \mu)/\sigma_{\bar{x}} = (62.35 - 66)/1.00 = -3.65$

d. $z = (\bar{x} - \mu)/\sigma_{\bar{x}} = (71.82 - 66)/1.00 = 5.82$

7.44 $\mu = 75$, $\sigma = 14$, and $n = 20$

$$\mu_{\bar{x}} = \mu = 75 \quad \text{and} \quad \sigma_{\bar{x}} = \sigma/\sqrt{n} = 14/\sqrt{20} = 3.13049517$$

a. For $\bar{x} = 68.5 : z = (68.5 - 75)/3.13049517 = -2.08$

For $\bar{x} = 77.3 : z = (77.3 - 75)/3.13049517 = .73$

$P(68.5 < \bar{x} < 77.3) = P(-2.08 < z < .73) = P(-2.08 < z < 0) + P(0 < z < .73) = .4812 + .2673 = .7485$

b. For $\bar{x} = 72.4 : z = (72.4 - 75)/3.13049517 = -.83$

$P(\bar{x} < 72.4) = P(z < -.83) = .5 - P(-.83 \le z \le 0) = .5 - .2967 = .2033$

7.45 $\mu = 48$, $\sigma = 8$, and $n = 16$

$$\mu_{\bar{x}} = \mu = 48 \quad \text{and} \quad \sigma_{\bar{x}} = \sigma/\sqrt{n} = 8/\sqrt{16} = 2.0$$

a. For $\bar{x} = 49.6 : z = (49.6 - 48)/2.0 = .80$

For $\bar{x} = 52.2 : z = (52.2 - 48)/2.0 = 2.10$

$P(49.6 < \bar{x} < 52.2) = P(.80 < z < 2.10) = .4821 - .2881 = .1940$

b. For $\bar{x} = 45.7 : z = (45.7 - 48)/2.0 = -2.3/2.0 = -1.15$

$P(\bar{x} > 45.7) = P(z > -1.15) = P(-1.15 < z < 0) + .5 = .3749 + .5 = .8749$

7.46 $\mu = 60$, $\sigma = 10$, and $n = 40$

$\mu_{\bar{x}} = \mu = 60$ and $\sigma_{\bar{x}} = \sigma/\sqrt{n} = 10/\sqrt{40} = 1.58113883$

a. For $\bar{x} = 62.2 : z = (62.2 - 60)/1.58113883 = 2.2/1.58113883 = 1.39$

$P(\bar{x} < 62.20) = P(z < 1.39) = .5 + P(0 < z < 1.39) = .5 + .4177 = .9177$

b   For $\bar{x} = 61.4 : z = (61.4 - 60)/1.58113883 = 1.4/1.58113883 = .89$

For $\bar{x} = 64.2 : z = (64.2 - 60)/1.58113883 = 4.2/1.58113883 = 2.66$

$P(61.4 < \bar{x} < 64.2) = P(.89 < z < 2.66) = P(0 < z < 2.66) - P(0 < z < .89) = .4961 - .3133 = .1828$

7.47 $\mu = 90$, $\sigma = 18$, and $n = 64$

$\mu_{\bar{x}} = \mu = 90$ and $\sigma_{\bar{x}} = \sigma/\sqrt{n} = 18/\sqrt{64} = 2.25$

a. For $\bar{x} = 82.3 : z = (82.3 - 90)/2.25 = -3.42$

$P(\bar{x} < 82.3) = P(z < -3.42) = .5 - P(-3.42 \leq z \leq 0) = .5 - .5 = .0000$ approximately

b. For $\bar{x} = 86.7 : z = (86.7 - 90)/2.25 = -1.47$

$P(\bar{x} > 86.7) = P(z > -1.47) = P(-1.47 < z < 0) + .5 = .4292 + .5 = .9292$

7.48 $\mu = 46$ miles per hour, $\sigma = 3$ miles per hour, and $n = 20$

$\sigma_{\bar{x}} = \sigma/\sqrt{n} = 3/\sqrt{20} = .67082039$ miles per hour

a. For $\bar{x} = 45 : z = (45 - 46)/.67082039 = -1.49$

$P(\bar{x} > 45) = P(z > -1.49) = .5 + P(-1.49 \leq z \leq 0) = .5 + .4319 = .9319$

b. For $\bar{x} = 45.5 : z = (45.5 - 46)/.67082039 = -.75$

$P(\bar{x} < 45.5) = P(z < -.75) = .5 - P(-.75 < z < 0) = .5 - .2734 = .2266$

c. For $\bar{x} = 44.5 : z = (44.5 - 46)/.67082039 = -2.24$

For $\bar{x} = 47 : z = (47 - 46)/.67082039 = 1.49$

$P(44.5 \le \bar{x} \le 47) = P(-2.24 \le z \le 1.49) = P(-2.24 \le z \le 0) + P(0 \le z \le 1.49)$

$= .4875 + .4319 = .9194$

7.49 $\mu = 3.02$, $\sigma = .29$, and $n = 20$

$\sigma_{\bar{x}} = \sigma/\sqrt{n} = .29/\sqrt{20} = .06484597$

a. For $\bar{x} = 3.10 : z = (3.10 - 3.02)/.06484597 = 1.23$

$P(\bar{x} \ge 3.10) = P(z \ge 1.23) = .5 - P(0 \le z \le 1.23) = .5 - .3907 = .1093$

b. For $\bar{x} = 2.90 : z = (2.90 - 3.02)/.06484597 = -1.85$

$P(\bar{x} \le 2.90) = P(z \le -1.85) = .5 - P(-1.85 \le z \le 0) = .5 - .4678 = .0322$

c. For $\bar{x} = 2.95 : z = (2.95 - 3.02)/.06484597 = -1.08$

For $\bar{x} = 3.11 : z = (3.11 - 3.02)/.06484597 = 1.39$

$P(2.95 \le \bar{x} \le 3.11) = P(-1.08 \le z \le 1.39) = P(-1.08 \le z \le 0) + P(0 \le z \le 1.39)$

$= .3599 + .4177 = .7776$

7.50 $\mu = 6.7$ minutes, $\sigma = 2.1$ minutes, and $n = 16$

$\sigma_{\bar{x}} = \sigma/\sqrt{n} = 2.1/\sqrt{16} = .525$ minute

a. For $\bar{x} = 7 : z = (7 - 6.7)/.525 = .57$

For $\bar{x} = 8 : z = (8 - 6.7)/.525 = 2.48$

$P(7 < \bar{x} < 8) = P(.57 < z < 2.48) = P(0 < z < 2.48) - P(0 < z < .57) = .4934 - .2157 = .2777$

b. $P(\bar{x}$ within 1 minute of $\mu) = P(5.7 \le \bar{x} \le 7.7)$

For $\bar{x} = 5.7 : z = (5.7 - 6.7)/.525 = -1.90$

For $\bar{x} = 7.7 : z = (7.7 - 6.7)/.525 = 1.90$

$P(5.7 \le \bar{x} \le 7.7) = P(-1.90 \le z \le 1.90) = P(-1.90 \le z \le 0) + P(0 \le z \le 1.90) = .4713 + .4713 = .9426$

c. $P(\bar{x}$ lower than μ by 1 minute or more) $= P(\bar{x} \le 5.7)$

For $\bar{x} = 5.7 : z = (5.7 - 6.7)/.525 = -1.90$

$P(\bar{x} \le 5.7) = P(z \le -1.90) = .5 - P(-1.90 < z < 0) = .5 - .4713 = .0287$

7.51 $\mu = \$18.96$ per hour, $\sigma = \$3.60$ per hour, and $n = 25$

$\sigma_{\bar{x}} = \sigma/\sqrt{n} = 3.60/\sqrt{25} = \$.72$ per hour

a. For $\bar{x} = 18 : z = (18 - 18.96)/.72 = -1.33$

For $\bar{x} = 20 : z = (20 - 18.96)/.72 = 1.44$

$P(18 < \bar{x} < 20) = P(-1.33 \le z \le 1.44) = P(-1.33 \le z \le 0) + P(0 \le z \le 1.44) = .4080 + .4251 = .8333$

b. $P(\bar{x}$ within $\$1.00$ per hour of $\mu) = P(17.963 \le \bar{x} \le 19.96)$

For $\bar{x} = 17.96 : z = (17.96 - 18.96)/.72 = -1.39$

For $\bar{x} = 19.96 : z = (19.96 - 18.96)/.72 = 1.39$

$P(17.96 \le \bar{x} \le 19.96) = P(-1.39 \le z \le 1.39) = P(-1.39 \le z \le 0) + P(0 \le z \le 1.39)$

$= .4177 + .4177 = .8354$

c. $P(\bar{x}$ greater than μ by $\$1.50$ per hour or more$) = P(\bar{x} \ge 20.46)$

For $\bar{x} = 20.46 : z = (20.46 - 18.96)/.72 = 2.08$

$P(\bar{x} \ge 20.46) = P(z \ge 2.08) = .5 - P(0 < z < 2.08) = .5 - .4812 = .0188$

7.52 $\mu = 8.4$ hours, $\sigma = 2.7$ hours, and $n = 45$

$\sigma_{\bar{x}} = \sigma/\sqrt{n} = 2.7/\sqrt{45} = .40249224$ hour

a. For $\bar{x} = 8 : z = (8 - 8.4)/.40249224 = -.99$

For $\bar{x} = 9 : z = (9 - 8.4)/.40249224 = 1.49$

$P(8 < \bar{x} < 9) = P(-.99 < z < 1.49) = P(-.99 < z < 0) + P(0 < z < 1.49) = .3389 + .4319 = .7708$

b. For $\bar{x} = 8 : z = (8 - 8.4)/.40249224 = -.99$

$P(\bar{x} < 8) = P(z < -.99) = .5 - P(-.99 \le z \le 0) = .5 - .3389 = .1611$

7.53 $\mu = \$1840$; $\sigma = \$453$;   and $n = 36$

$\sigma_{\bar{x}} = \sigma / \sqrt{n} = 453 / \sqrt{36} = \$75.5$

a. For $\bar{x} = 1750 : z = (1750 - 1840) / 75.5 = -1.19$

For $\bar{x} = 1950 : z = (1950 - 1840) / 75.5 = 1.46$

$P(1750 < \bar{x} < 1950) = P(-1.19 \leq z \leq 1.46) = P(-1.19 \leq z \leq 0) + P(0 \leq z \leq 1.46)$

$= .3830 + .4279 = .8109$

b. For $\bar{x} = 1700 : z = (1700 - 1840) / 75.5 = -1.85$

$P(\bar{x} < 1700) = P(z < -1.85) = .5 - P(-1.85 \leq z \leq 0) = .5 - .4678 = .0322$

7.54 $\mu = \$8367$, $\sigma = \$2400$, and $n = 225$

$\sigma_{\bar{x}} = \sigma / \sqrt{n} = 2400 / \sqrt{225} = \$160$

a. For $\bar{x} = 8100 : z = (8100 - 8367) / 160 = -1.67$

For $\bar{x} = 8500 : z = (8500 - 8367) / 160 = .83$

$P(8100 < \bar{x} < 8500) = P(-1.67 \leq z \leq .83) = P(-1.67 \leq z \leq 0) + P(0 \leq z \leq .83)$

$= .4525 + .2967 = .7492$

b. $P(\bar{x}$ within \$200 of $\mu) = P(8167 \leq \bar{x} \leq 8567)$

For $\bar{x} = 8167 : z = (8167 - 8367) / 160 = -1.25$

For $\bar{x} = 8567 : z = (8567 - 8367) / 160 = 1.25$

$P(8167 \leq \bar{x} \leq 8567) = P(-1.25 \leq z \leq 1.25) = P(-1.25 \leq z \leq 0) + P(0 \leq z \leq 1.25)$

$= .3944 + .3944 = .7888$

c. $P(\bar{x}$ greater than μ by \$300 or more$) = P(\bar{x} \geq 8667)$

For $\bar{x} = 8667 : z = (8667 - 8367) / 160 = 1.88$

$P(\bar{x} \geq 8667) = P(z \geq 1.88) = .5 - P(0 < z < 1.88) = .5 - .4699 = .0301$

7.55 $\mu = \$80$, $\sigma = \$25$, and $n = 75$

$\sigma_{\bar{x}} = \sigma / \sqrt{n} = 25 / \sqrt{75} = 2.88675135$

a. For $\bar{x} = 72 : z = (72 - 80) / 2.88675135 = -2.77$

For $\bar{x} = 77: z = (77 - 80)/2.88675135 = -1.04$

$P(72 \le \bar{x} \le 77) = P(-2.77 \le z \le -1.04) = P(-2.77 \le z \le 0) - P(-1.04 \le z \le 0)$

$= .4972 - .3508 = .1464$

b. $P(\mu - 6 \le \bar{x} \le \mu + 6) = P(74 \le \bar{x} \le 86)$

For $\bar{x} = 74: z = (74 - 80)/2.88675135 = -2.08$

For $\bar{x} = 86: z = (86 - 80)/2.88675135 = 2.08$

$P(74 \le \bar{x} \le 86) = P(-2.08 \le z \le 2.08) = .4812 + .4812 = .9624$

c. $P(\bar{x} \ge \mu + 5) = P(\bar{x} \ge 85)$

For $\bar{x} = 85: z = (85 - 80)/2.88675135 = 1.73$

$P(\bar{x} \ge 85) = P(z \ge 1.73) = .5 - P(0 < z < 1.73) = .5 - .4582 = .0418$

7.56 $\mu = \$12,450$, $\sigma = \$4300$, and $n = 50$

$\sigma_{\bar{x}} = 4300 / \sqrt{50} = \$608.11183182$

a. For $\bar{x} = \$11,500: z = (11,500 - 12,450)/608.11183182 = -1.56$

$P(\bar{x} > 11,500) = P(z \ge -1.56) = P(-1.56 < z < 0) + .5 = .4406 + .5 = .9406$

b. For $\bar{x} = 12,000: z = (12,000 - 12,450)/608.11183182 = -.74$

For $\bar{x} = 13,800: z = (13,800 - 12,450)/608.11183182 = 2.22$

$P(12,000 \le \bar{x} \le 13,800) = P(-.74 \le z \le 2.22) = P(-.74 \le z \le 0) + P(0 \le z \le 2.22)$

$= .2704 + .4868 = .7572$

c. $P(\mu - 1500 \le \bar{x} \le \mu + 1500) = P(10,950 \le \bar{x} \le 13,950)$

For $\bar{x} = 10,950: z = (10,950 - 12,450)/608.11183182 = -2.47$

For $\bar{x} = 13,950: z = (13,950 - 12,450)/608.11183182 = 2.47$

$P(10,950 \le \bar{x} \le 13,950) = P(-2.47 \le z \le 2.47) = .4932 + .4932 = .9864$

d. $P(\bar{x} > \mu + 1000) = P(\bar{x} > 13,450)$

For $\bar{x} = 13,450: z = (13,450 - 12,450)/608.11183182 = 1.64$

$P(\bar{x} > 13,450) = P(z > 1.64) = .5 - P(0 \le z \le 1.64) = .5 - .4495 = .0505$

7.57 $\mu = 68$ inches, $\sigma = 4$ inches, and $n = 100$

$\sigma_{\bar{x}} = \sigma / \sqrt{n} = 4 / \sqrt{100} = .4$ inch

a. For $\bar{x} = 67.8 : z = (67.8 - 68) / .4 = -.50$

$P(\bar{x} < 67.8) = P(z < -.50) = .5 - P(-.50 \le z \le 0) = .5 - .1915 = .3085$

b. For $\bar{x} = 67.5 : z = (67.5 - 68) / .4 = -1.25$

For $\bar{x} = 68.7 : z = (68.7 - 68) / .4 = 1.75$

$P(67.5 \le \bar{x} \le 68.7) = P(-1.25 \le z \le 1.75) = P(-1.25 \le z \le 0) + P(0 \le z \le 1.75)$

$= .3944 + .4599 = .8543$

c. $P(\bar{x}$ within .6 inches of $\mu) = P(67.4 \le \bar{x} \le 68.6)$

For $\bar{x} = 67.4 : z = (67.4 - 68) / .4 = -1.50$

For $\bar{x} = 68.6 : z = (68.6 - 68) / .4 = 1.50$

$P(67.4 \le \bar{x} \le 68.6) = P(-1.50 \le z \le 1.50) = .4322 + .4332 = .8664$

d. $P(\bar{x}$ is lower than μ by .5 inches or more$) = P(\bar{x} \le 67.5)$

For $\bar{x} = 67.5 : z = (67.5 - 68) / .4 = -1.25$

$P(\bar{x} \le 67.5) = P(z \le -1.25) = .5 - P(-1.25 \le z \le 0) = .5 - .3944 = .1056$

7.58 $\bar{x} = 2250$ hours, $\sigma = 150$ hours, and $n = 100$

$\sigma_{\bar{x}} = 150 / \sqrt{100} = 15.00$ hours

We are to find $P(\mu - 25 \le \bar{x} \le \mu + 25)$.

For $\bar{x} = \mu - 25 : z = (\mu - 25 - \mu) / 15.00 = -1.67$

For $\bar{x} = \mu + 25 : z = (\mu + 25 - \mu) / 15.00 = 1.67$

$P(\mu - 25 \le \bar{x} \le \mu + 25) = P(-1.67 \le z \le 1.67) = .4525 + .4525 = .9050$

7.59 $\bar{x} = 3$ inches, $\sigma = .1$ inch, and $n = 25$

$\sigma_{\bar{x}} = .1 / \sqrt{.25} = .02$ inch

For $\bar{x} = 2.95 : z = (2.95 - 3) / .02 = -2.50$

For $\bar{x} = 3.05 : z = (3.05 - 3)/.02 = 2.50$

$P(\bar{x} < 2.95) + P(\bar{x} > 3.05) = P(z < -2.50) + P(z > 2.50) = (.5 - .4938) + (.5 - .4938)$

$= .0062 + .0062 = .0124$

7.60 $p = 640/1000 = .64$ and $\hat{p} = 24/40 = .60$

7.61 $p = 600/5000 = .12$ and $\hat{p} = 18/120 = .15$

7.62   Number with characteristic in population $= 18,700 \times .30 = 5610$

Number with characteristic in sample $= 250 \times .25 = 62.5 \approx 62$ or 63

7.63   Number in population with characteristic $= 9500 \times .75 = 7125$

Number in sample with characteristic $= 400 \times .78 = 312$

7.64   a. The mean of $\hat{p}$ is: $\mu_{\hat{p}} = p$

b. The standard deviation of $\hat{p}$ is: $\sigma_{\hat{p}} = \sqrt{pq/n}$

c. The sampling distribution of $\hat{p}$ is approximately normal if np and nq are both greater than 5.

7.65   Sampling error $= \hat{p} - p = .66 - .71 = -.05$

7.66   Sampling error $= \hat{p} - p = .33 - .29 = .04$

7.67   The estimator of p is the sample proportion $\hat{p}$.

The sample proportion $\hat{p}$ is an unbiased estimator of p, since the mean of $\hat{p}$ is equal to p.

7.68   The sample proportion $\hat{p}$ is a consistent estimator of p, since $\sigma_{\hat{p}}$ decreases as the sample size is increased.

7.69 $\sigma_{\hat{p}} = \sqrt{pq/n}$, hence $\sigma_{\hat{p}}$ decreases as n increases.

7.70 $p = .63,\ q = 1 - p = 1 - .63 = .37$

a. $n = 100, \mu_{\hat{p}} = p = .63$, and $\sigma_{\hat{p}} = \sqrt{pq/n} = \sqrt{.63(.37)/100} = .048$

b. $n = 900, \mu_{\hat{p}} = p = .63$, and $\sigma_{\hat{p}} = \sqrt{pq/n} = \sqrt{.63(.37)/900} = .016$

7.71 $p = .21, \ q = 1 - p = 1 - .21 = .79$

     a. $n = 400, \mu_{\hat{p}} = p = .21$, and $\sigma_{\hat{p}} = \sqrt{pq / n} = \sqrt{.21(.79) / 400} = .020$

     b. $n = 750, \mu_{\hat{p}} = p = .21$, and $\sigma_{\hat{p}} = \sqrt{pq / n} = \sqrt{.21(.79) / 750} = .015$

7.72 $N = 4000, \ p = .12$, and $q = 1 - p = 1 - .12 = .88$

     a. $n / N = 800 / 4000 = .20 > .05$

$$\sigma_{\hat{p}} = \sqrt{\frac{pq}{n}} \sqrt{\frac{N-n}{N-1}} = \sqrt{\frac{.12(.88)}{800}} \sqrt{\frac{4000-800}{4000-1}} = .010$$

     b. $n / N = 30 / 4000 = .0075 < .05$, $\sigma_{\hat{p}} = \sqrt{pq / n} = \sqrt{.12(.88) / 30} = .059$

7.73 $N = 1400, \ p = .47$, and $q = 1 - p = 1 - .47 = .53$

     a. $n / N = 90 / 1400 = .064 > .05$

$$\sigma_{\hat{p}} = \sqrt{\frac{pq}{n}} \sqrt{\frac{N-n}{N-1}} = \sqrt{\frac{.47(.53)}{90}} \sqrt{\frac{1400-90}{1400-1}} = .051$$

     b. $n / N = 50 / 1400 = .036 < .05$, $\sigma_{\hat{p}} = \sqrt{pq / n} = \sqrt{.47(.53) / 50} = .071$

7.74   A sample is considered large enough to apply the central limit theorem if np and nq are both greater than 5.

7.75   a. $np = 400(.28) = 112$ and $nq = 400(.72) = 288$

          Since $np > 5$ and $nq > 5$, the central limit theorem applies.

     b. $np = 80(.05) = 4$; since $np < 5$, the central limit theorem does not apply.

     c. $np = 60(.12) = 7.2$ and $nq = 60(.88) = 52.8$

          Since $np > 5$ and $nq > 5$, the central limit theorem applies.

     d. $np = 100(.035) = 3.5$; since $np < 5$, the central limit theorem does not apply.

7.76   a. $np = 20(.45) = 9$ and $nq = 20(.55) = 11$

   Since $np > 5$ and $nq > 5$, the central limit theorem applies.

   b. $np = 75(.22) = 16.5$ and $nq = 75(.78) = 58.5$

   Since $np > 5$ and $nq > 5$, the central limit theorem applies.

   c. $np = 350(.01) = 3.5$; since $np < 5$, the central limit theorem does not apply.

   d. $np = 200(.022) = 4.4$; since $np < 5$, the central limit theorem does not apply.

7.77   a.   The proportion of these TV sets that are good is $4/6 = .667$

   b.   Total number of samples of size 5 is: $(_6C_5) = 6$

   c & d. Let: $G =$ good TV set and $D =$ defective TV set

   Let the six TV sets be denoted as: $1 = G, 2 = G, 3 = D, 4 = D, 5 = G$, and $6 = G$. The six possible samples, their sample proportions, and the sampling errors are given in the table below.

Sample	TV sets	$\hat{p}$	Sampling error
1, 2, 3, 4, 5	G, G, D, D, G	3/5=.60	$.60 - .667 = -.067$
1, 2, 3, 4, 6	G, G, D, D, G	3/5=.60	$.60 - .667 = .067$
1, 2, 3, 5, 6	G, G, D, G, G	4/5=.80	$.80 - .667 = .133$
1, 2, 4, 5, 6	G, G, D, G, G	4/5=.80	$.80 - .667 = .133$
1, 3, 4, 5, 6	G, D, D, G, G	3/5=.60	$.60 - .667 = -.067$
2, 3, 4, 5, 6	G, D, D, G, G	3/5=.60	$.60 - .667 = -.067$

$\hat{p}$	f	Relative Frequency
.60	4	4/6=.667
.80	2	2/6=.333
	$\sum f = 6$	

$\hat{p}$	$P(\hat{p})$
.60	.667
.80	.333

7.78   a.   Two of the five fires of this season were caused by arson. Hence, the proportion of fires caused by arson is: $p = 2/5 = .40$

   b.   The total number of samples of size 3 that can be selected is: $_5C_3 = 10$

   c & d. Let the five fires of the season be denoted as:

A = arson, B = accident, C = accident, D = arson, and E = accident

The following tables list all the possible samples of size 3, the sample proportions, the sampling errors and the sampling distribution of the sample proportion.

Sample	Employees			$\hat{p}$	Sampling error
A, B, C	arson,	accident,	accident	1/3 = .333	.333 − .40 = −.067
A, B, D	arson,	accident,	arson	2/3 = .667	.667 − .40 = .267
A, B, E	arson,	accident,	accident	1/3 = .333	.333 − .40 = −.067
A, C, D	arson,	accident,	arson	2/3 = .667	.667 − .40 = .267
A, C, E	arson,	accident,	accident	1/3 = .333	.333 − .40 = −.067
A, D, E	arson,	arson,	accident	2/3 = .667	.667 − .40 = .267
B, C, D	accident,	accident,	arson	1/3 = .333	.333 − .40 = −.067
B, C, E	accident,	accident,	accident	0/3 = .000	.000 − .40 = −.400
B, D, E	accident,	arson,	accident	1/3 = .333	.333 − .40 = −.067
C, D, E	accident,	arson,	accident	1/3 = .333	.333 − .40 = −.067

$\hat{p}$	f	Relative Frequency
.000	1	1/10 = .10
.333	6	6/10 = .60
.667	3	3/10 = .30

$\hat{p}$	$P(\hat{p})$
.000	.10
.333	.60
.667	.30

7.79 $n = 400$, $p = .45$, $q = 1 - p = 1 - .45 = .55$

$\mu_{\hat{p}} = p = .45$ and $\sigma_{\hat{p}} = \sqrt{pq/n} = \sqrt{.45(.55)/400} = .025$

$np = 400(.45) = 180$ and $nq = 400(.55) = 220$

Since np and nq are both greater than 5, the sampling distribution of $\hat{p}$ is approximately normal.

7.80 $n = 300$; $p = .37$; $q = 1 - p = 1 - .37 = .63$

$\mu_{\hat{p}} = p = .37$ and $\sigma_{\hat{p}} = \sqrt{pq/n} = \sqrt{(.37)(.63)/300} = .028$

$np = 300(.37) = 111$ and $nq = 300(.63) = 189$

Since np and nq are both greater than 5, the sampling distribution of $\hat{p}$ is approximately normal.

7.81 $n = 50$, $p = .28$, $q = 1 - p = 1 - .28 = .72$

$\mu_{\hat{p}} = p = .28$ and $\sigma_{\hat{p}} = \sqrt{pq/n} = \sqrt{.28(.72)/50} = .063$

$np = 50(.28) = 14$ and $nq = 50(.72) = 36$

Since np and nq are both greater than 5, the sampling distribution of $\hat{p}$ is approximately normal.

7.82 $n = 225;\ \ p = .51;,\ \ q = 1 - p = 1 - .51 = .49$

$\mu_{\hat{p}} = p = .51$ and $\sigma_{\hat{p}} = \sqrt{pq/n} = \sqrt{.51(.49)/225} = .033$

$np = 225(.51) = 115$ and $nq = 225(.49) = 110$

Since np and nq are both greater than 5, the sampling distribution of $\hat{p}$ is approximately normal.

7.83 $P(p - 2.0\sigma_{\hat{p}} \le \hat{p} \le p + 2.0\sigma_{\hat{p}}) = P(-2.00 \le z \le 2.00) = .4772 + .4722 = .9544$

Thus, 95.44% of the sample proportions will be within 2 standard deviations of the population proportion.

7.84 $P(p - 3.0\sigma_{\hat{p}} \le \hat{p} \le p + 3.0\sigma_{\hat{p}}) = P(-3.00 \le z \le 3.00) = .4987 + .4987 = .9974$

Thus, 99.74% of the sample proportions will be within 3 standard deviations of the population proportion.

7.85 $n = 100;\ \ p = .59;\ \ q = 1 - .59 = .41$

$\sigma_{\hat{p}} = \sqrt{pq/n} = \sqrt{.59(.41)/100} = .04918333$

a. $z = (\hat{p} - p)/\sigma_{\hat{p}} = (.56 - .59)/.04918333 = -.61$

b. $z = (\hat{p} - p)/\sigma_{\hat{p}} = (.68 - .59)/.04918333 = 1.83$

c. $z = (\hat{p} - p)/\sigma_{\hat{p}} = (.53 - .59)/.04918333 = -1.22$

d. $z = (\hat{p} - p)/\sigma_{\hat{p}} = (.65 - .59)/.04918333 = 1.22$

7.86 $n = 70;\ \ p = .25;\ \ q = 1 - .25 = .75$

$\sigma_{\hat{p}} = \sqrt{pq/n} = \sqrt{.25(.75)/70} = .05175492$

a. $z = (\hat{p} - p)/\sigma_{\hat{p}} = (.26 - .25)/.05175492 = .19$

b. $z = (\hat{p} - p)/\sigma_{\hat{p}} = (.32 - .25)/.05175492 = 1.35$

c. $z = (\hat{p} - p)/\sigma_{\hat{p}} = (.17 - .25)/.05175492 = -1.55$

d. $z = (\hat{p} - p)/\sigma_{\hat{p}} = (.20 - .25)/.05175492 = -.97$

7.87 $p = .616$, $q = 1 - p = 1 - .616 = .384$, and $n = 200$

$$\sigma_{\hat{p}} = \sqrt{pq/n} = \sqrt{.616(.384)/200} = .03439070$$

a. For $\hat{p} = .60: z = (.60 - .616)/.03439070 = -.47$

   For $\hat{p} = .66: z = (.66 - .616)/.03439070 = 1.28$

   $P(.60 \le \hat{p} \le .66) = P(-.47 < z < 1.28) = P(-.47 < z \le 0) + P(0 \le z < 1.28) = .1808 + .3997 = .5805$

b. For $\hat{p} = .64: z = (.64 - .616)/.03439070 = .70$

   $P(\hat{p} > .64) = P(z > .70) = .5 - P(0 \le z \le .70) = .5 - .2580 = .2420$

7.88 $p = .64$, $q = 1 - p = 1 - .64 = .36$, and $n = 50$

   $\sigma_{\hat{p}} = \sqrt{pq/n} = \sqrt{.64(.36)/50} = .06788225$

a. For $\hat{p} = .54: z = (.54 - .64)/.06788225 = -1.47$

   For $\hat{p} = .61: z = (.61 - .64)/.06788225 = -.44$

   $P(.54 < \hat{p} < .61) = P(-1.47 < z < -.44) = P(-1.47 < z < 0) - P(-.44 \le z \le 0)$

                  $= .4292 - .1700 = .2592$

b. For $\hat{p} = .71: z = (.71 - .64)/.06788225 = 1.03$

   $P(\hat{p} > .71) = P(z > 1.03) = .5 - P(0 \le z \le 1.03) = .5 - .3485 = .1515$

7.89 $p = .39;$, $q = 1 - .39 = .61;$ and $n = 300$

   $\sigma_{\hat{p}} = \sqrt{pq/n} = \sqrt{.39(.61)/300} = .02816026$

a. For $\hat{p} = .35: z = (.35 - .39)/.02816026 = -1.42$

   For $\hat{p} = .45: z = (.45 - .39)/.02816026 = 2.13$

   $P(.35 < \hat{p} < .45) = P(-1.42 < z < 2.13) = P(-1.42 < z < 0) + P(0 \le z \le 2.13) = .4222 + .4834 = .9056$

b. For $\hat{p} = .36: z = (.36 - .39)/.02816026 = -1.07$

   $P(\hat{p} > .36) = P(z > -1.07) = P(-1.07 < z < 0) + .5 = .3577 + .5 = .8577$

7.90 $p = .85$, $q = 1 - p = 1 - .85 = .15$, and $n = 100$

$\sigma_{\hat{p}} = \sqrt{pq / n} = \sqrt{.85(.15) / 100} = .03570714$

a. For $\hat{p} = .81 : z = (.81 - .85) / .03570714 = -1.12$

For $\hat{p} = .88 : z = (.88 - .85) / .03570714 = .84$

$P(.81 < \hat{p} < .88) = P(-1.12 < z < .84) = P(-1.12 < z < 0) + P(0 < z < .84)$

$= .3686 + .2995 = .6681$

b. For $\hat{p} = .87 : z = (.87 - .85) / .03570714 = .56$

$P(\hat{p} < .87) = P(z < .56) = .5 + P(0 < z < .56) = .5 + .2123 = .7123$

7.91 $p = .06$, $q = 1 - p = 1 - .06 = .94$, and $n = 100$

$\sigma_{\hat{p}} = \sqrt{pq / n} = \sqrt{.06(.94) / 100} = .02374868$

For $\hat{p} = .08 : z = (.08 - .06) / .02374868 = .84$

$P(\hat{p} \geq .08) = P(z \geq .84) = .5 - P(0 < z < .84) = .5 - .2995 = .2005$

7.92 $p = .80$, $q = 1 - .80 = .20$, and $n = 100$

$\sigma_{\hat{p}} = \sqrt{pq / n} = \sqrt{.80(.20) / 100} = .040$

a. The required probability is: $P(.80 - .05 \leq \hat{p} \leq .80 + .05) = P(.75 \leq \hat{p} \leq .85)$

For $\hat{p} = .75 : z = (.75 - .80) / .040 = -1.25$

For $\hat{p} = .85 : z = (.85 - .80) / .040 = 1.25$

$P(.75 \leq \hat{p} \leq .85) = P(-1.25 \leq z \leq 1.25) = P(-1.25 \leq z \leq 0) + P(0 \leq z \leq 1.25) = .3944 + .3944 = .7888$

b. We are to find the probability: $P(\hat{p} \leq p - .06) = P(\hat{p} \leq .80 - .06) = P(\hat{p} \leq .74)$.

For $\hat{p} = .74 : z = (.74 - .80) / .040 = -1.50$

$P(\hat{p} \leq .74) = P(z \leq -1.50) = .5 - P(-1.50 < z < 0) = .5 - .4332 = .0668$

c. $P(\hat{p} \geq p + .07) = P(\hat{p} \geq .80 + .07) = P(\hat{p} \geq .87)$

For $\hat{p} = .87 : z = (.87 - .80) / .040 = 1.75$

$P(\hat{p} \geq .87) = P(z \geq 1.75) = 5 - P(0 < z < 1.75) = .5 - .4599 = .0401$

7.93 $\mu = 750$ hours, $\sigma = 55$ hours, and $n = 25$

$\mu_{\bar{x}} = \mu = 750$ hours and $\sigma_{\bar{x}} = \sigma/\sqrt{n} = 55/\sqrt{25} = 11$ hours

The sampling distribution of $\bar{x}$ is normal because the population is normally distributed.

7.94 $\mu = 24$ minutes, $n = 100$, $\sigma = 7$ minutes,

$\mu_{\bar{x}} = \mu = 24$ minutes and $\sigma_{\bar{x}} = \sigma/\sqrt{n} = 7/\sqrt{100} = .7$ minutes

The sampling distribution of $\bar{x}$ is approximately normal because $n > 30$.

7.95 $\mu = 750$ hours, $\sigma = 55$ hours, and $n = 25$

$\sigma_{\bar{x}} = \sigma/\sqrt{n} = 55/\sqrt{25} = 11$ hours

a.   For $\bar{x} = 735 : z = (735 - 750)/11 = -1.36$

$P(\bar{x} > 735) = P(z > -1.36) = P(-1.36 < z < 0) + .5 = .4131 + .5 = .9131$

b.   For $\bar{x} = 725 : z = (725 - 750)/11 = -2.27$

For $\bar{x} = 740 : z = (740 - 750)/11 = -.91$

$P(725 < \bar{x} < 740) = P(-2.27 < z < -.91) = P(-2.27 < z < 0) - P(-.91 \le z \le 0)$

$= .4884 - .3186 = .1698$

c. $P(\bar{x}$ within 15 hours of $\mu) = P(735 \le \bar{x} \le 765)$

For $\bar{x} = 735 : z = (735 - 750)/11 = -1.36$

For $\bar{x} = 765 : z = (765 - 750)/11 = 1.36$

$P(735 \le \bar{x} \le 765) = P(-1.36 \le z \le 1.36) = .4131 + .4131 = .8262$

d. $P(\bar{x}$ is lower than μ by 20 hours or more$) = P(\bar{x} \le 730)$

For $\bar{x} = 730 : z = (730 - 750)/11 = -1.82$

$P(\bar{x} \le 730) = P(z \le -1.82) = .5 - P(-1.82 < z < 0) = .5 - .4656 = .0344$

7.96 $\mu = 24$ minutes, $n = 100$, $\sigma = 7$ minutes,

$\sigma_{\bar{x}} = \sigma/\sqrt{n} = 7/\sqrt{100} = .7$ minute

a. For $\bar{x} = 22 : z = (22 - 24)/.7 = -2.86$

$P(\bar{x} < 22) = P(z < -2.86) = .5 - P(-2.86 \le z \le 0) = .5 - .4979 = .0021$

b. For $\bar{x} = 23 : z = (23 - 24)/.7 = -1.43$

For $\bar{x} = 26 : z = (26 - 24)/.7 = 2.86$

$P(23 < \bar{x} < 26) = P(-1.43 < z < 2.86) = P(-1.43 < z < 0) + P(0 < z < 2.86) = .4236 + .4979 = .9215$

c. $P(\bar{x} \text{ within 1 minute of } \mu) = P(\mu - 1 \le \bar{x} \le \mu + 1) = P(23 \le \bar{x} \le 25)$

For $\bar{x} = 23 : z = (23 - 24)/.7 = -1.43$

For $\bar{x} = 25 : z = (25 - 24)/.7 = 1.43$

$P(23 \le \bar{x} \le 25) = P(-1.43 \le z \le 1.43) = P(-1.43 \le z \le 0) + P(0 \le z \le 1.43) = .4236 + .4236 = .8472$

d. $P(\bar{x} \text{ is greater than } \mu \text{ by 2 minutes or more}) = P(\bar{x} \ge \mu + 2) = P(\bar{x} \ge 26)$

For $\bar{x} = 26 : z = (26 - 24)/.7 = 2.86$

$P(\bar{x} \ge 26) = P(z \ge 2.86) = .5 - P(0 < z < 2.86) = .5 - .4979 = .0021$

7.97 $\mu = 16.87$ hours, $\sigma = 5$ hours, and $n = 100$

$\sigma_{\bar{x}} = \sigma/\sqrt{n} = 5/\sqrt{100} = .5$ hours

a. For $\bar{x} = 17 : z = (17 - 16.87)/.5 = .26$

$P(\bar{x} > 17) = P(z > .26) = .5 - P(0 \le z \le .26) = .5 - .1026 = .3974$

b. For $\bar{x} = 16.5 : z = (16.5 - 16.87)/.5 = -.74$

For $\bar{x} = 17.5 : z = (17.5 - 16.87)/.5 = 1.26$

$P(16.5 < \bar{x} < 17.5) = P(-.74 < z < 1.26) = P(-.74 < z < 0) + P(0 < z < 1.26) = .2704 + .3962 = .6666$

c. $P(\mu - .75 \le \bar{x} \le \mu + .75) = P(16.12 \le z \le 17.62)$

For $\bar{x} = 16.12 : z = (16.12 - 16.87)/.5 = -1.5$

For $\bar{x} = 17.62 : z = (17.62 - 16.87)/.5 = 1.5$

$P(16.12 \le \bar{x} \le 17.62) = P(-1.50 \le z \le 1.50) = P(-1.50 \le z \le 0) + P(0 \le z \le 1.50)$

$= .4332 + .4332 = .8664$

d. $P(\overline{x} \leq \mu - .75) = P(\overline{x} \leq 16.12)$

For $\overline{x} = 16.12 : z = (16.12 - 16.87)/.5 = -1.50$

$P(\overline{x} \leq 16.12) = P(z \leq -1.50) = .5 - P(-1.50 < z < 0) = .5 - .4332 = .0668$

7.98 $\mu = 64$ ounces, $\sigma = .4$ ounce, and $n = 16$

$\sigma_{\overline{x}} = \sigma/\sqrt{n} = .4/\sqrt{16} = .10$

For $\overline{x} = 63.75 : z = (63.75 - 64)/.10 = -2.50$

For $\overline{x} = 64.25 : z = (64.25 - 64)/.10 = 2.50$

$P(\overline{x} < 63.75) + P(\overline{x} > 64.25) = P(z < -2.50) + P(z > 2.50) = (.5 - .4938) + (.5 - .4938)$

$= .0062 + .0062 = .0124$

7.99 $p = .88, \ q = 1 - p = 1 - .88 = .12$, and $n = 80$

$\mu_{\hat{p}} = p = .88$, and $\sigma_{\hat{p}} = \sqrt{pq/n} = \sqrt{.88(.12)/80} = .036$

$np = 80(.88) = 70.4$ and $nq = 80(.12) = 9.6$

Since np and nq are both greater than 5, the sampling distribution of $\hat{p}$ is approximately normal.

7.100 $p = .70, \ q = 1 - .70 = .30,$ and $n = 400$

$\mu_{\hat{p}} = p = .70$, and $\sigma_{\hat{p}} = \sqrt{pq/n} = \sqrt{.70(.30)/400} = .023$

$np = 400(.70) = 280$ and $nq = 400(.30) = 120$

Since np and nq are both greater than 5, the sampling distribution of $\hat{p}$ is approximately normal.

7.101 $p = .70, q = 1 - p = 1 - .70 = .30$, and $n = 400$

$\sigma_{\hat{p}} = \sqrt{pq/n} = \sqrt{.70(.30)/400} = .02291288$

a. i. For $\hat{p} = .65 : z = (.65 - .70)/.02291288 = -2.18$

$P(\hat{p} < .65) = P(z < -2.18) = .5 - P(-2.18 \leq z \leq 0) = .5 - .4854 = .0146$

ii. For $\hat{p} = .73 : z = (.73 - .70)/.02291288 = 1.31$

For $\hat{p} = .76 : z = (.76 - .70)/.02291288 = 2.62$

$P(.73 < \hat{p} < .76) = P(1.31 < z < 2.62) = P(0 < z < 2.62) - P(0 < z < 1.31) = .4956 - .4049 = .0907$

b. $P(p-.06 \le \hat{p} \le p+.06) = P(.70-.06 \le \hat{p} \le .70+.06) = P(.64 \le \hat{p} \le .76)$

   For $\hat{p} = .64: z = (.64-.70)/.02291288 = -2.62$

   For $\hat{p} = .76: z = (.76-.70)/.02291288 = 2.62$

   $P(.64 \le \hat{p} \le .76) = P(-2.62 \le z \le 2.62) = .4956+.4956 = .9912$

c. $P(\hat{p} \ge p+.05) = P(\hat{p} \ge .70+.05) = P(\hat{p} \ge .75)$

   For $\hat{p} = .75: z = (.75-.70)/.02291288 = 2.18$

   $P(\hat{p} \ge .75) = P(z \ge 2.18) = .5 - P(0 < z < 2.18) = .5 - .4854 = .0146$

7.102 $p = .40, \quad q = 1-.40 = .60, \text{ and } n = 120$

   $\sigma_{\hat{p}} = \sqrt{pq/n} = \sqrt{.40(.60)/120} = .044721359$

a. $P(\hat{p} \text{ is within } .05 \text{ of } p) = P(.35 \le \hat{p} \le .45)$

   For $\hat{p} = .35: z = (.35-.40)/.044721359 = -1.12$

   For $\hat{p} = .45: z = (.45-.40)/.044721359 = 1.12$

   $P(.35 \le \hat{p} \le .45) = P(-1.12 \le z \le 1.12) = .3686+.3686 = .7372$

b. $P(\hat{p} \text{ is not within } .05 \text{ of } p) = 1 - P(\hat{p} \text{ is within } .05 \text{ of } p) = 1 - .7372 \text{ (from part a)} = .2628$

c. $P(\hat{p} \text{ is greater than } p \text{ by } .08 \text{ or more}) = P(\hat{p} \ge .48)$

   For $\hat{p} = .48: z = (.48-.40)/.044721359 = 1.79$

   $P(\hat{p} \ge .48) = P(z \ge 1.79) = .5 - P(0 < z < 1.79) = .5 - .4633 = .0367$

d. $P(\hat{p} \text{ is less than } p \text{ by } .06 \text{ or more}) = P(\hat{p} \le .34)$

   For $\hat{p} = .34: z = (.34-.40)/.044721359 = -1.34$

   $P(\hat{p} \le .34) = P(z \le -1.34) = .5 - P(-1.34 < z < 0) = .5 - .4099 = .0901$

7.103 $\sigma = \$105,000 \text{ and } n = 100$

   $\sigma_{\bar{x}} = \sigma/\sqrt{n} = 105,000/\sqrt{100} = \$10,500$

   The required probability is: $P(\mu-10,000 \le \bar{x} \le \mu+10,000)$

   For $\bar{x} = \mu-10,000: z = [(\mu-10,000)-\mu]/10,500 = -.95$

   For $\bar{x} = \mu+10,000: z = [(\mu+10,000)-\mu]/10,500 = .95$

$$P(\mu - 10{,}000 \le \bar{x} \le \mu + 10{,}000) = P(-.95 \le z \le .95) = .3289 + .3289 = .6578$$

7.104 Given $P(\bar{x} > 90) = .15$ and $P(\bar{x} < 65) = .30$

The corresponding z values are approximately $z = 1.04$ and $z = -.52$ respectively.

First we use $\bar{x} = \mu + z\sigma$ to find σ.

We have $90 = \mu + 1.04\sigma$ and $65 = \mu - .52\sigma$

Subtracting, gives $25 = 1.56\sigma$, so $\sigma = 16.0256$

Since $65 = \mu - .52\sigma$, we have $65 = \mu - .52(16.0256)$ so $\mu = 65 + .52(16.0256) \approx 73.33$.

7.105 $\mu = c, \sigma = .8$ ppm

We want $P(\mu - .5 \le \bar{x} \le \mu + .5) = .95$

$1.96\sigma_{\bar{x}} = .5$ or $\sigma_{\bar{x}} = .255$

$\sigma_{\bar{x}} = \sigma / \sqrt{n}$, so $n \ge \dfrac{\sigma^2}{(\sigma_{\bar{x}})^2} = \dfrac{(.8)^2}{(.255)^2} = 9.84$

Ten measurements are necessary.

7.106 a. $p = .60$, $np = 25(.6) = 15$, $nq = 25(.4) = 10$, so we can use the normal approximation to the binomial distribution.

$\mu = np = 15, \sigma = \sqrt{npq} = \sqrt{25(.6)(.4)} = 2.44948974$

$z = \dfrac{12.5 - 15}{2.44948974} = -1.02$

$P(\bar{x} > 12.5) = P(z > -1.02) = .8461$

b. $P(z > -1.65) = .9505$

$z = (\hat{p} - p) / \sigma_{\hat{p}}$, so $\sigma_{\hat{p}} = \dfrac{\hat{p} - p}{z} = \dfrac{.5 - .6}{-1.65} = .0606$

$\sigma_{\hat{p}} = \sqrt{pq / n}$, hence $n = pq /(\sigma_{\hat{p}})^2 = .6(.4) /(.0606)^2 = 65.34$

The reporter should take a sample of 66 voters or more.

7.107 a. $p = .53, n = 200$, and we assume that $n / N \le .05$

$\sigma_{\hat{p}} = \sqrt{pq / n} = \sqrt{(.53)(.47) / 200} = .03529164$

The shape of the sampling distribution is approximately normal.

$z = (\hat{p} - p) / \sigma_{\hat{p}} = \dfrac{.50 - .53}{.03529164} = -.85$

$$P(\hat{p} \geq .50) = P(z \geq -.85) = .5 + .3032 = .8023$$

b. $P(z > -1.65) = .9505$

$$z = (\hat{p} - p) / \sigma_{\hat{p}}, \text{ so } \sigma_{\hat{p}} = (\hat{p} - p) / z = \frac{.5 - .53}{-1.65} = .01818182$$

$$\sigma_{\hat{p}} = \sqrt{pq / n}, \text{ hence } n = pq / (\sigma_{\hat{p}})^2 = \frac{(.53)(.47)}{(.01818182)^2} = 753.53$$

The sample should include at least 754 voters.

7.108 $\mu = 290$ feet, $\sigma = 10$ feet, and $n = 3$

$$\sigma_{\bar{x}} = \sigma / \sqrt{n} = 10 / \sqrt{3} = 5.77350269 \text{ feet}$$

P (total length of three throws exceeds 885) = P(mean length of three throws exceeds 885/3)

$$= P(\bar{x} > 295)$$

For $\bar{x} = 295 : z = (295 - 290) / 5.77350269 = .87$

$$P(\bar{x} > 295) = P(z > .87) = .5 - P(0 \leq z \leq .87) = .5 - .3708 = .1922$$

7.109 $\mu = 160$ pounds, $\sigma = 25$ pounds, and $n = 35$

$$\sigma_{\bar{x}} = \sigma / \sqrt{n} = 25 / \sqrt{35} = 4.22577127$$

Since $n > 30$, $\bar{x}$ is approximately normally distributed.

P (sum of 35 weights exceeds 6000 pounds) = P (mean weight exceeds 6000/35) = $P(\bar{x} > 171.43)$

For $\bar{x} = 171.43 : z = (171.43 - 160) / 4.22577127 = 2.70$

$$P(\bar{x} > 171.43) = P(z > 2.70) = .5 - P(0 \leq z \leq 2.70) = .5 - .4965 = .0035$$

Self – Review Test for Chapter Seven

| 1. | b | 2. | b | 3. | a | 4. | a | 5. | b | 6. | b |

| 7. | c | 8. | a | 9. | a | 10. | a | 11. | c | 12. | a |

13.   According to the central limit theorem, for a large sample size, the sampling distribution of the sample mean is approximately normal irrespective of the shape of the population distribution. The mean and standard deviation of the sampling distribution of the sample mean are: $\mu_{\bar{x}} = \mu$ and $\sigma_{\bar{x}} = \sigma / \sqrt{n}$.

The sample size is usually considered to be large if $n \geq 30$.

From the same theorem, the sampling distribution of $\hat{p}$ is approximately normal for large samples. In the case of proportion, the sample is large if $np > 5$ and $nq > 5$.

14. $\mu = 145$ pounds and $\sigma = 18$ pounds

    a. $\mu_{\bar{x}} = \mu = 145$ pounds and $\sigma_{\bar{x}} = \sigma / \sqrt{n} = 18 / \sqrt{25} = 3.60$ pounds

    b. $\mu_{\bar{x}} = \mu = 145$ pounds and $\sigma_{\bar{x}} = \sigma / \sqrt{n} = 18 / \sqrt{100} = 1.80$ pounds

        In both cases the sampling distribution of $\bar{x}$ is approximately normal because the population has an approximate normal distribution.

15. $\mu = 11$ minutes and $\sigma = 2.7$ minutes

    a. $\mu_{\bar{x}} = \mu = 11$ minutes and $\sigma_{\bar{x}} = \sigma / \sqrt{n} = 2.7 / \sqrt{25} = .54$ minute

        Since the population has an unknown distribution and $n < 30$, we can draw no conclusion about the shape of the sampling distribution of $\bar{x}$.

    b. $\mu_{\bar{x}} = \mu = 11$ minutes and $\sigma_{\bar{x}} = \sigma / \sqrt{n} = 2.7 / \sqrt{75} = .312$ minutes

        Since $n > 30$, the sampling distribution of $\bar{x}$ is approximately normal.

16. $\mu = 42$ seconds, $\sigma = 10$ seconds, and $n = 50$

    $\sigma_{\bar{x}} = \sigma / \sqrt{n} = 10 / \sqrt{50} = 1.41421356$ seconds

    a. For $\bar{x} = 38: z = (38 - 42) / 1.4142136 = -2.83$

        For $\bar{x} = 41: z = (41 - 42) / 1.4142136 = -.71$

        $P(38 < \bar{x} < 41) = P(-2.83 < z < -.71) = P(-2.83 < z < 0) - P(-.71 \leq z \leq 0) = .4977 - .2611 = .2366$

    b. For $\bar{x} = 42 - 2 = 40: z = (40 - 42) / 1.4142136 = -1.41$

        For $\bar{x} = 42 + 2 = 44: z = (44 - 42) / 1.4142136 = 1.41$

        $P(40 < \bar{x} < 44) = P(-1.41 < z < 1.41) = P(-1.41 < z < 0) + P(0 \leq z \leq 1.41) = .4207 + .4207 = .8414$

    c. $P(\bar{x}$ greater than μ by 1 second or more)

For $\bar{x} = 43: z = (43 - 42)/1.4142136 = .71$

$P(\bar{x} \geq 43) = P(z \geq .71) = 5 - P(0 < z < .71) = .5 - .2611 = .2389$

d. For $\bar{x} = 39: z = (39 - 42)/1.4142136 = -2.12$

For $\bar{x} = 44: z = (44 - 42)/1.4142136 = 1.41$

$P(39 < \bar{x} < 44) = P(-2.12 < z < 1.41) = P(-2.12 < z < 0) + P(0 < z < 1.41) = .4830 + .4207 = .9037$

e. $\bar{x} = 43: z = (43 - 42)/1.4142136 = .71$

$P(\bar{x} \leq 43) = P(z \leq .71) = .5 + P(0 < z < .71) = .5 + .2611 = .7611$

17. $\mu = 16$ ounces, $\sigma = .18$ ounce, and $n = 16$

$\sigma_{\bar{x}} = \sigma/\sqrt{n} = .18/\sqrt{16} = .045$

a. i. For $\bar{x} = 15.90: z = (15.90 - 16)/.045 = -2.22$

For $\bar{x} = 15.95: z = (15.95 - 16)/.045 = -1.11$

$P(15.90 < \bar{x} < 15.95) = P(-2.22 < z < -1.11) = P(-2.22 < z < 0) - P(-1.11 < z < 0)$

$= .4868 - .3665 = .1203$

ii. For $\bar{x} = 15.95: z = (15.95 - 16)/.045 = -1.11$

$P(\bar{x} < 15.95) = P(z < -1.11) = .5 - P(-1.11 \leq z \leq 0) = .5 - .3665 = .1335$

iii. For $\bar{x} = 15.97: z = (15.97 - 16)/.045 = -.67$

$P(\bar{x} > 15.97) = P(z > -.67) = P(-.67 < z < 0) + .5 = .2486 + .5 = .7486$

b. $P(16 - .10 \leq \bar{x} \leq 16 + .10) = P(15.90 \leq \bar{x} \leq 16.10)$

For $\bar{x} = 15.90: z = (15.90 - 16)/.045 = -2.22$

For $\bar{x} = 16.10: z = (16.10 - 16)/.045 = 2.22$

$P(15.90 \leq \bar{x} \leq 16.10) = P(-2.22 \leq z \leq 2.22) = .4868 + .4868 = .9736$

c. $P(\bar{x} \leq 16 - .135) = P(\bar{x} \leq 15.865)$

For $\bar{x} = 15.865: z = (15.865 - 16)/.045 = -3.00$

$P(\bar{x} \leq 15.865) = P(z \leq -3.00) = .5 - P(-3.00 \leq z \leq 0) = .5 - .4987 = .0013$

18. $p = .07, \ q = 1 - .07 = .93$

    a. $n = 80$, $\mu_{\hat{p}} = p = .07$, and $\sigma_{\hat{p}} = \sqrt{pq/n} = \sqrt{.07(.93)/80} = .029$

        $np = 80(.07) = 5.6$ and $nq = 80(.93) = 74.4$

        Since np and nq are both greater than 5, the sampling distribution of $\hat{p}$ is approximately normal.

    b. $n = 200$, $\mu_{\hat{p}} = p = .07$, and $\sigma_{\hat{p}} = \sqrt{pq/n} = \sqrt{.07(.93)/200} = .018$

        $np = 200(.07) = 14$ and $nq = 200(.93) = 186$

        Since np and nq are both greater than 5, the sampling distribution of $\hat{p}$ is approximately normal.

    c. $n = 1000$, $\mu_{\hat{p}} = p = .07$, and $\sigma_{\hat{p}} = \sqrt{pq/n} = \sqrt{.07(.93)/1000} = .008$

        $np = 1000(.07) = 70$ and $nq = 1000(.93) = 930$

        Since np and nq are both greater than 5, the sampling distribution of $\hat{p}$ is approximately normal.

19. $p = .41, \quad q = 1 - .41 = .59$ and $n = 300$

    $\sigma_{\hat{p}} = \sqrt{pq/n} = \sqrt{.41(.59)/300} = .02839601$

    a. **i.**   For $\hat{p} = .45 : z = (.45 - .41)/.02839601 = 1.41$

          $P(\hat{p} > .45) = P(z > 1.41) = .5 - P(0 \le z \le 1.41) = .5 - .4207 = .0793$

      ii.   For $\hat{p} = .40 : z = (.40 - .41)/.02839601 = -.35$

          For $\hat{p} = .46 : z = (.46 - .41)/.02839601 = 1.76$

          $P(.40 < \hat{p} < .46) = P(-.35 < z < 1.76) = P(-.35 < z < 0) + P(0 < z < 1.76) = .1368 + .4608 = .5976$

      iii.   For $\hat{p} = .43 : z = (.43 - .41)/.02839601 = .70$

          $P(\hat{p} < .43) = P(z < .70) = .5 + P(0 < z < .70) = .5 + .2580 = .7580$

      iv.   For $\hat{p} = .36 : z = (.36 - .41)/.02839601 = -1.76$

          For $\hat{p} = .39 : z = (.39 - .41)/.02839601 = -.70$

          $P(.36 < \hat{p} < .39) = P(-1.76 < z < -.70) = P(-1.76 < z < 0) - P(-.70 \le z \le 0)$

             $= .4608 - .2580 = .2028$

b. $P(.41 - .04 \leq \hat{p} \leq .41 + .04) = P(.37 \leq \hat{p} \leq .45)$

   For $\hat{p} = .37: z = (.37 - .41)/.02839601 = -1.41$

   For $\hat{p} = .45: z = (.45 - .41)/.02839601 = 1.41$

   $P(.37 \leq \hat{p} \leq .45) = P(-1.41 \leq z \leq 1.41) = P(-1.41 \leq z \leq 0) + P(0 \leq z \leq 1.41) = .4207 + .4207 = .8414$

c. $P(\hat{p} < .41 - .05) = P(\hat{p} < .36)$

   For $\hat{p} = .36: z = (.36 - .41)/.02839601 = 1.76$

   $P(\hat{p} < .36) = P(z < -1.76) = .5 - P(-1.76 \leq z \leq 0) = .5 - .4608 = .0392$

d. $P(.41 - .03 \leq \hat{p} \leq .41 + .03) = P(.38 \leq \hat{p} \leq .44)$

   For $\hat{p} = .38: z = (.38 - .41)/.02839601 = -1.06$

   For $\hat{p} = .44: z = (.44 - .41)/.02839601 = 1.06$

   $P(.36 \leq \hat{p} \leq .44) = P(-1.06 \leq z \leq 1.06) = P(-1.06 \leq z \leq 0) + P(0 \leq z \leq 1.06) = .3554 - .3554 = .7108$

e. $P(\hat{p} > .41 + .03) = P(\hat{p} > .44)$

   For $\hat{p} = .44, z = (.44 - .41)/.02893601 = 1.06$

   $P(\hat{p} > .44) = P(z > 1.06) = .5 - P(0 \leq z \leq 1.06) = .5 - .3554 = .1446$

Chapter Eight

8.1   An estimator is a sample statistic used to estimate a population parameter. For example, the sample mean $\bar{x}$ is an estimator of the population mean μ and the sample proportion $\hat{p}$ is an estimator of the population proportion p. The value(s) assigned to a population parameter based on the value of a sample statistic is called an estimate.

8.2   The value of a sample statistic used to estimate a population parameter is called a point estimate. Under this procedure, we assign a single value to the population parameter being estimated. In interval estimation, an interval is constructed around the point estimate. This interval is likely to contain the corresponding population parameter.

8.3   The sample mean $\bar{x}$ is the point estimator of the population mean μ. The margin of error is $\pm 1.96\sigma_{\bar{x}}$ or $\pm 1.96 s_{\bar{x}}$.

8.4   The width of a confidence interval may be decreased by lowering the confidence level or by increasing the sample size. Since lowering the confidence level gives a less reliable estimate, it is preferable to increase the sample size.

8.5   The width of a confidence interval depends on $z\sigma/\sqrt{n}$ or $zs/\sqrt{n}$. Thus, when n increases, $z\sigma/\sqrt{n}$ (or $zs/\sqrt{n}$) decreases. As a result of this, the width of the confidence interval decreases.

Example: Suppose $\bar{x}=50$, $\sigma=12$, and $n=36$.

The 95% confidence interval for μ is: $\bar{x} \pm z\sigma_{\bar{x}} = 50 \pm 1.96(12)/\sqrt{36} = 50 \pm 3.92 = 46.08$ to 53.92

If $n = 100$, but all other values remain the same, the 95% confidence interval for μ is:

$\bar{x} \pm z\sigma_{\bar{x}} = 50 \pm 1.96(12)/\sqrt{100} = 50 \pm 2.35 = 47.65$ to 52.35

A comparison shows that the 95% confidence interval for μ is narrower when n is 100 than when n is 36.

8.6   The width of a confidence interval depends on $z\sigma_{\bar{x}}$ or $zs_{\bar{x}}$. A decrease in the confidence level decreases z, which reduces the width of the confidence interval.

Example: Suppose $\bar{x} = 50, \sigma = 12,$ and $n = 36$.

The 95% confidence interval for μ is: $\bar{x} \pm z\sigma_{\bar{x}} = 50 \pm 1.96(12)/\sqrt{36} = 50 \pm 3.92 = 46.08$ to 53.92

The 90% confidence interval for μ is : $\bar{x} \pm z\sigma_{\bar{x}} = 50 \pm 1.65(12)/\sqrt{36} = 50 \pm 3.30 = 46.70$ to 53.30

By comparison, the 90% confidence interval is narrower than the 95% confidence interval.

8.7 A confidence interval is an interval constructed around a point estimate. A confidence level indicates how confident we are that the confidence interval contains the population parameter.

8.8 The maximum error of estimate for μ is subtracted from and added to the sample statistic $\bar{x}$ to obtain a confidence interval for μ. It is given by $z\sigma_{\bar{x}}$ or $zs_{\bar{x}}$, where z is determined by the confidence level.

8.9 If we took all possible samples of a given size and constructed a 99% confidence interval for μ from each sample, we would expect about 99% of these confidence intervals would contain μ and 1% would not.

8.10 a. $z = 1.65$ b. $z = 1.96$ c. $z = 2.05$

   d. $z = 2.17$ e. $z = 2.33$ f. $z = 2.58$

8.11 $n = 64$, $\bar{x} = 24.5$, $s = 3.1$, and $s_{\bar{x}} = s/\sqrt{n} = 3.1/\sqrt{64} = .3875$

   a. Point estimate of μ is: $\bar{x} = 24.5$

   b. Margin of error $= \pm1.96 s_{\bar{x}} = \pm1.96(.3875) = \pm.760$

   c. The 99% confidence interval for μ is: $\bar{x} + zs_{\bar{x}} = 24.5 \pm 2.58(.3875) = 24.5 \pm 1.00 = 23.50$ to 25.50

   d. Maximum error of estimate for 99% confidence level is: $zs_{\bar{x}} = 2.58(.3875) = 1.00$

8.12 $n = 81$, $\bar{x} = 48.25$, $s = 4.8$, and $s_{\bar{x}} = s/\sqrt{n} = 4.8/\sqrt{81} = .53333333$

   a. Point estimate of μ is: $\bar{x} = 48.25$

   b. Margin of error $= \pm1.96 s_{\bar{x}} = \pm1.96(.53333333) = \pm1.05$

   c. The 95% confidence interval for μ is: $\bar{x} \pm zs_{\bar{x}} = 48.25 \pm 1.96(.53333333) = 48.25 \pm 1.05$

= 47.20 to 49.30.

d. Maximum error of estimate for 95% confidence level is: $zs_{\bar{x}} = 1.96(.53333333) = 1.05$

8.13 $n = 36$, $\bar{x} = 74.8$, $\sigma = 15.3$, and $\sigma_{\bar{x}} = \sigma / \sqrt{n} = 15.3 / \sqrt{36} = 2.55$

a. The 90% confidence interval for μ is: $\bar{x} \pm z\sigma_{\bar{x}} = 74.8 \pm 1.65(2.55) = 74.8 \pm 4.21 = 70.59$ to 79.01

b. The 95% confidence interval for μ is: $\bar{x} \pm z\sigma_{\bar{x}} = 74.8 \pm 1.96(2.55) = 74.8 \pm 5.00 = 69.80$ to 79.80

c. The 99% confidence interval for μ is: $\bar{x} \pm z\sigma_{\bar{x}} = 74.8 \pm 2.58(2.55) = 74.8 \pm 6.58 = 68.22$ to 81.38

d. Yes, the width of the confidence intervals increases as the confidence level increases. This occurs because as the confidence level increases, the value of z increases.

8.14 $n = 100$, $\bar{x} = 143.72$, $\sigma = 14.8$, and $\sigma_{\bar{x}} = \sigma / \sqrt{n} = 14.8 / \sqrt{100} = 1.48$

a. The 99% confidence interval for μ is: $\bar{x} \pm z\sigma_{\bar{x}} = 143.72 \pm 2.58(1.48) = 143.72 \pm 3.82 = 139.90$ to 147.54

b. The 95% confidence interval for μ is: $\bar{x} \pm z\sigma_{\bar{x}} = 143.72 \pm 1.96(1.48) = 143.72 \pm 2.90 = 140.82$ to 146.62

c. The 90% confidence interval for μ is: $\bar{x} \pm z\sigma_{\bar{x}} = 143.72 \pm 1.65(1.48) = 143.72 \pm 2.44 = 141.28$ to 146.16

d. Yes, the width of the confidence intervals decreases as the confidence level decreases. This occurs because as the confidence level decreases, the value of z decreases.

8.15 $\bar{x} = 81.90$ and $\sigma = 6.30$

a. $n = 36$, so $\sigma_{\bar{x}} = \sigma / \sqrt{n} = 6.30 / \sqrt{36} = 1.05$ The 99% confidence interval for μ is:

$\bar{x} \pm z\sigma_{\bar{x}} = 81.90 \pm 2.58(1.05) = 81.90 \pm 2.71 = 79.19$ to 84.61

b. $n = 81$, so $\sigma_{\bar{x}} = \sigma / \sqrt{n} = 6.30 / \sqrt{81} = .70$

The 99% confidence interval for μ is: $\bar{x} \pm z\sigma_{\bar{x}} = 81.90 \pm 2.58(.70) = 81.90 \pm 1.81 = 80.09$ to 83.71

    c. $n = 100$, so $\sigma_{\bar{x}} = \sigma / \sqrt{n} = 6.30 / \sqrt{100} = .63$

        The 99% confidence interval for μ is: $\bar{x} \pm z\sigma_{\bar{x}} = 81.90 \pm 2.58(.63) = 81.90 \pm 1.63 = 80.27$ to 83.53

    d. Yes, the width of the confidence intervals decreases as the sample size increases. This occurs because the standard deviation of the sample mean decreases as the sample size increases.

8.16 $\bar{x} = 48.52$ and $\sigma = 7.14$

    a. $n = 196$, so $\sigma_{\bar{x}} = \sigma / \sqrt{n} = 7.14 / \sqrt{196} = .51$

        The 95% confidence interval for μ is: $\bar{x} \pm z\sigma_{\bar{x}} = 48.52 \pm 1.96(.51) = 48.52 \pm 1.00 = 47.52$ to 49.52

    b. $n = 100$, so $\sigma_{\bar{x}} = \sigma / \sqrt{n} = 7.14 / \sqrt{100} = .714$

        The 95% confidence interval for μ is: $\bar{x} \pm z\sigma_{\bar{x}} = 48.52 \pm 1.96(.714) = 48.52 \pm 1.40 = 47.12$ to 49.92

    c. $n = 49$, so $\sigma_{\bar{x}} = \sigma / \sqrt{n} = 7.14 / \sqrt{49} = 1.02$

        The 95% confidence interval for μ is: $\bar{x} \pm z\sigma_{\bar{x}} = 48.52 \pm 1.96(1.02) = 48.52 \pm 2.00 = 46.52$ to 50.52

    d. Yes, the width of the confidence intervals increases as the sample size decreases. This occurs because the standard deviation of the sample mean increases as the sample size decreases.

8.17 a. $n = 100$, $\bar{x} = 55.32$, and $s = 8.4$, so $s_{\bar{x}} = s / \sqrt{n} = 8.4 / \sqrt{100} = .84$

        The 90% confidence interval for μ is: $\bar{x} \pm zs_{\bar{x}} = 55.32 \pm 1.65(.84) = 55.32 \pm 1.39 = 53.93$ to 56.71

    b. $n = 100$, $\bar{x} = 57.40$, and $s = 7.5$, so $s_{\bar{x}} = s / \sqrt{n} = 7.5 / \sqrt{100} = .75$

        The 90% confidence interval for μ is: $\bar{x} \pm zs_{\bar{x}} = 57.40 \pm 1.65(.75) = 57.40 \pm 1.24 = 56.16$ to 58.64

    c. $n = 100$, $\bar{x} = 56.25$, and $s = 7.9$, so $s_{\bar{x}} = s / \sqrt{n} = 7.9 / \sqrt{100} = .79$

        The 90% confidence interval for μ is: $\bar{x} \pm zs_{\bar{x}} = 56.25 \pm 1.65(.79) = 56.25 \pm 1.30 = 54.95$ to 57.55

    d. The confidence intervals of parts a and c cover μ but the confidence interval of part b does not.

8.18 a. $n = 400$, $\bar{x} = 92.45$, and $s = 12.20$, so $s_{\bar{x}} = s / \sqrt{n} = 12.20 / \sqrt{400} = .61$

        The 97% confidence interval for μ is: $\bar{x} \pm zs_{\bar{x}} = 92.45 \pm 2.17(.61) = 92.45 \pm 1.32 = 91.13$ to 93.77

b. $n = 400$, $\bar{x} = 91.75$, and $s = 14.50$, so $s_{\bar{x}} = s / \sqrt{n} = 14.50 / \sqrt{400} = .725$

The 97% confidence interval for μ is: $\bar{x} \pm z s_{\bar{x}} = 91.75 \pm 2.17(.725) = 91.75 \pm 1.57 = 90.18$ to 93.32

c. $n = 400$, $\bar{x} = 89.63$, and $s = 13.40$; so $s_{\bar{x}} = s / \sqrt{n} = 13.40 / \sqrt{400} = .67$

The 97% confidence interval for μ is: $\bar{x} \pm z s_{\bar{x}} = 89.63 \pm 2.17(.67) = 89.63 \pm 1.45 = 88.18$ to 91.08

d. The confidence intervals of parts b and c cover μ but the confidence interval of part a does not.

8.19 $n = 35$, $\bar{x} = \sum x / n = 1342 / 35 = 38.34$, and $\sigma = 2.65$

$\sigma_{\bar{x}} = \sigma / \sqrt{n} = 2.65 / \sqrt{35} = .44793176$

a. Point estimate of μ is: $\bar{x} = 38.34$

b. Margin of error $= \pm 1.96 \sigma_{\bar{x}} = \pm 1.96(.44793176) = \pm .88$

c. The 98% confidence interval for μ is: $\bar{x} \pm z \sigma_{\bar{x}} = 38.34 \pm 2.33(.44793176) = 38.34 \pm 1.04 = 37.30$ to 39.38

d. Maximum error of estimate $= z \sigma_{\bar{x}} = 2.33(.44793176) = 1.04$

8.20 $n = 32$, $\bar{x} = \sum x / n = 2543 / 32 = 79.47$, and $\sigma = 4.96$, so $\sigma_{\bar{x}} = \sigma / \sqrt{n} = 4.96 / \sqrt{32} = .87681241$

a. Point estimate of μ is: $\bar{x} = 79.47$

b. Margin of error $= \pm 1.96 \sigma_{\bar{x}} = \pm 1.96(.87681241) = \pm 1.72$

c. The 99% confidence interval for μ is: $\bar{x} \pm z s_{\bar{x}} = 79.47 \pm 2.58(.87681241) = 79.47 \pm 2.26 = 77.21$ to 81.73

d. Maximum error of estimate $= z \sigma_{\bar{x}} = 2.58(.87681241) = 2.26$

8.21 $n = 500$, $\bar{x} = \$355,000$, and $s = \$125,000$; so $s_{\bar{x}} = s / \sqrt{n} = 125,000 / \sqrt{500} = 5590.169944$

The 99% confidence interval for μ is:

$\bar{x} \pm z s_{\bar{x}} = 355,000 \pm 2.58(5590.169944) = 355,000 \pm 14,422.64 = \$340,577.36$ to $\$369,422.64$.

8.22 $n = 1000$, $\bar{x} = 16.87$ hours, and $s = 3.2$ hours; so $s_{\bar{x}} = s / \sqrt{n} = 3.2 / \sqrt{1000} = .10119289$

The 99% confidence interval for μ is: $\bar{x} \pm z s_{\bar{x}} = 16.87 \pm 2.58(.10119289) = 16.87 \pm .20$

$= 16.67$ to 17.07 hours.

8.23 $n = 100$, $\bar{x} = \$4081$, and $s = \$845$; so $s_{\bar{x}} = s / \sqrt{n} = 845 / \sqrt{100} = \$84.50$

The 99% confidence interval for μ is: $\bar{x} \pm zs_{\bar{x}} = 4081 \pm 2.58(84.5) = 4081 \pm 218.01 = \$3862.99$ to $\$4299.01$

8.24 $n = 1000,\quad \bar{x} = \$145,000,\ \text{and}\ s = \$24,000;\ \text{so} \qquad s_{\bar{x}} = s/\sqrt{n} = 24,000/\sqrt{1000} = 758.9466384$

The 95% confidence interval for μ is:

$\bar{x} \pm zs_{\bar{x}} = 145,000 \pm 1.96(758.9466384) = 145,000 \pm 1487.54 = \$143,512.46$ to $\$146,487.54$

8.25 $n = 60,\ \bar{x} = \$961,000,\ \text{and}\ s = \$180,000,\ \text{so} \qquad s_{\bar{x}} = s/\sqrt{n} = 180,000/\sqrt{60} = 23237.90008$

a. Point estimate of $\mu = \bar{x} = \$961,000$

   Margin of error $= \pm 1.96 s_{\bar{x}} = \pm 1.96(23,237.90008) = \pm \$45,546.28$

b. The 90% confidence interval for μ is:

   $\bar{x} \pm zs_{\bar{x}} = 961,000 \pm 1.65(23,237.90008) = 961,000 \pm 38,342.54 = \$922,657.46$ to $\$999,342.54$

8.26 $n = 500,\quad \bar{x} = \$7182,\ \text{and}\ s = \$2100;\ \text{so} \qquad s_{\bar{x}} = s/\sqrt{n} = 2100/\sqrt{500} = \$93.91485506$

a. Point estimate of $\mu = \bar{x} = \$7182$

   Margin of error $= \pm 1.96 s_{\bar{x}} = \pm 1.96(93.91485506) = \pm \$184.07$

b. The 98% confidence interval for μ is:

   $\bar{x} \pm zs_{\bar{x}} = 7182 \pm 2.33(93.91485506) = 7182 \pm 218.82 = \$6963.18$ to $\$7400.82$

8.27 $n = 1200,\quad \bar{x} = 2.3\ \text{visits, and}\ s = .44\ \text{visits, so} \qquad s_{\bar{x}} = s/\sqrt{n} = .44/\sqrt{1200} = .01270171$

a. The 97% confidence interval for μ is:

   $\bar{x} \pm zs_{\bar{x}} = 2.3 \pm 2.17(.01270171) = 2.3 \pm .03 = 2.27$ to 2.33 visits.

b. The sample mean of 2.3 visits is an estimate of μ based on a random sample. Because of sampling error, this estimate might differ from the true mean, μ, so we make an interval estimate to allow for this uncertainty and sampling error.

8.28 $n = 50,\quad \bar{x} = 70\ \text{hours, and}\ s = 16\ \text{hours, so} \qquad s_{\bar{x}} = s/\sqrt{n} = 16/\sqrt{50} = 2.26274170\ \text{hours}$

a. The 95% confidence interval for μ is: $\bar{x} \pm zs_{\bar{x}} = 70 \pm 1.96(2.26274170) = 70 \pm 4.43 = 65.57$ to 74.43 hours

b. The sample mean of 70 hours is an estimate of μ based on a random sample. Because of sampling error, this estimate might differ from the true mean, μ, so we make an interval estimate to allow for this uncertainty and sampling error.

8.29 $n = 40$, $\bar{x} = 36.02$ inches and $\sigma = .10$ inches, so $\sigma_{\bar{x}} = \sigma / \sqrt{n} = .10 / \sqrt{40} = .01581139$ inches

The 99% confidence interval for μ is:

$\bar{x} \pm z\sigma_{\bar{x}} = 36.02 \pm 2.58(.01581139) = 36.02 \pm .04 = 35.98$ to 36.06 inches

Since the upper limit, 36.06, is greater than 36.05, the machine needs an adjustment.

8.30 $n = 35$, $\bar{x} = 31.94$ ounces, and, $\sigma = .15$ ounces, so $\sigma_{\bar{x}} = \sigma / \sqrt{n} = .15 / \sqrt{35} = .02535463$ ounces.

The 99% confidence interval for μ is:

$\bar{x} \pm z\sigma_{\bar{x}} = 31.94 \pm 2.58(.02535463) = 31.94 \pm .07 = 31.87$ to 32.01 ounces.

Since the upper limit, 32.01, is less than 32.15, and the lower limit, 31.87, is greater than 31.85, the machine does not need an adjustment.

8.31 a. $n = 70$, $\bar{x} = \$420$, and $s = \$110$, so $s_{\bar{x}} = s / \sqrt{n} = 110 / \sqrt{70} = \$13.1475147$

The 99% confidence interval for μ is:

$\bar{x} \pm zs_{\bar{x}} = 420 \pm 2.58(13.1475147) = 420 \pm 33.92 = \$386.08$ to $\$453.92$

b. The width of the confidence interval obtained in part a may be reduced by:

1. Lowering the confidence level
2. Increasing the sample size

The second alternative is better because lowering the confidence level lowers the probability that the confidence interval contains μ.

8.32 a. $n = 40$, $\bar{x} = \$1575$, and $s = \$215$, so $s_{\bar{x}} = s / \sqrt{n} = 215 / \sqrt{40} = \$33.99448485$

The 97% confidence interval for μ is:

$\bar{x} \pm zs_{\bar{x}} = 1575 \pm 2.17(33.99448485) = 1575 \pm 73.77 = \$1501.23$ to $\$1648.77$

b. The width of the confidence interval obtained in part a may be reduced by:

1. Lowering the confidence level
2. Increasing the sample size

The second alternative is better because lowering the confidence level lowers the probability that the confidence interval contains μ.

8.33 To estimate μ, the mean commuting time:

1. Take a random sample of 30 or more students from your school
2. Determine the commuting time for each student
3. Find the mean $\bar{x}$ and the standard deviation s of this sample
4. Find $s_{\bar{x}} = s / \sqrt{n}$
5. The 99% confidence interval for μ is obtained by using the formula $\bar{x} \pm 2.58 s_{\bar{x}}$

8.34 To estimate μ, the mean age of cars:

1. Take a random sample of 30 or more U.S. car owners
2. Determine age of each car
3. Find the mean $\bar{x}$ and the standard deviation s of this sample
4. Find $s_{\bar{x}} = s / \sqrt{n}$
5. The 95% confidence interval for μ is obtained by using the formula $\bar{x} \pm 1.96 s_{\bar{x}}$

8.35 The following are the similarities between the standard normal distribution and the t distribution:

1. Both distributions are symmetric (bell–shaped) about the mean.
2. Neither distribution meets the horizontal axis.
3. Total area under each of these curves is 1.0 or 100%.
4. The mean of both of these distributions is zero.

The following are the main differences between the standard normal distribution and the t distribution:

1. The t distribution has a lower height and a wider spread than the normal distribution.
2. The standard deviation of the standard normal distribution is 1 and that of the t distribution is $\sqrt{df /(df - 2)}$, which is always greater than 1.
3. The t distribution has only one parameter, the degrees of freedom, whereas the (standard) normal distribution has two parameters, μ and σ.

8.36 The normal distribution has two parameters: μ and σ. Given the values of these parameters for a normal distribution, we can find the area under the normal curve between any two points. The t distribution has only one parameter: the degrees of freedom. The shape of the t distribution curve is determined by the degrees of freedom.

8.37   The number of degrees of freedom is defined as the number of observations that can be chosen freely. As an example, suppose the mean score of five students in an examination is 81. Consequently, the sum of these five scores is 81(5) = 405. Now, how many scores, out of five, can we choose freely so that the sum of these five scores is 405? The answer is that we are free to choose $5 - 1 = 4$ scores. Suppose we choose 87, 73, 69, and 94 as the four scores. Given these four scores and the information that the mean of the five scores is 81, the fifth score is: $405 - 87 - 73 - 69 - 94 = 82$

Thus, once we have chosen four scores, the fifth score is automatically determined. Hence, the number of degrees of freedom for this example are: $df = n - 1 = 5 - 1 = 4$

We subtract 1 from n to obtain the degrees of freedom because we lose one degree of freedom to calculate the mean.

8.38   The t distribution is used to make a confidence interval for the population mean μ if the following three assumptions hold true.

1. The population from which the sample is taken is (approximately) normally distributed.
2. The sample size is small, that is, $n < 30$.
3. The population standard deviation σ is not known.

8.39   a. $t = 1.782$ b. $df = n - 1 = 26 - 1 = 25$ and $t = -2.060$

c. $t = -3.579$ d. $df = n - 1 = 24 - 1 = 23$ and $t = 2.807$

8.40   a. $df = n - 1 = 21 - 1 = 20$ and $t = -1.325$ b. $df = n - 1 = 14 - 1 = 13$ and $t = 2.160$

c. $t = 3.686$ d. $t = -2.771$

8.41   a. Area in the right tail = .01   b. Area in the left tail = .005

c. Area in the left tail = .025   d. Area in the right tail = .01

8.42   a. Area in the left tail = .05   b. Area in the right tail = .005

c. Area in the right tail = .10   d. Area in the left tail = .01

8.43   a. $\alpha / 2 = .5 - .99 / 2 = .005$ and $t = 3.012$

b. $df = n - 1 = 26 - 1 = 25$, $\alpha / 2 = .5 - .95 / 2 = .025$ and $t = 2.060$

c. $\alpha / 2 = .5 - .90 / 2 = .05$ and $t = 1.746$

8.44   a. $df = n - 1 = 22 - 1 = 21$, $\alpha / 2 = .5 - .95 / 2 = .025$ and $t = 2.080$

b. $\alpha / 2 = .5 - .90 / 2 = .05$ and $t = 1.812$

c. $df = 24 - 1 = 23$, $\alpha / 2 = .5 - .99 / 2 = .005$ and $t = 2.807$

8.45 From the sample data: $n = 12, \sum x = 153, \sum x^2 = 2087$

$$\bar{x} = \sum x / n = 153 / 12 = 12.75$$

$$s = \sqrt{\frac{\sum x^2 - (\sum x)^2 / n}{n-1}} = \sqrt{\frac{2087 - (153)^2 / 12}{12-1}} = 3.51942661$$

$$s_{\bar{x}} = s / \sqrt{n} = 3.51942661 / \sqrt{12} = 1.01597095$$

a. Point estimate of μ is: $= 12.75$

b. $\alpha / 2 = .01 / 2 = .005$, $df = n - 1 = 12 - 1 = 11$, and $t = 3.106$

The 99% confidence interval for μ is:

$\bar{x} \pm t s_{\bar{x}} = 12.75 \pm 3.106(1.01597095) = 12.75 \pm 3.16 = 9.59$ to 15.91.

c. Maximum error of estimate is: $t s_{\bar{x}} = 3.106(1.01597095) = 3.16$

8.46 From the sample data: $n = 10$, $\sum x = 441$, $\sum x^2 = 19,977$

$$\bar{x} = \sum x / n = 441 / 10 = 44.10$$

$$s = \sqrt{\frac{\sum x^2 - (\sum x)^2 / n}{n-1}} = \sqrt{\frac{19,977 - (441)^2 / 10}{10-1}} = 7.66594199$$

$$s_{\bar{x}} = s / \sqrt{n} = 7.66594199 / \sqrt{10} = 2.42418371$$

a. Point estimate of μ is: $\bar{x} = 44.10$

b. $\alpha / 2 = .5 - .95 / 2 = .025$, $df = n - 1 = 10 - 1 = 9$, and $t = 2.262$

The 95% confidence interval for μ is:

$\bar{x} \pm t s_{\bar{x}} = 44.10 \pm 2.262(2.42418371) = 44.10 \pm 5.48 = 38.62$ to 49.58

c. Maximum error of estimate is: $t s_{\bar{x}} = 2.262(2.42418371) = 5.48$

8.47 a. $df = n - 1 = 16 - 1 = 15$, $\alpha / 2 = .5 - (.95 / 2) = .025$ and $t = 2.131$

$$s_{\bar{x}} = s / \sqrt{n} = 8.9 / \sqrt{16} = 2.225$$

The 95% confidence interval for μ is: $\bar{x} \pm t s_{\bar{x}} = 68.50 \pm 2.131(2.225) = 68.50 \pm 4.74 = 63.76$ to 73.24.

b. $\alpha / 2 = .5 - (.90 / 2) = .05$, $df = 15$, and $t = 1.753$

The 90% confidence interval for μ is: $\bar{x} \pm t s_{\bar{x}} = 68.50 \pm 1.753(2.225) = 68.50 \pm 3.90 = 64.60$ to 72.40

The width of the 90% confidence interval for μ is smaller than that of the 95% confidence interval. This is so because the value of t for a 90% confidence level is smaller than that for a 95% confidence level (with *df* remaining the same).

c. $s_{\bar{x}} = s / \sqrt{n} = 8.9 / \sqrt{25} = 1.78$

$df = n - 1 = 25 - 1 = 24$, $\alpha / 2 = .5 - (.95 / 2) = .025$ and $t = 2.064$

The 95% confidence interval for μ is: $\bar{x} \pm t s_{\bar{x}} = 68.50 \pm 2.064(1.78) = 68.50 \pm 3.67 = 64.83$ to 72.17

The width of the 95% confidence interval for μ is smaller with $n = 25$ than that of the 95% confidence interval with $n = 16$. This is so because the value of the standard deviation of the sample mean decreases as the sample size increases.

8.48   a. $df = n - 1 = 27 - 1 = 26$, $\alpha / 2 = .5 - (.95 / 2) = .025$ and $t = 2.056$

$s_{\bar{x}} = s / \sqrt{n} = 4.9 / \sqrt{27} = .94300544$

The 95% confidence interval for μ is: $\bar{x} \pm t s_{\bar{x}} = 25.5 \pm 2.056(.94300544) = 25.5 \pm 1.94 = 23.56$ to 27.44

b. $df = 27 - 1 = 26, \alpha / 2 = .5 - (.99 / 2) = .005$ and $t = 2.779$

The 99% confidence interval for μ is: $\bar{x} \pm t s_{\bar{x}} = 25.5 \pm 2.779(.94300544) = 25.5 \pm 2.62 = 22.88$ to 28.12

The width of the 99% confidence interval for μ is larger than that of the 95% confidence interval. This is so because the value of t for a 99% confidence level is larger than that for a 95% confidence level.

c. $s_{\bar{x}} = s / \sqrt{n} = 4.9 / \sqrt{20} = 1.09567331$

$df = n - 1 = 20 - 1 = 19$, $\alpha / 2 = .5 - (.95 / 2) = .025$ and $t = 2.093$

The 95% confidence interval for μ is: $\bar{x} \pm t s_{\bar{x}} = 25.5 \pm 2.093(1.09567331) = 25.5 \pm 2.29 = 23.21$ to 27.79

The width of the 95% confidence interval for μ is larger with $n = 20$ than that of the 95% confidence interval with $n = 27$. This is so because the value of the standard deviation of the sample mean is larger for a smaller sample size.

8.49 $n = 16$, $\bar{x} = 31$ minutes, and $s = 7$ minutes; then $s_{\bar{x}} = s / \sqrt{n} = 7 / \sqrt{16} = 1.75$ minutes

$df = n - 1 = 16 - 1 = 15$, $\alpha / 2 = .5 - (.99 / 2) = .005$ and $t = 2.947$

The 99% confidence interval for μ is: $\bar{x} \pm t s_{\bar{x}} = 31 \pm 2.947(1.75) = 31 \pm 5.16 = 25.84$ to 36.16 minutes

8.50 $n = 20$, $\bar{x} = 41.2$ bushels per acre, and $s = 3$ bushels per acre

$s_{\bar{x}} = s/\sqrt{n} = 3/\sqrt{20} = .67082039$ bushels

$df = n-1 = 20-1 = 19, \quad \alpha/2 = .5-(.90/2) = .05$ and $t = 1.729$

The 90% confidence interval for μ is:

$\bar{x} \pm ts_{\bar{x}} = 41.2 \pm 1.729(.67082039) = 41.2 \pm 1.16 = 40.04$ to 42.36 bushels per acre.

8.51 $n = 25, \quad \bar{x} = \$5.70,$, and $s = \$1.05$; so $s_{\bar{x}} = s/\sqrt{n} = 1.05/\sqrt{25} = \$.21$

        $df = n-1 = 25-1 = 24, \quad \alpha/2 = .5-(.95/2) = .025$ and $t = 2.064$

        The 95% confidence interval for μ is: $\bar{x} \pm ts_{\bar{x}} = 5.70 \pm 2.064(.21) = 5.70 \pm .43 = \$5.27$ to \$6.13

8.52 $n = 16, \quad \bar{x} = 31.98$ ounces, and $s = .26$ ounce; then $s_{\bar{x}} = s/\sqrt{n} = .26/\sqrt{16} = .0650$ ounce

        $df = n-1 = 16-1 = 15, \quad \alpha/2 = .5-(.95/2) = .025$ and $t = 2.131$

        The 95% confidence interval for μ is: $\bar{x} \pm ts_{\bar{x}} = 31.98 \pm 2.131(.0650) = 31.98 \pm .14 = 31.84$ to 32.12 ounces

8.53 $n = 25, \quad \bar{x} = \$143$, and $s = \$28$;   then $s_{\bar{x}} = s/\sqrt{n} = 28/\sqrt{25} = 5.60$

        $df = n-1 = 25-1 = 24, \quad \alpha/2 = .5-(.98/2) = .01$ and $t = 2.492$

        The 98% confidence interval for μ is: $\bar{x} \pm ts_{\bar{x}} = 143 \pm 2.492(5.60) = 143 \pm 13.96 = \$129.04$ to \$156.96

8.54 $n = 25, \quad \bar{x} = 22$ minutes, and $s = 6$ minutes; so $s_{\bar{x}} = s/\sqrt{n} = 6/\sqrt{25} = 1.2$ minutes.

        $df = n-1 = 25-1 = 24, \quad \alpha/2 = .5-(.99/2) = .005$ and $t = 2.797$

        The 99% confidence interval for μ is: $\bar{x} \pm ts_{\bar{x}} = 22 \pm 2.797(1.2) = 22 \pm 3.36 = 18.64$ to 25.36 minutes

8.55   a. $n = 16, \quad \bar{x} = 26.4$ MPG, and $s = 2.3$ MPG;   then $s_{\bar{x}} = s/\sqrt{n} = 2.3/\sqrt{16} = .575$

            $df = n-1 = 16-1 = 15, \quad \alpha/2 = .5-(.99/2) = .005$ and $t = 2.947$

            The 99% confidence interval for μ is: $\bar{x} + ts_{\bar{x}} = 26.4 \pm 2.947(.575) = 26.4 \pm 1.69 = 24.71$ to 28.09 MPG

        b.   The width of the confidence interval obtained in part a can be reduced by:

           1.   Lowering the confidence level

           2.   Increasing the sample size

              The second alternative is better because by lowering the confidence level we will simply lower the probability that our confidence interval includes μ.

8.56   a. $n = 20, \quad \bar{x} = 23$ hours, and $s = 3.75$ hours, so $s_{\bar{x}} = s/\sqrt{n} = 3.75/\sqrt{20} = .83852549$

$df = n-1 = 20-1 = 19, \quad \alpha/2 = 5-(.98/2) = .01 \text{ and } t = 2.539$

The 98% confidence interval for μ is:

$\bar{x} + ts_{\bar{x}} = 23 \pm 2.539(.83852549) = 23 \pm 2.13 = 20.87 \text{ to } 25.13 \text{ hours}$

b. The width of the confidence interval constructed in part a may be reduced by:

1. Lowering the confidence level

2. Increasing the sample size

   The second alternative is better because lowering the confidence level lowers the probability that our confidence interval contains μ.

8.57 $n = 10, \sum x = 742, \sum x^2 = 55,310$

$\bar{x} = \sum x/n = 742/10 = 74.20 \text{ mph}$

$s = \sqrt{\dfrac{\sum x^2 - (\sum x)^2/n}{n-1}} = \sqrt{\dfrac{55,310 - (742)^2/10}{10-1}} = 5.30827446$

$s_{\bar{x}} = s/\sqrt{n} = 5.30827446/\sqrt{10} = 1.67862377 \text{ mph}$

$df = n-1 = 10-1 = 9, \quad \alpha/2 = .5-(.90/2) = .05 \text{ and } t = 1.833$

The 90% confidence interval for μ is:

$\bar{x} \pm ts_{\bar{x}} = 74.20 \pm 1.833(1.67862377) = 74.20 \pm 3.08 = 71.12 \text{ to } 77.28 \text{ mph}$

8.58 $n = 9, \quad \sum x = 72, \quad \sum x^2 = 708, \text{ and } \bar{x} = \sum x/n = 72/9 = 8 \text{ hours; then}$

$s = \sqrt{\dfrac{\sum x^2 - (\sum x)^2/n}{n-1}} = \sqrt{\dfrac{708 - (72)^2/9}{9-1}} = 4.06201920 \text{ hours and}$

$s_{\bar{x}} = s/\sqrt{n} = 4.06201920/\sqrt{9} = 1.35400640 \text{ hours.}$

$df = n-1 = 9-1 = 8, \quad \alpha/2 = .5-(.95/2) = .025 \text{ and } t = 2.306$

The 95% confidence interval for μ is: $\bar{x} \pm ts_{\bar{x}} = 8 \pm 2.306(1.354006401) = 8 \pm 3.12 = 4.88 \text{ to } 11.12 \text{ hours}$

8.59 To estimate the mean time taken by a cashier to serve customers at a supermarket, we will follow the following procedure. (Assume that service times are normally distributed.)

1. Take a random sample of 10 customers served by this cashier

2. Collect the information on serving time for these customers

3. Calculate the sample mean, the standard deviation, and $s_{\bar{x}}$

4. Choose the confidence level and determine the *t* value

5. Make the confidence interval for μ using the *t* distribution

8.60 To estimate the mean waiting time μ (assuming that waiting times are normally distributed):

1. Take a random sample of 15 customers at the bank

2. Using a timing device, record the waiting time for each of the 15 customers

3. Find the mean and the standard deviation of these 15 waiting times

4. Calculate $s_{\bar{x}} = s / \sqrt{15}$, and t for 14 df and the chosen confidence level

5. The confidence interval for μ is $\bar{x} \pm t s_{\bar{x}}$

8.61 The normal distribution will be used to make a confidence interval for the population proportion if the sampling distribution of the sample proportion is (approximately) normal. We know from Chapter 7 (Section 7.7.3) that the sampling distribution of the sample proportion is (approximately) normal if:

$$np > 5 \text{ and } nq > 5$$

Thus, the normal distribution can be used to make a confidence interval for the population proportion if:

$$np > 5 \text{ and } nq > 5$$

8.62 The sample proportion $\hat{p}$ is the point estimator of p. The margin of error for a point estimate of p is $\pm 1.96 s_{\hat{p}}$, where $s_{\hat{p}} = \sqrt{\hat{p}\hat{q}/n}$.

8.63   a. $n = 50$, $\hat{p} = .25$, $\hat{q} = 1 - \hat{p} = 1 - .25 = .75$, $n\hat{p} = 50(.25) = 12.5$, and $n\hat{q} = 50(.75) = 37.50$

           Since $n\hat{p} > 5$ and $n\hat{q} > 5$, the sample size is large enough to use the normal distribution.

      b. $n = 160$, $\hat{p} = .03$, $\hat{q} = 1 - .03 = .97$, $n\hat{p} = 160(.03) = 4.8$, and $n\hat{q} = 160(.97) = 155.2$

           Since $n\hat{p} < 5$, the sample size is not large enough to use the normal distribution.

      c. $n = 400$, $\hat{p} = .65$, $\hat{q} = 1 - \hat{p} = 1 - .65 = .35$, $n\hat{p} = 400(.65) = 260$, and $n\hat{q} = 400(.35) = 140$

           Since $n\hat{p} > 5$ and $n\hat{q} > 5$, the sample size is large enough to use the normal distribution.

      d. $n = 75$, $\hat{p} = .06$, $\hat{q} = 1 - \hat{p} = 1 - .06 = .94$, $n\hat{p} = 75(.06) = 4.5$, and $n\hat{q} = 75(.94) = 70.5$

           Since $n\hat{p} < 5$, the sample size is not large enough to use the normal distribution.

8.64   a. $n = 120$, $\hat{p} = .04$, $\hat{q} = 1 - \hat{p} = 1 - .04 = .96$, $n\hat{p} = 120(.04) = 4.8$, and $n\hat{q} = 120(.96) = 115.2$

           Since $n\hat{p} < 5$, the sample size is not large enough to use the normal distribution.

      b. $n = 60$, $\hat{p} = .08$, $\hat{q} = 1 - \hat{p} = 1 - .08 = .92$, $n\hat{p} = 60(.08) = 4.8$, and $n\hat{q} = 60(.92) = 55.2$

Since $n\hat{p} < 5$, the sample size is not large enough to use the normal distribution.

c. $n = 40$, $\hat{p} = .50$, $\hat{q} = 1 - \hat{p} = 1 - .50 = .50$, $n\hat{p} = 40(.50) = 20$, and $n\hat{q} = 40(.50) = 20$

Since $n\hat{p} > 5$ and $n\hat{q} > 5$, the sample size is large enough to use the normal distribution.

d. $n = 900$, $\hat{p} = .15$, $\hat{q} = 1 - \hat{p} = 1 - .15 = .85$, $n\hat{p} = 900(.15) = 135$, and $n\hat{q} = 900(.85) = 765$

Since $n\hat{p} > 5$ and $n\hat{q} > 5$, the sample size is large enough to use the normal distribution.

8.65   a. $n = 400$, $\hat{p} = .63$, and $\hat{q} = 1 - .63 = .37$; then $s_{\hat{p}} = \sqrt{\hat{p}\hat{q}/n} = \sqrt{.63(.37)/400} = .02414022$

The 95% confidence interval for p is: $\hat{p} \pm zs_{\hat{p}} = .63 \pm 1.96(.02414022) = .63 \pm .047 = .583$ to $.677$

b. $n = 400$, $\hat{p} = .59$, and $\hat{q} = .41$;   then $s_{\hat{p}} = \sqrt{.59(.41)/400} = .02459167$

The 95% confidence interval for p is: $\hat{p} \pm zs_{\hat{p}} = .59 \pm 1.96(.02459167) = .59 \pm .048 = .542$ to $.638$

c. $n = 400$, $\hat{p} = .67$, and $\hat{q} = .33$;   then $s_{\hat{p}} = \sqrt{.67(.33)/400} = .02351064$

The 95% confidence interval for p is: $\hat{p} \pm zs_{\hat{p}} = .67 \pm 1.96(.02351064) = .67 \pm .046 = .624$ to $.716$

d.   The confidence intervals of parts a and c cover p, but the confidence interval of part b does not.

8.66   a. $n = 900$, $\hat{p} = .32$, and $\hat{q} = 1 - \hat{p} = 1 - .32 = .68$;   then $s_{\hat{p}} = \sqrt{\hat{p}\hat{q}/n} = \sqrt{.32(.68)/900} = .01554921$

The 90% confidence interval for p is: $\hat{p} \pm zs_{\hat{p}} = .32 \pm 1.65(.01554921) = .32 \pm .026 = .294$ to $.346$

b. $n = 900$, $\hat{p} = .36$, and $\hat{q} = .64$;   then $s_{\hat{p}} = \sqrt{.36(.64)/900} = .0160$

The 90% confidence interval for p is: $\hat{p} \pm zs_{\hat{p}} = .36 \pm 1.65(.0160) = .36 \pm .026 = .334$ to $.386$

c. $n = 900$, $\hat{p} = .30$, and $\hat{q} = .70$;   then $s_{\hat{p}} = \sqrt{.30(.70)/900} = .01527525$

The 90% confidence interval for p is: $\hat{p} \pm zs_{\hat{p}} = .30 \pm 1.65(.01527525) = .30 \pm .025 = .275$ to $.325$

d.   The confidence intervals of parts a and b cover p, but the confidence interval of part c does not.

8.67 $n = 500$, $\hat{p} = .68$, and $\hat{q} = 1 - \hat{p} = 1 - .68 = .32$; $s_{\hat{p}} = \sqrt{\hat{p}\hat{q}/n} = \sqrt{.68(.32)/500} = .02086145$

a. The 90% confidence interval for p is: $\hat{p} \pm z s_{\hat{p}} = .68 \pm 1.65(.02086145) = .68 \pm .034 = .646$ to $.714$

b. The 95% confidence interval for p is: $\hat{p} \pm z s_{\hat{p}} = .68 \pm 1.96(.02086145) = .68 \pm .041 = .639$ to $.721$

c. The 99% confidence interval for p is: $\hat{p} \pm z s_{\hat{p}} = .68 \pm 2.58(.02086145) = .68 \pm .054 = .626$ to $.734$

d. Yes, the width of the confidence intervals increases as the confidence level increases. This occurs because as the confidence level increases, the value of z increases.

8.68 $n = 200,$ $\hat{p} = .27,$ and $\hat{q} = 1 - \hat{p} = 1 - .27 = .73$; then $s_{\hat{p}} = \sqrt{\hat{p}\hat{q}/n} = \sqrt{.27(.73)/200} = .03139267$

a. The 99% confidence interval for p is: $\hat{p} \pm z s_{\hat{p}} = .27 \pm 2.58(.03139267) = .27 \pm .081 = .189$ to $.351$

b. The 97% confidence interval for p is: $\hat{p} \pm z s_{\hat{p}} = .27 \pm 2.17(.03139267) = .27 \pm .068 = .202$ to $.338$

c. The 90% confidence interval for p is: $\hat{p} \pm z s_{\hat{p}} = .27 \pm 1.65(.03139267) = .27 \pm .052 = .218$ to $.322$

d. Yes, the width of the confidence intervals decreases as the confidence level decreases. This occurs because as the confidence level decreases, the value of z decreases.

8.69 $\hat{p} = .73$, and $\hat{q} = 1 - \hat{p} = 1 - .73 = .27$

a. $n = 100$ and $s_{\hat{p}} = \sqrt{\hat{p}\hat{q}/n} = \sqrt{.73(.27)/100} = .04439595$

The 99% confidence interval for p is: $\hat{p} \pm z s_{\hat{p}} = .73 \pm 2.58(.04439595) = .73 \pm .115 = .615$ to $.845$

b. $n = 600$ and $s_{\hat{p}} = \sqrt{.73(.27)/600} = .01812457$

The 99% confidence interval for p is: $\hat{p} \pm z s_{\hat{p}} = .73 \pm 2.58(.01812457) = .73 \pm .047 = .683$ to $.777$

c. $n = 1500$ and $s_{\hat{p}} = \sqrt{.73(.27)/1500} = .01146298$

The 99% confidence interval for p is: $\hat{p} \pm z s_{\hat{p}} = .73 \pm 2.58(.01146298) = .73 \pm .030 = .700$ to $.760$

d. Yes, the width of the confidence intervals decreases as the sample size increases. This occurs because increasing the sample size decreases the standard deviation of the sample proportion.

8.70 $\hat{p} = .31$ and $\hat{q} = 1 - \hat{p} = 1 - .31 = .69$

a. $n = 1200$ and $s_{\hat{p}} = \sqrt{\hat{p}\hat{q}/n} = \sqrt{.31(.69)/1200} = .01335103$

   The 95% confidence interval for p is: $\hat{p} \pm z s_{\hat{p}} = .31 \pm 1.96(.01335103) = .31 \pm .026 = .284$ to $.336$

b. $n = 500$, $s_{\hat{p}} = \sqrt{\hat{p}\hat{q}/n} = \sqrt{.31(.69)/500} = .02068333$

   The 95% confidence interval for p is: $\hat{p} \pm z s_{\hat{p}} = .31 \pm 1.96(.02068333) = .31 \pm .041 = .269$ to $.351$

c. $n = 80$, $s_{\hat{p}} = \sqrt{\hat{p}\hat{q}/n} = \sqrt{.31(.69)/80} = .05170832$

   The 95% confidence interval for p is: $\hat{p} \pm z s_{\hat{p}} = .31 \pm 1.96(.05170832) = .31 \pm .101 = .209$ to $.411$

d. Yes, the width of the confidence intervals increases as the sample size decreases. This occurs because decreasing the sample size increases the standard deviation of the sample proportion.

8.71 $n = 1004$, $\hat{p} = .36$, and $\hat{q} = 1 - .36 = .64$; ;   then $s_{\hat{p}} = \sqrt{\hat{p}\hat{q}/n} = \sqrt{.36(.64)/1004} = .01514867$

a. Point estimate of $p = \hat{p} = .36$

   Margin of error $= \pm 1.96 s_{\hat{p}} = \pm 1.96(.01514867) = \pm .030$

b. The 99% confidence interval for p is: $\hat{p} \pm z s_{\hat{p}} = .36 \pm 2.58(.01514867) = .36 \pm .039 = .321$ to $.399$

8.72 $n = 617$, $\hat{p} = .25$, and $\hat{q} = 1 - .25 = .75$;   then $s_{\hat{p}} = \sqrt{\hat{p}\hat{q}/n} = \sqrt{.25(.75)/617} = .01742434$

a. Point estimate of $p = \hat{p} = .25$

   Margin of error $= \pm 1.96 s_{\hat{p}} = \pm 1.96(.01743243) = \pm .034$

b. The 95% confidence interval for p is: $\hat{p} \pm z s_{\hat{p}} = .25 \pm 1.96(.01743243) = .25 \pm .034 = .216$ to $.284$

8.73 $n = 611$, $\hat{p} = .36$, and $\hat{q} = 1 - .36 = .64$;   therefore $s_{\hat{p}} = \sqrt{\hat{p}\hat{q}/n} = \sqrt{.36(.64)/611} = .01941872$

a. Point estimate of $p = \hat{p} = .36$

Margin of error $= \pm 1.96 s_{\hat{p}} = \pm 1.96(.01941872) = \pm.038$

b. The 95% confidence interval for p is: $\quad \hat{p} \pm zs_{\hat{p}} = .36 \pm 1.96(.01941872) = .36 \pm .038 = .322$ to $.398$

8.74 $n = 240$, $\quad \hat{p} = 96/240 = .40$, and $\hat{q} = 1 - .40 = .60$; then $\quad s_{\hat{p}} = \sqrt{\hat{p}\hat{q}/n} = \sqrt{.40(.60)/240} = .03162278$

a. Point estimate of $p = \hat{p} = .40$ or 40%

Margin of error $= \pm 1.96 s_{\hat{p}} = \pm 1.96(.03162278) = \pm.062$ or 6.2%

b. The 97% confidence interval for p is:

$\hat{p} \pm zs_{\hat{p}} = .40 \pm 2.17(.03162278) = .40 \pm .069 = .331$ to $.469$ or 33.1% to 46.9%.

8.75 $n = 50$, $\quad \hat{p} = 35/50 = .70$, and $\hat{q} = 1 - .70 = .30$; then $\quad\quad s_{\hat{p}} = \sqrt{\hat{p}\hat{q}/n} = \sqrt{.70(.30)/50} = .06480741$

a. The 98% confidence interval for p is:

$\hat{p} \pm zs_{\hat{p}} = .70 \pm 2.33(.06480741) = .70 \pm .151 = .549$ to $.851$ or 54.9% to 85.1%.

b. The width of the confidence interval constructed in part a may be reduced by:

1. Lowering the confidence level

2. Increasing the sample size

The second alternative is better because lowering the confidence level lowers the probability that the confidence interval contains p.

8.76   a. $n = 50$, $\quad \hat{p} = .38$, and $\hat{q} = 1 - .38 = .62$; then $\quad\quad s_{\hat{p}} = \sqrt{\hat{p}\hat{q}/n} = \sqrt{.38(.62)/50} = .06864401$

The 99% confidence interval for p is:

$\hat{p} \pm zs_{\hat{p}} = .38 \pm 2.58(.06864401) = .38 \pm .177 = .203$ to $.557$ or 20.3% to 55.7%.

b. The width of the confidence interval constructed in part a may be reduced by:

1. Lowering the confidence level

2. Increasing the sample size

The second alternative is better because lowering the confidence level lowers the probability that the confidence interval contains p.

8.77 $n = 1200$, $\hat{p} = .17$, and $\hat{q} = 1 - .17 = .83$; therefore $s_{\hat{p}} = \sqrt{\hat{p}\hat{q}/n} = \sqrt{.17(.83)/1200} = .01084358$

    a. The 99% confidence interval for p is:

       $\hat{p} \pm zs_{\hat{p}} = .17 \pm 2.58(.01084358) = .17 \pm .028 = .142$ to $.198$ or 14.2% to 19.8%.

    b. The sample proportion of .17 is an estimate of p based on a random sample. Because of sampling error, this estimate might differ from the true proportion, p, so we make an interval estimate to allow for this uncertainty and sampling error.

8.78 $n = 400$, $\hat{p} = .64$, and $\hat{q} = 1 - .64 = .36$;   then $s_{\hat{p}} = \sqrt{\hat{p}\hat{q}/n} = \sqrt{.64(.36)/400} = .024$

    a. The 98% confidence interval for p is: $\hat{p} \pm zs_{\hat{p}} = .64 \pm 2.33(.024) = .64 \pm .056 = .584$ to $.696$

    b. The sample proportion of .64 is an estimate of p based on a random sample. Because of sampling error, this estimate might differ from the true proportion p, so we make an interval estimate to allow for this uncertainty and sampling error.

8.79   Nine of the 15 judges in the sample are in favor of the death penalty. Hence,

    $n = 15$, $\hat{p} = 9/15 = .60$, and $\hat{q} = 1 - .60 = .40$. Therefore $s_{\hat{p}} = \sqrt{\hat{p}\hat{q}/n} = \sqrt{.60(.40)/15} = .12649111$

    a. Point estimate of $p = \hat{p} = .60$

       Margin of error $= \pm 1.96 s_{\hat{p}} = \pm 1.96(.12649111) = \pm .248$

    b. The 95% confidence interval for p is:

       $\hat{p} \pm zs_{\hat{p}} = .60 \pm 1.96(.12649111) = .60 \pm .248 = .352$ to $.848$ or 35.2% to 84.8%

8.80   Eight of the twenty–four policy holders in the sample have tried alternative treatments. Hence,

    $n = 24$, $\hat{p} = 8/24 = .333$, and $\hat{q} = 1 - .333 = .667$. Therefore $s_{\hat{p}} = \sqrt{\hat{p}\hat{q}/n} = \sqrt{.333(.667)/24} = .09620096$

    a. Point estimate of $p = \hat{p} = .333$

       Margin of error $= \pm 1.96 z_{\hat{p}} = \pm 1.96(.09620096) = \pm .189$

b. The 99% confidence interval for p is:

$\hat{p} \pm zs_{\hat{p}} = .333 \pm 2.58(.09620096) = .333 \pm .248 = .085$ to $.581$ or 8.5% to 58.1%

8.81 To estimate the proportion of students who hold off campus jobs:

1. Take a random sample of 40 students at your college.

2. Determine the number of students in this sample who hold off campus jobs

3. Calculate $\hat{p}$ and $s_{\hat{p}}$

4. Choose the confidence level and find the required value of z from the normal distribution table

5. Obtain the confidence interval for p using the formula $\hat{p} \pm zs_{\hat{p}}$

8.82 To estimate the percentage of students who are satisfied with campus food services:

1. Take a random sample of 30 students who use campus food services.

2. Determine the number of students in this sample who are satisfied with the campus food services

3. Calculate $\hat{p}$ and $s_{\hat{p}}$

4. Choose the confidence level and find the required value of z from the normal distribution table

5. Obtain the confidence interval for p using the formula $\hat{p} \pm zs_{\hat{p}}$

To obtain a confidence interval for the percentage, multiply the upper and lower limits of the confidence interval for p by 100.

8.83 a. $\sigma = 12.5$, $E = 2.50$, and $z = 2.58$ for 99% confidence level

$$n = \frac{z^2 \sigma^2}{E^2} = \frac{(2.58)^2 (12.5)^2}{(2.50)^2} = 166.41 \approx 167$$

b. $\sigma = 12.5$, $E = 3.20$, and $z = 2.05$ for 96% confidence level

$$n = \frac{z^2 \sigma^2}{E^2} = \frac{(2.05)^2 (12.5)^2}{(3.20)^2} = 64.13 \approx 65$$

8.84 a. $\sigma = 14.50$, $E = 5.50$, and $z = 2.33$ for 98% confidence level

$$n = \frac{z^2 \sigma^2}{E^2} = \frac{(2.33)^2 (14.50)^2}{(5.50)^2} = 37.73 \approx 38$$

b. $\sigma = 14.50$ $E = 4.25$, and $z = 1.96$ for 95% confidence level

$$n = \frac{z^2 \sigma^2}{E^2} = \frac{(1.96)^2 (14.50)^2}{(4.25)^2} = 44.72 \approx 45$$

8.85 a. $E = 2.3$, $\sigma = 15.40$, and $z = 2.58$ for 99% confidence level

$$n = \frac{z^2\sigma^2}{E^2} = \frac{(2.58)^2(15.40)^2}{(2.3)^2} = 298.42 \approx 299$$

b. $E = 4.1$, $\sigma = 23.45$, and $z = 1.96$ for 95% confidence level

$$n = \frac{z^2\sigma^2}{E^2} = \frac{(1.96)^2(23.45)^2}{(4.1)^2} = 125.67 \approx 126$$

c. $E = 25.9$, $\sigma = 122.25$, and $z = 1.65$ for 90% confidence level

$$n = \frac{z^2\sigma^2}{E^2} = \frac{(1.65)^2(122.25)^2}{(25.9)^2} = 60.65 \approx 61$$

8.86   a. $E = .17, \sigma = .90$, and $z = 2.58$ for 99% confidence level

$$n = \frac{z^2\sigma^2}{E^2} = \frac{(2.58)^2(.90)^2}{(.17)^2} = 186.56 \approx 187$$

b. $E = 1.45$, $\sigma = 5.82$, and $z = 1.96$ for 95% confidence level

$$n = \frac{z^2\sigma^2}{E^2} = \frac{(1.96)^2(5.82)^2}{(1.45)^2} = 61.89 \approx 62$$

c. $E = 5.65$, $\sigma = 18.20$, and $z = 1.65$ for 90% confidence level

$$n = \frac{z^2\sigma^2}{E^2} = \frac{(1.65)^2(18.20)^2}{(5.65)^2} = 28.25 \approx 29$$

8.87 $\sigma = \$11$, $E = \$2$, and $z = 1.96$ for 95% confidence level

$$n = z^2\sigma^2 / E^2 = (1.96)^2(11)^2 / (2)^2 = 116.21 \approx 117$$

8.88 $\sigma = .20$ ounce, $E = .04$ ounce, and $z = 2.58$ for 99% confidence level

$$n = \frac{z^2\sigma^2}{E^2} = \frac{(2.58)^2(.20)^2}{(.04)^2} = 166.41 \approx 167$$

8.89 $\sigma = \$31$, $E = \$3$, and $z = 1.65$ for 90% confidence level $n = \frac{z^2\sigma^2}{E^2} = \frac{(1.65)^2(31)^2}{(3)^2} = 290.70 \approx 291$

8.90 $\sigma = 2.5$ hours, $E = .75$ hours, and $z = 2.33$ for 98% confidence level

$$n = \frac{z^2 \sigma^2}{E^2} = \frac{(2.33)^2 (2.5)^2}{(.75)^2} = 60.32 \approx 61$$

8.91   a. $E = .035,$ $\hat{p} = .29,$ $\hat{q} = 1 - .29 = .71,$ and $z = 2.58$ for 99% confidence level

$$n = \frac{z^2 \hat{p}\hat{q}}{E^2} = \frac{(2.58)^2 (.29)(.71)}{(.035)^2} = 1118.82 \approx 1119$$

b. $E = .035,$ $z = 2.58,$ and $p = q = .50,$ for most conservative sample size

$$n = \frac{z^2 pq}{E^2} = \frac{(2.58)^2 (.50)(.50)}{(.035)^2} = 1358.45 \approx 1359$$

8.92   a. $E = .045,$ $\hat{p} = .53,$ $\hat{q} = 1 - .53 = .47,$ and $z = 2.33$ for 98% confidence level

$$n = \frac{z^2 \hat{p}\hat{q}}{E^2} = \frac{(2.33)^2 (.53)(.47)}{(.045)^2} = 667.82 \approx 668$$

b. $E = .045,$ $z = 2.33,$ and $p = q = .50$ for most conservative sample size

$$n = \frac{z^2 pq}{E^2} = \frac{(2.33)^2 (.50)(.50)}{(.045)^2} = 670.23 \approx 671$$

8.93   a. $E = .03,$ $z = 2.58,$ and $p = q = .50,$ for most conservative sample size

$$n = \frac{z^2 pq}{E^2} = \frac{(2.58)^2 (.50)(.50)}{(.03)^2} = 1849$$

b. $E = .04, z = 1.96,$ $p = q = .50$ for most conservative sample size

$$n = \frac{z^2 pq}{E^2} = \frac{(1.96)^2 (.50)(.50)}{(.04)^2} = 600.25 \approx 601$$

c. $E = .01,$ $z = 1.65,$ $p = q = .50,$ for most conservative sample size

$$n = \frac{z^2 pq}{E^2} = \frac{(1.65)^2 (.50)(.50)}{(.01)^2} = 6806.25 \approx 6807$$

8.94   a. $E = .03,$ $\hat{p} = .32,$ $\hat{q} = 1 - \hat{p} = 1 - .32 = .68,$ and $z = 2.58$

$$n = \frac{z^2 \hat{p}\hat{q}}{E^2} = \frac{(2.58)^2(.32)(.68)}{(.03)^2} = 1609.37 \approx 1610$$

b. $E = .04,\quad \hat{p} = .78,\ \hat{q} = 1 - \hat{p} = 1 - .78 = .22,\ \text{and}\ z = 1.96$

$$n = \frac{z^2 \hat{p}\hat{q}}{E^2} = \frac{(1.96)^2(.78)(.22)}{(.04)^2} = 412.01 \approx 413$$

c. $E = .02,\ \hat{p} = .64,\ \hat{q} = 1 - \hat{p} = 1 - .64 = .36,\ \text{and}\ z = 1.65$

$$n = \frac{z^2 \hat{p}\hat{q}}{E^2} = \frac{(1.65)^2(.64)(.36)}{(.02)^2} = 1568.16 \approx 1569$$

8.95 $E = .02,\quad z = 2.58,\quad p = q = .50$

$n = z^2 pq / E^2 = (2.58)^2(.50)(.50) / (.02)^2 = 4160.25 \approx 4161$

8.96 $E = .02,\quad \hat{p} = .93,\quad \hat{q} = 1 - .93 = .07,\ \text{and}\ z = 2.58$

$$n = z^2 \hat{p}\hat{q} / E^2 = \frac{(2.58)^2(.93)(.07)}{(.02)^2} = 1083.33 \approx 1084$$

8.97 $E = .03,\quad \hat{p} = .76,\quad \hat{q} = 1 - .76 = .24,\ \text{and}\ z = 2.58$

$$n = z^2 \hat{p}\hat{q} / E^2 = \frac{(2.58)^2(.76)(.24)}{(.03)^2} = 1349.03 \approx 1350$$

8.98 $E = .03,\quad z = 2.58,\quad p = q = .50$

$$n = z^2 pq / E^2 = \frac{(2.58)^2(.50)(.50)}{(.03)^2} = 1849$$

8.99 $n = 100,\quad \bar{x} = \$273,,\ \text{and}\ s = \$60;$ therefore $s_{\bar{x}} = s / \sqrt{n} = 60 / \sqrt{100} = 6$

a. Point estimate of $\mu = \bar{x} = \$273$

Margin of error $= \pm 1.96 s_{\bar{x}} = \pm 1.96(6) = \pm\$11.76$

b. The 95 % confidence interval for μ is: $\bar{x} \pm z s_{\bar{x}} = 273 \pm 1.96(6) = 273 \pm 11.76 = \$261.24$ to $\$284.76$

8.100 $n = 100,$ $\bar{x} = \$2640,$ and $s = \$578;$ then $s_{\bar{x}} = s / \sqrt{n} = 578 / \sqrt{100} = \$57.80$

      a.   Point estimate of $\mu = \bar{x} = \$2640$

           Margin of error $= \pm 1.96 s_{\bar{x}} = \pm 1.96 (57.80) = \pm \$113.29$

      b.   The 97% confidence interval for μ is:

      $\bar{x} \pm z s_{\bar{x}} = 2640 \pm 2.17 (57.80) = 2640 \pm 125.43 = \$2514.57$ to $\$2765.43$

8.101 $n = 36,$ $\bar{x} = 24.015$ inches, and $\sigma = .06$ inches; so $\sigma_{\bar{x}} = \sigma / \sqrt{n} = .06 / \sqrt{36} = .01$

      The 99% confidence interval for μ is:

      $\bar{x} \pm z \sigma_{\bar{x}} = 24.015 \pm 2.58 (.01) = 24.015 \pm .026 = 23.989$ to 24.041 inches.

      Since the upper limit of the confidence interval is 24.041, which is greater than 24.025, the machine needs an adjustment.

8.102 $n = 50,$ $\bar{x} = 3.99$ inches, and $\sigma = .04$ inches; so $\sigma_{\bar{x}} = \sigma / \sqrt{n} = .04 / \sqrt{50} = .00565685$

      The 98% confidence interval for μ is:

      $\bar{x} \pm z \sigma_{\bar{x}} = 3.99 \pm 2.33 (.00565685) = 3.99 \pm .013 = 3.977$ to 4.003 inches.

      Since the lower limit of the confidence interval is less than 3.98 inches, the machine needs an adjustment.

8.103 $n = 32,$ $\sum x = 1779,$ $\sum x^2 = 142,545$

      $\bar{x} = \sum x / n = 1779 / 32 = 55.59$ minutes

      $s = \sqrt{\dfrac{\sum x^2 - \left(\sum x\right)^2 / n}{n-1}} = \sqrt{\dfrac{142,545 - (1779)^2 / 32}{32-1}} = 37.52148578$ minutes

      $s_{\bar{x}} = s / \sqrt{n} = 37.52148578 / \sqrt{32} = 6.63292426$ minutes

      a.   Point estimate of $\mu = \bar{x} = 55.59$ minutes

           Margin of error $= \pm 1.96 s_{\bar{x}} = \pm 1.96 (6.63292426) = \pm 13.00$ minutes

      b.   The 98% confidence interval for μ is:

           $\bar{x} \pm z s_{\bar{x}} = 55.59 \pm 2.33 (6.632924) = 55.59 \pm 15.45 = 40.14$ to 71.04 minutes

8.104 $n = 36,$ $\sum x = 547,$ $\sum x^2 = 9265$

      $\bar{x} = \sum x / n = 547 / 36 = 15.19$ hours

$$s = \sqrt{\frac{\sum x^2 - (\sum x)^2 / n}{n-1}} = \sqrt{\frac{9265 - (547)^2 / 36}{36-1}} = 5.21984917$$

$$s_{\bar{x}} = s / \sqrt{n} = 5.21984917 / \sqrt{36} = .86997486 \text{ hour}$$

    a. Point estimate of $\mu = \bar{x} = 15.19$ hours

       Margin of error $= \pm 1.96 s_{\bar{x}} = \pm 1.96(.86997486) = \pm 1.71$ hours

    b. The 99% confidence interval for μ is:

       $\bar{x} \pm z s_{\bar{x}} = 15.19 \pm 2.58(.86997486) = 15.19 \pm 2.24 = 12.95$ to 17.43 hours

8.105 $n = 25$, $\bar{x} = \$685$, and $s = \$74$; therefore $s_{\bar{x}} = s / \sqrt{n} = 74 / \sqrt{25} = \$14.80$

       $df = n-1 = 25-1 = 24$, $\alpha/2 = .5 - (.99/2) = .005$ and $t = 2.797$

       The 99% confidence interval for μ is: $\bar{x} \pm t s_{\bar{x}} = 685 \pm 2.797(14.80) = 685 \pm 41.40 = \$643.60$ to $\$726.40$.

8.106 $n = 18$, $\bar{x} = 24$ minutes, and $s = 4.5$ minutes; so $s_{\bar{x}} = s / \sqrt{n} = 4.5 / \sqrt{18} = 1.06066017$ minutes

       $df = n-1 = 18-1 = 17$, $\alpha/2 = .5 - (.95/2) = .025$ and $t = 2.110$

       The 95% confidence interval for μ is:

       $\bar{x} \pm t s_{\bar{x}} = 24 \pm 2.110(1.06066017) = 24 \pm 2.24 = 21.76$ to 26.24 minutes

8.107 $n = 20$, $\bar{x} = 9.75$ hours, and $s = 2.2$ hours; so $s_{\bar{x}} = s / \sqrt{n} = 2.2 / \sqrt{20} = .49193496$ hour

       $df = n-1 = 20-1 = 19$, $\alpha/2 = .5 - (.90/2) = .05$ and $t = 1.729$

       The 90% confidence interval for μ is:

       $\bar{x} \pm t s_{\bar{x}} = 9.75 \pm 1.729(.49193496) = 9.75 \pm .85 = 8.90$ to 10.60 hours

8.108 $n = 20$, $\bar{x} = 4.5$ hours, and $s = .75$ hours; so $s_{\bar{x}} = s / \sqrt{n} = .75 / \sqrt{20} = .16770510$ hour

       $df = n-1 = 20-1 = 19$, $\alpha/2 = .5 - (.98/2) = .01$ and $t = 2.539$

       The 98% confidence interval for μ is:

       $\bar{x} \pm t s_{\bar{x}} = 4.5 \pm 2.539(.16770510) = 4.5 \pm .43 = 4.07$ to 4.93 hours.

8.109 $n = 12$, $\sum x = 26.80$, and $\sum x^2 = 62.395$. This means $\bar{x} = \sum x / n = 26.80/12 = 2.23$ hours,

$$s = \sqrt{\frac{\sum x^2 - (\sum x)^2 / n}{n-1}} = \sqrt{\frac{62.395 - (26.80)^2 / 12}{12-1}} = .48068764 \text{ hours, and}$$

$s_{\bar{x}} = s / \sqrt{n} = .48068764 / \sqrt{12} = .13876257$ hours

$df = n - 1 = 12 - 1 = 11, \quad \alpha / 2 = .5 - (.95 / 2) = .025$ and $t = 2.201$

The 95% confidence interval for μ is:

$\bar{x} \pm t s_{\bar{x}} = 2.23 \pm 2.201(.13876257) = 2.23 \pm .31 = 1.92$ to 2.54 hours.

8.110 $n = 10, \sum x = 1514$, and $\sum x^2 = 229{,}646$; Then $\quad \bar{x} = \sum x / n = 1514 / 10 = 151.40$ calories

$s = \sqrt{\dfrac{\sum x^2 - (\sum x)^2 / n}{n - 1}} = \sqrt{\dfrac{229{,}646 - (1514)^2 / 10}{10 - 1}} = 6.88315173$ calories, and

$s_{\bar{x}} = s / \sqrt{n} = 6.88315173 / \sqrt{10} = 2.17664369$ calories

$df = n - 1 = 10 - 1 = 9, \quad \alpha / 2 = .5 - (.99 / 2) = .005$ and $t = 3.250$

The 99% confidence interval for μ is:

$\bar{x} \pm t s_{\bar{x}} = 151.40 \pm 3.250(2.17664369) = 151.40 \pm 7.07 = 144.33$ to 158.47 calories

8.111 $n = 50, \quad \hat{p} = .12$, and $\hat{q} = 1 - .12 = .88$; therefore $\quad s_{\hat{p}} = \sqrt{\hat{p}\hat{q} / n} = \sqrt{(.12)(.88) / 50} = .04595650$

a. Point estimate for $p = \hat{p} = .12$ or 12%

   Margin of error $= \pm 1.96 s_{\hat{p}} = \pm 1.96(.04595650) = \pm .090$ or $\pm 9.0\%$

b. The 99% confidence interval for p is:

   $\hat{p} \pm z s_{\hat{p}} = .12 \pm 2.58(.04595650) = .12 \pm .119 = .001$ to .239 or .1% to 23.9%

8.112 $n = 500, \quad \hat{p} = .44$, and $\hat{q} = 1 - .44 = .56$; therefore $\quad s_{\hat{p}} = \sqrt{\hat{p}\hat{q} / n} = \sqrt{(.44)(.56) / 500} = .02219910$

a. Point estimate of $p = \hat{p} = .44$ or 44%

   Margin of error $= \pm 1.96 s_{\hat{p}} = \pm 1.96(.02219910) = \pm .044$ or 4.4%

b. The 90% confidence interval for p is:

   $\hat{p} \pm z s_{\hat{p}} = .44 \pm 1.65(.02219910) = .44 \pm .037 = .403$ to .477 or 40.3% to 47.7%

8.113 $n = 20, \quad \hat{p} = 8 / 20 = .40$, and $\hat{q} = 1 - .40 = .60$; hence, $s_{\hat{p}} = \sqrt{\hat{p}\hat{q} / n} = \sqrt{(.40)(.60) / 20} = .10954451$

The 99% confidence interval for p is:

$\hat{p} \pm z s_{\hat{p}} = .40 \pm 2.58(.10954451) = .40 \pm .283 = .117$ to .683 or 11.7% to 68.3%

8.114 $n = 16$, $\hat{p} = 5/16 = .313$, , and $\hat{q} = 1 - .313 = .687$; hence $s_{\hat{p}} = \sqrt{\hat{p}\hat{q}/n} = \sqrt{(.313)(.687)/16} = .11592859$

The 97% confidence interval for p is:

$\hat{p} \pm z s_{\hat{p}} = .313 \pm 2.17(.11592859) = .313 \pm .252 = .061$ to $.565$ or 6.1% to 56.5%

8.115 $\sigma = 3$ hours, $E = 1.2$ hours, and $z = 2.58$; hence $n = \dfrac{z^2 \sigma^2}{E^2} = \dfrac{(2.58)^2 (3)^2}{(1.2)^2} = 41.60 \approx 42$

8.116 $\sigma = \$31,500$, $E = \$3,500$, and $z = 1.65$; therefore $n = \dfrac{z^2 \sigma^2}{E^2} = \dfrac{(1.65)^2 (31,500)^2}{(3500)^2} = 220.52 \approx 221$

8.117 $E = .05$, $p = q = .50$, and $z = 1.96$; hence $n = \dfrac{z^2 \, pq}{E^2} = \dfrac{(1.96)^2 (.50)(.50)}{(.05)^2} = 384.16 \approx 385$

8.118 $E = .05$, $\hat{p} = .63$, $\hat{q} = 1 - .63 = .37$, and $z = 1.96$; then $n = \dfrac{z^2 \hat{p}\hat{q}}{E^2} = \dfrac{(1.96)^2 (.63)(.373)}{(.05)^2} = 358.19 \approx 359$

8.119 $n > 30$; 95% confidence interval: \$8.46 to \$9.86

95% confidence interval for a large sample: $\bar{x} \pm 1.96 s_{\bar{x}}$

   a. $\bar{x} = \dfrac{\$9.86 + \$8.46}{2} = \$9.16$

   b. $\bar{x} = 1.96$, $s_{\bar{x}} = 9.86$, hence $s_{\bar{x}} = \dfrac{9.86 - \bar{x}}{1.96} = \dfrac{9.86 - 9.16}{1.96} = .35714286$

Confidence level = 99%, i.e. $z = 2.58$ so the 99% confidence interval is $\bar{x} \pm 2.58 s_{\bar{x}} = \$8.24$ to \$10.08

8.120   a.   Let $x =$ time devoted to commercials per hour on Channel 66.

A sample of 30 twenty – minute time intervals yielded a mean of 4.68 minutes of commercials with a standard deviation of 1.30 minutes. This is equivalent to:

$\bar{x} = 3(4.68) = 14.04$ minutes and $s = 3(1.30) = 3.90$ minutes for a 60–minute time interval.

$s_{\bar{x}} = s/\sqrt{n} = 3.90/\sqrt{30} = .71203932$ minute.

The 90% confidence interval for μ is:

$\bar{x} \pm z s_{\bar{x}} = 14.04 \pm 1.65(.71203932) = 14.04 \pm 1.17 = 12.87$ to 15.21 minutes per hour

b. $n = 30$, $\hat{p} = 7/30$

The point estimate of μ is $(7/30) \times 60$ minutes per hour $= 14$ minutes per hour

c. Both estimates yield approximately 14 minutes per hour.

d. $s_{\hat{p}} = \sqrt{\hat{p}\hat{q}/n} = \sqrt{\left(\dfrac{7}{30}\right)\left(\dfrac{23}{30}\right)/30} = .07722022$ and $\hat{p} = 7/30 = .233$

So the 90% confidence interval for $\hat{p}$ is $\hat{p} = \pm 1.65 s_{\hat{p}} = .233 \pm 1.65(.07722022) = .106$ to $.360$. Hence

the 90% confidence interval for $\mu = p \times 60$ minutes per hour $= 6.36$ to 21.60 minutes per hour.

e. Maximum error of Waldo's estimate:

$z \times s_{\hat{p}} \times 60$ minutes per hour $= 1.65 \times (.07722022) \times (60)$ minutes per hour $= 7.64$ minutes per hour

Maximum error of class estimate: $z \times s_{\bar{x}} = 1.65 \times (.71203932) = 1.17$ minutes per hour

f. We need a new $s_{\hat{p}}$ such that $s_{\bar{x}} = s_{\hat{p}} \cdot 60$ minutes per hour; hence,

$.71203932 = \sqrt{\dfrac{(7/30) \cdot (23/30)}{n}} \cdot 60$, so $n = \sqrt{\dfrac{(7/30) \cdot (23/30)}{(.71203932/60)^2}} = 1270.2 \approx 1271$.

It does not seem to be feasible to achieve the same accuracy as that of the class using Waldo's method.

8.121 Let: $p_1 =$ proportion of 12–18 year old females who expect a female president within 10 years

$p_2 =$ proportion of 12–18 year old females who expect a female president within 15 years

$p_3 =$ proportion of 12–18 year old females who expect a female president within 20 years

$p_4 =$ proportion of 12–18 year old females who do not expect a female president not within their
lifetime

$n = 1250$, $\hat{p}_1 = .40$, $\hat{p}_2 = .25$, $\hat{p}_3 = .21$, $\hat{p}_4 = .14$

$n\hat{p}$ and $n\hat{q}$ exceed 5 for all these proportions, so the sample is considered large.

$s_{\hat{p}_1} = \sqrt{\hat{p}_1 \hat{q}_1 / n} = \sqrt{.40(.60)/1250} = .01385641$

The 95% confidence interval for p_1 is:

$\hat{p}_1 \pm z s_{\hat{p}_1} = .40 \pm 1.96(..01385641) = .40 \pm .027 = .373$ to $.427$ or 37.3% to 42.7%

$s_{\hat{p}_2} = \sqrt{\hat{p}_2 \hat{q}_2 / n} = \sqrt{.25(.75)/1250} = .01224745$

The 95% confidence interval for p_2 is:

$\hat{p}_2 \pm z s_{\hat{p}_2} = .25 \pm 1.96(.01224745) = .25 \pm .024 = .226$ to $.274$ or 22.6% to 27.4%

$$s_{\hat{p}_3} = \sqrt{\hat{p}_3 \hat{q}_3 / n} = \sqrt{.21(.79)/1250} = .01152042$$

The 95% confidence interval for p_3 is:

$$\hat{p}_3 \pm zs_{\hat{p}_3} = .21 \pm 1.96(.01152042) = .21 \pm .023 = .187 \text{ to } .233 \text{ or } 18.7\% \text{ to } 23.3\%$$

$$s_{\hat{p}_4} = \sqrt{\hat{p}_4 \hat{q}_4 / n} = \sqrt{.14(.86)/1250} = .00981428$$

The 95% confidence interval for p_4 is:

$$\hat{p}_4 \pm zs_{\hat{p}_4} = .14 \pm 1.96(.00981428) = .14 \pm .019 = .121 \text{ to } .159 \text{ or } 12.1\% \text{ to } 15.9\%$$

A confidence interval is a range of numbers (in this particular case proportions or percentages) which give an estimate for the true value (i.e. proportion of 12–18 year old females who feel this way). The 95% means that we are 95% confident that this interval actually contains the true value. A single percentage that we assign as an estimate would almost always differ from the true value, hence a range with the associated confidence level is more informative. We assume that the 1200 people are a random sample of 12–18 year old females.

8.122 $n = 100$, and $\hat{p} = 10/100 = .10$

a. $s_{\hat{p}} = \sqrt{\dfrac{.10(.90)}{100}} = .030$

The 95% confidence interval for p is:

$$\hat{p} \pm 1.96 s_{\hat{p}} = .10 \pm 1.96(.030) = .10 \pm .059 = .041 \text{ to } .159 \text{ or } 4.1\% \text{ to } 15.9\%$$

b.   No, 18% does not lie within the confidence interval in part a. This suggests that the vaccine is effective to some degree.

c.   There are several things that may have distorted the outcome of this experiment: we only point out two. It is not mentioned, although it is reasonable to assume that all the dogs that were vaccinated did not have Lyme disease to begin with. Assuming this, the vaccinated dogs were exposed only 1 year to possible tick bites, whereas the other dogs in the area probably were exposed much longer. Also, the owners who agreed to participate in this experiment may be more concerned about their dogs' health and restrict the area that the dogs can access, thereby decreasing the exposure to the ticks as well.

8.123 $s = 4.1$ miles, $E = 1.0$ and $z = 1.96$.

σ may be estimated by s.   Hence, $n = z^2 s^2 / E^2 = (1.96)^2 (4.1)^2 /(1.0)^2 = 64.58 \approx 65$

Thus, an additional 65–20=45 observations must be taken.

8.124 The major problem with this procedure is that the sample is not drawn from the whole target population. This introduces nonsampling error.

    i. Households with no cars would be excluded from this sample. Furthermore, households with more than 2 cars could be counted more than once in the sample. Both of these problems would result in an upwardly biased estimate of p.

    ii. Because of the characteristics of this particular gas station, the clientele may not be typical of all gas station clientele (with respect to multiple car ownership). Other problems may be present, such as the next 200 gasoline customers may not be typical of all customers, or there may be dependency between gasoline customers from the same household. If the attendant took a random sample of 200 households, and determined how many of these households owned more than 2 cars, the sampling error would still be present, but could be reduced by taking a larger sample.

8.125 a. Here, $\sigma = 170$, $E = 100$, and $1 - \alpha = .99$, and $z = 2.58$

Thus, the required sample size is $n = \dfrac{z^2 \sigma^2}{E^2} = \dfrac{(2.58)^2 (170)^2}{(100)^2} = 19.24$ or 20 days

Note that since $n < 30$, we must assume that the number of cars passing each day is approximately normally distributed. Or we may take a large $(n \geq 30)$ sample.

b. Since $n = z^2 \sigma^2 / E^2$, then, $z = E \sqrt{n} / \sigma = 100 \sqrt{20} / 272 = 1.64$ which corresponds to a confidence level of approximately 90%.

c. Since, $n = \dfrac{z^2 \sigma^2}{E^2}$, then, $E = \dfrac{z\sigma}{\sqrt{n}} = \dfrac{2.58(130)}{\sqrt{20}} = 75.00$

Thus, they can be 99% confident that their point estimate is within 75 cars of the true average.

8.126 No, the student's analysis does not make sense. The relevant parameter, p, is the proportion of all U.S. senators in favor of the bill. The value, .55, is not a sample proportion: instead it is the population proportion, p. Since, $p = .55$ is known there is no need to estimate it.

Self – Review Test for Chapter Eight

1.   a. Estimation means assigning values to a *population parameter* based on the value of a *sample statistic*.

    b. An estimator is the *sample statistic* used to estimate a *population parameter*.

    c. The value of a *sample statistic* is called the point estimate of the corresponding *population parameter*.

2. b   **3.** a   **4.** a   **5.** d   **6.** b

7. $n = 36$, $\bar{x} = \$159,000$, and $s = \$27,000$; hence $s_{\bar{x}} = s/\sqrt{n} = 27,000/\sqrt{36} = \$4500$

     a. Point estimate of $\mu = \bar{x} = \$159,000$

     Margin of error $= \pm1.96 s_{\bar{x}} = \pm1.96(4500) = \pm\$8820$

     b. The 99% confidence interval for μ is:

        $\bar{x} \pm z s_{\bar{x}} = 159,000 \pm 2.58(4500) = 159,000 \pm 11,610 = \$147,390$ to $\$170,610$

8. $n = 25$, $\bar{x} = \$410,425$, and $s = \$74,820$

     $df = n - 1 = 25 - 1 = 24$, $\quad \alpha/2 = .5 - (.95/2) = .025$ and $t = 2.064$

     $s_{\bar{x}} = s/\sqrt{n} = 74,820/\sqrt{25} = \$14,964$

     The 95% confidence interval for μ is:

     $\bar{x} \pm t s_{\bar{x}} = 410,425 \pm 2.064(14,964) = 410,425 \pm 30,885.70 = \$379,539.30$ to $\$441,310.70$

9. $n = 612$, $\hat{p} = .41$, and $\hat{q} = 1 - .41 = .59$,   so $s_{\hat{p}} = \sqrt{\hat{p}\hat{q}/n} = \sqrt{(.41)(.59)/612} = .01988118$

     a. Point estimate of $p = \hat{p} = .41$

     Margin of error $= \pm1.96 s_{\hat{p}} = \pm1.96(.01988118) = \pm.039$

     b. The 95% confidence interval for p is: $\hat{p} \pm z s_{\hat{p}} = .41 \pm 1.96(.01988118) = .41 \pm .039 = .371$ to $.449$

10. $E = .65$ houses, $\sigma = 2.2$ houses, and $z = 1.96$, then $n = \dfrac{z^2 \sigma^2}{E^2} = \dfrac{(1.96)^2 (2.2)^2}{(.65)^2} = 44.01 \approx 45.$

11. $E = .05$, $\quad z = 1.65$, $\quad p = q = .50$, then $n = \dfrac{z^2 pq}{E^2} = \dfrac{(1.65)^2 (.50)(.50)}{(.05)^2} = 272.25 \approx 273.$

12. $E = .05$, $\quad \hat{p} = .70$, $\hat{q} = 1 - \hat{p} = 1 - .70 = .30$, and $z = 1.65$; so $n = \dfrac{z^2 \hat{p}\hat{q}}{E^2} = \dfrac{(1.65)^2 (.70)(.30)}{(.05)^2} = 228.69 \approx 229.$

13. The width of the confidence interval can be reduced by:

     1. Lowering the confidence level

2. Increasing the sample size

   The second alternative is better because by lowering the confidence level we will simply lower the probability that our confidence interval includes μ.

14. To estimate the mean number of hours that all students at your college work per week:

1. Take a random sample of 12 students from your college who hold jobs

2. Record the number of hours each of these students worked last week

3. Calculate $\bar{x}$, s, and $s_{\bar{x}}$ from these data

4. After choosing the confidence level, find the value for the t distribution with $11 df$ and for an area of $\alpha/2$ in the right tail.

5. Obtain the confidence interval for μ by using the formula $\bar{x} \pm t s_{\bar{x}}$

   You are assuming that the hours worked by all students at your college have a normal distribution.

15. To estimate the proportion of people who are happy with their current jobs:

1. Take a random sample of 35 workers

2. Determine whether or not each worker is happy with his or her job

3. Calculate $\hat{p}, \hat{q}$, and $s_{\hat{p}}$

4. Choose the confidence level and find the required value of z from the normal distribution table

5. Obtain the confidence interval for p by using the formula $\hat{p} \pm z s_{\hat{p}}$

Chapter Nine

9.1   a. The null hypothesis is a claim about a population parameter that is assumed to be true until it is declared false.

          b. An alternative hypothesis is a claim about a population parameter that will be true if the null hypothesis is false.

          c. The critical point(s) divides the whole area under a distribution curve into rejection and non-rejection regions.

          d. The significance level, denoted by α, is the probability of making a Type I error, that is, the probability of rejecting the null hypothesis when it is actually true.

          e. The nonrejection region is the area where the null hypothesis is not rejected.

          f. The rejection region is the area where the null hypothesis is rejected.

          g. A hypothesis test is a two-tailed test if the rejection regions are in both tails of the distribution curve; it is a left-tailed test if the rejection region is in the left tail; and it is a right-tailed test if the rejection region is in the right tail.

          h. Type I error: A type I error occurs when a true null hypothesis is rejected. The probability of committing a Type I error, denoted by α is: $\alpha = P(H_0 \text{ is rejected} \mid H_0 \text{ is true})$

             Type II error: A Type II error occurs when a false null hypothesis is not rejected. The probability of committing a Type II error, denoted by β, is: $\beta = P(H_0 \text{ is not rejected} \mid H_0 \text{ is false})$

9.2   Table 9.2 on page 402 of the text shows the four possible outcomes – two correct decisions and two errors.

9.3   A hypothesis test is a two-tailed test if the sign in the alternative hypothesis is "$\neq$"; it is a left-tailed test if the sign in the alternative hypothesis is "$<$" (less than); and it is a right-tailed test if the sign in the alternative hypothesis is "$>$" (greater than). Table 9.3 on page 405 of the text describes these relationships.

9.4   a. Right-tailed test because the rejection region is in the right tail of the distribution curve.

          b. Two-tailed test because the rejection region lies in both tails of the distribution curve.

          c. Left-tailed test because the rejection region lies in the left tail of the distribution curve.

9.5   a. Left-tailed test   b. Right-tailed test   c. Two-tailed test

9.6   The null hypothesis is initially assumed to be true.

9.7   a. Type II error   b. Type I error

9.8   a. Type II error   b. Type I error

9.9   a. H_0: $\mu = 20$ hours; H_1: $\mu \neq 20$ hours;   a two-tailed test
     b. H_0: $\mu = 10$ hours; H_1: $\mu > 10$ hours;   a right-tailed test
     c. H_0: $\mu = 3$ years; H_1: $\mu \neq 3$ years;   a two-tailed test
     d. H_0: $\mu = \$1000$; H_1: $\mu < \$1000$;   a left-tailed test
     e. H_0: $\mu = 12$ minutes; H_1: $\mu > 12$ minutes;   a right-tailed test

9.10   a. H_0: $\mu = 9.5$ hours; H_1: $\mu \neq 9.5$ hours;   a two-tailed test
      b. H_0: $\mu = \$105$; H_1: $\mu < \$105$;   a left-tailed test
      c. H_0: $\mu = \$29,000$; H_1: $\mu > \$29,000$;   a right-tailed test
      d. H_0: $\mu = 2.9$; H_1: $\mu < 2.9$;   a left-tailed test
      e. H_0: $\mu = 175$; H_1: $\mu > 175$;   a right-tailed test

9.11   For a two-tailed test, the p–value is twice the area in the tail of the sampling distribution curve beyond the observed value of the sample test statistic.

For a one-tailed test, the p–value is the area in the tail of the sampling distribution curve beyond the observed value of the sample test statistic.

9.12   a. Step 1: H_0: $\mu = 23$; H_1: $\mu \neq 23$;   A two-tailed test

      Step 2:   Since $n > 30$, use the normal distribution.

      Step 3: $s_{\bar{x}} = s / \sqrt{n} = 5 / \sqrt{50} = .70710678$

            $z = (\bar{x} - \mu) / s_{\bar{x}} = (21.25 - 23) / .70710678 = -2.47$

      From the normal distribution table, area to the left of $z = -2.47$ is $.5 - .4932 = .0068$ approximately. Hence, p–value $= 2(.0068) = .0136$

     b. Step 1: H_0: $\mu = 15$; H_1: $\mu < 15$;   A left-tailed test.

      Step 2:   Since $n > 30$, use the normal distribution.

      Step 3: $s_{\bar{x}} = s / \sqrt{n} = 5.5 / \sqrt{80} = .61491869$

            $z = (\bar{x} - \mu) / s_{\bar{x}} = (13.25 - 15) / .61491869 = -2.85$

From the normal distribution table, area to the left of $z = -2.85$ is $.5 - .4978 = .0022$ approximately.

Hence, p–value $= .0022$

c. Step 1: H_0: $\mu = 38$; H_1: $\mu > 38$; A right-tailed test.

Step 2: Since $n > 30$, use the normal distribution.

Step 3: $s_{\bar{x}} = s / \sqrt{n} = 7.2 / \sqrt{35} = 1.21702213$

$z = (\bar{x} - \mu) / s_{\bar{x}} = (40.25 - 38) / 1.21702213 = 1.85$

From the normal distribution table, area to the right of $z = 1.85$ is $.5 - .4678 = .0322$ approximately.

Hence, p–value $= .0322$

9.13 a. Step 1: H_0: $\mu = 46$; H_1: $\mu \neq 46$; A two-tailed test.

Step 2: Since $n > 30$, use the normal distribution.

Step 3: $s_{\bar{x}} = s / \sqrt{n} = 9.7 / \sqrt{40} = 1.53370467$

$z = (\bar{x} - \mu) / s_{\bar{x}} = (49.60 - 46) / 1.53370467 = 2.35$

From the normal distribution table, area to the right of $z = 2.35$ is $.5 - .4906 = .0094$ approximately.

Hence, p–value $= 2(.0094) = .0188$

b. Step 1: H_0: $\mu = 26$; H_1: $\mu < 26$; A left-tailed test.

Step 2: Since $n > 30$, use the normal distribution.

Step 3: $s_{\bar{x}} = s / \sqrt{n} = 4.3 / \sqrt{33} = .74853392$

$z = (\bar{x} - \mu) / s_{\bar{x}} = (24.3 - 26) / .74853392 = -2.27$

From the normal distribution table, area to the left of $z = -2.27$ is $.5 - .4884 = .0116$ approximately.

Hence, p–value $= .0116$

c. Step 1: H_0: $\mu = 18$; H_1: $\mu > 18$; A right-tailed test.

Step 2: Since $n > 30$, use the normal distribution.

Step 3: $s_{\bar{x}} = s / \sqrt{n} = 7.8 / \sqrt{55} = 1.05175179$

$z = (\bar{x} - \mu) / s_{\bar{x}} = (20.50 - 18) / 1.05175179 = 2.38$

From the normal distribution table, area to the right of $z = 2.38$ is $.5 - .4913 = .0087$ approximately.

Hence, p–value $= .0087$

9.14 a. Step 1: H_0: $\mu = 29$; H_1: $\mu \neq 29$; A two-tailed test.

Step 2: Since $n > 30$, use the normal distribution.

Step 3: $s_{\bar{x}} = s / \sqrt{n} = 8 / \sqrt{60} = 1.03279556$

$z = (\bar{x} - \mu) / s_{\bar{x}} = (31.4 - 29) / 1.03279556 = 2.32$

The area under the standard normal curve to the right of $z = 2.32$ is $.5 - .4898 = .0102$.

Hence, p–value $= 2(.0102) = .0204$

    b. For $\alpha = .05$, reject H_0, since p–value $< .05$.

    c. For $\alpha = .01$, do not reject H_0, since p–value $> .01$.

9.15   a. Step 1: H_0: $\mu = 72$; H_1: $\mu > 72$;   A right-tailed test.

              Step 2:   Since $n > 30$, use the normal distribution.

              Step 3: $s_{\bar{x}} = s / \sqrt{n} = 6 / \sqrt{36} = 1.0$

$$z = (\bar{x} - \mu) / s_{\bar{x}} = (74.07 - 72) / 1.0 = 2.07$$

        The area under the standard normal curve to the right of $z = 2.07$ is $.5 - .4808 = .0192$.

        Hence p–value $= .0192$

    b. For $\alpha = .01$, do not reject H_0, since p–value $> .01$.

    c. For $\alpha = .025$, reject H_0, since p–value $< .025$.

9.16 H_0: $\mu = 10$ ounces; H_1: $\mu < 10$ ounces;   A left-tailed test.

        $s_{\bar{x}} = s / \sqrt{n} = .15 / \sqrt{36} = .025$

        Test statistic: $z = (\bar{x} - \mu) / s_{\bar{x}} = (9.955 - 10) / .025 = -1.80$

        The area under the standard normal curve to the left of $z = -1.80$ is $.5 - .4641 = .0359$

        Hence, p–value $= .0359$

        For $\alpha = .01$, do not reject H_0, since p–value $> .01$.

9.17 H_0: $\mu = 16.3$ weeks; H_1: $\mu > 16.3$ weeks;   A right-tailed test.

        $s_{\bar{x}} = s / \sqrt{n} = 4.2 / \sqrt{400} = .21$

        Test statistic: $z = (\bar{x} - \mu) / s_{\bar{x}} = (16.9 - 16.3) / .21 = 2.86$

        The area under the standard normal curve to the right of $z = 2.86$ is $.5 - .4979 = .0021$ approximately.

        Hence, p–value $= .0021$

        For $\alpha = .02$, reject H_0, since p–value $< .02$.

9.18 H_0: $\mu = 45$ months; H_1: $\mu < 45$ months;   A left-tailed test.

        $s_{\bar{x}} = s / \sqrt{n} = 4.5 / \sqrt{36} = .75$

        Test statistic: $z = (\bar{x} - \mu) / s_{\bar{x}} = (43.75 - 45) / .75 = -1.67$

        The area under the standard normal curve to the left of $z = -1.67$ is $.5 - .4525 = .0475$

        Hence, p–value $= .0475$.

        For $\alpha = .025$, do not reject H_0, since p–value $> .025$.

9.19 H_0: $\mu \geq 14$ hours; H_1 : $\mu < 14$ hours;   A left-tailed test.

        $s_{\bar{x}} = s / \sqrt{n} = 3.0 / \sqrt{200} = .21213203$

Test statistic: $z = (\bar{x} - \mu) / s_{\bar{x}} = (13.75 - 14) / .21213203 = -1.18$

The area under the standard normal curve to the left of $z = -1.18$ is $.5 - .3810 = .1190$

Hence, p–value $= .1190$.

For $\alpha = .05$, do not reject H_0, since p–value $> .05$.

9.20 a. H_0: $\mu = \$226$; H_1: $\mu \neq \$226$;   A two-tailed test.

$s_{\bar{x}} = s / \sqrt{n} = 77 / \sqrt{250} = \$4.86990760$

$z = (\bar{x} - \mu) / s_{\bar{x}} = (238 - 226) / 4.86990760 = 2.46$

The area under the standard normal curve to the right of $z = 2.46$ is $.5 - .4931 = .0069$.

Hence, p–value $= 2(.0069) = .0138$

b. For $\alpha = .01$, do not reject H_0, since p–value $> .01$.

For $\alpha = .02$, reject H_0, since p–value $< .02$.

9.21 a. H_0: $\mu = 10$ minutes; H_1: $\mu < 10$ minutes;   A left-tailed test.

$s_{\bar{x}} = s / \sqrt{n} = 3.75 / \sqrt{100} = .375$

$z = (\bar{x} - \mu) / s_{\bar{x}} = (9.25 - 10) / .375 = -2.00$

The area under the standard normal curve to the left of $z = -2.00$ is $.5 - .4772 = .0228$.

Hence, p–value $= .0228$

b. Do not reject H_0 for $\alpha = .02$, since p–value $> .02$.

Reject H_0 for $\alpha = .05$, since p–value $< .05$.

9.22 a. H_0: $\mu = 36$ inches; H_1: $\mu \neq 36$ inches;   A two-tailed test.

$\sigma_{\bar{x}} = \sigma / \sqrt{n} = .05 / \sqrt{40} = .00790569$

$z = (\bar{x} - \mu) / \sigma_{\bar{x}} = (36.015 - 36) / .00790569 = 1.90$

The area under the standard normal curve to the right of $z = 1.90$ is $.5 - .4713 = .0287$.

Hence, p–value $= 2(.0287) = .0574$

b. For $\alpha = .02$, do not reject H_0, since p–value $> .02$. Thus, the inspector will not stop the machine.

For $\alpha = .10$, reject H_0, since p–value $< .10$. Thus, the inspector will stop the machine.

9.23 a. H_0: $\mu = 32$ ounces; H_1: $\mu \neq 32$ ounces;   A two-tailed test.

$\sigma_{\bar{x}} = \sigma / \sqrt{n} = .15 / \sqrt{35} = .02535463$

$z = (\bar{x} - \mu) / \sigma_{\bar{x}} = (31.90 - 32) / .02535463 = -3.94$

The area under the standard normal curve to the left of $z = -3.94$ is approximately $.5 - .5 = 0$.

Hence, p–value $= 2(.00) = .00$ approximately

b. For $\alpha = .01$, reject H_0, since p–value $< .01$.

For $\alpha = .05$, reject H_0, since p–value $< .05$.

Thus in either case, the inspector will stop the machine.

9.24 The five steps of a test of hypothesis are:

1. State the null and alternative hypotheses.

2. Select the distribution to use.

3. Determine the rejection and nonrejection regions.

4. Calculate the value of the test statistic.

5. Make a decision.

9.25 The level of significance in a test of hypothesis is the probability of making a Type I error. It is the area under the probability distribution curve where we reject H_0.

9.26 Yes, rejecting H_0 is equivalent to stating that the evidence from the sample is strong that H_1 is true.

9.27 The critical value of z separates the rejection region from the nonrejection region and is found from a table such as the standard normal distribution table. The observed value of z is the value calculated for a sample statistic such as $\bar{x}$.

9.28 a. The rejection region lies to the left of $z = -1.96$ and to the right of $z = 1.96$.

The nonrejection region lies between $z = -1.96$ and $z = 1.96$.

b. The rejection region lies to the left of $z = -2.33$.

The nonrejection region lies to the right of $z = -2.33$.

c. The rejection region lies to the right of $z = 2.05$.

The nonrejection region lies to the left of $z = 2.05$.

9.29 a. The rejection region lies to the left of $z = -2.58$ and to the right of $z = 2.58$.

The nonrejection region lies between $z = -2.58$ and $z = 2.58$.

b. The rejection region lies to the left of $z = -2.58$.

The nonrejection region lies to the right of $z = -2.58$.

c. The rejection region lies to the right of $z = 1.96$.

The nonrejection region lies to the left of $z = 1.96$.

9.30 If H_0 is rejected, the difference between the observed value of $\bar{x}$ and the hypothesized value of μ is "statistically significant".

9.31   If H_0 is not rejected, the difference between the hypothesized value of μ and the observed value of $\bar{x}$ is "statistically not significant".

9.32   a. .025   b. .05   c. .01

9.33   a. .10   b. .02   c. .005

9.34 $n = 120$, $\bar{x} = 32$, $s = 6$, and $s_{\bar{x}} = s/\sqrt{n} = 6/\sqrt{120} = .54772256$

    a. Critical value: $z = 1.65$

       Observed value: $z = (\bar{x} - \mu)/s_{\bar{x}} = (32 - 28)/.54772256 = 7.30$

    b. Critical values: $z = -1.96$ and 1.96

       Observed value: $z = (\bar{x} - \mu)/s_{\bar{x}} = (32 - 28)/.54772256 = 7.30$

9.35 $n = 90$, $\bar{x} = 15$, $s = 4$, and $s_{\bar{x}} = s/\sqrt{n} = 4/\sqrt{90} = .42163702$

    a. Critical value: $z = -2.33$

       Observed value: $z = (\bar{x} - \mu)/s_{\bar{x}} = (15 - 20)/.42163702 = -11.86$

    b. Critical value: $z = -2.58$ and 2.58.

       Observed value: $z = (\bar{x} - \mu)/s_{\bar{x}} = (15 - 20)/.42163702 = -11.86$

9.36   a. The rejection region lies to the left of $z = -1.65$.

       The nonrejection region lies to the right of $z = -1.65$.

    b. The rejection region lies to the left of $z = -1.96$ and to the right of $z = 1.96$.

       The nonrejection region lies between $z = -1.96$ and $z = 1.96$.

    c. The rejection region lies to the right of $z = 1.65$.

       The nonrejection region lies to the left of $z = 1.65$.

9.37   a. The rejection region lies to the left of $z = -2.33$.

       The nonrejection region lies to the right of $z = -2.33$.

    b. The rejection region lies to the left of $z = -2.58$ and to the right of $z = 2.58$.

       The nonrejection region lies between $z = -2.58$ and $z = 2.58$.

    c. The rejection region lies to the right of $z = 2.33$.

       The nonrejection region lies to the left of $z = 2.33$.

9.38   a. $n = 64$, $\bar{x} = 98$, $s = 12$, and $s_{\bar{x}} = s/\sqrt{n} = 12/\sqrt{64} = 1.50$

       Critical values: $z = -2.58$ and 2.58

Test statistic: $z = (\bar{x} - \mu) / s_{\bar{x}} = (98 - 100) / 1.50 = -1.33$; Do not reject H_0.

b. $n = 64$, $\bar{x} = 104$, $s = 10$, and $s_{\bar{x}} = s / \sqrt{n} = 10 / \sqrt{64} = 1.25$

Critical values: $z = -2.58$ and 2.58

Test statistic: $z = (\bar{x} - \mu) / s_{\bar{x}} = (104 - 100) / 1.25 = 3.20$; Reject H_0.

Comparing parts a and b shows that two samples selected from the same population can yield opposite conclusions on the same test of hypothesis.

9.39   a. $n = 100$, $\bar{x} = 43$, $s = 5$, and $s_{\bar{x}} = s / \sqrt{n} = 5 / \sqrt{100} = .50$

Critical value: $z = -1.96$

Test statistic: $z = (\bar{x} - \mu) / s_{\bar{x}} = (43 - 45) / .50 = -4.00$; Reject H_0.

b. $n = 100$, $\bar{x} = 43.8$, $s = 7$, and $s_{\bar{x}} = s / \sqrt{n} = 7 / \sqrt{100} = .70$

Critical value: $z = -1.96$

Test statistic: $z = (\bar{x} - \mu) / s_{\bar{x}} = (43.8 - 45) / .70 = -1.71$; Do not reject H_0.

Comparing parts a and b shows that two samples selected from the same population can yield opposite conclusions on the same test of hypothesis.

9.40   a.   Step 1: $H_0: \mu = 25$; $H_1: \mu \neq 25$;   A two-tailed test.

           Step 2:   Since $n > 30$, use the normal distribution.

           Step 3:   For $\alpha = .01$, the critical values of z are -2.58 and 2.58.

           Step 4: $s_{\bar{x}} = s / \sqrt{n} = 3 / \sqrt{81} = .33333333$

                    $z = (\bar{x} - \mu) / s_{\bar{x}} = (28.5 - 25) / .33333333 = 10.50$

           Step 5:   Reject H_0.

b.   Step 1: $H_0: \mu = 12$; $H_1: \mu < 12$;   A left-tailed test.

           Step 2:   Since $n > 30$, use the normal distribution.

           Step 3:   For $\alpha = .05$, the critical value of z is -1.65.

           Step 4: $\sigma_{\bar{x}} = \sigma / \sqrt{n} = 4.5 / \sqrt{45} = .67082039$

                    $z = (\bar{x} - \mu) / \sigma_{\bar{x}} = (11.25 - 12) / .67082039 = -1.12$

           Step 5:   Do not reject H_0.

c.   Step 1: $H_0: \mu = 40$; $H_1: \mu > 40$;   A right-tailed test.

           Step 2:   Since $n > 30$, use the normal distribution.

           Step 3:   For $\alpha = .10$, the critical value of z is 1.28.

           Step 4: $s_{\bar{x}} = s / \sqrt{n} = 7 / \sqrt{100} = .70$

                    $z = (\bar{x} - \mu) / s_{\bar{x}} = (47 - 40) / .70 = 10.00$

           Step 5:   Reject H_0.

9.41   a.   Step 1: $H_0: \mu = 80;$ $H_1: \mu \neq 80;$ A two-tailed test.

   Step 2:   Since $n > 30$, use the normal distribution.

   Step 3:   For $\alpha = .10$, the critical values of z are -1.65 and 1.65.

   Step 4: $\sigma_{\bar{x}} = \sigma / \sqrt{n} = 15 / \sqrt{33} = 2.61116484$

   $z = (\bar{x} - \mu) / \sigma_{\bar{x}} = (76.5 - 80) / 2.61116484 = -1.34$

   Step 5:   Do not reject H_0.

   b.   Step 1: $H_0: \mu = 32;$ $H_1: \mu < 32;$ A left-tailed test.

   Step 2:   Since $n > 30$, use the normal distribution.

   Step 3:   For $\alpha = .01$, the critical value of z is -2.33.

   Step 4: $s_{\bar{x}} = s / \sqrt{n} = 7.4 / \sqrt{75} = .85447840$

   $z = (\bar{x} - \mu) / s_{\bar{x}} = (26.5 - 32) / .85447840 = -6.44$

   Step 5:   Reject H_0.

   c.   Step 1: $H_0: \mu = 55;$ $H_1: \mu > 55;$ A right-tailed test.

   Step 2:   Since $n > 30$, use the normal distribution.

   Step 3:   For $\alpha = .05$, the critical value of z is 1.65.

   Step 4: $s_{\bar{x}} = s / \sqrt{n} = 4 / \sqrt{40} = .63245553$

   $z = (\bar{x} - \mu) / s_{\bar{x}} = (60.5 - 55) / .63245553 = 8.70$

   Step 5:   Reject H_0.

9.42   Step 1: $H_0: \mu = \$940;$ $H_1: \mu \neq \$940;$ A two-tailed test.

   Step 2:   Since $n > 30$, use the normal distribution.

   Step 3:   For $\alpha = .01$, the critical values of z are -2.58 and 2.58.

   Step 4: $s_{\bar{x}} = s / \sqrt{n} = 330 / \sqrt{324} = 18.33333333$

   $z = (\bar{x} - \mu) / s_{\bar{x}} = (1005 - 940) / 18.33333333 = 3.55$

   Step 5:   Reject H_0. Conclude this year's holiday expenditures differs from the 2001 level of $940.

9.43   Step 1: $H_0: \mu = 16.3$ weeks; $H_1: \mu > 16.3$ weeks;   A right-tailed test.

   Step 2:   Since $n > 30$, use the normal distribution.

   Step 3:   For $\alpha = .02$, the critical value of z is 2.05.

   Step 4: $s_{\bar{x}} = s / \sqrt{n} = 4.2 / \sqrt{400} = .21$

   $z = (\bar{x} - \mu) / s_{\bar{x}} = (16.9 - 16.3) / .21 = 2.86$

   Step 5:   Reject H_0. Conclude the mean duration of unemployment exceeds 16.3 weeks.

9.44 $H_0: \mu = \$1,055,500;$ $\qquad$ $H_1: \mu \neq 1,055,500;$ $\qquad$ A two-tailed test.

For $\alpha = .01$, the critical values of z are -2.58 and 2.58.

$s_{\bar{x}} = s / \sqrt{n} = 220,000 / \sqrt{36} = 36,666.66667$

Test statistic: $z = (\bar{x} - \mu) / s_{\bar{x}} = (1,130,000 - 1,055,500) / 36,666.66667 = 2.03$

Do not reject H_0. Do not conclude that the mean expenditure on women's athletic scholarships for all institutions in NCAA Division I has changed.

9.45 $H_0: \mu = \$2.4$ million; $\qquad$ $H_1: \mu > \$2.4$ million; $\qquad$ A right-tailed test.

For $\alpha = .025$, the critical value of z is 1.96.

$s_{\bar{x}} = s / \sqrt{n} = .5 / \sqrt{32} = ..08838835$

Test statistic: $z = (\bar{x} - \mu) / s_{\bar{x}} = (2.6 - 2.4) / .08838835 = 2.26$

Reject H_0. Conclude the mean annual cash compensation of CEOs exceeds \$2.4 million.

9.46   a. $H_0: \mu = 394$ minutes; $\qquad$ $H_1: \mu > 394$ minutes; $\qquad$ A right-tailed test.

For $\alpha = .02$, the critical value of z is 2.05.

$s_{\bar{x}} = s / \sqrt{n} = 81 / \sqrt{295} = 4.71600233$

Test statistic: $z = (\bar{x} - \mu) / s_{\bar{x}} = (402 - 394) / 4.71600233 = 1.70$

Do not reject H_0. Do not conclude that the mean time women talk on their cell phones currently exceeds 394 minutes per month.

b. The Type I error in this case would be to conclude that the mean time women talk on their cell phones currently exceeds 394 minutes per month when it actually does not. $P(\text{Type I error}) = .02$

9.47   a. $H_0: \mu = 38.1$ years; $\qquad$ $H_1: \mu < 38.1$ years; $\qquad$ A left-tailed test.

For $\alpha = .01$, the critical value of z is -2.33.

$s_{\bar{x}} = s / \sqrt{n} = 8 / \sqrt{700} = .30237158$

Test statistic: $z = (\bar{x} - \mu) / s_{\bar{x}} = (37 - 38.1) / .30237158 = -3.64$

Reject H_0. Conclude that the mean age of motor cycle owners is less than 38.1 years.

b. The Type I error in this case would be to conclude that the mean age of motor cycle owners is less than 38.1 years when it is actually equal to 38.1 years . $P(\text{Type I error}) = \alpha = .01$

9.48   a. $H_0: \mu = 11$ hours; $\qquad$ $H_1: \mu < 11$ hours; $\qquad$ A left-tailed test.

For $\alpha = .01$, the critical value of z is -2.33.

$s_{\bar{x}} = s / \sqrt{n} = 2.2 / \sqrt{100} = .2200$

Test statistic: $z = (\bar{x} - \mu) / s_{\bar{x}} = (9 - 11) / .2200 = -9.09$

Reject H_0. Conclude that adult males spend less than 11 hours watching sports on television.

b. If $\alpha = 0$, there can be no rejection region. Thus, we cannot reject H_0.

9.49   a. H_0: $\mu \geq \$35,000$; H_1: $\mu < \$35,000$;   A left-tailed test.

For $\alpha = .01$, the critical value of z is -2.33.

$s_{\bar{x}} = s / \sqrt{n} = 5400 / \sqrt{150} = \$440.90815370$

Test statistic: $z = (\bar{x} - \mu) / s_{\bar{x}} = (33,400 - 35,000) / 440.90815370 = -3.63$

Reject H_0 and conclude that the company should not open a restaurant in this area.

b.   If $\alpha = 0$, there can be no rejection region. Thus, we cannot reject H_0.

9.50   a. H_0: $\mu = 10$ ounces; H_1: $\mu < 10$ ounces;   A left-tailed test.

For $\alpha = .025$, the critical value of z is -1.96.

$s_{\bar{x}} = s / \sqrt{n} = .15 / \sqrt{36} = .025$

Test statistic: $z = (\bar{x} - \mu) / s_{\bar{x}} = (9.955 - 10) / .025 = -1.80$

Do not reject H_0. Conclude the mean weight of cheddar cheese packages is not less than 10 ounces.

b.   For $\alpha = .05$, the critical value of z is -1.65. From part a, the value of the test statistic is -1.88.

Reject H_0 and conclude that the mean weight of cheddar cheese packages is less than 10 ounces.

The decisions in parts a and b are different. The results of this sample are not very conclusive, since raising the significance level from .025 to .05 reverses the decision.

9.51   a. H_0: $\mu \geq 8$ hours; H_1: $\mu < 8$ hours;   A left-tailed test

For $\alpha = .01$, the critical value of z is -2.33.

$s_{\bar{x}} = s / \sqrt{n} = 2.1 / \sqrt{200} = .14849242$

Test statistic: $z = (\bar{x} - \mu) / s_{\bar{x}} = (7.68 - 8) / .14849242 = -2.15$

Do not reject H_0. Conclude that the claim is true.

b.   For $\alpha = .025$, the critical value of z is -1.96. From part a, the value of the test statistic is -2.15.

Hence, we reject H_0 and conclude that the claim is false.

The decisions in parts a and b are different. The results of this sample are not very conclusive, since raising the significance level from .01 to .025 reverses the decision.

9.52 H_0: $\mu = 36$ inches; H_1: $\mu \neq 36$ inches;   A two-tailed test.

For $\alpha = .01$, the critical values of z are -2.58 and 2.58.

$\sigma_{\bar{x}} = \sigma / \sqrt{n} = .05 / \sqrt{40} = .00790569$

Test statistic: $z = (\bar{x} - \mu) / \sigma_{\bar{x}} = (36.015 - 36) / .00790569 = 1.90$

Do not reject H_0; the machine does not need to be adjusted.

9.53 H_0: $\mu = 32$ ounces; H_1: $\mu \neq 32$ ounces;   A two-tailed test.

For $\alpha = .02$, the critical values of z are -2.33 and 2.33.

$\sigma_{\bar{x}} = \sigma / \sqrt{n} = .15 / \sqrt{35} = .02535463$

Test statistic: $z = (\bar{x} - \mu) / \sigma_{\bar{x}} = (31.90 - 32) / .02535463 = -3.94$

Reject H_0; the machine needs to be adjusted.

9.54   To make a hypothesis test about the mean net weight of All Taste cereal boxes, first we will take a sample of 30 or more such cereal boxes and measure the net weight of each of these boxes. Then, using the formulas learned in Chapter 3 for the mean and standard deviation for sample data, we will calculate the sample mean and sample standard deviation. Then, we will determine the significance level and make the test of hypothesis.

9.55   To make a hypothesis test about the mean textbook costs of college freshmen, first we will take a sample of 30 or more freshmen and collect the information on how much each of them spends on textbooks. Then, using the formulas learned in Chapter 3 for the mean and standard deviation for sample data, we will calculate the sample mean and sample standard deviation. Then we will determine the significance level and make the test of hypothesis.

9.56   To use the t distribution:

1. The sample size should be small ($n < 30$).
2. The population being sampled must be (approximately) normal.
3. The population standard deviation σ must be unknown.

9.57   a. Area in each tail $= \alpha / 2 = .02 / 2 = .01$ and $df = n - 1 = 20 - 1 = 19$

   Rejection region: $t < -2.539$ and $t > 2.539$

   Nonrejection region: $-2.539 < t < 2.539$

   b. Area in the left tail $= \alpha = .01$ and $df = n - 1 = 16 - 1 = 15$

   Rejection region: $t < -2.602$

   Nonrejection region: $t > -2.602$

   c. Area in the right tail $= \alpha = .05$ and $df = n - 1 = 18 - 1 = 17$

   Rejection region: $t > 1.740$

   Nonrejection region: $t < 1.740$

9.58   a. Area in each tail $= \alpha / 2 = .01 / 2 = .005$ and $df = n - 1 = 15 - 1 = 14$

   Rejection region: $t < -2.977$ and $t > 2.977$

   Nonrejection region: $-2.977 < t < 2.977$

    b. Area in the left tail = α = .005 and $df = n - 1 = 25 - 1 = 24$

       Rejection region: $t < -2.797$

       Nonrejection region $t > -2.797$

    c. Area in the right tail = α = .025 and $df = n - 1 = 22 - 1 = 21$

       Rejection region: $t > 2.080$

       Nonrejection region: $t < 2.080$

9.59   a. Area in the right tail = α = .01 and $df = n - 1 = 25 - 1 = 24$

       Critical value: $t = 2.492$

$$s_{\bar{x}} = s / \sqrt{n} = 7.5 / \sqrt{25} = 1.50$$

       Observed value: $t = (\bar{x} - \mu) / s_{\bar{x}} = (58.5 - 55) / 1.50 = 2.333$

    b. Area in each tail = $\alpha / 2 = .01 / 2 = .005$ and $df = n-1 = 25-1 = 24$

       Critical values: $t = -2.797$ and 2.797

       Observed value: $t = (\bar{x} - \mu) / s_{\bar{x}} = (58.5 - 55) / 1.50 = 2.333$

9.60   a. Area in the left tail: α = .05 and $df = n - 1 = 16 - 1 = 15$

       Critical value: $t = -1.753$

$$s_{\bar{x}} = s / \sqrt{n} = 8 / \sqrt{16} = 2.00$$

       Observed value: $t = (\bar{x} - \mu) / s_{\bar{x}} = (42.4 - 46) / 2.00 = -1.800$

    b. Area in each tail: $\alpha / 2 = .05 / 2 = .025$ and $df = n-1 = 16 - 1 = 15$

       Critical values: $t = -2.131$ and 2.131

       Observed value: $t = (\bar{x} - \mu) / s_{\bar{x}} = (42.4 - 46) / 2.00 = -1.800$

9.61   a. The rejection region lies to the left of $t = -2.539$.

       The nonrejection region lies to the right of $t = -2.539$.

    b. The rejection region lies to the left of $t = -2.861$ and to the right of $t = 2.861$.

       The nonrejection region lies between $t = -2.861$ and $t = 2.861$

    c. The rejection region lies to the right of $t = 2.539$.

       The nonrejection region lies to the left of $t = 2.539$.

9.62   a. The rejection region lies to the left of $t = -1.721$.

       The nonrejection region lies to the right of $t = -1.721$.

    b. The rejection region lies to the left of $t = -2.080$ and to the right of $t = 2.080$.

       The nonrejection region lies between $t = -2.080$ and $t = 2.080$.

    c. The rejection region lies to the right of $t = 1.721$.

       The nonrejection region lies to the left of $t = 1.721$.

9.63   a.   Step 1: $H_0: \mu = 80$; $H_1: \mu \neq 80$;   A two-tailed test.

   Step 2:   The sample size is small ($n < 30$), the population is normally distributed and the population standard deviation σ, is unknown. Hence, we use the t distribution to make the test.

   Step 3: $df = n - 1 = 25 - 1 = 24$ and $\alpha / 2 = .01 / 2 = .005$

   Critical values: $t = -2.797$ and 2.797

   Step 4: $s_{\bar{x}} = s / \sqrt{n} = 8 / \sqrt{25} = 1.60$

   $t = (\bar{x} - \mu) / s_{\bar{x}} = (77 - 80) / 1.60 = -1.875$

   Step 5:   Do not reject H_0.

   b.   Steps 1, 2, and 3 are the same as for part a.

   Step 4: $s_{\bar{x}} = s / \sqrt{n} = 6 / \sqrt{25} = 1.20$

   $t = (\bar{x} - \mu) / s_{\bar{x}} = (86 - 80) / 1.20 = 5.000$

   Step 5: Reject H_0.

   Comparing parts a and b shows that two samples selected from the same population can yield opposite conclusions on the same test of hypothesis.

9.64   a.   Step 1: $H_0: \mu = 40$; $H_1: \mu > 40$;   A right-tailed test.

   Step 2:   The sample size is small ($n < 30$), the population is normally distributed, and the population standard deviation σ, is unknown. Hence, we use t distribution to make the test.

   Step 3: $df = n - 1 = 16 - 1 = 15$ and $\alpha = .025$

   Critical value: $t = 2.131$

   Step 4: $s_{\bar{x}} = s / \sqrt{n} = 5 / \sqrt{16} = 1.25$

   $t = (\bar{x} - \mu) / s_{\bar{x}} = (45 - 40) / 1.25 = 4.000$

   Step 5:   Reject H_0.

   b.   Steps 1, 2, and 3 are the same as for part a.

   Step 4: $s_{\bar{x}} = s / \sqrt{n} = 7 / \sqrt{16} = 1.75$

   $t = (\bar{x} - \mu) / s_{\bar{x}} = (41.9 - 40) / 1.75 = 1.086$

   Step 5:   Do not reject H_0.

   Comparing parts a and b shows that two samples selected from the same population can yield opposite conclusions on the same test of hypothesis.

9.65   a. $H_0: \mu = 24;$ $H_1: \mu \neq 24;$ A two-tailed test.

         $df = n - 1 = 25 - 1 = 24$ and $\alpha / 2 = .01 / 2 = .005$

         The critical values of t are -2.797 and 2.797.

         $s_{\bar{x}} = s / \sqrt{n} = 4.9 / \sqrt{25} = .98$

         $t = (\bar{x} - \mu) / s_{\bar{x}} = (28.5 - 24) / .98 = 4.592$ Reject H_0.

   b. $H_0: \mu = 30;$ $H_1: \mu < 30:$ A left-tailed test.

         $df = n - 1 = 16 - 1 = 15$ and $\alpha = .025$

         The critical value of t is -2.131.

         $s_{\bar{x}} = s / \sqrt{n} = 6.6 / \sqrt{16} = 1.65$

         $t = (\bar{x} - \mu) / s_{\bar{x}} = (27.5 - 30) / 1.65 = -1.515$ Do not reject H_0.

   c. $H_0: \mu = 18,$ $H_1: \mu > 18;$ A right tailed test.

         $df = n - 1 = 20 - 1 = 19$ and $\alpha = .10$

         The critical value of t is 1.328.

         $s_{\bar{x}} = s / \sqrt{n} = 8 / \sqrt{20} = 1.78885438$

         $t = (\bar{x} - \mu) / s_{\bar{x}} = (22.5 - 18) / 1.78885438 = 2.516$ Reject H_0.

9.66   a. $H_0: \mu = 60;$ $H_1: \mu \neq 60;$ A two-tailed test.

         $df = n - 1 = 14 - 1 = 13$ and $\alpha / 2 = .05 / 2 = .025$

         The critical values of t are -2.160 and 2.160.

         $s_{\bar{x}} = s / \sqrt{n} = 9 / \sqrt{14} = 2.40535118$

         $t = (\bar{x} - \mu) / s_{\bar{x}} = (57 - 60) / 2.40535118 = -1.247$ Do not reject H_0.

   b. $H_0: \mu = 35;$ $H_1: \mu < 35;$ A left-tailed test.

         $df = n - 1 = 24 - 1 = 23$ and $\alpha = .005$

         The critical value of t is -2.807.

         $s_{\bar{x}} = s / \sqrt{n} = 5.4 / \sqrt{24} = 1.10227038$

         $t = (\bar{x} - \mu) / s_{\bar{x}} = (28 - 35) / 1.10227038 = -6.351$ Reject H_0.

   c. $H_0: \mu = 47;$ $H_1: \mu > 47;$ A right-tailed test.

         $df = n - 1 = 18 - 1 = 17$ and $\alpha = .001$

         The critical value of t is 3.646.

         $s_{\bar{x}} = s / \sqrt{n} = 6 / \sqrt{18} = 1.41421356$

         $t = (\bar{x} - \mu) / s_{\bar{x}} = (50 - 47) / 1.41421356 = 2.121$ Do not reject H_0.

9.67 $H_0: \mu = 69.5$ inches; $H_1: \mu \neq 69.5$ inches;   A two-tailed test.

         $df = n - 1 = 25 - 1 = 24$ and $\alpha / 2 = .01 / 2 = .005$

The critical values of t are -2.797 and 2.797.

$s_{\bar{x}} = s/\sqrt{n} = 2.1/\sqrt{25} = .420$

$t = (\bar{x} - \mu)/s_{\bar{x}} = (70.25 - 69.5)/.420 = 1.786$ Do not reject H_0.

9.68 $H_0: \mu = \$60$; $H_1: \mu > \$60$;   A right-tailed test.

$df = n - 1 = 25 - 1 = 24$ and $\alpha = .05$

The critical value of t is 1.711.

$s_{\bar{x}} = s/\sqrt{n} = 2/\sqrt{25} = .40$

$t = (\bar{x} - \mu)/s_{\bar{x}} = (63.50 - 60)/.40 = 8.750$

Reject H_0. Conclude the current mean cost of a visit to the doctor exceeds $60.

9.69 $H_0: \mu \leq 7$ hours; $H_1: \mu > 7$ hours;   A right-tailed test.

$df = n - 1 = 20 - 1 = 19$ and $\alpha = .025$

The critical value of t is 2.093.

$s_{\bar{x}} = s/\sqrt{n} = 2.3/\sqrt{20} = .51429563$

$t = (\bar{x} - \mu)/s_{\bar{x}} = (10.5 - 7)/.51429563 = 6.805$ Reject H_0. The president's claim is not true.

9.70 $H_0: \mu = \$850$; $H_1: \mu < \$850$;   A left-tailed test.

$df = n - 1 = 25 - 1 = 24$ and $\alpha = .01$

The critical value of t is -2.492.

$s_{\bar{x}} = s/\sqrt{n} = 230/\sqrt{25} = \$46.00$

$t = (\bar{x} - \mu)/s_{\bar{x}} = (780 - 850)/46.00 = -1.522$

Do not reject H_0. The mean balance of checking accounts at this bank is not less than $850.

9.71 $H_0: \mu \leq 30$ calories; $H_1: \mu > 30$ calories;   A right-tailed test.

$df = n - 1 = 16 - 1 = 15$ and $\alpha = .05$

The critical value of t is 1.753.

$s_{\bar{x}} = s/\sqrt{n} = 3/\sqrt{16} = .750$

$t = (\bar{x} - \mu)/s_{\bar{x}} = (32 - 30)/.750 = 2.667$ Reject H_0. The manufacturer's claim is false.

9.72 $H_0: \mu = \$43,070$; $H_1: \mu > \$43,070$;   A right-tailed test.

$df = n - 1 = 28 - 1 = 27$ and $\alpha = .01$

The critical value of t is 2.473.

$s_{\bar{x}} = s/\sqrt{n} = 3200/\sqrt{28} = 604.7431568$

$t = (\bar{x} - \mu) / s_{\bar{x}} = (45{,}000 - 43{,}070) / 604.7431568 = 3.191$

Reject H_0. Conclude that the mean starting salary offer exceeds \$43,070.

9.73   a. H_0: $\mu \le 45$ minutes; H_1: $\mu > 45$ minutes;   A right-tailed test.

$df = n - 1 = 20 - 1 = 19$ and $\alpha = .01$

The critical value of t is 2.539.

$s_{\bar{x}} = s / \sqrt{n} = 3 / \sqrt{20} = .67082039$

$t = (\bar{x} - \mu) / s_{\bar{x}} = (49.50 - 45) / .67082039 = 6.708$

Reject H_0. The mean drying time for these paints is more than 45 minutes.

b.   The Type I error would occur if the mean drying time for these paints is 45 minutes or less, but we conclude otherwise. The probability of such an error is .01 here.

9.74   a. H_0: $\mu \ge \$150$; H_1: $\mu < \$150$;   A left-tailed test.

$df = n - 1 = 25 - 1 = 24$ and $\alpha = .01$

The critical value of t is –2.492.

$s_{\bar{x}} = s / \sqrt{n} = 28 / \sqrt{25} = \$5.60$

$t = (\bar{x} - \mu) / s_{\bar{x}} = (139 - 150) / 5.60 = -1.964$

Do not reject H_0. Do not conclude that the mean waiter's earnings from tips is less than \$150.

b.   The Type I error would occur if the mean waiter's earnings from tips is \$150 or more but we concluded that it was less than \$150. The probability of this error is $\alpha = .01$.

9.75   a.   If $\alpha = 0$, there is no rejection region. Hence, the decision must be: "Do not reject H_0".

b. H_0: $\mu \ge 1200$ words; H_1: $\mu < 1200$ words;   A left-tailed test.

$df = n - 1 = 25 - 1 = 24$ and $\alpha = .05$

Critical value: $t = -1.711$

$s_{\bar{x}} = s / \sqrt{n} = 85 / \sqrt{25} = 17.00$

$t = (\bar{x} - \mu) / s_{\bar{x}} = (1125 - 1200) / 17.00 = -4.412$

Reject H_0. Conclude that the claim of the business school is false.

9.76   a.   If $\alpha = 0$, there is no rejection region, so we could not reject H_0. Thus, we could not conclude that workers work less than 43.5 hours per week.

b. H_0: $\mu \ge 43.5$ hours; H_1: $\mu < 43.5$ hours;   A left-tailed test.

$df = n - 1 = 24 - 1 = 23$ and $\alpha = .05$

The critical value of t is –1.714.

$s_{\bar{x}} = s / \sqrt{n} = 1.3 / \sqrt{24} = .26536139$

$$t = (\bar{x} - \mu) / s_{\bar{x}} = (41.7 - 43.5) / .26536139 = -6.78$$

Reject H_0. Conclude that the claim that production workers in the mining industry work less than 43.5 hours per week is true.

9.77   From the given data: $n = 10$, $\sum x = 257$, and $\sum x^2 = 7341$

$$\bar{x} = \sum x / n = 257 / 10 = 25.70 \text{ hours.}$$

$$s = \sqrt{\frac{\sum x^2 - (\sum x)^2 / n}{n-1}} = \sqrt{\frac{7341 - (257)^2 / 10}{10-1}} = 9.04372097$$

H_0: $\mu = 18$ hours; H_1: $\mu \neq 18$ hours;   A two-tailed test.

$df = n - 1 = 10 - 1 = 9$ and $\alpha / 2 = .05 / 2 = .025$

Critical values: $t = -2.262$ and $t = 2.262$

$$s_{\bar{x}} = s / \sqrt{n} = 9.04372097 / \sqrt{10} = 2.85987568$$

$$t = (\bar{x} - \mu) / s_{\bar{x}} = (25.70 - 18) / 2.85987568 = 2.692$$

Reject H_0. Conclude that the claim of the earlier study is false.

9.78 $n = 12$, $\sum x = 885$, and $\sum x^2 = 78,045$

$$\bar{x} = \sum x / n = 885 / 12 = \$73.75$$

$$s = \sqrt{\frac{\sum x^2 - (\sum x)^2 / n}{n-1}} = \sqrt{\frac{78,045 - (885)^2 / 12}{12-1}} = 34.08045294$$

H_0: $\mu = \$65$; H_1: $\mu > \$65$;   A right-tailed test.

$df = n - 1 = 12 - 1 = 11$ and $\alpha = .01$

Critical value: $t = 2.718$

$$s_{\bar{x}} = s / \sqrt{n} = 34.08045294 / \sqrt{12} = 9.83817934$$

$$t = (\bar{x} - \mu) / s_{\bar{x}} = (73.75 - 65) / 9.83817934 = .889$$

Do not reject H_0. We cannot conclude that the mean amount spent after the campaign is greater than $65.

9.79   To make a hypothesis test about the mean amount spent on gas by all customers at the given gas station, first we will take a sample of less than 30 customers and collect the information on how much they spent on gas at this gas station. Then, using the formulas learned in Chapter 3 for the mean and standard deviation for sample data, we will calculate the sample mean and sample standard deviation. Assuming the amounts spent on gas by this station's customers are normally distributed, we will determine the significance level and make the test. Our hypotheses would be:

H_0: $\mu = \$10.90$ versus H_1: $\mu \neq \$10.90$

9.80 We should test the machine for several ($n < 30$) one–hour periods, recording the number of bolts produced in each period. Then, we would use the formulas from Chapter 3 to calculate the sample mean and sample standard deviation for the production data. Our hypotheses would be: H_0: $\mu \geq 88$, versus H_1: $\mu < 88$, where μ denotes the true mean number of bolts produced per hour by this machine. Assuming the number of bolts produced per hour by this machine is normally distributed, we would select a level of significance and carry out the hypothesis test of the mean (small sample).

9.81 In order to use the normal distribution in a test of hypothesis about a population proportion, both np and nq must be greater than 5, where p is the value of the population proportion in the null hypothesis and $q = 1 - p$.

9.82 a. No; $np = 40(.11) = 4.4 < 5$; not a large sample
   b. Yes; $np = 100(.73) = 73 > 5$, and $nq = 100(.27) = 27 > 5$
   c. No; $np = 80(.05) = 4 < 5$
   d. Yes; $np = 50(.14) = 7 > 5$, and $nq = 50(.86) = 43 > 5$

9.83 a. Yes; $np = 30(.65) = 19.5 > 5$; and $nq = 30(.35) = 10.5 > 5$
   b. No; $np = 70(.05) = 3.5 < 5$
   c. No; $np = 60(.06) = 3.6 < 5$
   d. Yes; $np = 900(.17) = 153 > 5$, and $nq = 900(.83) = 747 > 5$

9.84 a. The rejection region lies to the left of $z = -1.65$ and to the right of $z = 1.65$.
   The nonrejection region lies between $z = -1.65$ and $z = 1.65$.
   b. The rejection region lies to the left of $z = -2.33$.
   The nonrejection region lies to the right of $z = -2.33$.
   c. The rejection region lies to the right of $z = 1.65$.
   The nonrejection region lies to the left of $z = 1.65$.

9.85 a. The rejection region lies to the left of $z = -1.96$ and to the right of $z = 1.96$.
   The nonrejection region lies between $z = -1.96$ and $z = 1.96$.
   b. The rejection region lies to the left of $z = -2.05$.
   The nonrejection region lies to the right of $z = -2.05$.
   c. The rejection region lies to the right of $z = 1.96$.
   The nonrejection region lies to the left of $z = 1.96$.

9.86   a. Area in the right tail = α = .05

Critical value: $z = 1.65$

$$\sigma_{\hat{p}} = \sqrt{\frac{pq}{n}} = \sqrt{\frac{.30(.70)}{500}} = .02049390$$

Observed value of $z = \dfrac{\hat{p} - p}{\sigma_{\hat{p}}} = \dfrac{.38 - .30}{.02049390} = 3.90$

b. Area in each tail = $\alpha / 2 = .05 / 2 = .025$

Critical values: $z = -1.96$ and 1.96

Observed value of z is 3.90 as in part a.

9.87   a. Area in the left tail = α = .01

Critical value: $z = -2.33$

$$\sigma_{\hat{p}} = \sqrt{\frac{pq}{n}} = \sqrt{\frac{.63(.37)}{200}} = .03413942$$

Observed value of $z = \dfrac{\hat{p} - p}{\sigma_{\hat{p}}} = \dfrac{.60 - .63}{.03413942} = -.88$

b. Area in each tail = $\alpha / 2 = .01 / 2 = .005$

Critical values: $z = -2.58$ and 2.58

Observed value of z is $-.88$ as in part a.

9.88   a. The rejection region lies to the left of $z = -1.65$.

The nonrejection region lies to the right of $z = -1.65$.

b. The rejection region lies to the left of $z = -1.96$ and to the right of $z = 1.96$.

The nonrejection region lies between $z = -1.96$ and $z = 1.96$.

c. The rejection region lies to the right of $z = 1.65$.

The nonrejection region lies to the left of $z = 1.65$.

9.89   a. The rejection region lies to the left of $z = -2.33$.

The nonrejection region lies to the right of $z = -2.33$.

b. The rejection region lies to the left of $z = -2.58$ and to the right of $z = 2.58$.

The nonrejection region lies between $z = -2.58$ and $z = 2.58$.

c. The rejection region lies to the right of $z = 2.33$.

The nonrejection region lies to the left of $z = 2.33$.

9.90   a. Step 1: $H_0: p = .70$; $H_1: p \neq .70$;   A two-tailed test

Step 2: $np = 600(.70) = 420$ and $nq = 600(.30) = 180$

Since $np > 5$ and $nq > 5$, use the normal distribution.

Step 3:   For α = .01, the critical values of z are –2.58 and 2.58.

Step 4: $\sigma_{\hat{p}} = \sqrt{\dfrac{pq}{n}} = \sqrt{\dfrac{.70(.30)}{600}} = .01870829$

$z = \dfrac{\hat{p} - p}{\sigma_{\hat{p}}} = \dfrac{.68 - .70}{.01870829} = -1.07$

Step 5:   Do not reject Ho.

b. Steps 1, 2, and 3 are identical to part a.

Step 4: $z = \dfrac{\hat{p} - p}{\sigma_{\hat{p}}} = \dfrac{.76 - .70}{.01870829} = 3.21$

Step 5:   Reject H_0.

The results of parts a and b show that two different samples from the same population can yield opposite decisions on a test of the same hypothesis.

9.91   a. Step 1: $H_0: p = .45$; $H_1: p < .45$;   A left-tailed test.

Step 2: $np = 400(.45) = 180$ and $nq = 400(.55) = 220$

Since $np > 5$ and $nq > 5$, use the normal distribution.

Step 3:   For α = .025, the critical value of z is –1.96.

Step 4: $\sigma_{\hat{p}} = \sqrt{\dfrac{pq}{n}} = \sqrt{\dfrac{.45(.55)}{400}} = .02487469$

$z = \dfrac{\hat{p} - p}{\sigma_{\hat{p}}} = \dfrac{.42 - .45}{.02487469} = -1.21$

Step 5:   Do not reject H_0.

b. Steps 1, 2, and 3 are identical to part a.

Step 4: $z = \dfrac{\hat{p} - p}{\sigma_{\hat{p}}} = \dfrac{.39 - .45}{.02487469} = -2.41$

Step 5:   Reject H_0.

The results of parts a and b show that two different samples from the same population can yield opposite decisions on a test of the same hypothesis.

9.92   a. Step 1: $H_0: p = .45$; $H_1: p \neq .45$;   A two-tailed test.

Step 2: $np = 100(.45) = 45$ and $nq = 100(1 - .45) = 55$

Since $np > 5$ and $nq > 5$, use the normal distribution.

Step 3:   For α = .10, the critical values of z are –1.65 and 1.65.

Step 4: $p = .45$ and $q = 1 - p = 1 - .45 = .55$

$$\sigma_{\hat{p}} = \sqrt{\frac{pq}{n}} = \sqrt{\frac{.45(.55)}{100}} = .04974937$$

$$z = \frac{\hat{p} - p}{\sigma_{\hat{p}}} = \frac{.49 - .45}{.04974937} = .80$$

Step 5: Do not reject the null hypothesis.

b. Step 1: H_0: $p = .72$; H_1: $p < .72$;   A left-tailed test.

Step 2: $np = 700(.72) = 504$ and $nq = 700(1 - .72) = 196$

Since $np > 5$ and $nq > 5$, use the normal distribution.

Step 3: For $\alpha = .05$, the critical value of z is -1.65.

Step 4: $p = .72$ and $q = 1 - p = 1 - .72 = .28$

$$\sigma_{\hat{p}} = \sqrt{\frac{pq}{n}} = \sqrt{\frac{.72(.28)}{700}} = .01697056$$

$$z = \frac{\hat{p} - p}{\sigma_{\hat{p}}} = \frac{.64 - .72}{.01697056} = -4.71$$

Step 5: Reject the null hypothesis.

c. Step 1: H_0: $p = .30$; H_1: $p > .30$;   A right-tailed test.

Step 2: $np = 200(.30) = 60$ and $nq = 200(1 - .30) = 140$

Since $np > 5$, and $nq > 5$, use the normal distribution.

Step 3: For $\alpha = .01$, the critical value of z is 2.33.

Step 4: $p = .30$ and $q = 1 - p = 1 - .30 = .70$

$$\sigma_{\hat{p}} = \sqrt{\frac{pq}{n}} = \sqrt{\frac{.30(.70)}{200}} = .03240370$$

$$z = \frac{\hat{p} - p}{\sigma_{\hat{p}}} = \frac{.33 - .30}{.03240370} = .93$$

Step 5: Do not reject the null hypothesis.

9.93   a. Step 1: H_0: $p = .57$; H_1: $p \neq .57$;   A two-tailed test.

Step 2: $np = 800(.57) = 456$ and $nq = 800(1 - .57) = 344$

Since $np > 5$ and $nq > 5$, use the normal distribution.

Step 3: For $\alpha = .05$, the critical values of z are -1.96 and 1.96.

Step 4: $p = .57$ and $q = 1 - p = 1 - .57 = .43$

$$\sigma_{\hat{p}} = \sqrt{\frac{pq}{n}} = \sqrt{\frac{.57(.43)}{800}} = .01750357$$

$$z = \frac{\hat{p} - p}{\sigma_{\hat{p}}} = \frac{.50 - .57}{.01750357} = -4.00$$

Step 5: Reject the null hypothesis.

b. Step 1: H_0: $p = .26$; H_1: $p < .26$;   A left-tailed test.

Step 2: $np = 400(.26) = 104$ and $nq = 400(1 - .26) = 296$

          Since $np > 5$ and $nq > 5$, use the normal distribution.

Step 3:   For $\alpha = .01$, the critical value of z is -2.33.

Step 4: $p = .26$ and $q = 1 - p = 1 - .26 = .74$

$$\sigma_{\hat{p}} = \sqrt{\frac{pq}{n}} = \sqrt{\frac{.26(.74)}{400}} = .02193171$$

$$z = \frac{\hat{p} - p}{\sigma_{\hat{p}}} = \frac{.23 - .26}{.02193171} = -1.37$$

Step 5:   Do not reject the null hypothesis.

c. Step 1: H_0: $p = .84$; H_1: $p > .84$;   A right-tailed test.

Step 2: $np = 250(.84) = 210$ and $nq = 250(1 - .84) = 40$

          Since $np > 5$, and $nq > 5$, use the normal distribution.

Step 3:   For $\alpha = .025$, the critical value of is 1.96.

Step 4: $p = .84$ and $q = 1 - p = 1 - .84 = .16$

$$\sigma_{\hat{p}} = \sqrt{\frac{pq}{n}} = \sqrt{\frac{.84(.16)}{250}} = .02318620$$

$$z = \frac{\hat{p} - p}{\sigma_{\hat{p}}} = \frac{.85 - .84}{.02318620} = .43$$

Step 5:   Do not reject the null hypothesis.

9.94 H_0: $p = .209$; H_1: $p < .209$;   A left-tailed test.

For $\alpha = .05$, the critical value of z is -1.65.

$$\sigma_{\hat{p}} = \sqrt{pq/n} = \sqrt{(.209)(.791)/250} = .02571529$$

$$\hat{p} = .181$$

$$z = \frac{\hat{p} - p}{\sigma_{\hat{p}}} = \frac{.181 - .209}{.02571529} = -1.09$$

Do not reject H_0. Do not conclude that the percentage of registered voters who did not vote in November 2002 because they were too busy was less than 20.9%.

9.95 H_0: $p = .49$; H_1: $p > .49$;   A right-tailed test.

For $\alpha = .025$, the critical value of z is 1.96.

$$\sigma_{\hat{p}} = \sqrt{pq/n} = \sqrt{(.49)(.51)/200} = .03534827$$

$$\hat{p} = .52$$

$$z = \frac{\hat{p} - p}{\sigma_{\hat{p}}} = \frac{.52 - .49}{.03534827} = .85$$

Do not reject H_0. Do not conclude that the current percentage of management and professional jobs held by women exceeds 49%.

9.96 H_0: $p = .47$; H_1: $p \neq .47$;   A two-tailed test.

For $\alpha = .05$, the critical values of z are -1.96 and 1.96.

$\hat{p} = 430 / 1000 = .43$

$$\sigma_{\hat{p}} = \sqrt{pq / n} = \sqrt{(.47)(.53)/1000} = .01578290$$

$$z = \frac{\hat{p} - p}{\sigma_{\hat{p}}} = \frac{.43 - .47}{.01578290} = -2.53$$

Reject H_0. Conclude that the proportion of Americans who dream of owning their own business is different from 47%.

9.97 H_0: $p = .45$; H_1: $p > .45$;   A right-tailed test.

For $\alpha = .01$, the critical value of z is 2.33.

$\hat{p} = 248 / 500 = .496$

$$\sigma_{\hat{p}} = \sqrt{pq / n} = \sqrt{(.45)(.55)/500} = .02224860$$

$$z = \frac{\hat{p} - p}{\sigma_{\hat{p}}} = \frac{.496 - .45}{.02224860} = 2.07$$

Do not reject H_0. Do not conclude that the proportion of employers monitoring their employees use of company phones exceeds 45%.

9.98   a. H_0: $p = .40$; H_1: $p \neq .40$;   A two-tailed test.

For $\alpha = .02$, the critical values of z are -2.33 and 2.33.

$\hat{p} = 352 / 800 = .44$

$$\sigma_{\hat{p}} = \sqrt{pq / n} = \sqrt{(.40)(.60)/800} = .01732051$$

$$z = \frac{\hat{p} - p}{\sigma_{\hat{p}}} = \frac{.44 - .40}{.01732051} = 2.31$$

Do not reject H_0. Do not conclude that the current percentage of Americans who consider themselves somewhat or very overweight differs from 40%.

b. The Type I error would occur if the current percentage of Americans who consider themselves somewhat or very overweight was actually 40%, but we concluded that it differed from 40%.

$P(\text{Type I error}) = \alpha = .02$

9.99   a. $H_0: p = .32;$ $\qquad\qquad$ $H_1: p > .32;$ $\qquad\qquad$ A right-tailed test.

For $\alpha = .025$, the critical value of z is 1.96.

$\hat{p} = 396 \, / \, 1100 = .36$

$\sigma_{\hat{p}} = \sqrt{pq \, / \, n} = \sqrt{(.32)(.68) \, / \, 1100} = .01406479$

$z = \dfrac{\hat{p} - p}{\sigma_{\hat{p}}} = \dfrac{.36 - .32}{.01406479} = 2.84$

Reject H_0. Conclude that more than 32% of households earning \$75,000 or more would struggle to pay an unexpected bill of \$5000.

b.   The Type I error would occur if we concluded that more than 32% of households earning \$75,000 or more would struggle to pay an unexpected bill of \$5000, when actually it is not true. The probability of making this error is $\alpha = .025$.

9.100   a. $H_0: p \geq .35;$ $\qquad\qquad$ $H_1: p < .35;$ $\qquad\qquad$ A left-tailed test.

For $\alpha = .025$, the critical value of z is -1.96.

$\hat{p} = 112 \, / \, 400 = .28$

$\sigma_{\hat{p}} = \sqrt{\dfrac{pq}{n}} = \sqrt{\dfrac{.35(.65)}{400}} = .02384848$

$z = \dfrac{\hat{p} - p}{\sigma_{\hat{p}}} = \dfrac{.50 - .47}{.02384848} = -2.94$

Reject H_0. Conclude that the company should not market the yogurt.

b.   If $\alpha = 0$, there is no rejection region, so we could not reject H_0. Thus, we would conclude that the company should market the yogurt.

9.101   a. $H_0: p \geq .60;$ $\qquad\qquad$ $H_1: p < .60;$ $\qquad\qquad$ A left-tailed test.

For $\alpha = .01$, the critical value of z is -2.33.

$\hat{p} = 208 \, / \, 400 = .52$

$\sigma_{\hat{p}} = \sqrt{\dfrac{pq}{n}} = \sqrt{\dfrac{.60(.40)}{400}} = .02449490$

$z = \dfrac{\hat{p} - p}{\sigma_{\hat{p}}} = \dfrac{.52 - .60}{.02449490} = -3.27$

Reject H_0. Conclude that the company's claim is not true.

b.   If $\alpha = 0$, there is no rejection region, so we could not reject H_0. Thus, we would conclude that the company's claim is true.

9.102   a. $H_0: p \leq .05;$ $\qquad\qquad$ $H_1: p > .05;$ $\qquad\qquad$ A right-tailed test.

For $\alpha = .025$, the critical value of $z = 1.96$.

$$\hat{p} = 17/200 = .085$$

$$\sigma_{\hat{p}} = \sqrt{\frac{pq}{n}} = \sqrt{\frac{.05(.95)}{200}} = .01541104$$

$$z = \frac{\hat{p} - p}{\sigma_{\hat{p}}} = \frac{.085 - .05}{.01541104} = 2.27$$

Reject H_0. Conclude that the production process should be stopped.

b.   If $\alpha = .01$, the critical value of z is 2.33. Since the value of the test statistic is 2.27, we do not reject H_0. Thus, our decision differs from that of part a; we conclude that the process should not be stopped. The decisions in parts a and b are different. The results of this sample are not very conclusive, since lowering the significance level from 2.5% to 1% reverses the decision.

9.103   a. $H_0: p \le .07$; $H_1: p > .07$;   A right-tailed test.

For $\alpha = .02$, the critical value of $z = 2.05$.

$$\hat{p} = 22/200 = .11$$

$$\sigma_{\hat{p}} = \sqrt{\frac{pq}{n}} = \sqrt{\frac{.07(.93)}{200}} = .01804162$$

$$z = \frac{\hat{p} - p}{\sigma_{\hat{p}}} = \frac{.11 - .07}{.01804162} = 2.22$$

Reject H_0. Conclude that the machine should be stopped.

b.   If $\alpha = .01$, the critical value of z is 2.33. Since the value of the test statistic is 2.22, we do not reject H_0. Thus, our decision differs from that of part a; we conclude that the machine should not be stopped. The decisions in parts a and b are different. The results of this sample are not very conclusive, since lowering the significance level from 2% to 1% reverses the decision.

9.104   Take a (large) random sample of the bank's customers and find the percentage of customers in this sample who are satisfied. The hypotheses are: $H_0: p = .75$; $H_1: p \ne .75$

Choose a level of significance and determine the critical values of z. Using $\hat{p}$ from the sample data, calculate $z = \dfrac{\hat{p} - .75}{\sigma_{\hat{p}}}$. If H_0 is rejected, conclude that p has changed; if H_0 is not rejected, conclude that p is still .75.

9.105   Select a random sample of 40 students from your school and ask them whether or not they hold off–campus jobs. From this data calculate $\hat{p}$.

Using hypotheses $H_0: p = .65$ and $H_1: p \ne .65$, choose a level of significance and find the critical values from the table of the normal distribution.

Then compute $\sigma_{\hat{p}} = \sqrt{\dfrac{(.65)(.35)}{40}}$ and $z = \dfrac{\hat{p} - .65}{\sigma_{\hat{p}}}$ and make a decision.

9.106 H_0: $\mu = 120$; H_1: $\mu > 120$

$s_{\bar{x}} = s / \sqrt{n} = 15 / \sqrt{81} = 1.66666667$

$z = (\bar{x} - \mu) / s_{\bar{x}} = (123.5 - 120) / 1.66666667 = 2.10$

a. For $\alpha = .025$, the critical value of z is 1.96. Hence, reject H_0.

b. P(Type I error) $= \alpha = .025$

c. Area to the right of $z = 2.10$ is $.5 - .4821 = .0179$. Hence, p–value $= .0179$

   If $\alpha = .01$, do not reject H_0 since p–value $> .01$.

   If $\alpha = .05$, reject H_0 since p–value $< .05$.

9.107 H_0: $\mu = 40$; H_1: $\mu \neq 40$

$s_{\bar{x}} = s / \sqrt{n} = 6 / \sqrt{64} = .75$

$z = (\bar{x} - \mu) / s_{\bar{x}} = (38.4 - 40) / .75 = -2.13$

a. For $\alpha = .02$, the critical values of z are -2.33 and 2.33. Hence, do not reject H_0.

b. P(Type I error) $= \alpha = .02$

c. Area to the left of $z = -2.13$ is $.5 - .4834 = .0166$. Hence, p–value $= 2(.0166) = .0332$

   If $\alpha = .01$, do not reject H_0 since p–value $> .01$.

   If $\alpha = .05$, reject H_0 since p–value $< .05$.

9.108 H_0: $p = .82$; H_1: $p \neq .82$

$\sigma_{\hat{p}} = \sqrt{\dfrac{pq}{n}} = \sqrt{\dfrac{.82(.18)}{600}} = .01568439$

$z = \dfrac{\hat{p} - p}{\sigma_{\hat{p}}} = \dfrac{.86 - .82}{.01568439} = 2.55$

a. For $\alpha = .02$, the critical values of z are -2.33 and 2.33. Reject H_0.

b. P(Type I error) $= \alpha = .02$

c. Area to the right of $z = 2.55$ is $.5 - .4946 = .0054$. Hence, p–value $= 2(.0054) = .0108$

   If $\alpha = .025$, reject H_0, since p–value $< .025$.

   If $\alpha = .005$, do not reject H_0, since p–value $> .005$.

9.109 H_0: $p = .44$; H_1: $p < .44$

$\sigma_{\hat{p}} = \sqrt{\dfrac{pq}{n}} = \sqrt{\dfrac{.44(.56)}{450}} = .02339991$

$$z = \frac{\hat{p} - p}{\sigma_{\hat{p}}} = \frac{.39 - .44}{.02339991} = -2.14$$

a. For $\alpha = .02$, the critical value of z is -2.05. Reject H_0.

b. $P(\text{Type I error}) = \alpha = .02$

c. Area to the left of -2.14 is $.5 - .4838 = .0162$. Hence, p–value $= .0162$

If $\alpha = .01$, do not reject H_0, since p–value $> .01$.

If $\alpha = .025$, reject H_0, since p–value $< .025$.

9.110 $H_0: \mu = \$1770;$ $\qquad\qquad$ $H_1: \mu \neq \$1770$

$s_{\bar{x}} = s / \sqrt{n} = 460 / \sqrt{300} = 26.55811238$

$z = (\bar{x} - \mu) / s_{\bar{x}} = (1840 - 1770) / 26.55811238 = 2.64$

Area to the right of $z = 2.64$ is $.5 - .4959 = .0041$. Hence, p–value $= 2(.0041) = .0082$

Reject H_0 for $\alpha = .01$, since p–value $< .01$.

9.111 $H_0: \mu = 1245$ cubic feet; $\quad$ $H_1: \mu < 1245$ cubic feet

$s_{\bar{x}} = s / \sqrt{n} = 250 / \sqrt{100} = 25.00$

$z = (\bar{x} - \mu) / s_{\bar{x}} = (1175 - 1245) / 25.00 = -2.80$

Area to the left of $z = -2.80$ is $.5 - .4974 = .0026$. Hence, p–value $= .0026$

Reject H_0 for $\alpha = .025$, since p–value $< .025$.

9.112 a. $H_0: \mu \geq 50;$ $\qquad$ $H_1: \mu < 50;$ $\qquad\qquad$ A left-tailed test.

For $\alpha = .025$, the critical value of z is -1.96.

$s_{\bar{x}} = s / \sqrt{n} = 4 / \sqrt{36} = .66666667$

$z = (\bar{x} - \mu) / s_{\bar{x}} = (48 - 50) / .66666667 = -3.00$

Reject H_0. Conclude the site worker's claim is false.

b. The Type I error would be to conclude that the site worker's claim is false when it is actually true. This probability is $.025$.

c. If $\alpha = 0$, there is no rejection region so we cannot reject H_0. Thus, our conclusion of part a is changed.

9.113 a. $H_0: \mu = 180$ months; $\quad$ $H_1: \mu < 180$ months; $\qquad$ A left-tailed test.

For $\alpha = .02$, the critical value of z is -2.05.

$s_{\bar{x}} = s / \sqrt{n} = 27 / \sqrt{60} = 3.48568501$

$z = (\bar{x} - \mu) / s_{\bar{x}} = (171 - 180) / 3.48568501 = -2.58$

Reject H_0. The sample supports the alternative hypothesis that the current mean sentence for such crimes is less than 180 months.

  b. The Type I error would be to conclude that the current mean sentence for such crimes is less than 180 months when in fact it is not. $P(\text{Type I error}) = .02$

  c. If $\alpha = 0$, there is no rejection region and, hence, we cannot reject H_0. Thus, our conclusion would change.

9.114 a. H_0: $\mu = \$1,121,703$; H_1: $\mu \neq \$1,121,703$;   A two-tailed test.

  For $\alpha = .05$, the critical values of z are -1.96 and 1.96.

  $s_{\bar{x}} = s/\sqrt{n} = 180,000/\sqrt{36} = \$30,000$

  $z = (\bar{x} - \mu)/s_{\bar{x}} = (1,190,000 - 1,121,703)/30,000 = 2.28$

  Reject H_0. Conclude that the mean sale price of homes in this zip code is different from $\$1,121,703$

  b. For $\alpha = .01$, the critical values of z are -2.58 and 2.58.

  Do not reject H_0. The sample does not support the alternative hypothesis.

The results of parts a and b show that the sample does not support the alternative hypothesis very strongly, since lowering the significance level from .05 to .01 reverses the conclusion.

9.115 a. H_0: $\mu \leq 2400$ square feet; H_1: $\mu > 2400$ square feet;   A right-tailed test.

  For $\alpha = .05$, the critical value of z is 1.65.

  $s_{\bar{x}} = s/\sqrt{n} = 472/\sqrt{50} = 66.75088014$ square feet

  $z = (\bar{x} - \mu)/s_{\bar{x}} = (2540 - 2400)/66.75088014 = 2.10$

  Reject H_0. The sample supports the alternative hypothesis that the real estate agents' claim is false.

  b. For $\alpha = .01$, the critical value of z is 2.33.

  From part a, the observed value of z is 2.10. Hence, do not reject H_0.

The results of parts a and b show that the sample does not support the alternative hypothesis very strongly, since lowering the significance level from .05 to .01 reverses the conclusion.

9.116 H_0: $\mu \leq 15$ minutes; H_1: $\mu > 15$ minutes;   A right-tailed test.

For $\alpha = .01$, the critical value of z is 2.33.

  $s_{\bar{x}} = s/\sqrt{n} = 2.4/\sqrt{36} = .40$ minutes

  $z = (\bar{x} - \mu)/s_{\bar{x}} = (15.75 - 15)/.40 = 1.88$

Do not reject H_0. At the 1% level of significance, the difference between the sample mean and the hypothesized value of the population mean is small enough to attribute to chance. Thus, the journalist's conclusion is unfair to the restaurant.

9.117 H_0: $\mu = 8$ minutes; H_1: $\mu < 8$ minutes;   A left-tailed test.

For $\alpha = .025$, the critical value of z is -1.96.

$s_{\bar{x}} = s / \sqrt{n} = 2.1 / \sqrt{32} = .37123106$ minute

$z = (\bar{x} - \mu) / s_{\bar{x}} = (7.5 - 8) / .37123106 = -1.35$

Do not reject H_0. At the .025 level of significance, the difference between the sample mean and the hypothesized value of the population mean is small enough to attribute to chance. Thus, the manager's claim is not justified.

9.118 $H_0: \mu = \$30,765;$ $H_1: \mu > \$30,765;$ A right-tailed test.

$df = n - 1 = 20 - 1 = 19$ and $\alpha = .01$

The critical value of t is 2.539.

$s_{\bar{x}} = s / \sqrt{n} = 1300 / \sqrt{20} = 290.6888371$

$t = (\bar{x} - \mu) / s_{\bar{x}} = (31,400 - 30,765) / 290.6888371 = 2.184$

Do not reject H_0. Do not conclude the mean resale price of Mercedes CLKs exceeds \$30,765.

9.119 $H_0: \mu = 25$ minutes; $H_1: \mu \neq 25$ minutes;   A two-tailed test.

$df = n - 1 = 16 - 1 = 15$ and $\alpha / 2 = .01 / 2 = .005$

The critical values of t are -2.947 and 2.947.

$s_{\bar{x}} = s / \sqrt{n} = 4.8 / \sqrt{16} = 1.20$

$t = (\bar{x} - \mu) / s_{\bar{x}} = (27.5 - 25) / 1.20 = 2.083$

Do not reject H_0.

9.120   a. $H_0: \mu = 36$ minutes; $H_1: \mu < 36$ minutes;   A left-tailed test.

$df = n - 1 = 25 - 1 = 24$ and $\alpha = .025$

The critical value of t is -2.064.

$s_{\bar{x}} = s / \sqrt{n} = 8.5 / \sqrt{25} = 1.7$ minutes

$t = (\bar{x} - \mu) / s_{\bar{x}} = (32.1 - 36) / 1.7 = -2.294$

Reject H_0. Conclude that the mean duration of all such visits is less than 36 minutes.

b.   If $\alpha = 0$, there is no rejection region, so we cannot reject H_0. Thus, there is no need to go through the five steps of hypothesis testing.

9.121   a. $H_0: \mu = 114$ minutes; $H_1: \mu < 114$ minutes;   A left-tailed test.

$df = n - 1 = 25 - 1 = 24$ and $\alpha = .01$

The critical value of t is -2.492.

$s_{\bar{x}} = s / \sqrt{n} = 11 / \sqrt{25} = 2.20$ minutes

$t = (\bar{x} - \mu) / s_{\bar{x}} = (109 - 114) / 2.20 = -2.273$

Do not reject H_0. The mean time currently spent by all adults with their families is not less than 114 minutes a day.

b. If $\alpha = 0$, there is no rejection region, so there is no need to go through the five steps of hypothesis testing. We cannot reject H_0.

9.122 $H_0: \mu \leq 2$ hours; $H_1: \mu > 2$ hours;   A right-tailed test.

df $= n - 1 = 12 - 1 = 11$ and $\alpha = .01$

The critical value of t is 2.718.

$\sum x = 26.80$, and $\sum x^2 = 62.395$

$\bar{x} = \sum x / n = 26.80 / 12 = 2.233$

$$s = \sqrt{\frac{\sum x^2 - (\sum x)^2 / n}{n - 1}} = \sqrt{\frac{62.395 - (26.80)^2 / 12}{12 - 1}} = .48068764$$

$s_{\bar{x}} = s / \sqrt{n} = .48068764 / \sqrt{12} = .13876257$

$t = (\bar{x} - \mu) / s_{\bar{x}} = (2.233 - 2) / .13876257 = 1.679$ Do not reject H_0.

9.123 $H_0: \mu \leq 150$ calories; $H_1: \mu > 150$ calories

$df = n - 1 = 10 - 1 = 9$ and $\alpha = .025$

The critical value of t is 2.262.

$\sum x = 1527$, and $\sum x^2 = 233,663$

$\bar{x} = \sum x / n = 1527 / 10 = 152.7$

$$s = \sqrt{\frac{\sum x^2 - (\sum x)^2 / n}{n - 1}} = \sqrt{\frac{233,663 - (1527)^2 / 10}{10 - 1}} = 7.37940076$$

$s_{\bar{x}} = s / \sqrt{n} = 7.37940076 / \sqrt{10} = 2.33357142$

$t = (\bar{x} - \mu) / s_{\bar{x}} = (152.70 - 150) / 2.33357142 = 1.157$ Do not reject H_0.

9.124   a. $H_0: p = .15$; $H_1: p \neq .15$;   A two-tailed test.

For $\alpha = .02$, the critical values of z are -2.33 and 2.33

$$\sigma_{\hat{p}} = \sqrt{\frac{pq}{n}} = \sqrt{\frac{.15(.85)}{600}} = .01457738$$

$$z = \frac{\hat{p} - p}{\sigma_{\hat{p}}} = \frac{.17 - .15}{.01457738} = 1.37$$ Do not reject H_0.

b. The Type I error would be to conclude that the percentage of all people who could name the Chief Supreme Court Justice was different from 15% when it is actually equal to 15%.

9.125   a. H_0: $p = .32$; H_1: $p > .32$;   A right-tailed test.

For $\alpha = .02$, the critical value of $z = 2.05$.

$$\sigma_{\hat{p}} = \sqrt{\frac{pq}{n}} = \sqrt{\frac{.32(.68)}{850}} = .016$$

$$z = \frac{\hat{p} - p}{\sigma_{\hat{p}}} = \frac{.35 - .32}{.016} = 1.88$$

Do not reject H_0. Do not conclude that the current percentage of faculty who hold this opinion exceeds 32%.

b.   The Type I error would occur if we concluded that the current percentage of the faculty who hold this opinion exceeds 32%, when in fact it does not. P(Type I error) = .02.

9.126 H_0: $p = .40$; H_1: $p \neq .40$;   A two-tailed test.

For $\alpha = .01$, the critical values of z are -2.58 and 2.58

$\hat{p} = 259 / 700 = .37$

$$\sigma_{\hat{p}} = \sqrt{\frac{pq}{n}} = \sqrt{\frac{.40(.60)}{700}} = .01851640$$

$$z = \frac{\hat{p} - p}{\sigma_{\hat{p}}} = \frac{.37 - .40}{.01851640} = -1.62$$

Do not reject H_0. We cannot conclude that the percentage of people who buy national brand coffee is different from 40%.

9.127 H_0: $p = .56$; H_1: $p < .56$;   A left-tailed test.

For $\alpha = .02$, the critical value of z is -2.05.

$\hat{p} = .50$

$$\sigma_{\hat{p}} = \sqrt{\frac{pq}{n}} = \sqrt{\frac{.56(.44)}{150}} = .04052982$$

$$z = \frac{\hat{p} - p}{\sigma_{\hat{p}}} = \frac{.50 - .56}{.04052982} = -1.48$$

Do not reject H_0. Do not conclude that the current percentage of attorneys who take work on vacation is less than 56%.

9.128   a. H_0: $p = .80$; H_1: $p < .80$;   A left-tailed test.

For $\alpha = .01$, the critical value of $z = -2.33$.

$$\sigma_{\hat{p}} = \sqrt{\frac{pq}{n}} = \sqrt{\frac{.80(.20)}{40}} = .06324555$$

$$z = \frac{\hat{p} - p}{\sigma_{\hat{p}}} = \frac{.75 - .80}{.06324555} = -.79$$

Do not reject H_0. Do not conclude that the company's claim is false.

b. If $\alpha = 0$, there is no rejection region; thus, we cannot reject H_0 and cannot conclude that the company's claim is false.

9.129   a. $H_0: p \geq .90$; $H_1: p < .90$;   A left-tailed test.

For $\alpha = .02$, the critical value of z is -2.05.

$$\sigma_{\hat{p}} = \sqrt{\frac{pq}{n}} = \sqrt{\frac{.90(.10)}{90}} = .03162278$$

$$\hat{p} = 75 / 90 = .833$$

$$z = \frac{\hat{p} - p}{\sigma_{\hat{p}}} = \frac{.833 - .90}{.03162278} = -2.12$$

Reject H_0. Conclude that the company's policy is not maintained.

b. If $\alpha = 0$, there is no rejection region; thus, we cannot reject H_0 and cannot conclude that the company's policy is not maintained.

9.130 $H_0: p = .32$; $H_1: p > .32$;   A right-tailed test.

$$\sigma_{\hat{p}} = \sqrt{\frac{pq}{n}} = \sqrt{\frac{.32(.68)}{850}} = .016$$

$$z = \frac{\hat{p} - p}{\sigma_{\hat{p}}} = \frac{.35 - .32}{.016} = 1.88$$

The area to the right of $z = 1.88$ under the normal curve is $.5 - .4699 = .0301$

Hence, p–value $= .0301$

If $\alpha = .05$, reject H_0 since p–value $< .05$.

If $\alpha = .01$, do not reject H_0 since p–value $> .01$.

9.131 $H_0: p \geq .90$; $H_1: p < .90$;   A left-tailed test.

$$\sigma_{\hat{p}} = \sqrt{\frac{pq}{n}} = \sqrt{\frac{.90(.10)}{90}} = .03162278$$

$$\hat{p} = 75 / 90 = .833$$

$$z = \frac{\hat{p} - p}{\sigma_{\hat{p}}} = \frac{.833 - .90}{.03162278} = -2.12$$

The area to the left of $z = -2.12$ under the normal curve is $.5 - .4830 = .0170$.

Hence, p–value $= .0170$

If $\alpha = .05$, reject H_0 since p–value $< .05$.

If $\alpha = .01$, do not reject H_0 since p–value $> .01$.

9.132 a. P(a student guessing randomly wins all five bets) $= \left(\dfrac{18}{38}\right)^5 = .0238$

b. The significance level of α represents the probability of a Type I error, i.e., concluding that a student is not guessing when in fact the student is guessing. From part a we know that the probability that we reject H_0 despite the fact that the student is guessing is .0238.

c. $\mu = np = 100 \cdot (.0238) = 2.38$. So we would expect 2 students to win all five bets even if all the students are guessing.

9.133 Since $np > 5$ and $nq > 5$, the distribution of $\hat{p}$ is approximately normal.

For $\alpha = .05$, the critical value of z is 1.65.

$$\sigma_{\hat{p}} = \sqrt{\frac{pq}{n}} = \sqrt{\frac{.04(.96)}{130}} = .01718676$$

Now $z = \dfrac{\hat{p} - p}{\sigma_{\hat{p}}}$, so $1.65 = \dfrac{\hat{p} - .04}{.01718676}$

Solving for $\hat{p}$ yields $\hat{p} = 1.65(.01718676) + .04 = .0684$

Thus, we would reject H_0 if $\hat{p} > .0684$.

Hence, $.0684 = \dfrac{c}{n} = \dfrac{c}{130}$

Then, $c = 130(.0684) = 8.89 \approx 9$

Therefore, reject H_0 and shut down the machine if the number of defectives in a sample of 130 parts is 9 or more.

9.134 $H_0 ; p \le .15;$ $\qquad\qquad H_1 : p > .15$

Let x be the number of late arrivals in a random sample of 50 flights. The significance level, α, is the probability of rejecting H_0 when H_0 is true. In this case, $\alpha = P(x > 9 \mid p = .15)$.

If H_0 is true, x is a binomial random variable with $n = 50$ and $p = .15$. Since np and nq are both greater than 5, we may use a normal approximation.

$\mu = np = 50(.15) = 7.5$ and $\sqrt{npq} = \sqrt{50(.15)(.85)} = 2.52487623$

Correcting for continuity, $\alpha = P(x > 8.5)$.

For $x = 8.5$: $z = (8.5 - 7.5) / 2.52487623 = .40$ Thus, $\alpha = P(x > 8.5) = P(z > .40) = .5 - .1554 = .3446$ approximately. Hence, the significance level for this test is approximately .34.

9.135 First, we must be sure that the cure rate of the new therapy is not lower than that of the old therapy. Let p be the proportion of all patients cured with the new therapy. We must test the hypotheses:

$H_0: p = .60;$ $H_1: p < .60$

For $\alpha = .01$, the critical value of z is -2.33.

$$\sigma_{\hat{p}} = \sqrt{\frac{pq}{n}} = \sqrt{\frac{.60(.40)}{200}} = .03464102$$

From the data: $\hat{p} = x / n = 108 / 200 = .54$

Test statistic: $z = \dfrac{\hat{p} - p}{\sigma_{\hat{p}}} = \dfrac{.54 - .60}{.03464102} = -1.73$

Do not reject H_0. Thus, we cannot conclude that the new therapy has a lower cure rate. Next, we must see if the new therapy is effective in reducing the number of visits. Let μ be the mean number of visits required for all patients using the new therapy regime. We must test the hypotheses:

$H_0: \mu = 140$ visits; $H_1: \mu < 140$ visits

For $\alpha = .01$, the critical value of z is -2.33.

$$s_{\bar{x}} = s / \sqrt{n} = 38 / \sqrt{200} = 2.68700577$$

Test statistic: $z = (\bar{x} - \mu) / s_{\bar{x}} = (132 - 140) / 2.68700577 = -2.98$

Reject H_0. Conclude that the new therapy regime requires fewer visits on average. Based on the results of these two hypothesis tests, the health care provider should support the new therapy regime.

9.136   Let $\mu = $ mean life of these light bulbs

$H_0: \mu = 750$ hours; $H_1: \mu < 750$ hours;

$\sigma = 50$ and $n = 64$

$$\sigma_{\bar{x}} = \sigma / \sqrt{n} = 50 / \sqrt{64} = 6.25$$

First decision rule: Reject H_0 if $\bar{x} < 735$

For $\bar{x} = 735$: $z = (\bar{x} - \mu) / \sigma_{\bar{x}} = (735 - 750) / 6.25 = -2.40$

Thus, $\alpha = P(\bar{x} < 735 \mid \mu = 750) = P(z < -2.40) = .5 - .4918 = .0082$

Using this decision rule, the probability of concluding that GE has printed too high an average length of life on the package when in fact they have not is .0082.

Second decision rule: "Reject H_0 if $\bar{x} < 700$"

For $\bar{x} = 700$: $z = (\bar{x} - \mu) / \sigma_{\bar{x}} = (700 - 750) / 6.25 = -8.00$

Thus, $\alpha = P(\bar{x} < 700 \mid \mu = 750) = P(z < -8.00) = .5 - .5 = .0000$ approximately

Using this decision rule, there is virtually no chance of concluding that GE has printed too high an average length of life on the package when in fact they have not.

9.137   a. Let p be the proportion of all people receiving the new vaccine who contract the disease within a year. Then the appropriate hypotheses are: $H_0: p = .30;$ $H_1: p < .30$

b. Let x be the number of people in a sample of 100 inoculated with the new vaccine who contract the disease within a year. Then, under H_0, x is a binomial random variable with $n = 100$ and $p = .30$.

Hence, $\mu = np = 100(.30) = 30$

$\sigma = \sqrt{npq} = \sqrt{100(.30)(.70)} = 4.58257569$

If 84 or more of the 100 people in the sample do not contract the disease, then $x < 16$. Using a normal approximation, and correcting for continuity, we need $P(x < 16.5)$.

For $x = 16.5$: $z = \dfrac{16.5 - 30}{4.58257569} = -2.95$

Thus, $\alpha = P(\bar{x} < 16.5 \mid p = .30) = P(z < -2.95) = .5 - .4984 = .0016$

c. Let x be the number of people in a sample of 20 inoculated with the new vaccine who contract the disease within a year. Then, under H_0, x is a binomial random variable with $n = 20$ and $p = .30$. Using Table IV, Appendix C:

$\alpha = P(x < 3 \mid p = .30) = P(0) + P(1) + P(2) = .0008 + .0068 + .0278 = .0354$

9.138 a. Let μ = the average thickness of the ice on the pond

b. The appropriate hypotheses are:

H_0: $\mu \leq 5$ inches and H_1: $\mu > 5$ inches

c. A Type I error would be to allow skating when the ice is too thin.

A Type II error would be to prohibit skating when the ice is thick enough.

d. The significance level should be small, because committing a Type I error could result in serious injury or death. On the other hand, committing a Type II error would merely cause inconvenience. A significance level of $\alpha = .005$ might be appropriate. This would mean that the probability of allowing skating when the ice is too thin would be .005, or 1/200.

9.139 The following are two possible experiments we might conduct to investigate the effectiveness of middle taillights.

I. Let: p_1 = proportion of all collisions involving cars built since 1984 that were rear–end collisions.

Let: p_2 = proportion of all collisions involving cars built before 1984 that were rear–end collisions. (p_2 would be known)

We would test H_0: $p_1 = p_2$ versus H_1: $p_1 < p_2$.

We would take a random sample of collisions involving cars built since 1984 and determine the number that were rear–end collisions. We would have to assume the following:

i. The only change in cars built since 1984 that would reduce rear–end collisions is the middle taillight.

ii. None of the cars built before 1984 had middle taillights.

iii. People's driving habits, traffic volume, and other variables that might affect rear–end collisions have not changed appreciably since 1984.

II. Let: μ_1 = mean number of rear–end collisions per 1000 cars built since 1984

μ_2 = mean number of rear–end collisions per 1000 cars built before 1984 (μ_2 would be known)

We would test H_0: $\mu_1 = \mu_2$ versus H_1: $\mu_1 < \mu_2$.

We could take several random samples of 1000 cars built since 1984. We would determine the number of rear–end collisions in each sample of 1000 cars.

We would find the mean and standard deviation of these numbers and use them to form the test statistic.

We would require the same assumptions as those listed for the test in part 1. Also, if we took less than 30 samples, we would have to assume that the number of rear–end collisions per 1000 cars has a normal distribution.

9.140   a. H_0: $p = .50$; H_1: $p > .50$

b. Let x be the number of heads obtained in 10 tosses of the coin. Then x is a binomial random variable with $n = 10$ and (when H_0 is true) $p = .50$

Using the referee's decision rule, $\alpha = P(\text{reject } H_0 \mid H_0 \text{ true}) = P(x > 8 \mid p = .50) = P(8) + P(9) + P(10)$

$= .0439 + .0098 + .0010$ (from Table IV of Appendix C) $= .0547$

Self–Review Test for Chapter Nine

1. a	2. b	3. a	4. b	5. a	6. a	7. a	8. b
9. c	10. a	11. c	12. b	13. d	14. c	15. a	16. b

17.   a. Step 1: H_0: $\mu = \$85,900$; H_1: $\mu > 85,900$;   A right-tailed test.

Step 2:   Since $n > 30$, use the normal distribution.

Step 3: $s_{\bar{x}} = s / \sqrt{n} = 27,000 / \sqrt{36} = \$4500$

$z = (\bar{x} - \mu) / s_{\bar{x}} = (95,000 - 85,900) / 4500 = 2.02$

The area to the right of $z = 2.02$ under the normal curve is $.5 - .4783 = .0217$. Hence, p–value $= .0217$

b. If $\alpha = .01$, do not reject H_0, since p–value $> .01$.

If $\alpha = .05$, reject H_0, since p–value $< .05$.

18.   a. Step 1: H_0: $\mu = 185$ minutes; H_1: $\mu < 185$ minutes

Step 2:   Since $n > 30$, use the normal distribution.

Step 3:   For $\alpha = .01$, the critical value of z is -2.33.

Step 4 $s_{\bar{x}} = s / \sqrt{n} = 12 / \sqrt{36} = 2$

$z = (\bar{x} - \mu) / s_{\bar{x}} = (179 - 185) / 2 = -3.00$

Step 5:   Reject H_0. Conclude that the mean durations of games have decreased after the meeting.

b. The Type I error would be to conclude that the mean durations of games have decreased after the meeting when they are actually equal to the duration of games before the meeting.

P(Type I error) $= \alpha = .01$

c. If $\alpha = 0$, there is no rejection region, so do not reject H_0.

d. From part a, $z = -3.00$. The area to the left of $z = -3.00$ under the normal curve is $.5 - .4987 = .0013$. Hence, p–value $= .0013$.   For $\alpha = .01$, reject H_0, since p–value $< .01$.

19. a. $H_0: \mu \geq 31$ months; $H_1: \mu < 31$ months

The critical value of t is: -2.131

$s_{\bar{x}} = s / \sqrt{n} = 7.2 / \sqrt{16} = 1.80$

$t = (\bar{x} - \mu) / s_{\bar{x}} = (25 - 31) / 1.80 = -3.333$ Reject H_0. Conclude that the editor's claim is false.

b. The Type I error would be to conclude that the editor's claim is false when it is actually true.

P(Type I error) $= \alpha = .025$

c. For $\alpha = .001$, the critical value of t is -3.733. Since the observed value of $t = -2.778$ is greater than -3.733, we would not reject H_0.

20. a. $H_0: p = .50$; $H_1: p < .50$;   A left-tailed test.

For $\alpha = .05$, the critical value of z is -1.65.

$\hat{p} = 450 / 1000 = .45$

$\sigma_{\hat{p}} = \sqrt{\dfrac{pq}{n}} = \sqrt{\dfrac{.50(.50)}{1000}} = .015811388$

$z = \dfrac{\hat{p} - p}{\sigma_{\hat{p}}} = \dfrac{.45 - .50}{.015811388} = -3.16$ Reject H_0. Conclude that the less than 50% have a will.

b. The Type I error would be to conclude that the percentage of adults with wills was less than 50 % when it is actually 50%. P(Type I error) $= \alpha = .05$

c. If $\alpha = 0$, there is no rejection region, so do not reject H_0.

21. a. Referring to Problem 20.

The area to the left of $z = -3.16$ under the normal curve is $.5 - .5000 = .0000$.

Hence, p–value $= .0000$

b. If $\alpha = .05$, reject H_0, since p–value $< .05$

If $\alpha = .01$, reject H_0, since p–value $< .01$

Chapter Ten

10.1 The two samples are independent if they are drawn from two different populations and the elements of the two samples are not related. As an example, suppose we want to estimate the difference between the salaries of male and female university professors. To do so, we will select two samples from two different populations, one from all male university professors and the second from all female university professors. These two populations will include different elements that are not related.

In two dependent samples, the elements of one sample are related to the elements of the second sample. To test if a certain course that claims to reduce stress, does indeed decrease stress, we will take a sample of people who are suffering from stress. We will measure the stress level for these people before they take this course and then after they finish it. Based on these results we will make a decision. Notice, that in this example, we have the same group of people for two samples of data, one before taking the course and the second after completing the course.

10.2 The sampling distribution of $\bar{x}_1 - \bar{x}_2$ for large and independent samples is approximately normal. The mean of $\bar{x}_1 - \bar{x}_2$ is $\mu_1 - \mu_2$, where μ_1 and μ_2 are the population means. The standard deviation of

$\bar{x}_1 - \bar{x}_2$ is $\sqrt{\dfrac{\sigma_1^2}{n_1} + \dfrac{\sigma_2^2}{n_2}}$, where σ_1 and σ_2 are the population standard deviations and n_1 and n_2 are the

sample sizes.

10.3 a. The point estimate of $\mu_1 - \mu_2$ is $\bar{x}_1 - \bar{x}_2 = 5.56 - 4.80 = .76$

$$s_{\bar{x}_1-\bar{x}_2} = \sqrt{\frac{s_1^2}{n_1} + \frac{s_2^2}{n_2}} = \sqrt{\frac{(1.65)^2}{240} + \frac{(1.58)^2}{270}} = .14349103$$

Margin of error $= \pm 1.96 s_{\bar{x}_1-\bar{x}_2} = \pm 1.96(.14349103) = \pm.28$

b. The z value for the 99% confidence level is 2.58.

The 99% confidence interval for $\mu_1 - \mu_2$ is:

$$(\bar{x}_1 - \bar{x}_2) \pm z s_{\bar{x}_1-\bar{x}_2} = .76 \pm 2.58\,(.14349103) = 0.76 \pm .37 = .39 \text{ to } 1.13$$

10.4   a. The point estimate of $\mu_1 - \mu_2$ is $\bar{x}_1 - \bar{x}_2 = 22.0 - 27.6 = -5.6$

$$s_{\bar{x}_1 - \bar{x}_2} = \sqrt{\frac{s_1^2}{n_1} + \frac{s_2^2}{n_2}} = \sqrt{\frac{(4.9)^2}{300} + \frac{(4.5)^2}{250}} = .40128959$$

Margin of error $= \pm 1.96 s_{\bar{x}_1 - \bar{x}_2} = \pm 1.96(.40128959) = \pm.79$

b. The z value for the 95% confidence level is 1.96.

The 95% confidence interval for $\mu_1 - \mu_2$ is:

$$(\bar{x}_1 - \bar{x}_2) \pm z s_{\bar{x}_1 - \bar{x}_2} = -5.6 \pm 1.96\,(.40128959) = -5.6 \pm .79 = -6.39 \text{ to } -4.81$$

10.5 $H_0: \mu_1 - \mu_2 = 0;$ $H_1: \mu_1 - \mu_2 \neq 0$

For $\alpha = .05$, the critical values of z are -1.96 and 1.96.

From Exercise 10.3, $s_{\bar{x}_1 - \bar{x}_2} = .14349103$

$$z = \frac{(\bar{x}_1 - \bar{x}_2) - (\mu_1 - \mu_2)}{s_{\bar{x}_1 - \bar{x}_2}} = \frac{(5.56 - 4.80) - 0}{.14349103} = 5.30 \quad \text{Reject } H_0.$$

10.6 $H_0: \mu_1 - \mu_2 = 0;$ $H_1: \mu_1 - \mu_2 \neq 0$

For $\alpha = .01$, the critical values of z are -2.58 and 2.58.

From Exercise 10.4, $s_{\bar{x}_1 - \bar{x}_2} = .40128959$

$$z = \frac{(\bar{x}_1 - \bar{x}_2) - (\mu_1 - \mu_2)}{s_{\bar{x}_1 - \bar{x}_2}} = \frac{(22.0 - 27.6) - 0}{.40128959} = -13.96 \quad \text{Reject } H_0.$$

10.7 $H_0: \mu_1 - \mu_2 = 0;$ $H_1: \mu_1 - \mu_2 > 0$

For $\alpha = .01$, the critical value of z is 2.33.

From Exercise 10.3, $s_{\bar{x}_1 - \bar{x}_2} = .14349103$

$$z = \frac{(\bar{x}_1 - \bar{x}_2) - (\mu_1 - \mu_2)}{s_{\bar{x}_1 - \bar{x}_2}} = \frac{(5.56 - 4.80) - 0}{.14349103} = 5.30 \quad \text{Reject } H_0.$$

10.8 $H_0: \mu_1 - \mu_2 = 0;$ $H_1: \mu_1 - \mu_2 < 0$

For $\alpha = .05$, the critical value of z is -1.65.

From Exercise 10.4, $s_{\bar{x}_1 - \bar{x}_2} = .40128959$

$$z = \frac{(\bar{x}_1 - \bar{x}_2) - (\mu_1 - \mu_2)}{s_{\bar{x}_1 - \bar{x}_2}} = \frac{(22.0 - 27.6) - 0}{.40128959} = -13.96 \quad \text{Reject } H_0.$$

10.9   a. The point estimate of $\mu_1 - \mu_2$ is $\bar{x}_1 - \bar{x}_2 = 6.7 - 6.3 = .4$ correct answer.

$$s_{\bar{x}_1-\bar{x}_2} = \sqrt{\frac{s_1^2}{n_1} + \frac{s_2^2}{n_2}} = \sqrt{\frac{(.7)^2}{400} + \frac{(.8)^2}{600}} = .04787136$$

Margin of error $= \pm 1.96 \, s_{\bar{x}_1-\bar{x}_2} = \pm 1.96(.04787136) = \pm .09$ correct answer.

b. The z value for the 98% confidence level is 2.33.

   The 98% confidence interval for $\mu_1 - \mu_2$ is:

   $$(\bar{x}_1 - \bar{x}_2) \pm zs_{\bar{x}_1-\bar{x}_2} = .4 \pm 2.33 \, (.04787136) = .4 \pm .11 = .29 \text{ to } .51 \text{ correct answer.}$$

c. $H_0: \mu_1 - \mu_2 = 0;$ $H_1: \mu_1 - \mu_2 > 0$

   For $\alpha = .05$, the critical value of z is 1.65.

   $$z = \frac{(\bar{x}_1 - \bar{x}_2) - (\mu_1 - \mu_2)}{s_{\bar{x}_1-\bar{x}_2}} = \frac{(.4) - 0}{.04787136} = 8.36$$

   Reject H_0. Conclude that the mean correct response rate for Germans exceeds the Swedish one.

10.10   a. The point estimate of $\mu_1 - \mu_2$ is $\bar{x}_1 - \bar{x}_2 = 101 - 92 = 9$ hours.

   $$s_{\bar{x}_1-\bar{x}_2} = \sqrt{\frac{s_1^2}{n_1} + \frac{s_2^2}{n_2}} = \sqrt{\frac{(15)^2}{35} + \frac{(10)^2}{35}} = 3.04724700$$

   Margin of error $= \pm 1.96 \, s_{\bar{x}_1-\bar{x}_2} = \pm 1.96(3.04724700) = \pm 5.97$ hours

b. The z value for the 97% confidence level is 2.17.

   The 97% confidence interval for $\mu_1 - \mu_2$ is:

   $$(\bar{x}_1 - \bar{x}_2) \pm zs_{\bar{x}_1-\bar{x}_2} = 9 \pm 2.17 \, (3.04724700) = 9 \pm 6.61 = 2.39 \text{ to } 15.61 \text{ hours}$$

c. $H_0: \mu_1 - \mu_2 = 0;$ $H_1: \mu_1 - \mu_2 \neq 0$

   For $\alpha = .02$, the critical values of z are -2.33 and 2.33.

   $$z = \frac{(\bar{x}_1 - \bar{x}_2) - (\mu_1 - \mu_2)}{s_{\bar{x}_1-\bar{x}_2}} = \frac{9 - 0}{3.04724700} = 2.95$$

   Reject H_0. Conclude that the mean repellent time for deer with the two repellents is different.

10.11   a. The point estimate of $\mu_1 - \mu_2$ is $\bar{x}_1 - \bar{x}_2 = \$85.69 - \$84.58 = \$1.11$.

   $$s_{\bar{x}_1-\bar{x}_2} = \sqrt{\frac{s_1^2}{n_1} + \frac{s_2^2}{n_2}} = \sqrt{\frac{(18.5)^2}{1000} + \frac{(18)^2}{1100}} = .79799465$$

   Margin of error $= \pm 1.96 \, s_{\bar{x}_1-\bar{x}_2} = \pm 1.96(.79799465) = \pm \$1.56$

b. The z value for the 90% confidence level is 1.65.

The 90% confidence interval for $\mu_1 - \mu_2$ is:

$$(\bar{x}_1 - \bar{x}_2) \pm z s_{\bar{x}_1 - \bar{x}_2} = 1.11 \pm 1.65 \,(.79799465) = 1.11 \pm 1.32 = -\$.21 \text{ to } \$2.43$$

c. $H_0: \mu_1 - \mu_2 = 0;$ $\qquad H_1: \mu_1 - \mu_2 > 0.$

For $\alpha = .01$, the critical value of z is 2.33.

$$z = \frac{(\bar{x}_1 - \bar{x}_2) - (\mu_1 - \mu_2)}{s_{\bar{x}_1 - \bar{x}_2}} = \frac{(1.11) - 0}{.79799465} = 1.39$$

Do not reject H_0. Do not conclude that the mean hotel room rate in 2001 exceeded that of 2000.

10.12 a. The point estimate of $\mu_1 - \mu_2$ is $\bar{x}_1 - \bar{x}_2 = 21 - 16 = 5$ miles.

$$s_{\bar{x}_1 - \bar{x}_2} = \sqrt{\frac{s_1^2}{n_1} + \frac{s_2^2}{n_2}} = \sqrt{\frac{(5.2)^2}{500} + \frac{(4.4)^2}{400}} = .32012498$$

Margin of error $= \pm 1.96 \, s_{\bar{x}_1 - \bar{x}_2} = \pm 1.96 (.32012498) = \pm .63$ miles

b. The z value for the 97% confidence level is 2.17.

The 97% confidence interval for $\mu_1 - \mu_2$ is:

$$(\bar{x}_1 - \bar{x}_2) \pm z s_{\bar{x}_1 - \bar{x}_2} = 5 \pm 2.17 \,(.32012498) = 5 \pm .69 = 4.31 \text{ to } 5.69 \text{ miles}$$

c. $H_0: \mu_1 - \mu_2 = 0;$ $\qquad H_1: \mu_1 - \mu_2 \neq 0$

For $\alpha = .02$, the critical values of z are -2.33 and 2.33.

$$z = \frac{(\bar{x}_1 - \bar{x}_2) - (\mu_1 - \mu_2)}{s_{\bar{x}_1 - \bar{x}_2}} = \frac{(5) - 0}{.32012498} = 15.62$$

Reject H_0. Conclude that the mean commuting distance is different for males and females.

10.13 a. $s_{\bar{x}_1 - \bar{x}_2} = \sqrt{\dfrac{s_1^2}{n_1} + \dfrac{s_2^2}{n_2}} = \sqrt{\dfrac{(1.20)^2}{45} + \dfrac{(1.85)^2}{50}} = .31693848$

For the 98% confidence level, $z = 2.33$.

The 98% confidence interval for $\mu_1 - \mu_2$ is:

$$(\bar{x}_1 - \bar{x}_2) \pm z s_{\bar{x}_1 - \bar{x}_2} = (6.4 - 9.3) \pm 2.33 \,(.31693848) = -2.9 \pm .74 = -3.64 \text{ to } -2.16 \text{ days}$$

b. $H_0: \mu_1 - \mu_2 = 0;$ $\qquad H_1: \mu_1 - \mu_2 < 0$

For $\alpha = .025$, the critical value of z is -1.96.

$$z = \frac{(\bar{x}_1 - \bar{x}_2) - (\mu_1 - \mu_2)}{s_{\bar{x}_1 - \bar{x}_2}} = \frac{(-2.9) - 0}{.31693848} = -9.15$$

Reject H_0. The mean number of days missed per year by mothers working for companies that provide daycare facilities on premises is less than that of mothers working for companies that do not provide such facilities.

c. The Type I error would be to conclude that the mean number of days missed by the first group of mothers is less than that the mean for the second group when the two means are actually equal. P(Type I error) $= \alpha = .025$.

10.14 a. $s_{\bar{x}_1 - \bar{x}_2} = \sqrt{\dfrac{s_1^2}{n_1} + \dfrac{s_2^2}{n_2}} = \sqrt{\dfrac{(41,000)^2}{44} + \dfrac{(39,000)^2}{37}} = \$8905.765187$

For the 99% confidence level, $z = 2.58$

The 99% confidence interval for $\mu_1 - \mu_2$ is: $(\bar{x}_1 - \bar{x}_2) \pm z s_{\bar{x}_1 - \bar{x}_2} =$

$(211,193 - 197,512) \pm 2.58 (8905.765187) = 13,681 \pm 22,976.87 = -\$9295.87$ to $\$36,657.87$

b. H_0: $\mu_1 - \mu_2 = 0$; H_1: $\mu_1 - \mu_2 \neq 0$

For $\alpha = .01$, the critical values of z are -2.58 and 2.58.

$$z = \frac{(\bar{x}_1 - \bar{x}_2) - (\mu_1 - \mu_2)}{s_{\bar{x}_1 - \bar{x}_2}} = \frac{13,681 - 0}{8905.765187} = 1.54$$

Do not reject H_0. Do not conclude the mean sale price of homes is different.

c. The Type I error would be to conclude the mean sale price of homes is different when in fact it is not. P(Type I error) $= \alpha = .01$.

10.15 a. $s_{\bar{x}_1 - \bar{x}_2} = \sqrt{\dfrac{s_1^2}{n_1} + \dfrac{s_2^2}{n_2}} = \sqrt{\dfrac{(2.7)^2}{100} + \dfrac{(2.1)^2}{108}} = .33724373$

For the 95% confidence level, $z = 1.96$.

The 95% confidence interval for $\mu_1 - \mu_2$ is:

$(\bar{x}_1 - \bar{x}_2) \pm z s_{\bar{x}_1 - \bar{x}_2} = (19 - 15.5) \pm 1.96 (.33724373) = 3.5 \pm .66 = 2.84$ to 4.16 minutes

b. H_0: $\mu_1 - \mu_2 = 0$; H_1: $\mu_1 - \mu_2 > 0$.

For $\alpha = .025$, the critical value of z is 1.96.

$$z = \frac{(\bar{x}_1 - \bar{x}_2) - (\mu_1 - \mu_2)}{s_{\bar{x}_1 - \bar{x}_2}} = \frac{3.5 - 0}{.33724373} = 10.38$$

Reject H_0. Conclude the mean average time doctors spend with patients is less than last year.

c. If $\alpha = 0$, there is no rejection region, so do not reject H_0.

10.16   a. $s_{\bar{x}_1 - \bar{x}_2} = \sqrt{\dfrac{s_1^2}{n_1} + \dfrac{s_2^2}{n_2}} = \sqrt{\dfrac{(800)^2}{45} + \dfrac{(1000)^2}{51}} = 183.9295119$

For the 99% confidence level, $z = 2.58$.

The 99% confidence interval for $\mu_1 - \mu_2$ is:

$(\bar{x}_1 - \bar{x}_2) \pm z s_{\bar{x}_1 - \bar{x}_2} = (3300 - 3850) \pm 2.58\,(183.9295119) = -550 \pm 474.54 = -\$1024.54 \text{ to } -\$75.46$

b. $H_0: \mu_1 - \mu_2 = 0;$ $H_1: \mu_1 - \mu_2 \ne 0$

For $\alpha = .01$, the critical values of z are -2.58 and 2.58.

$z = \dfrac{(\bar{x}_1 - \bar{x}_2) - (\mu_1 - \mu_2)}{s_{\bar{x}_1 - \bar{x}_2}} = \dfrac{-550 - 0}{183.9295119} = -2.99$

Reject H_0. Conclude the mean repair costs for the two types of cars are different.

c. If $\alpha = 0$, there is no rejection region, so do not reject H_0.

10.17   a. $s_{\bar{x}_1 - \bar{x}_2} = \sqrt{\dfrac{s_1^2}{n_1} + \dfrac{s_2^2}{n_2}} = \sqrt{\dfrac{(1.2)^2}{200} + \dfrac{(1.5)^2}{300}} = .12124356$

For the 97% confidence level, $z = 2.17$

The 97% confidence interval for $\mu_1 - \mu_2$ is:

$(\bar{x}_1 - \bar{x}_2) \pm z s_{\bar{x}_1 - \bar{x}_2} = (4.5 - 4.75) \pm 2.17\,(.12124356) = -.25 \pm .26 = -.51 \text{ to } .01 \text{ minute}$

b. $H_0: \mu_1 - \mu_2 = 0;$ $H_1: \mu_1 - \mu_2 < 0$

For $\alpha = .025$, the critical value of z is -1.96.

$z = \dfrac{(\bar{x}_1 - \bar{x}_2) - (\mu_1 - \mu_2)}{s_{\bar{x}_1 - \bar{x}_2}} = \dfrac{(-.25) - 0}{.12124356} = -2.06$

Reject H_0. Thus, we conclude that the bank's claim is true.

c. The area to the left of $z = -2.06$ is $.5 - .4803 = .0197$. Hence, p–value $= .0197$.

If $\alpha = .01$, do not reject H_0 since p–value $> .01$.

If $\alpha = .05$, reject H_0 since p–value $< .05$.

10.18   a. $s_{\bar{x}_1-\bar{x}_2} = \sqrt{\dfrac{s_1^2}{n_1}+\dfrac{s_2^2}{n_2}} = \sqrt{\dfrac{(5.5)^2}{50}+\dfrac{(6.4)^2}{40}} = 1.27632284$

For the 98% confidence level, $z = 2.33$.

The 98% confidence interval for $\mu_1 - \mu_2$ is:

$(\bar{x}_1 - \bar{x}_2) \pm zs_{\bar{x}_1-\bar{x}_2} = (141 - 144) \pm 2.33\,(1.27632284) = -3 \pm 2.97 = -5.97$ to $-.03$ calories

b. $H_0: \mu_1 - \mu_2 = 0;$ $H_1: \mu_1 - \mu_2 < 0$

For $\alpha = .01$, the critical value of z is -2.33.

$z = \dfrac{(\bar{x}_1 - \bar{x}_2) - (\mu_1 - \mu_2)}{s_{\bar{x}_1-\bar{x}_2}} = \dfrac{(-3) - 0}{1.27632284} = -2.35$

Reject H_0. Conclude the dairy's claim is true.

c.   The area to the left of $z = -2.35$ is $.5 - .4906 = .0094$. Hence, p–value $= .0094$.

If $\alpha = .005$, do not reject H_0, since p–value $> .005$.

If $\alpha = .025$, reject H_0, since p–value $< .025$.

10.19   1.   The populations from which the two samples are drawn are (approximately) normally distributed.

2.   The samples are small ($n_1 < 30$ and $n_2 < 30$) and independent.

3.   The standard deviations of the two populations are unknown but equal.

10.20   a.   The point estimate of $\mu_1 - \mu_2$ is $\bar{x}_1 - \bar{x}_2 = 12.50 - 14.60 = -2.10$.

b.   Since $n_1 < 30$ and $n_2 < 30$, we will use the t distribution to make the confidence interval for $\mu_1 - \mu_2$.

First we calculate the values of s_p and $s_{\bar{x}_1-\bar{x}_2}$ using the formulas discussed in Section 10.2 of

Chapter 10 of the text.

$s_p = \sqrt{\dfrac{(n_1-1)s_1^2 + (n_2-1)s_2^2}{n_1+n_2-2}} = \sqrt{\dfrac{(25-1)(3.75)^2 + (20-1)(3.10)^2}{25+20-2}} = 3.47780337$

$s_{\bar{x}_1-\bar{x}_2} = s_p\sqrt{\dfrac{1}{n_1}+\dfrac{1}{n_2}} = 3.47780337\sqrt{\dfrac{1}{25}+\dfrac{1}{20}} = 1.04334101$

$df = n_1 + n_2 - 2 = 25 + 20 - 2 = 43$

Area in each tail of the t curve $= .5 - (.95/2) = .025$

The t value for $df = 43$ and $.025$ area in the right tail is 2.017.

The 95% confidence interval for $\mu_1 - \mu_2$ is:

$(\bar{x}_1 - \bar{x}_2) \pm ts_{\bar{x}_1-\bar{x}_2} = (12.50 - 14.60) \pm 2.017\,(1.04334101) = -2.10 \pm 2.10 = -4.20$ to 0

10.21 a. The point estimate of $\mu_1 - \mu_2$ is $\bar{x}_1 - \bar{x}_2 = 33.75 - 28.50 = 5.25$.

b. Since $n_1 < 30$ and $n_2 < 30$, we will use the t distribution to make the confidence interval for $\mu_1 - \mu_2$. First we calculate the values of s_p and $s_{\bar{x}_1 - \bar{x}_2}$ using the formulas discussed in Section 10.2 of Chapter 10 of the text.

$$s_p = \sqrt{\frac{(n_1 - 1)s_1^2 + (n_2 - 1)s_2^2}{n_1 + n_2 - 2}} = \sqrt{\frac{(18 - 1)(5.25)^2 + (20 - 1)(4.55)^2}{18 + 20 - 2}} = 4.89305063$$

$$s_{\bar{x}_1 - \bar{x}_2} = s_p \sqrt{\frac{1}{n_1} + \frac{1}{n_2}} = 4.89305063 \sqrt{\frac{1}{18} + \frac{1}{20}} = 1.58971861$$

$df = n_1 + n_2 - 2 = 18 + 20 - 2 = 36$

Area in each tail of the t curve $= .5 - (.99/2) = .005$

The t value for $df = 36$ and .005 area in the right tail is 2.719.

The 99% confidence interval for $\mu_1 - \mu_2$ is:

$$(\bar{x}_1 - \bar{x}_2) \pm t s_{\bar{x}_1 - \bar{x}_2} = (33.75 - 28.50) \pm 2.719 \,(1.58971861) = 5.25 \pm 4.32 = .93 \text{ to } 9.57$$

10.22 Step 1: $H_0: \mu_1 - \mu_2 = 0$; $H_1: \mu_1 - \mu_2 \neq 0$.   A two tailed test.

Step 2: Since $n_1 < 30$ and $n_2 < 30$, use the t distribution.

Step 3: $df = n_1 + n_2 - 2 = 25 + 20 - 2 = 43$

Area in each tail of the t curve $= .05/2 = .025$

The critical values of t are -2.017 and 2.017.

Step 4: From the solution to Exercise 10.20, $s_{\bar{x}_1 - \bar{x}_2} = 1.04334101$.

$$t = \frac{(\bar{x}_1 - \bar{x}_2) - (\mu_1 - \mu_2)}{s_{\bar{x}_1 - \bar{x}_2}} = \frac{(12.50 - 14.60) - 0}{1.04334101} = -2.013$$

Step 5: Do not reject the null hypothesis. The two means are not different.

10.23 Step 1: $H_0: \mu_1 - \mu_2 = 0$; $H_1: \mu_1 - \mu_2 \neq 0$.   A two tailed test.

Step 2: Since $n_1 < 30$ and $n_2 < 30$, use the t distribution.

Step 3: $df = n_1 + n_2 - 2 = 18 + 20 - 2 = 36$

Area in each tail of the t curve $= .01/2 = .005$

The critical values of t are -2.719 and 2.719.

Step 4: From the solution to Exercise 10.21, $s_{\bar{x}_1 - \bar{x}_2} = 1.58971861$.

$$t = \frac{(\bar{x}_1 - \bar{x}_2) - (\mu_1 - \mu_2)}{s_{\bar{x}_1 - \bar{x}_2}} = \frac{(33.75 - 28.50) - 0}{1.58971861} = 3.302$$

Step 5: Reject the null hypothesis. The two means are different.

10.24 $H_0: \mu_1 - \mu_2 = 0;$ $H_1: \mu_1 - \mu_2 < 0;$ A left tailed test.

$df = n_1 + n_2 - 2 = 25 + 20 - 2 = 43$ and   Area in each tail of the t curve = .01

The critical value of t is -2.416.

From the solution to Exercise 10.20, $s_{\bar{x}_1 - \bar{x}_2} = 1.04334101$.

$$t = \frac{(\bar{x}_1 - \bar{x}_2) - (\mu_1 - \mu_2)}{s_{\bar{x}_1 - \bar{x}_2}} = \frac{(12.50 - 14.60) - 0}{1.04334101} = -2.013$$

Do not reject the null hypothesis. Hence, μ_1 is not less than μ_2.

10.25 $H_0: \mu_1 - \mu_2 = 0;$ $H_1: \mu_1 - \mu_2 > 0.$ A right tailed test.

$df = n_1 + n_2 - 2 = 18 + 20 - 2 = 36$ and   Area in each tail of the t curve = .05

The critical value of t is 1.688.

From the solution to Exercise 10.21, $s_{\bar{x}_1 - \bar{x}_2} = 1.58971861$.

$$t = \frac{(\bar{x}_1 - \bar{x}_2) - (\mu_1 - \mu_2)}{s_{\bar{x}_1 - \bar{x}_2}} = \frac{(33.75 - 28.50) - 0}{1.58971861} = 3.302$$

Reject the null hypothesis. Hence, μ_1 is greater than μ_2.

10.26   From Sample 1: $n_1 = 13$, $\Sigma x = 385$, $\Sigma x^2 = 11,701$

$\bar{x}_1 = \Sigma x / n_1 = 385 / 13 = 29.62$

$$s_1 = \sqrt{\frac{\Sigma x^2 - (\Sigma x)^2 / n_1}{n_1 - 1}} = \sqrt{\frac{11,701 - (385)^2 / 13}{13 - 1}} = 4.99230177$$

From Sample 2: $n_2 = 12$, $\Sigma x = 306$, $\Sigma x^2 = 7918$

$\bar{x}_2 = \Sigma x / n_2 = 306 / 12 = 25.50$

$$s_2 = \sqrt{\frac{\Sigma x^2 - (\Sigma x)^2 / n_2}{n_2 - 1}} = \sqrt{\frac{7918 - (306)^2 / 12}{12 - 1}} = 3.23334895$$

a.   The point estimate of $\mu_1 - \mu_2$ is $\bar{x}_1 - \bar{x}_2 = 29.62 - 25.50 = 4.12$

b. $s_p = \sqrt{\dfrac{(n_1 - 1)s_1^2 + (n_2 - 1)s_2^2}{n_1 + n_2 - 2}} = \sqrt{\dfrac{(13 - 1)(4.99230177)^2 + (12 - 1)(3.23334895)^2}{13 + 12 - 2}} = 4.24303482$

$s_{\bar{x}_1 - \bar{x}_2} = s_p \sqrt{\dfrac{1}{n_1} + \dfrac{1}{n_2}} = 4.24303482 \sqrt{\dfrac{1}{13} + \dfrac{1}{12}} = 1.69857333$

$df = n_1 + n_2 - 2 = 13 + 12 - 2 = 23$

The t value for the 98% confidence level is 2.500.

The 98% confidence interval for $\mu_1 - \mu_2$ is:

$$(\bar{x}_1 - \bar{x}_2) \pm t s_{\bar{x}_1 - \bar{x}_2} = 4.12 \pm 2.500\,(1.69857333) = 4.12 \pm 4.25 = -.13 \text{ to } 8.37$$

c. $H_0: \mu_1 - \mu_2 = 0;$ $H_1: \mu_1 - \mu_2 > 0$

For $\alpha = .01$ and $df = 23$, the critical value of t is 2.500.

$$t = \frac{(\bar{x}_1 - \bar{x}_2) - (\mu_1 - \mu_2)}{s_{\bar{x}_1 - \bar{x}_2}} = \frac{4.12 - 0}{1.69857333} = 2.426$$

Do not reject the null hypothesis. Do not conclude that μ_1 is greater than μ_2.

10.27   From Sample 1: $n_1 = 11$, $\Sigma x = 118$, $\Sigma x^2 = 1360$

$$\bar{x}_1 = \Sigma x / n_1 = 118 / 11 = 10.73$$

$$s_1 = \sqrt{\frac{\Sigma x^2 - (\Sigma x)^2 / n_1}{n_1 - 1}} = \sqrt{\frac{1360 - (118)^2 / 11}{11 - 1}} = 3.06890564$$

From Sample 2: $n_2 = 12$, $\Sigma x = 193$, $\Sigma x^2 = 3213$

$$\bar{x}_2 = \Sigma x / n_2 = 193 / 12 = 16.08$$

$$s_2 = \sqrt{\frac{\Sigma x^2 - (\Sigma x)^2 / n_2}{n_2 - 1}} = \sqrt{\frac{3213 - (193)^2 / 12}{12 - 1}} = 3.14666731$$

a.   The point estimate of $\mu_1 - \mu_2$ is $\bar{x}_1 - \bar{x}_2 = 10.73 - 16.08 = -5.35$

b. $s_p = \sqrt{\dfrac{(n_1 - 1)s_1^2 + (n_2 - 1)s_2^2}{n_1 + n_2 - 2}} = \sqrt{\dfrac{(11 - 1)(3.06890564)^2 + (12 - 1)(3.14666731)^2}{11 + 12 - 2}} = 3.10988045$

$$s_{\bar{x}_1 - \bar{x}_2} = s_p \sqrt{\frac{1}{n_1} + \frac{1}{n_2}} = 3.10988045 \sqrt{\frac{1}{11} + \frac{1}{12}} = 1.29813735$$

$df = n_1 + n_2 - 2 = 11 + 12 - 2 = 21$

The t value for the 99% confidence level is 2.831.

The 99% confidence interval for $\mu_1 - \mu_2$ is:

$$(\bar{x}_1 - \bar{x}_2) \pm t s_{\bar{x}_1 - \bar{x}_2} = -5.35 \pm 2.831\,(1.29813735) = -5.35 \pm 3.68 = -9.03 \text{ to } -1.67$$

c. $H_0: \mu_1 - \mu_2 = 0;$ $H_1: \mu_1 - \mu_2 < 0$

For $\alpha = .025$ and $df = 21$, the critical value of t is -2.080.

$$t = \frac{(\bar{x}_1 - \bar{x}_2) - (\mu_1 - \mu_2)}{s_{\bar{x}_1 - \bar{x}_2}} = \frac{-5.35 - 0}{1.29813735} = -4.121$$

Reject H_0. Conclude that μ_1 is less than μ_2.

10.28 $s_p = \sqrt{\dfrac{(n_1-1)s_1^2+(n_2-1)s_2^2}{n_1+n_2-2}} = \sqrt{\dfrac{(25-1)(17.50)^2+(20-1)(14.40)^2}{25+20-2}} = 16.20353105$

$s_{\bar{x}_1-\bar{x}_2} = s_p\sqrt{\dfrac{1}{n_1}+\dfrac{1}{n_2}} = 16.20353105\sqrt{\dfrac{1}{25}+\dfrac{1}{20}} = 4.86105932$

$df = n_1+n_2-2 = 25+20-2 = 43$

a. The *t* value for the 99% confidence level is 2.695.

The 99% confidence interval for $\mu_1-\mu_2$ is:

$(\bar{x}_1-\bar{x}_2) \pm ts_{\bar{x}_1-\bar{x}_2} = (80-96) \pm 2.695\,(4.86105932) = -16 \pm 13.10 = -\$29.10 \text{ to } -\$2.90$

b. $H_0: \mu_1-\mu_2 = 0$; $H_1: \mu_1-\mu_2 < 0$

For $df = 43$ and the area in the left tail of the *t* curve = .05, the critical value of *t* is –1.681.

$t = \dfrac{(\bar{x}_1-\bar{x}_2)-(\mu_1-\mu_2)}{s_{\bar{x}_1-\bar{x}_2}} = \dfrac{(80-96)-0}{14.86105932} = -3.291$

Reject H_0. Conclude that the mean amount spent by all male customers at this supermarket is less than that of all female customers.

10.29 $s_p = \sqrt{\dfrac{(n_1-1)s_1^2+(n_2-1)s_2^2}{n_1+n_2-2}} = \sqrt{\dfrac{(20-1)(1.02)^2+(25-1)(1.34)^2}{20+25-2}} = 1.20909345$

$s_{\bar{x}_1-\bar{x}_2} = s_p\sqrt{\dfrac{1}{n_1}+\dfrac{1}{n_2}} = 1.20909345\sqrt{\dfrac{1}{20}+\dfrac{1}{25}} = .36272804$

$df = n_1+n_2-2 = 20+25-2 = 43$

a. The *t* value for 99% confidence level is 2.695.

The 99% confidence interval for $\mu_1-\mu_2$ is:

$(\bar{x}_1-\bar{x}_2) \pm ts_{\bar{x}_1-\bar{x}_2} = (10.6-11.57) \pm 2.695\,(.36272804) = -.97 \pm .98 = -\$1.95 \text{ to } -\$.01$

b. $H_0: \mu_1-\mu_2 = 0$; $H_1: \mu_1-\mu_2 < 0$.

For $df = 43$ and the area in the right tail of the *t* curve = .05, the critical value of *t* is –1.681

$t = \dfrac{(\bar{x}_1-\bar{x}_2)-(\mu_1-\mu_2)}{s_{\bar{x}_1-\bar{x}_2}} = \dfrac{(10.6-11.57)-0}{.36272804} = -2.674$

Reject H_0. Conclude that the average hourly earnings of cleaners is less than bellhops.

10.30 $s_p = \sqrt{\dfrac{(n_1-1)s_1^2+(n_2-1)s_2^2}{n_1+n_2-2}} = \sqrt{\dfrac{(27-1)(2.2)^2+(18-1)(2.5)^2}{27+18-2}} = 2.32323952$

$$s_{\bar{x}_1-\bar{x}_2} = s_p \sqrt{\frac{1}{n_1}+\frac{1}{n_2}} = 2.32323952 = \sqrt{\frac{1}{27}+\frac{1}{18}} = .70693927$$

$df = n_1 + n_2 - 2 = 27 + 18 - 2 = 43$

a. Area in each tail of the t curve $= .5 - (.98/2) = .01$

The t value for $df = 43$ and .01 area in the right tail is 2.416.

The 98% confidence interval for $\mu_1 - \mu_2$ is:

$(\bar{x}_1 - \bar{x}_2) \pm ts_{\bar{x}_1-\bar{x}_2} = (72 - 68) \pm 2.416 (.70693927) = 4 \pm 1.71 = 2.29$ to 5.71

b. $H_0: \mu_1 - \mu_2 = 0$; $H_1: \mu_1 - \mu_2 > 0$

For $df = 43$ and the area in the right tail of the t curve $= .01$, the critical value of t is 2.416.

$$t = \frac{(\bar{x}_1 - \bar{x}_2) - (\mu_1 - \mu_2)}{s_{\bar{x}_1-\bar{x}_2}} = \frac{(72-68)-0}{.70693927} = 5.658$$

Reject H_0. The mean speed of cars driven by all men is higher than that of all women.

10.31 $$s_p = \sqrt{\frac{(n_1-1)s_1^2+(n_2-1)s_2^2}{n_1+n_2-2}} = \sqrt{\frac{(25-1)(7)^2+(23-1)(6.2)^2}{25+23-2}} = 6.62944683$$

$$s_{\bar{x}_1-\bar{x}_2} = s_p \sqrt{\frac{1}{n_1}+\frac{1}{n_2}} = 6.62944683\sqrt{\frac{1}{25}+\frac{1}{23}} = 1.91541987$$

$df = n_1 + n_2 - 2 = 25 + 23 - 2 = 46$

a. Area in each tail of the t curve $= .5 - (.90/2) = .05$

The t value for $df = 46$ and .05 area in the right tail is 1.679.

The 90% confidence interval for $\mu_1 - \mu_2$ is:

$(\bar{x}_1 - \bar{x}_2) \pm ts_{\bar{x}_1-\bar{x}_2} = (29 - 22) \pm 1.679 (1.91541987) = 7 \pm 3.22 = 3.78$ to 10.22 hours

b. $H_0: \mu_1 - \mu_2 = 0$; $H_1: \mu_1 - \mu_2 \neq 0$

For $df = 46$ and the area in each tail of the t curve $= .05/2 = .025$

The critical values of t are -2.013 and 2.013.

$$t = \frac{(\bar{x}_1 - \bar{x}_2) - (\mu_1 - \mu_2)}{s_{\bar{x}_1-\bar{x}_2}} = \frac{(29-22)-0}{1.91541987} = 3.655$$

Reject H_0. Conclude that mean free time for tenth and twelfth–grades are different.

10.32 $$s_p = \sqrt{\frac{(n_1-1)s_1^2+(n_2-1)s_2^2}{n_1+n_2-2}} = \sqrt{\frac{(25-1)(.75)^2+(28-1)(.59)^2}{25+28-2}} = .67007023$$

$$s_{\bar{x}_1 - \bar{x}_2} = s_p \sqrt{\frac{1}{n_1} + \frac{1}{n_2}} = .67007023 \sqrt{\frac{1}{25} + \frac{1}{28}} = .18437806$$

$df = n_1 + n_2 - 2 = 25 + 28 - 2 = 51$

a. The t value for the 98% confidence level is 2.402.

The 98% confidence interval for $\mu_1 - \mu_2$ is:

$$(\bar{x}_1 - \bar{x}_2) \pm t s_{\bar{x}_1 - \bar{x}_2} = (7.6 - 8.1) \pm 2.402 \, (.18437806) = -.5 \pm .44 = -.94 \text{ to} - .06$$

b. H_0: $\mu_1 - \mu_2 = 0$; H_1: $\mu_1 - \mu_2 \neq 0$.

For $df = 51$ and the area in each tail of the t curve $= .01/2 = .005$

The critical values of t are -2.676 and 2.676.

$$t = \frac{(\bar{x}_1 - \bar{x}_2) - (\mu_1 - \mu_2)}{s_{\bar{x}_1 - \bar{x}_2}} = \frac{(7.6 - 8.1) - 0}{.18437806} = -2.712$$

Reject H_0. The mean satisfaction indexes for all customers of the two supermarkets are different.

10.33 $$s_p = \sqrt{\frac{(n_1 - 1)s_1^2 + (n_2 - 1)s_2^2}{n_1 + n_2 - 2}} = \sqrt{\frac{(25 - 1)(11)^2 + (22 - 1)(9)^2}{25 + 22 - 2}} = 10.11599394$$

$$s_{\bar{x}_1 - \bar{x}_2} = s_p \sqrt{\frac{1}{n_1} + \frac{1}{n_2}} = 10.11599394 \sqrt{\frac{1}{25} + \frac{1}{22}} = 2.95716900$$

$df = n_1 + n_2 - 2 = 25 + 22 - 2 = 45$

a. Area in each tail of the t curve $= .5 - (.99/2) = .005$

The t value for $df = 45$ and .005 area in the right tail is 2.690.

The 99% confidence interval for $\mu_1 - \mu_2$ is:

$$(\bar{x}_1 - \bar{x}_2) \pm t s_{\bar{x}_1 - \bar{x}_2} = (44 - 49) \pm 2.690 \, (2.95716900) = -5 \pm 7.95 = -12.95 \text{ to } 2.95 \text{ minutes}$$

b. H_0: $\mu_1 - \mu_2 = 0$; $\qquad H_1$: $\mu_1 - \mu_2 < 0$

For $df = 45$ and the area in the left tail of the t curve $= .01$, the critical value of t is -2.412.

$$t = \frac{(\bar{x}_1 - \bar{x}_2) - (\mu_1 - \mu_2)}{s_{\bar{x}_1 - \bar{x}_2}} = \frac{(44 - 49) - 0}{2.95716900} = -1.691$$

Do not reject H_0. Do not conclude that relief time for Brand A is less than for Brand B.

10.34 $$s_p = \sqrt{\frac{(n_1 - 1)s_1^2 + (n_2 - 1)s_2^2}{n_1 + n_2 - 2}} = \sqrt{\frac{(25 - 1)(1800)^2 + (27 - 1)(1700)^2}{25 + 27 - 2}} = 1748.713813$$

$$s_{\bar{x}_1 - \bar{x}_2} = s_p \sqrt{\frac{1}{n_1} + \frac{1}{n_2}} = 1748.713813 \sqrt{\frac{1}{25} + \frac{1}{27}} = 485.3650783$$

$df = n_1 + n_2 - 2 = 25 + 27 - 2 = 50$

a. Area in each tail of the t curve $= .5 - (.95/2) = .025$

   The t value for $df = 50$ and $.025$ area in the right tail is 2.009.

   The 95% confidence interval for $\mu_1 - \mu_2$ is: $(\bar{x}_1 - \bar{x}_2) \pm ts_{\bar{x}_1 - \bar{x}_2} =$

   $(21,375 - 18,921) \pm 2.009 \,(485.3650783) = 2454 \pm 975.10 = \$1478.90$ to $\$3429.10$

b. $H_0: \mu_1 - \mu_2 = 0;$ $\qquad H_1: \mu_1 - \mu_2 > 0$

   For $df = 50$ and the area in the right tail of the t curve $= .05$, the critical value of t is 1.676.

   $t = \dfrac{(\bar{x}_1 - \bar{x}_2) - (\mu_1 - \mu_2)}{s_{\bar{x}_1 - \bar{x}_2}} = \dfrac{2454 - 0}{485.3650783} = 5.056$

   Reject H_0. Conclude the mean resale price of Acura MDXs exceeds that of Toyota Sequoias.

10.35 $s_p = \sqrt{\dfrac{(n_1 - 1)s_1^2 + (n_2 - 1)s_2^2}{n_1 + n_2 - 2}} = \sqrt{\dfrac{(10-1)(6)^2 + (10-1)(5)^2}{10 + 10 - 2}} = 5.52268051$

$s_{\bar{x}_1 - \bar{x}_2} = s_p \sqrt{\dfrac{1}{n_1} + \dfrac{1}{n_2}} = 5.52268051 \sqrt{\dfrac{1}{10} + \dfrac{1}{10}} = 2.46981781$

$df = n_1 + n_2 - 2 = 10 + 10 - 2 = 18$

a. Area in each tail of the t curve $= .5 - (.99/2) = .005$

   The t value for $df = 18$ and $.005$ area in the right tail is 2.878.

   The 99% confidence interval for $\mu_1 - \mu_2$ is:

   $(\bar{x}_1 - \bar{x}_2) \pm ts_{\bar{x}_1 - \bar{x}_2} = (203 - 187) \pm 2.878\,(2.46981781) = 16 \pm 7.11 = 8.89$ to 23.11 seconds

b. $H_0: \mu_1 - \mu_2 = 0;$ $\qquad H_1: \mu_1 - \mu_2 > 0$

   For $df = 18$ and the area in the left tail of the t curve $= .01$, the critical value of t is 2.552.

   $t = \dfrac{(\bar{x}_1 - \bar{x}_2) - (\mu_1 - \mu_2)}{s_{\bar{x}_1 - \bar{x}_2}} = \dfrac{(203 - 187) - 0}{2.46981781} = 6.478$

   Reject H_0. Conclude that the mean time taken by the Piranha is greater than that of the Crocodile to shred 100 sheets of paper.

c. If $\alpha = 0$, there is no rejection region, so do not reject H_0.

10.36 $s_{\bar{x}_1 - \bar{x}_2} \sqrt{\dfrac{(3.90)^2}{24} + \dfrac{(5.15)^2}{16}} = 1.51373916$

$$df = \frac{\left(\dfrac{s_1^2}{n_1} + \dfrac{s_2^2}{n_2}\right)^2}{\dfrac{(s_1^2/n_1)^2}{n_1 - 1} + \dfrac{(s_2^2/n_2)^2}{n_2 - 1}} = \frac{\left[\dfrac{(3.90)^2}{24} + \dfrac{(5.15)^2}{16}\right]^2}{\dfrac{[(3.90)^2/24]^2}{24-1} + \dfrac{[(5.15)^2/16]^2}{16-1}} = 26.17 \approx 26$$

Area in each tail of the *t* curve = .5 – (.95/2) = .025

The *t* value for *df* = 26 and .025 area in the right tail is 2.056.

The 95% confidence interval for $\mu_1 - \mu_2$ is:

$$(\bar{x}_1 - \bar{x}_2) \pm t s_{\bar{x}_1 - \bar{x}_2} = (20.50 - 22.60) \pm 2.056\,(1.51373916) = -2.10 \pm 3.11 = -5.21 \text{ to } 1.01$$

10.37 $\quad s_{\bar{x}_1 - \bar{x}_2} = \sqrt{\dfrac{s_1^2}{n_1} + \dfrac{s_2^2}{n_2}} = \sqrt{\dfrac{(3.55)^2}{15} + \dfrac{(5.40)^2}{19}} = 1.54107220$

Using the formula for degrees of freedom given on page 471 of the text, we obtain: *df* = 31

Area in each tail of the *t* curve = .5 – (.99/2) = .005

The *t* value for *df* = 31 and .005 area in the right tail is 2.744.

The 95% confidence interval for $\mu_1 - \mu_2$ is:

$$(\bar{x}_1 - \bar{x}_2) \pm t s_{\bar{x}_1 - \bar{x}_2} = (52.61 - 43.75) \pm 2.744\,(1.54107720) = 8.86 \pm 4.23 = 4.63 \text{ to } 13.09 .$$

10.38 $\quad H_0: \mu_1 - \mu_2 = 0; \qquad H_1: \mu_1 - \mu_2 \neq 0$

Using the formula for degrees of freedom given on page 471 of the text, we obtain: *df* = 26

Area in each tail of the *t* curve = (.05/2) = .025

The critical values of *t* are –2.056 and 2.056.

From the solution to Exercise 10.36, $s_{\bar{x}_1 - \bar{x}_2} = 1.51373916$

$$t = \frac{(\bar{x}_1 - \bar{x}_2) - (\mu_1 - \mu_2)}{s_{\bar{x}_1 - \bar{x}_2}} = \frac{(20.50 - 22.60) - 0}{1.51373916} = -1.387$$

Do not reject H_0. Do not conclude the two population means are different.

10.39 $\quad H_0: \mu_1 - \mu_2 = 0; \qquad H_1: \mu_1 - \mu_2 \neq 0$

Using the formula for degrees of freedom given on page 471 of the text, we obtain: *df* = 31

Area in each tail of the *t* curve = (.01/2) = .005

The critical values of *t* are –2.744 and 2.744.

From the solution to Exercise 10.37, $s_{\bar{x}_1 - \bar{x}_2} = 1.54107220$

$$t = \frac{(\bar{x}_1 - \bar{x}_2) - (\mu_1 - \mu_2)}{s_{\bar{x}_1 - \bar{x}_2}} = \frac{(52.61 - 43.75) - 0}{1.54107219} = 5.749$$

Reject H_0. Conclude the two population means are different.

10.40 $H_0: \mu_1 - \mu_2 = 0;$ $H_1: \mu_1 - \mu_2 < 0$

Using the formula for degrees of freedom given on page 471 of the text, we obtain: $df = 26$

Area in the left tail of the t curve $= .01$

The critical value of t is -2.479.

From the solution to Exercise 10.36, $s_{\bar{x}_1 - \bar{x}_2} = 1.51373916$

$$t = \frac{(\bar{x}_1 - \bar{x}_2) - (\mu_1 - \mu_2)}{s_{\bar{x}_1 - \bar{x}_2}} = \frac{(20.50 - 22.60) - 0}{1.51373916} = -1.387$$

Do not reject H_0. Do not conclude that μ_1 is less than μ_2.

10.41 $H_0: \mu_1 - \mu_2 = 0;$ $H_1: \mu_1 - \mu_2 > 0$

Using the formula for degrees of freedom given on page 473 of the text, we obtain: $df = 31$

Area in the right each tail of the t curve $= .05$

The critical value of t is 1.696.

From the solution to Exercise 10.37, $s_{\bar{x}_1 - \bar{x}_2} = 1.54107220$

$$t = \frac{(\bar{x}_1 - \bar{x}_2) - (\mu_1 - \mu_2)}{s_{\bar{x}_1 - \bar{x}_2}} = \frac{(52.61 - 43.75) - 0}{1.54107220} = 5.749$$

Reject H_0. Conclude that μ_1 is greater than μ_2.

10.42 $s_{\bar{x}_1 - \bar{x}_2} = \sqrt{\dfrac{s_1^2}{n_1} + \dfrac{s_2^2}{n_2}} = \sqrt{\dfrac{(17.5)^2}{25} + \dfrac{(14.40)^2}{20}} = \$4.75583852$

Using the formula for degrees of freedom given on page 471 of the text, we obtain: $df = 42$

a.   Area in each tail of the t curve $= .5 - (.99/2) = .005$

The t value for $df = 42$ and $.005$ area in the right tail is 2.698.

The 99% confidence interval for $\mu_1 - \mu_2$ is:

$(\bar{x}_1 - \bar{x}_2) \pm ts_{\bar{x}_1 - \bar{x}_2} = (80 - 96) \pm 2.698\,(4.75583852) = -16 \pm 12.83 = -\$28.83$ to $-\$3.17$

b. $H_0: \mu_1 - \mu_2 = 0;$ $H_1: \mu_1 - \mu_2 < 0$

Area in the left tail of the t curve $= .05$

The critical value of t is -1.682.

$$t = \frac{(\bar{x}_1 - \bar{x}_2) - (\mu_1 - \mu_2)}{s_{\bar{x}_1 - \bar{x}_2}} = \frac{(80 - 96) - 0}{4.75583852} = -3.364$$

Reject H_0. The mean amount spent by all male customers at this supermarket is less than that spent by all female customers.

10.43 $s_{\bar{x}_1 - \bar{x}_2} = \sqrt{\dfrac{s_1^2}{n_1} + \dfrac{s_2^2}{n_2}} = \sqrt{\dfrac{(1.02)^2}{20} + \dfrac{(1.34)^2}{25}} = .35191476$

Using the formula for degrees of freedom given on page 471 of the text, we obtain: $df = 42$

a. Area in each tail of the t curve $= .5 - (.99/2) = .005$

The t value for $df = 43$ and .005 area in the right tail is 2.698.

The 99% confidence interval for $\mu_1 - \mu_2$ is:

$(\bar{x}_1 - \bar{x}_2) \pm t s_{\bar{x}_1 - \bar{x}_2} = (10.60 - 11.57) \pm 2.698 \, (.35191476) = -.97 \pm .95 = -\$1.92 \text{ to } -\$.02$

b. $H_0: \mu_1 - \mu_2 = 0;$ $H_1: \mu_1 - \mu_2 < 0$

$df = 42$ and   Area in the left tail of the t curve $= .05$

The critical value of t is -1.682.

$t = \dfrac{(\bar{x}_1 - \bar{x}_2) - (\mu_1 - \mu_2)}{s_{\bar{x}_1 - \bar{x}_2}} = \dfrac{(10.60 - 11.57) - 0}{.35191476} = -2.756$

Reject H_0. Conclude that the average hourly earnings of cleaners are less than bellhops.

10.44 $s_{\bar{x}_1 - \bar{x}_2} = \sqrt{\dfrac{s_1^2}{n_1} + \dfrac{s_2^2}{n_2}} = \sqrt{\dfrac{(2.2)^2}{27} + \dfrac{(2.5)^2}{18}} = .72559044$

Using the formula for degrees of freedom given on page 471 of the text, we obtain: $df = 33$

a. Area in each tail of the t curve $= .5 - (.98/2) = .01$

The t value for $df = 33$ and .01 area in the right tail is 2.445.

The 98% confidence interval for $\mu_1 - \mu_2$ is:

$(\bar{x}_1 - \bar{x}_2) \pm t s_{\bar{x}_1 - \bar{x}_2} = (72 - 68) \pm 2.445 \, (.72559044) = 4 \pm 1.77 = 2.23 \text{ to } 5.77 \text{ miles per hour}$

b. $H_0: \mu_1 - \mu_2 = 0;$ $H_1: \mu_1 - \mu_2 > 0.$

$df = 33$ and   Area in the right tail of the t curve $= .01$

The critical value of t is 2.445.

$t = \dfrac{(\bar{x}_1 - \bar{x}_2) - (\mu_1 - \mu_2)}{s_{\bar{x}_1 - \bar{x}_2}} = \dfrac{(72 - 68) - 0}{.72559044} = 5.513$

Reject H_0. The mean speed of cars driven by all men drivers on the given highway is higher than that of the cars driven by all women drivers.

10.45 $s_{\bar{x}_1 - \bar{x}_2} = \sqrt{\dfrac{s_1^2}{n_1} + \dfrac{s_2^2}{n_2}} = \sqrt{\dfrac{(7)^2}{25} + \dfrac{(6.2)^2}{23}} = 1.90559816$

Using the formula for degrees of freedom given on page 473 of the text, we obtain: $df = 45$

a. $df = 45$ and   Area in each tail of the t curve $= .5 - (.90/2) = .05$

The t value for $df = 45$ and .05 area in the right tail is 1.679.

The 95% confidence interval for $\mu_1 - \mu_2$ is:

$$(\bar{x}_1 - \bar{x}_2) \pm t s_{\bar{x}_1 - \bar{x}_2} = (29 - 22) \pm 1.679\,(1.90559816) = 7 \pm 3.20 = 3.80 \text{ to } 10.20$$

b. $H_0: \mu_1 - \mu_2 = 0;$ $\qquad$ $H_1: \mu_1 - \mu_2 \neq 0$

$df = 45$ $\qquad$ and $\qquad$ Area in each tail of the t curve $= (.05/2) = .025$

The critical values of t are -2.014 and 2.014.

$$t = \frac{(\bar{x}_1 - \bar{x}_2) - (\mu_1 - \mu_2)}{s_{\bar{x}_1 - \bar{x}_2}} = \frac{(29 - 22) - 0}{1.90559816} = 3.673$$

Reject H_0. Conclude that μ_1 is not equal to μ_2.

10.46 $\quad s_{\bar{x}_1 - \bar{x}_2} = \sqrt{\dfrac{s_1^2}{n_1} + \dfrac{s_2^2}{n_2}} = \sqrt{\dfrac{(.75)^2}{25} + \dfrac{(.59)^2}{28}} = .18690143$

Using the formula for degrees of freedom given on page 471 of the text, we obtain: $df = 45$

a. Area in each tail of the t curve $= .5 - (.98/2) = .01$

The t value for $df = 45$ and $.01$ area in the right tail is 2.412.

The 98% confidence interval for $\mu_1 - \mu_2$ is:

$$(\bar{x}_1 - \bar{x}_2) \pm t s_{\bar{x}_1 - \bar{x}_2} = (7.6 - 8.1) \pm 2.412\,(.18690143) = -.5 \pm .45 = -.95 \text{ to } -.05$$

b. $H_0: \mu_1 - \mu_2 = 0;$ $\qquad$ $H_1: \mu_1 - \mu_2 \neq 0$

$df = 45$ $\qquad$ and $\qquad$ Area in the right tail of the t curve $= .01/2 = .005$

The critical values of t are -2.690 and 2.690.

$$t = \frac{(\bar{x}_1 - \bar{x}_2) - (\mu_1 - \mu_2)}{s_{\bar{x}_1 - \bar{x}_2}} = \frac{(7.6 - 8.1) - 0}{.18690143} = -2.675$$

Do not reject H_0. The mean satisfaction indexes for all customers for the two supermarkets are the same.

10.47 $\quad s_{\bar{x}_1 - \bar{x}_2} = \sqrt{\dfrac{s_1^2}{n_1} + \dfrac{s_2^2}{n_2}} = \sqrt{\dfrac{(11)^2}{25} + \dfrac{(9)^2}{22}} = 2.91921534$

Using the formula for degrees of freedom given on page 471 of the text, we obtain: $df = 44$

a. Area in each tail of the t curve $= .5 - (.99/2) = .005$

The t value for $df = 44$ and $.005$ area in the right tail is 2.692.

The 99% confidence interval for $\mu_1 - \mu_2$ is:

$$(\bar{x}_1 - \bar{x}_2) \pm t s_{\bar{x}_1 - \bar{x}_2} = (44 - 49) \pm 2.692\,(2.91921534) = -5 \pm 7.86 = -12.86 \text{ to } 2.86 \text{ minutes}$$

b. $H_0: \mu_1 - \mu_2 = 0;$ $\qquad$ $H_1: \mu_1 - \mu_2 < 0.$

$df = 44$ and   Area in the left tail of the t curve $= .01$

The critical value of t is -2.414.

$$t = \frac{(\bar{x}_1 - \bar{x}_2) - (\mu_1 - \mu_2)}{s_{\bar{x}_1 - \bar{x}_2}} = \frac{(44 - 49) - 0}{2.91921534} = -1.713$$

Do not reject H_0. The mean relief time for Brand A is not less than the mean relief time for Brand B.

10.48 $s_{\bar{x}_1 - \bar{x}_2} = \sqrt{\dfrac{s_1^2}{n_1} + \dfrac{s_2^2}{n_2}} = \sqrt{\dfrac{(1800)^2}{25} + \dfrac{(1700)^2}{27}} = 486.4535302$

Using the formula for degrees of freedom given on page 471 of the text, we obtain: $df = 49$

a. Area in each tail of the t curve $= .5 - (.95/2) = .025$

The t value for $df = 49$ and $.025$ area in the right tail is 2.010.

The 95% confidence interval for $\mu_1 - \mu_2$ is:

$(\bar{x}_1 - \bar{x}_2) \pm ts_{\bar{x}_1 - \bar{x}_2} = (21{,}375 - 18{,}921) \pm 2.010\,(486.4535302) = 2454 \pm 977.77 = \$1476.23$ to $\$3431.77$

b. H_0: $\mu_1 - \mu_2 = 0$; H_1: $\mu_1 - \mu_2 > 0$

$df = 49$ and   Area in the right tail of the t curve $= .05$

The critical value of t is 1.677.

$$t = \frac{(\bar{x}_1 - \bar{x}_2) - (\mu_1 - \mu_2)}{s_{\bar{x}_1 - \bar{x}_2}} = \frac{2454 - 0}{486.4535302} = 5.045$$

Reject H_0. The mean resale price of Acura MDXs exceeds that of Toyota Sequoias.

10.49 $s_{\bar{x}_1 - \bar{x}_2} = \sqrt{\dfrac{s_1^2}{n_1} + \dfrac{s_2^2}{n_2}} = \sqrt{\dfrac{(6)^2}{10} + \dfrac{(5)^2}{10}} = 2.46981781$

Using the formula for degrees of freedom given on page 471 of the text, we obtain: $df = 17$

a. Area in each tail of the t curve $= .5 - (.99/2) = .005$

The t value for $df = 17$ and $.005$ area in the right tail is 2.898.

The 99% confidence interval for $\mu_1 - \mu_2$ is:

$(\bar{x}_1 - \bar{x}_2) \pm ts_{\bar{x}_1 - \bar{x}_2} = (203 - 187) \pm 2.898\,(2.46981781) = 16 \pm 7.16 = 8.84$ to 23.16 second

b H_0: $\mu_1 - \mu_2 = 0$; H_1: $\mu_1 - \mu_2 > 0$

$df = 17$ and   Area in the right tail of the t curve $= .01$

The critical value of t is 2.567.

$$t = \frac{(\bar{x}_1 - \bar{x}_2) - (\mu_1 - \mu_2)}{s_{\bar{x}_1 - \bar{x}_2}} = \frac{(203 - 187) - 0}{2.46981781} = 6.478$$

Reject H_0. Conclude that the Piranha shreds 100 sheets of paper slower than the Crocodile.

c. If $\alpha = 0$, there is no rejection region, so do not reject H_0.

10.50 The paired samples procedure is used when two samples are drawn in such a way that they include the same elements and two data values are obtained from each element, one for each sample.

10.51 a. $s_{\bar{d}} = s_d / \sqrt{n} = 13.5 / \sqrt{11} = 4.07040315$

  $df = n - 1 = 11 - 1 = 10$ and   Area in each tail of the t curve $= .5 - (.99/2) = .005$

  The t value for $df = 10$ and .005 area in the right tail is 3.169.

  The 99% confidence interval for μ_d is:

  $\bar{d} \pm t s_{\bar{d}} = 25.4 \pm 3.169 \,(4.07040315) = 25.4 \pm 12.90 = 12.50$ to 38.30

 b. $s_{\bar{d}} = s_d / \sqrt{n} = 4.8 / \sqrt{23} = 1.00086919$

  $df = n - 1 = 23 - 1 = 22$ and   Area in each tail of the t curve $= .5 - (.95/2) = .025$

  The t value for $df = 22$ and .025 area in the right tail is 2.074.

  The 95% confidence interval for μ_d is:

  $\bar{d} \pm t s_{\bar{d}} = 13.2 \pm 2.074 \,(1.00086919) = 13.2 \pm 2.08 = 11.12$ to 15.28

 c. $s_{\bar{d}} = s_d / \sqrt{n} = 11.7 / \sqrt{18} = 2.75771645$

  $df = n - 1 = 18 - 1 = 17$ and   Area in each tail of the t curve $= .5 - (.90/2) = .05$

  The t value for $df = 17$ and .05 area in the right tail is 1.740.

  The 90% confidence interval for μ_d is:

  $\bar{d} \pm t s_{\bar{d}} = 34.6 \pm 1.740 \,(2.75771645) = 34.6 \pm 4.80 = 29.80$ to 39.40

10.52 a. $s_{\bar{d}} = s_d / \sqrt{n} = 6.3 / \sqrt{12} = 1.81865335$

  $df = n - 1 = 12 - 1 = 1$ and   Area in each tail of the t curve $= .5 - (.99/2) = .005$

  The t value for $df = 11$ and .005 area in the right tail is 3.106.

  The 99% confidence interval for μ_d is:

  $\bar{d} \pm t s_{\bar{d}} = 17.5 \pm 3.106 \,(1.81865335) = 17.5 \pm 5.65 = 11.85$ to 23.15

 b. $s_{\bar{d}} = s_d / \sqrt{n} = 14.7 / \sqrt{27} = 2.82901632$

  $df = n - 1 = 27 - 1 = 26$ and   Area in each tail of the t curve $= .5 - (.95/2) = .025$

  The t value for $df = 26$ and .025 area in the right tail is 2.056.

  The 95% confidence interval for μ_d is:

$\overline{d} \pm ts_{\overline{d}} = 55.9 \pm 2.056 \,(2.82901632) = 55.9 \pm 5.82 = 50.08$ to 61.72

c. $s_{\overline{d}} = s_d / \sqrt{n} = 8.3 / \sqrt{16} = 2.075$

$df = n - 1 = 16 - 1 = 15$ and Area in each tail of the t curve $= .5 - (.90/2) = .05$

The t value for $df = 15$ and $.05$ area in the right tail is 1.753.

The 90% confidence interval for μ_d is:

$\overline{d} \pm ts_{\overline{d}} = 29.3 \pm 1.753 \,(2.075) = 29.3 \pm 3.64 = 25.66$ to 32.94

10.53 a. $H_0: \mu_d = 0;$ $H_1: \mu_d \neq 0$

$df = n - 1 = 9 - 1 = 8$ and Area in each tail of the t curve $= .10/2 = .05$

The critical values of t are -1.860 and 1.860.

$s_{\overline{d}} = s_d / \sqrt{n} = 2.5 / \sqrt{9} = 8.33333333$

$t = \dfrac{\overline{d} - \mu_d}{s_{\overline{d}}} = \dfrac{6.7 - 0}{.833333333} = 8.040$ Reject the null hypothesis.

b. $H_0: \mu_d = 0;$ $H_1: \mu_d > 0$

$df = n - 1 = 22 - 1 = 21$ and Area in the right tail of the t curve $= .05$

The critical value of t is 1.721.

$s_{\overline{d}} = s_d / \sqrt{n} = 6.4 / \sqrt{22} = 1.36448458$

$t = \dfrac{\overline{d} - \mu_d}{s_{\overline{d}}} = \dfrac{14.8 - 0}{1.36448458} = 10.847$ Reject the null hypothesis.

c. $H_0: \mu_d = 0;$ $H_1: \mu_d < 0$

$df = n - 1 = 17 - 1 = 16$ and Area in the left tail of the t curve $= .01$

The critical value of t is -2.583.

$s_{\overline{d}} = s_d / \sqrt{n} = 4.8 / \sqrt{17} = 1.16417100$

$t = \dfrac{\overline{d} - \mu_d}{s_{\overline{d}}} = \dfrac{-9.3 - 0}{1.16417100} = -7.989$ Reject the null hypothesis.

10.54 a. $H_0: \mu_d = 0;$ $H_1: \mu_d \neq 0$

$df = n - 1 = 26 - 1 = 25$ and Area in each tail of the t curve $= .05/2 = .025$

The critical values of t are -2.060 and 2.060.

$$s_{\bar{d}} = s_d / \sqrt{n} = 3.9 / \sqrt{26} = .76485293$$

$$t = \frac{\bar{d} - \mu_d}{s_{\bar{d}}} = \frac{9.6 - 0}{.76485293} = 12.551 \qquad\qquad \text{Reject the null hypothesis.}$$

b. $H_0: \mu_d = 0;$ $\qquad\qquad\qquad$ $H_1: \mu_d > 0$

$\quad df = n - 1 = 15 - 1 = 14$ $\qquad$ and $\qquad\qquad$ Area in the right tail of the t curve = .01

$\quad$ The critical value of t is 2.624.

$$s_{\bar{d}} = s_d / \sqrt{n} = 4.7 / \sqrt{15} = 1.21353478$$

$$t = \frac{\bar{d} - \mu_d}{s_{\bar{d}}} = \frac{8.8 - 0}{1.21353478} = 7.252 \qquad\qquad \text{Reject the null hypothesis.}$$

c. $H_0: \mu_d = 0;$ $\qquad\qquad\qquad$ $H_1: \mu_d < 0$

$\quad df = n - 1 = 20 - 1 = 19$ $\qquad$ and $\qquad\qquad$ Area in the left tail of the t curve = .10

$\quad$ The critical value of t is -1.328.

$$s_{\bar{d}} = s_d / \sqrt{n} = 2.3 / \sqrt{20} = .51429563$$

$$t = \frac{\bar{d} - \mu_d}{s_{\bar{d}}} = \frac{-7.4 - 0}{.51429563} = -14.389 \qquad\qquad \text{Reject the null hypothesis.}$$

10.55

Before (B)	8	5	4	9	6	9	5	
After (A)	10	8	5	11	6	7	9	
$d = B - A$	-2	-3	-1	-2	0	2	-4	$\sum d = -10$
d^2	4	9	1	4	0	4	16	$\sum d^2 = 38$

$$\bar{d} = \sum d / n = -10/7 = -1.43$$

$$s_d = \sqrt{\frac{\sum d^2 - \frac{(\sum d)^2}{n}}{n-1}} = \sqrt{\frac{38 - \frac{(-10)^2}{7}}{7-1}} = 1.98805959$$

$$s_{\bar{d}} = s_d / \sqrt{n} = 1.98805959 / \sqrt{7} = .75141590$$

a. $df = n - 1 = 7 - 1 = 6$ $\quad$ and $\qquad\qquad$ Area in each tail of the t curve $= .5 - (.95/2) = .025$

$\quad$ The t value for $df = 6$ and .025 area in the right tail is 2.447.

$\quad$ The 95% confidence interval for μ_d is:

$$\bar{d} \pm t s_{\bar{d}} = -1.43 \pm 2.447 (.75141590) = -1.43 \pm 1.84 = -3.27 \text{ to } .41$$

b. $H_0: \mu_d = 0;$ $\qquad\qquad$ $H_1: \mu_d < 0$

$df = n - 1 = 7 - 1 = 6$ and   Area in the left tail of the t curve = .01

The critical value of t is –3.143.

$$t = \frac{\bar{d} - \mu_d}{s_{\bar{d}}} = \frac{-1.43 - 0}{.75141590} = -1.903$$

Do not reject the null hypothesis. Attending the given course does not increase the mean score.

10.56

Before (B)	103	97	111	95	102	96	108	
After (A)	100	95	104	101	96	91	101	
$d = B–A$	3	2	7	–6	6	5	7	$d = 24$
d^2	9	4	49	36	36	25	49	$d^2 = 208$

$\bar{d} = \sum d / n = 24 / 7 = 3.43$

$$s_d = \sqrt{\frac{\sum d^2 - \frac{(\sum d)^2}{n}}{n - 1}} = \sqrt{\frac{208 - \frac{(24)^2}{7}}{7 - 1}} = 4.57737708$$

$s_{\bar{d}} = s_d / \sqrt{n} = 4.57737708 / \sqrt{7} = 1.73008592$

a. $df = n - 1 = 7 - 1 = 6$ and   Area in each tail of the t curve = .5 – (.99/2) = .005

The t value for $df = 6$ and .005 area in the right tail is 3.707.

The 99% confidence interval for μ_d is:

$\bar{d} \pm t s_{\bar{d}} = 3.43 \pm 3.707 (1.73008592) = 3.43 \pm 6.41 = -2.98$ to 9.84 minutes

b. $H_0: \mu_d = 0$; $H_1: \mu_d < 0$

$df = n - 1 = 7 - 1 = 6$ and   Area in the right tail of the t curve = .025

The critical value of t is 2.447.

$$t = \frac{\bar{d} - \mu_d}{s_{\bar{d}}} = \frac{3.43 - 0}{1.73008592} = 1.983$$

Do not reject the null hypothesis. Do not conclude the dietary supplements cause riders to be faster.

10.57

Before (B)	81	75	89	91	65	70	90	64	
After (A)	97	72	93	110	78	69	115	72	
$d = B–A$	–16	3	–4	–19	–13	1	–25	–8	$d = -81$
d^2	256	9	16	361	169	1	625	64	$d^2 = 1501$

$\bar{d} = \sum d / n = -81/8 = -10.13$

$$s_d = \sqrt{\frac{\sum d^2 - \frac{(\sum d)^2}{n}}{n-1}} = \sqrt{\frac{1501 - \frac{(-81)^2}{8}}{8-1}} = 9.86244681$$

$$s_{\bar{d}} = s_d / \sqrt{n} = 9.86244681 / \sqrt{8} = 3.48690151$$

a. $df = n - 1 = 8 - 1 = 7$

    Area in each tail of the t curve $= .5 - (.90/2) = .05$

    The t value for $df = 7$ and .05 area in the right tail is 1.895.

    The 90% confidence interval for μ_d is:

$$\bar{d} \pm t s_{\bar{d}} = -10.13 \pm 1.895\,(3.48690151) = -10.13 \pm 6.61 = -16.74 \text{ to } -3.52$$

b. $H_0: \mu_d = 0$; $H_1: \mu_d < 0$.

    $df = n - 1 = 8 - 1 = 7$

    Area in the left tail of the t curve $= .05$

    The critical value of t is -1.895.

$$t = \frac{\bar{d} - \mu_d}{s_{\bar{d}}} = \frac{-10.13 - 0}{3.48690151} = -2.905$$

    Reject the null hypothesis. Attending the given course increases the mean writing speed.

10.58

Before (B)	180	195	177	221	208	199	
After (A)	183	187	161	204	197	189	
$d = B{-}A$	-3	8	16	17	11	10	$d = 59$
d^2	9	64	256	289	121	100	$d^2 = 839$

$$\bar{d} = \sum d / n = 59/6 = 9.83$$

$$s_d = \sqrt{\frac{\sum d^2 - \frac{(\sum d)^2}{n}}{n-1}} = \sqrt{\frac{839 - \frac{(-59)^2}{6}}{6-1}} = 7.19490561$$

$$s_{\bar{d}} = s_d / \sqrt{n} = 7.19490561 / \sqrt{6} = 2.93730791$$

a. $df = n - 1 = 6 - 1 = 5$

    Area in each tail of the t curve $= .5 - (.95/2) = .025$

    The t value for $df = 5$ and .025 area in the right tail is 2.571.

    The 95% confidence interval for μ_d is:

$$\bar{d} \pm t s_{\bar{d}} = 9.83 \pm 2.571\,(2.93730791) = 9.83 \pm 7.55 = 2.28 \text{ to } 17.38$$

b. $H_0: \mu_d = 0$; $H_1: \mu_d > 0$

$df = n - 1 = 6 - 1 = 5$ and   Area in the right tail of the t curve = .01

The critical value of t is 3.365.

$$t = \frac{\overline{d} - \mu_d}{s_{\overline{d}}} = \frac{9.83 - 0}{2.93730791} = 3.347$$

Do not reject the null hypothesis. The mean weight loss for all persons due to the given special program is not greater than zero.

10.59

Without (B)	24.60	28.30	18.90	23.70	15.40	29.50	
With (A)	26.30	31.70	18.20	25.30	18.30	30.90	
$d = B - A$	−1.70	−3.40	.70	−1.60	−2.90	−1.40	$\sum d = -10.30$
d^2	2.89	11.56	.49	2.56	8.41	1.96	$\sum d^2 = 27.87$

$\overline{d} = \sum d / n = -10.30 / 6 = -1.72$

$$s_d = \sqrt{\frac{\sum d^2 - \frac{(\sum d)^2}{n}}{n-1}} = \sqrt{\frac{27.87 - \frac{(-10.3)^2}{6}}{6-1}} = 1.42746862$$

$s_{\overline{d}} = s_d / \sqrt{n} = 1.42746862 / \sqrt{6} = .58276162$

a. $df = n - 1 = 6 - 1 = 5$ and   Area in each tail of the t curve = $.5 - (.99/2) = .005$

The t value for $df = 5$ and .005 area in the right tail is 4.032.

The 99% confidence interval for μ_d is:

$\overline{d} \pm t s_{\overline{d}} = -1.72 \pm 4.032 (.58276162) = -1.72 \pm 2.35 = -4.07$ to .63 miles per gallon

b. $H_0: \mu_d = 0;$ $H_1: \mu_d < 0$

$df = n - 1 = 6 - 1 = 5$

Area in the left tail of the t curve = .025

The critical value of t is −2.571

$$t = \frac{\overline{d} - \mu_d}{s_{\overline{d}}} = \frac{-1.72 - 0}{.58276162} = -2.951$$

Reject the null hypothesis. The use of gasoline additive increases the gasoline mileage.

10.60

Machine I	23	26	19	24	27	22	20	18	
Machine II	21	24	23	25	24	28	24	23	
$d = I - II$	2	2	−4	−1	3	−6	−4	−5	$\sum d = -13$
d^2	4	4	16	1	9	36	16	25	$\sum d^2 = 111$

$$\bar{d} = \sum d / n = -13/8 = -1.63$$

$$s_d = \sqrt{\frac{\sum d^2 - \frac{(\sum d)^2}{n}}{n-1}} = \sqrt{\frac{111 - \frac{(-13)^2}{8}}{8-1}} = 3.58319490$$

$$s_{\bar{d}} = s_d / \sqrt{n} = 3.58319490 / \sqrt{8} = 1.26685071$$

a. $df = n - 1 = 8 - 1 = 7$

Area in each tail of the t curve $= .5 - (.98/2) = .01$

The t value for $df = 7$ and .01 area in the right tail is 2.998.

The 98% confidence interval for μ_d is:

$$\bar{d} \pm t s_{\bar{d}} = -1.63 \pm 2.998\,(1.26685071) = -1.63 \pm 3.80 = -5.43 \text{ to } 2.17 \text{ minutes}$$

b. $H_0: \mu_d = 0;$ $H_1: \mu_d \neq 0$

For $df = 7$ andarea in each tail of the t distribution curve $= .05/2 = .025$, critical values of t are
-2.365 and 2.365.

$$t = \frac{\bar{d} - \mu_d}{s_{\bar{d}}} = \frac{-1.63 - 0}{1.26685071} = -1.287$$

Do not reject the null hypothesis. The mean assembly time is the same for both machines.

10.61 The sampling distribution of $\hat{p}_1 - \hat{p}_2$ for two large samples is approximately normal, with its mean

equal to $p_1 - p_2$, and standard deviation equal to $\sqrt{\dfrac{p_1 q_1}{n_1} + \dfrac{p_2 q_2}{n_2}}$, where p_1 and p_2 are the population

proportions, $q_1 = 1 - p_1$, $q_2 = 1 - p_2$ and n_1 and n_2 are the sample sizes.

10.62 Two samples are considered large enough for the sampling distribution of the difference between two
sample proportions to be approximately normal when $n_1 p_1, n_1 q_1, n_2 p_2$, and $n_2 q_2$ are all greater than
5.

10.63 $s_{\hat{p}_1 - \hat{p}_2} = \sqrt{\dfrac{\hat{p}_1 \hat{q}_1}{n_1} + \dfrac{\hat{p}_2 \hat{q}_2}{n_2}} = \sqrt{\dfrac{(.55)(.45)}{300} + \dfrac{(.62)(.38)}{200}} = .04475489$

The z value for a 99% confidence level is 2.58.

The 99% confidence interval for $p_1 - p_2$ is:

$$(\hat{p}_1 - \hat{p}_2) \pm z s_{\hat{p}_1 - \hat{p}_2} = (.55 - .62) \pm 2.58\,(.04475489) = -.07 \pm .115 = -.185 \text{ to } .045$$

10.64 $s_{\hat{p}_1 - \hat{p}_2} = \sqrt{\dfrac{\hat{p}_1 \hat{q}_1}{n_1} + \dfrac{\hat{p}_2 \hat{q}_2}{n_2}} = \sqrt{\dfrac{(.81)(.19)}{100} + \dfrac{(.77)(.23)}{150}} = .05215042$

The z value for a 95% confidence level is 1.96.

The 95% confidence interval for $p_1 - p_2$ is:

$(\hat{p}_1 - \hat{p}_2) \pm zs_{\hat{p}_1 - \hat{p}_2} = (.81 - .77) \pm 1.96\,(.05215042) = .04 \pm .102 = -.062$ to $.142$

10.65 H_0: $p_1 - p_2 = 0$; $\qquad H_1$: $p_1 - p_2 \neq 0$

For $\alpha = .01$, the critical values of z are -2.58 and 2.58

$\bar{p} = \dfrac{n_1 \hat{p}_1 + n_2 \hat{p}_2}{n_1 + n_2} = \dfrac{300(.55) + 200(.62)}{300 + 200} = .578 \qquad$ and $\qquad \bar{q} = 1 - \bar{p} = 1 - .578 = .422$

$s_{\hat{p}_1 - \hat{p}_2} = \sqrt{\bar{p}\,\bar{q}\left(\dfrac{1}{n_1} + \dfrac{1}{n_2}\right)} = \sqrt{(.578)(.422)\left(\dfrac{1}{300} + \dfrac{1}{200}\right)} = .04508474$

$z = \dfrac{(\hat{p}_1 - \hat{p}_2) - (p_1 - p_2)}{s_{\hat{p}_1 - \hat{p}_2}} = \dfrac{(.55 - .62) - 0}{.04508474} = -1.55$

Do not reject H_0; the population proportions are not different.

10.66 H_0: $p_1 - p_2 = 0$; $\qquad H_1$: $p_1 - p_2 \neq 0$

For $\alpha = .05$, the critical values of z are -1.96 and 1.96.

$\bar{p} = \dfrac{n_1 \hat{p}_1 + n_2 \hat{p}_2}{n_1 + n_2} = \dfrac{100(.81) + 150(.77)}{100 + 150} = .786 \qquad$ and $\qquad \bar{q} = 1 - \bar{p} = 1 - .786 = .214$

$s_{\hat{p}_1 - \hat{p}_2} = \sqrt{\bar{p}\,\bar{q}\left(\dfrac{1}{n_1} + \dfrac{1}{n_2}\right)} = \sqrt{(.786)(.214)\left(\dfrac{1}{100} + \dfrac{1}{150}\right)} = .05294714$

$z = \dfrac{(\hat{p}_1 - \hat{p}_2) - (p_1 - p_2)}{s_{\hat{p}_1 - \hat{p}_2}} = \dfrac{(.81 - .77) - 0}{.05294714} = .76$

Do not reject the null hypothesis. The two population proportions are not different.

10.67 H_0: $p_1 - p_2 = 0$; $\qquad H_1$: $p_1 - p_2 < 0$

For $\alpha = .01$, the critical value of $z = -2.33$.

From the solution to exercise 10.65, $s_{\hat{p}_1 - \hat{p}_2} = .04508474$. Hence,

$z = \dfrac{(\hat{p}_1 - \hat{p}_2) - (p_1 - p_2)}{s_{\hat{p}_1 - \hat{p}_2}} = \dfrac{(.55 - .62) - 0}{.04508474} = -1.55$

Do not reject H_0. The proportion of the first population is not less than the proportion of the second population.

10.68 H_0: $p_1 - p_2 = 0$; H_1: $p_1 - p_2 > 0$

For $\alpha = .025$, the critical value of $z = 2.05$.

From the solution to exercise 10.66, $s_{\hat{p}_1 - \hat{p}_2} = .05294714$. Hence,

$$z = \frac{(\hat{p}_1 - \hat{p}_2) - (p_1 - p_2)}{s_{\hat{p}_1 - \hat{p}_2}} = \frac{(.81 - .77) - 0}{.05294714} = .76$$

Do not reject the null hypothesis. The proportion of the first population is not greater than the proportion of the second population.

10.69 $\hat{p}_1 = x_1 / n_1 = 305 / 500 = .61$

$\hat{p}_2 = x_2 / n_2 = 348 / 600 = .58$

a. The estimate of $p_1 - p_2$ is: $\hat{p}_1 - \hat{p}_2 = .61 - .58 = .03$

b. $s_{\hat{p}_1 - \hat{p}_2} = \sqrt{\dfrac{\hat{p}_1 \hat{q}_1}{n_1} + \dfrac{\hat{p}_2 \hat{q}_2}{n_2}} = \sqrt{\dfrac{(.61)(.39)}{500} + \dfrac{(.58)(.42)}{600}} = .02969512$

The z value for a 97% confidence level is 2.17.

The 97% confidence interval for $p_1 - p_2$ is:

$(\hat{p}_1 - \hat{p}_2) \pm z s_{\hat{p}_1 - \hat{p}_2} = (.61 - .58) \pm 2.17\,(.02969512) = .03 \pm .064 = -.034$ to $.094$

c. H_0: $p_1 - p_2 = 0$; H_1: $p_1 - p_2 > 0$

For $\alpha = .025$, the rejection region lies to the right of $z = 1.96$.

The nonrejection region lies to the left of $z = 1.96$.

d. $\bar{p} = \dfrac{x_1 + x_2}{n_1 + n_2} = \dfrac{305 + 348}{500 + 600} = .594$; $\bar{q} = 1 - \bar{p} = 1 - .594 = .406$

$$s_{\hat{p}_1 - \hat{p}_2} = \sqrt{\bar{p}\bar{q}\left(\frac{1}{n_1} + \frac{1}{n_2}\right)} = \sqrt{(.594)(.406)\left(\frac{1}{500} + \frac{1}{600}\right)} = .02973664$$

$$z = \frac{(\hat{p}_1 - \hat{p}_2) - (p_1 - p_2)}{s_{\hat{p}_1 - \hat{p}_2}} = \frac{(.61 - .58) - 0}{.02973664} = 1.01$$

e. Do not reject H_0.

10.70 $\hat{p}_1 = x_1 / n_1 = 290 / 1000 = .29$ and $\hat{p}_2 = x_2 / n_2 = 396 / 1200 = .33$

a. The estimate of $p_1 - p_2$ is: $\hat{p}_1 - \hat{p}_2 = .29 - .33 = -.04$

b. $s_{\hat{p}_1-\hat{p}_2} = \sqrt{\dfrac{\hat{p}_1\hat{q}_1}{n_1} + \dfrac{\hat{p}_2\hat{q}_2}{n_2}} = \sqrt{\dfrac{(.29)(.71)}{1000} + \dfrac{(.33)(.67)}{1200}} = .01975222$

The z value for a 98% confidence level is 2.33.

The 98% confidence interval for $p_1 - p_2$ is:

$(\hat{p}_1 - \hat{p}_2) \pm zs_{\hat{p}_1-\hat{p}_2} = -.04 \pm 2.33\,(.01975222) = -.04 \pm .046 = -.086 \text{ to } .006$

c. $H_0: p_1 - p_2 = 0;\qquad H_1: p_1 - p_2 < 0$

For $\alpha = .01$, the rejection region lies to the left of $z = -2.33$.

The nonrejection region lies to the right of $z = -2.33$.

d. $\bar{p} = \dfrac{x_1 + x_2}{n_1 + n_2} = \dfrac{290 + 396}{1000 + 1200} = .312;\ \bar{q} = 1 - \bar{p} = 1 - .312 = .688$

$s_{\hat{p}_1-\hat{p}_2} = \sqrt{\bar{p}\,\bar{q}\left(\dfrac{1}{n_1} + \dfrac{1}{n_2}\right)} = \sqrt{(.312)(.688)\left(\dfrac{1}{1000} + \dfrac{1}{1200}\right)} = .01983774$

$z = \dfrac{(\hat{p}_1 - \hat{p}_2) - (p_1 - p_2)}{s_{\hat{p}_1-\hat{p}_2}} = \dfrac{(.29 - .33) - 0}{.01983774} = -2.02$

e. Do not reject H_0.

10.71 a. $s_{\hat{p}_1-\hat{p}_2} = \sqrt{\dfrac{\hat{p}_1\hat{q}_1}{n_1} + \dfrac{\hat{p}_2\hat{q}_2}{n_2}} = \sqrt{\dfrac{(.334)(.666)}{750} + \dfrac{(.359)(.641)}{738}} = .02466589$

The z value for a 99% confidence level is 2.58.

The 99% confidence interval for $p_1 - p_2$ is:

$(\hat{p}_1 - \hat{p}_2) \pm zs_{\hat{p}_1-\hat{p}_2} = (.334 - .359) \pm 2.58\,(.02466589) = -.025 \pm .064 = -.089 \text{ to } .039$

b. $H_0: p_1 - p_2 = 0;\qquad H_1: p_1 - p_2 < 0$

For $\alpha = .01$, the critical value of z is -2.33.

$\bar{p} = \dfrac{n_1\hat{p}_1 + n_2\hat{p}_2}{n_1 + n_2} = \dfrac{750(.334) + 738(.359)}{750 + 738} = .346 \qquad \text{and} \qquad \bar{q} = 1 - \bar{p} = 1 - .346 = .654$

$s_{\hat{p}_1-\hat{p}_2} = \sqrt{\bar{p}\,\bar{q}\left(\dfrac{1}{n_1} + \dfrac{1}{n_2}\right)} = \sqrt{(.346)(.654)\left(\dfrac{1}{750} + \dfrac{1}{738}\right)} = .02466434$

$z = \dfrac{(\hat{p}_1 - \hat{p}_2) - (p_1 - p_2)}{s_{\hat{p}_1-\hat{p}_2}} = \dfrac{(.334 - .359) - 0}{.02466434} = -1.01$

Reject H_0. Conclude that the proportion of all ninth–grade girls who consider themselves overweight is not less than the proportion of all twelfth–grade girls who think they are overweight.

10.72   a. $\hat{p}_1 = .054$ and $\hat{p}_2 = .026$

$$s_{\hat{p}_1 - \hat{p}_2} = \sqrt{\frac{\hat{p}_1 \hat{q}_1}{n_1} + \frac{\hat{p}_2 \hat{q}_2}{n_2}} = \sqrt{\frac{(.054)(.946)}{1004} + \frac{(.026)(.974)}{935}} = .00882978$$

The z value for a 98% confidence level is 2.33.

The 98% confidence interval for $p_1 - p_2$ is:

$$(\hat{p}_1 - \hat{p}_2) \pm z s_{\hat{p}_1 - \hat{p}_2} = (.054 - .026) \pm 2.33\,(.00882978) = .028 \pm .021 = .007 \text{ to } .049$$

b. H_0: $p_1 - p_2 = 0$; H_1: $p_1 - p_2 \neq 0$

For $\alpha = .01$, the critical values of z are -2.58 and 2.58.

$$\bar{p} = \frac{n_1 \hat{p}_1 + n_2 \hat{p}_2}{n_1 + n_2} = \frac{1004(.054) + 935(.026)}{1004 + 935} = .040$$

$$s_{\hat{p}_1 - \hat{p}_2} = \sqrt{\bar{p}\,\bar{q}\left(\frac{1}{n_1} + \frac{1}{n_2}\right)} = \sqrt{(.040)(.960)\left(\frac{1}{1004} + \frac{1}{935}\right)} = .00890598$$

$$z = \frac{(\hat{p}_1 - \hat{p}_2) - (p_1 - p_2)}{s_{\hat{p}_1 - \hat{p}_2}} = \frac{(.054 - .026) - 0}{.00890598} = 3.14$$

Reject H_0. Conclude that the proportion of overdue accounts is different in the U.S. and Canada.

10.73   a. $s_{\hat{p}_1 - \hat{p}_2} = \sqrt{\dfrac{\hat{p}_1 \hat{q}_1}{n_1} + \dfrac{\hat{p}_2 \hat{q}_2}{n_2}} = \sqrt{\dfrac{(.35)(.65)}{501} + \dfrac{(.30)(.70)}{500}} = .02956504$

The z value for a 95% confidence level is 1.96.

The 95% confidence interval for $p_1 - p_2$ is:

$$(\hat{p}_1 - \hat{p}_2) \pm z s_{\hat{p}_1 - \hat{p}_2} = (.35 - .30) \pm 1.96\,(.02956504) = .05 \pm .058 = -.008 \text{ to } .108$$

b. H_0: $p_1 - p_2 = 0$; H_1: $p_1 - p_2 > 0$.

For $\alpha = .025$, the critical value of z is 1.96.

$$\bar{p} = \frac{n_1 \hat{p}_1 + n_2 \hat{p}_2}{n_1 + n_2} = \frac{501(.35) + 500(.30)}{501 + 500} = .325 \quad \text{and} \quad \bar{q} = 1 - p = 1 - .325 = .675$$

$$s_{\hat{p}_1 - \hat{p}_2} = \sqrt{\bar{p}\,\bar{q}\left(\frac{1}{n_1} + \frac{1}{n_2}\right)} = \sqrt{(.325)(.675)\left(\frac{1}{501} + \frac{1}{500}\right)} = .02960784$$

$$z = \frac{(\hat{p}_1 - \hat{p}_2) - (p_1 - p_2)}{s_{\hat{p}_1 - \hat{p}_2}} = \frac{(.35 - .30) - 0}{.02960784} = 1.69$$

Do not reject H_0. Do not conclude that the proportion of all women in this age group who have tattoos exceeds the corresponding proportion of men.

10.74 a. $\hat{p}_1 = x_1 / n_1 = 53 / 400 = .133$ and $\hat{p}_2 = x_2 / n_2 = 51 / 470 = .109$

The point estimate of $p_1 - p_2$ is: $\hat{p}_1 - \hat{p}_2 = .133 - .109 = .024$

b. $s_{\hat{p}_1 - \hat{p}_2} = \sqrt{\dfrac{\hat{p}_1 \hat{q}_1}{n_1} + \dfrac{\hat{p}_2 \hat{q}_2}{n_2}} = \sqrt{\dfrac{(.133)(.867)}{400} + \dfrac{(.109)(.891)}{470}} = .02224666$

The z value for a 95% confidence level is 1.96.

The 95% confidence interval for $p_1 - p_2$ is:

$(\hat{p}_1 - \hat{p}_2) \pm z s_{\hat{p}_1 - \hat{p}_2} = (.133 - .109) \pm 1.96 (.02224666) = .024 \pm .044 = -.020$ to $.068$

c. H_0: $p_1 - p_2 = 0$; H_1: $p_1 - p_2 \neq 0$.

For $\alpha = .05$, the critical values of z are -1.96 and 1.96.

$\bar{p} = \dfrac{x_1 + x_2}{n_1 + n_2} = \dfrac{53 + 51}{400 + 470} = .120$

$s_{\hat{p}_1 - \hat{p}_2} = \sqrt{\bar{p}\,\bar{q}\left(\dfrac{1}{n_1} + \dfrac{1}{n_2}\right)} = \sqrt{(.120)(.880)\left(\dfrac{1}{400} + \dfrac{1}{470}\right)} = .02210613$

$z = \dfrac{(\hat{p}_1 - \hat{p}_2) - (p_1 - p_2)}{s_{\hat{p}_1 - \hat{p}_2}} = \dfrac{(.133 - .109) - 0}{.02210613} = 1.09$

Do not reject H_0. Do not conclude that the proportion of cars failing inspection at the two stations are different.

10.75 a. $\hat{p}_1 = x_1 / n_1 = 246 / 600 = .41$

$\hat{p}_2 = x_2 / n_2 = 266 / 700 = .38$

The point estimate of $p_1 - p_2$ is: $\hat{p}_1 - \hat{p}_2 = .41 - .38 = .03$

b. $s_{\hat{p}_1 - \hat{p}_2} = \sqrt{\dfrac{\hat{p}_1 \hat{q}_1}{n_1} + \dfrac{\hat{p}_2 \hat{q}_2}{n_2}} = \sqrt{\dfrac{(.41)(.59)}{600} + \dfrac{(.38)(.62)}{700}} = .02719813$

The z value for a 95% confidence level is 1.96.

The 95% confidence interval for $p_1 - p_2$ is:

$(\hat{p}_1 - \hat{p}_2) \pm z s_{\hat{p}_1 - \hat{p}_2} = (.41 - .38) \pm 1.96 (.02719813) = .03 \pm .053 = -.023$ to $.083$

c. H_0: $p_1 - p_2 = 0$; H_1: $p_1 - p_2 \neq 0$

For $\alpha = .05$, the critical values of z are -1.96 and 1.96.

$$\bar{p} = \frac{x_1 + x_2}{n_1 + n_2} = \frac{246 + 266}{600 + 700} = .394$$

$$s_{\hat{p}_1 - \hat{p}_2} = \sqrt{\bar{p}\bar{q}\left(\frac{1}{n_1} + \frac{1}{n_2}\right)} = \sqrt{(.394)(.606)\left(\frac{1}{600} + \frac{1}{700}\right)} = .02718513$$

$$z = \frac{(\hat{p}_1 - \hat{p}_2) - (p_1 - p_2)}{s_{\hat{p}_1 - \hat{p}_2}} = \frac{(.41 - .38) - 0}{.02718513} = 1.10$$

Do not reject H_0. Conclude that the proportion of all men and women who prefer national brand products are not different.

10.76 a. $\hat{p}_1 = x_1 / n_1 = 160 / 500 = .32$

        $\hat{p}_2 = x_2 / n_2 = 66 / 300 = .22$

        The point estimate of $p_1 - p_2$ is: $\hat{p}_1 - \hat{p}_2 = .32 - .22 = .10$

    b. $s_{\hat{p}_1 - \hat{p}_2} = \sqrt{\frac{\hat{p}_1 \hat{q}_1}{n_1} + \frac{\hat{p}_2 \hat{q}_2}{n_2}} = \sqrt{\frac{(.32)(.68)}{500} + \frac{(.22)(.78)}{300}} = .03173641$

        The z value for a 99% confidence level is 2.58.

        The 99% confidence interval for $p_1 - p_2$ is:

        $(\hat{p}_1 - \hat{p}_2) \pm z s_{\hat{p}_1 - \hat{p}_2} = (.32 - .22) \pm 2.58\,(.03173641) = .10 \pm .082 = .018$ to $.182$

    c. $H_0: p_1 - p_2 = 0;$ $H_1: p_1 - p_2 \neq 0.$

        For $\alpha = .01$, the critical values of z are -2.58 and 2.58.

$$\bar{p} = \frac{x_1 + x_2}{n_1 + n_2} = \frac{160 + 66}{500 + 300} = .283$$

$$s_{\hat{p}_1 - \hat{p}_2} = \sqrt{\bar{p}\,\bar{q}\left(\frac{1}{n_1} + \frac{1}{n_2}\right)} = \sqrt{(.283)(.717)\left(\frac{1}{500} + \frac{1}{300}\right)} = .03289669$$

$$z = \frac{(\hat{p}_1 - \hat{p}_2) - (p_1 - p_2)}{s_{\hat{p}_1 - \hat{p}_2}} = \frac{(.32 - .22) - 0}{.03289669} = 3.04$$

        Reject H_0. Conclude that the proportions of all men and women who play the lottery often are different.

10.77 a. $\hat{p}_1 = x_1 / n_1 = 364 / 400 = .91$

        $\hat{p}_2 = x_2 / n_2 = 279 / 300 = .93$

        $s_{\hat{p}_1 - \hat{p}_2} = \sqrt{\frac{\hat{p}_1 \hat{q}_1}{n_1} + \frac{\hat{p}_2 \hat{q}_2}{n_2}} = \sqrt{\frac{(.91)(.09)}{400} + \frac{(.93)(.07)}{300}} = .02053655$

The z value for a 97% confidence level is 2.17.

The 97% confidence interval for $p_1 - p_2$ is:

$(\hat{p}_1 - \hat{p}_2) \pm zs_{\hat{p}_1 - \hat{p}_2} = (.91 - .93) \pm 2.17\,(.02053655) = -.02 \pm .045 = -.065$ to $.025$

b. H_0: $p_1 - p_2 = 0$; H_1: $p_1 - p_2 < 0$.

For $\alpha = .025$, the critical value of z is -1.96.

$\bar{p} = \dfrac{x_1 + x_2}{n_1 + n_2} = \dfrac{364 + 279}{400 + 300} = .919$

$s_{\hat{p}_1 - \hat{p}_2} = \sqrt{\bar{p}\,\bar{q}\left(\dfrac{1}{n_1} + \dfrac{1}{n_2}\right)} = \sqrt{(.919)(.081)\left(\dfrac{1}{400} + \dfrac{1}{300}\right)} = .02083813$

$z = \dfrac{(\hat{p}_1 - \hat{p}_2) - (p_1 - p_2)}{s_{\hat{p}_1 - \hat{p}_2}} = \dfrac{(.91 - .93) - 0}{.02083813} = -.96$

Do not reject H_0. Conclude that the proportion of all orders mailed within 72 hours from the West Coast warehouse is not less than the corresponding proportion for the East Coast warehouse.

b. The area to the left of $z = -.96$ under the normal curve is $.5 - .3315 = .1685$. Hence, p–value = $.1685$.

10.78 a. $\hat{p}_1 = x_1\,/\,n_1 = 240\,/\,800 = .30$

$\hat{p}_2 = x_2\,/\,n_2 = 279\,/\,900 = .31$

$s_{\hat{p}_1 - \hat{p}_2} = \sqrt{\dfrac{\hat{p}_1\hat{q}_1}{n_1} + \dfrac{\hat{p}_2\hat{q}_2}{n_2}} = \sqrt{\dfrac{(.30)(.70)}{800} + \dfrac{(.31)(.69)}{900}} = .02236441$

The z value for a 98% confidence level is 2.33.

The 98% confidence interval for $p_1 - p_2$ is:

$(\hat{p}_1 - \hat{p}_2) \pm zs_{\hat{p}_1 - \hat{p}_2} = (.30 - .31) \pm 2.33\,(.02236441) = -.01 \pm .052 = -.062$ to $.042$

b. H_0: $p_1 - p_2 = 0$; H_1: $p_1 - p_2 \neq 0$.

For $\alpha = .01$, the critical values of z are -2.58 and 2.58.

$\bar{p} = \dfrac{x_1 + x_2}{n_1 + n_2} = \dfrac{240 + 279}{800 + 900} = .305$

$s_{\hat{p}_1 - \hat{p}_2} = \sqrt{\bar{p}\bar{q}\left(\dfrac{1}{n_1} + \dfrac{1}{n_2}\right)} = \sqrt{(.305)(.695)\left(\dfrac{1}{800} + \dfrac{1}{900}\right)} = .02237178$

$z = \dfrac{(\hat{p}_1 - \hat{p}_2) - (p_1 - p_2)}{s_{\hat{p}_1 - \hat{p}_2}} = \dfrac{(.30 - .31) - 0}{.02237178} = -.45$

Do not reject H_0. Conclude that the proportion of all sales at the two stores for which at least one item is returned are not different.

c.   The area to the left of $z = -.45$ under the normal curve is $.5 - .1736 = .3264$.
Hence, p–value $= 2(.3264) = .6528$.

10.79   a. $s_{\bar{x}_1 - \bar{x}_2} = \sqrt{\dfrac{s_1^2}{n_1} + \dfrac{s_2^2}{n_2}} = \sqrt{\dfrac{(2800)^2}{33} + \dfrac{(3600)^2}{35}} = 779.6547132$

The z value for the 99% confidence level is 2.58.

The 99% confidence interval for $\mu_1 - \mu_2$ is: $(\bar{x}_1 - \bar{x}_2) \pm z s_{\bar{x}_1 - \bar{x}_2} =$

$(11{,}500 - 17{,}200) \pm 2.58(779.6547132) = -5700 \pm 2011.51 = -\$7711.51 \text{ to } -\$3688.49$

b. $H_0: \mu_1 - \mu_2 = 0;$ $H_1: \mu_1 - \mu_2 < 0.$
For $\alpha = .01$, the critical value of z is -2.33.

$z = \dfrac{(\bar{x}_1 - \bar{x}_2) - (\mu_1 - \mu_2)}{s_{\bar{x}_1 - \bar{x}_2}} = \dfrac{(11{,}500 - 17{,}200) - 0}{779.6547132} = -7.31$

Reject H_0. Conclude that the mean malpractice premium is less for anesthesiologists than pediatricians.

c.   If $\alpha = 0$, there is no rejection region, so we cannot reject H_0.

10.80 $s_{\bar{x}_1 - \bar{x}_2} = \sqrt{\dfrac{s_1^2}{n_1} + \dfrac{s_2^2}{n_2}} = \sqrt{\dfrac{(6)^2}{500} + \dfrac{(4)^2}{600}} = .31411251$

a.   The z value for the 98% confidence level is 2.33.

The 98% confidence interval for $\mu_1 - \mu_2$ is:

$(\bar{x}_1 - \bar{x}_2) \pm z s_{\bar{x}_1 - \bar{x}_2} = (51 - 53) \pm 2.33(.31411251) = -2 \pm .73 = -2.73 \text{ to } -1.27 \text{ hours}$

b. $H_0: \mu_1 - \mu_2 = 0;$ $H_1: \mu_1 - \mu_2 \neq 0.$
For $\alpha = .01$, the critical values of z are -2.58 and 2.58.

$z = \dfrac{(\bar{x}_1 - \bar{x}_2) - (\mu_1 - \mu_2)}{s_{\bar{x}_1 - \bar{x}_2}} = \dfrac{(51 - 53) - 0}{.31411251} = -6.37$

Reject H_0. The mean hours worked is different for men and women.

c.   If $\alpha = 0$, there is no rejection region, so we cannot reject H_0. Thus, our conclusion would change.

10.81 $s_{\bar{x}_1 - \bar{x}_2} = \sqrt{\dfrac{s_1^2}{n_1} + \dfrac{s_2^2}{n_2}} = \sqrt{\dfrac{(1.80)^2}{40} + \dfrac{(1.35)^2}{45}} = .34856850$

a. The *z* value for the 99% confidence level is 2.58

The 99% confidence interval for $\mu_1 - \mu_2$ is:

$(\bar{x}_1 - \bar{x}_2) \pm z s_{\bar{x}_1 - \bar{x}_2} = (11{-}9) \pm 2.58(.34856850) = 2 \pm .90 = 1.10$ to 2.90 policies

b. $H_0: \mu_1 - \mu_2 = 0$; $H_1: \mu_1 - \mu_2 > 0$.

For $\alpha = .01$, the critical value of *z* is 2.33.

$z = \dfrac{(\bar{x}_1 - \bar{x}_2) - (\mu_1 - \mu_2)}{s_{\bar{x}_1 - \bar{x}_2}} = \dfrac{(11-9)-0}{.34856850} = 5.74$

Reject H_0. Conclude that persons with a business degree are better salespersons than those with a degree in another area.

10.82 $s_{\bar{x}_1 - \bar{x}_2} = \sqrt{\dfrac{s_1^2}{n_1} + \dfrac{s_2^2}{n_2}} = \sqrt{\dfrac{(30)^2}{1500} + \dfrac{(35)^2}{2000}} = 1.10113578$

a. The *z* value for the 95% confidence level is 1.96

The 95% confidence interval for $\mu_1 - \mu_2$ is:

$(\bar{x}_1 - \bar{x}_2) \pm z s_{\bar{x}_1 - \bar{x}_2} = (388{-}505) \pm 1.96(1.10113578) = -117 \pm 2.16 = -\$119.16$ to $-\$114.84$

b. $H_0: \mu_1 - \mu_2 = 0$; $H_1: \mu_1 - \mu_2 < 0$.

For $\alpha = .025$, the critical value of *z* is -1.96.

$z = \dfrac{(\bar{x}_1 - \bar{x}_2) - (\mu_1 - \mu_2)}{s_{\bar{x}_1 - \bar{x}_2}} = \dfrac{(388-505)-0}{1.10113578} = -106.25$

Reject H_0. The weekly earnings of non–union female workers are less than those of female union members.

10.83 $s_p = \sqrt{\dfrac{(n_1-1)s_1^2 + (n_2-1)s_2^2}{n_1 + n_2 - 2}} = \sqrt{\dfrac{(28-1)(.43)^2 + (24-1)(.38)^2}{28 + 24 - 2}} = .40776219$

$s_{\bar{x}_1 - \bar{x}_2} = s_p \sqrt{\dfrac{1}{n_1} + \dfrac{1}{n_2}} = .40776219 \sqrt{\dfrac{1}{28} + \dfrac{1}{24}} = .11342897$

a. $H_0: \mu_1 - \mu_2 = 0$; $H_1: \mu_1 - \mu_2 \neq 0$.

$df = 28 + 24 - 2 = 50$ and   Area in each tail of the t curve $= (.05/2) = .025$

The critical values of t are -2.009 and 2.009.

$$t = \frac{(\bar{x}_1 - \bar{x}_2) - (\mu_1 - \mu_2)}{s_{\bar{x}_1 - \bar{x}_2}} = \frac{(2.62 - 2.74) - 0}{.11342897} = -1.058$$

Do not reject the null hypothesis. The mean GPAs of all such male and all such female college students are not different.

b. $df = 28 + 24 - 2 = 50$ and   Area in each tail of the t curve $= .5 - (.90/2) = .05$

The t value for $df = 50$ and .05 area in the right tail is 1.676.

The 90% confidence interval for $\mu_1 - \mu_2$ is:

$$(\bar{x}_1 - \bar{x}_2) \pm ts_{\bar{x}_1 - \bar{x}_2} = (2.62 - 2.74) \pm 1.676\,(.11342897) = -.12 \pm .19 = -.31 \text{ to } .07$$

10.84 $$s_p = \sqrt{\frac{(n_1 - 1)s_1^2 + (n_2 - 1)s_2^2}{n_1 + n_2 - 2}} = \sqrt{\frac{(25 - 1)(.35)^2 + (25 - 1)(.45)^2}{25 + 25 - 2}} = .40311289$$

$$s_{\bar{x}_1 - \bar{x}_2} = s_p \sqrt{\frac{1}{n_1} + \frac{1}{n_2}} = .40311289 \sqrt{\frac{1}{25} + \frac{1}{25}} = .11401754$$

a. $df = 25 + 25 - 2 = 48$ and   Area in each tail of the t curve $= .5 - (.90/2) = .05$

The t value for $df = 48$ and .05 area in the right tail is 1.677.

The 90% confidence interval for $\mu_1 - \mu_2$ is:

$$(\bar{x}_1 - \bar{x}_2) \pm ts_{\bar{x}_1 - \bar{x}_2} = (9.43 - 9.94) \pm 1.677\,(.11401754) = -.51 \pm .19 = -\$.70 \text{ to } -\$.32$$

b. $H_0: \mu_1 - \mu_2 = 0$; $H_1: \mu_1 - \mu_2 \neq 0$.

For $\alpha = .05$ and $df = 48$, the critical values of t are -2.011 and 2.011.

$$t = \frac{(\bar{x}_1 - \bar{x}_2) - (\mu_1 - \mu_2)}{s_{\bar{x}_1 - \bar{x}_2}} = \frac{(9.43 - 9.94) - 0}{.11401754} = -4.473$$

Reject H_0. The mean price of a pizza is different in the two cities.

10.85 $$s_p = \sqrt{\frac{(n_1 - 1)s_1^2 + (n_2 - 1)s_2^2}{n_1 + n_2 - 2}} = \sqrt{\frac{(25 - 1)(14)^2 + (20 - 1)(12)^2}{25 + 20 - 2}} = 13.15383046$$

$$s_{\bar{x}_1 - \bar{x}_2} = s_p \sqrt{\frac{1}{n_1} + \frac{1}{n_2}} = 13.15383046 \sqrt{\frac{1}{25} + \frac{1}{20}} = 3.94614914$$

a. $df = 25 + 20 - 2 = 43$ and   Area in each tail of the t curve $= .5 - (.99/2) = .005$

The t value for $df = 43$ and .005 area in the right tail is 2.695.

The 99% confidence interval for $\mu_1 - \mu_2$ is:

$$(\bar{x}_1 - \bar{x}_2) \pm ts_{\bar{x}_1 - \bar{x}_2} = (97 - 89) \pm 2.695\,(3.94614914) = 8 \pm 10.63 = -\$2.63 \text{ to } \$18.63$$

b. $H_0: \mu_1 - \mu_2 = 0$; $H_1: \mu_1 - \mu_2 > 0$

$df = 25 + 20 - 2 = 43$ and   Area in right tail of the t curve = .01

The critical value of t is 2.416.

$$t = \frac{(\bar{x}_1 - \bar{x}_2) - (\mu_1 - \mu_2)}{s_{\bar{x}_1 - \bar{x}_2}} = \frac{(97 - 89) - 0}{3.94614914} = 2.027$$

Do not reject the null hypothesis. The mean monthly insurance premium paid by drivers insured by company A is not higher than the mean monthly insurance premium paid by drivers insured by company B.

10.86 $s_p = \sqrt{\dfrac{(n_1 - 1)s_1^2 + (n_2 - 1)s_2^2}{n_1 + n_2 - 2}} = \sqrt{\dfrac{(15 - 1)(1.1)^2 + (17 - 1)(1)^2}{15 + 17 - 2}} = 1.04785495$

$$s_{\bar{x}_1 - \bar{x}_2} = s_p \sqrt{\frac{1}{n_1} + \frac{1}{n_2}} = 1.04785495 \sqrt{\frac{1}{15} + \frac{1}{17}} = .37119838$$

a. $df = 15 + 17 - 2 = 30$ and   Area in each tail of the t curve = $.5 - (.99/2) = .005$

The t value for $df = 30$ and .005 area in the right tail is 2.750.

The 99% confidence interval for $\mu_1 - \mu_2$ is:

$$(\bar{x}_1 - \bar{x}_2) \pm ts_{\bar{x}_1 - \bar{x}_2} = (5.1 - 4.2) \pm 2.750\,(.37119838) = .9 \pm 1.02 = -.12 \text{ to } 1.92 \text{ minutes}$$

b. $H_0: \mu_1 - \mu_2 = 0$; $H_1: \mu_1 - \mu_2 > 0$.

$df = 15 + 17 - 2 = 30$ and   Area in right tail of the t distribution curve = .05

The critical value of t is 1.697.

$$t = \frac{(\bar{x}_1 - \bar{x}_2) - (\mu_1 - \mu_2)}{s_{\bar{x}_1 - \bar{x}_2}} = \frac{.9 - 0}{.37119838} = 2.425$$

Reject the null hypothesis. The mean evacuation time is faster at closing than during fire drills.

10.87 $s_{\bar{x}_1 - \bar{x}_2} = \sqrt{\dfrac{s_1^2}{n_1} + \dfrac{s_2^2}{n_2}} = \sqrt{\dfrac{(.43)^2}{28} + \dfrac{(.38)^2}{24}} = .11233983$

Using the formula for degrees of freedom given on page 471 of the text, we obtain: $df = 49$

a. $H_0: \mu_1 - \mu_2 = 0$; $H_1: \mu_1 - \mu_2 \neq 0$.

Area in each tail of the t distribution curve = $.05/2 = .025$

The critical values of t are -2.010 and 2.010.

$$t = \frac{(\bar{x}_1 - \bar{x}_2) - (\mu_1 - \mu_2)}{s_{\bar{x}_1 - \bar{x}_2}} = \frac{(2.62 - 2.74) - 0}{.11233983} = -1.068$$

Do not reject the null hypothesis. The mean GPAs of all such male and all such female college students are not different.

b. Area in each tail of the t distribution curve $= .5 - (.90/2) = .05$

The t value for $df = 49$ and $.05$ area in the right tail is 1.677.

The 90% confidence interval for $\mu_1 - \mu_2$ is:

$$(\bar{x}_1 - \bar{x}_2) \pm t s_{\bar{x}_1 - \bar{x}_2} = (2.62 - 2.74) \pm 1.677\,(.11233983) = -.12 \pm .19 = -.31 \text{ to } .07$$

10.88 $\quad s_{\bar{x}_1 - \bar{x}_2} = \sqrt{\dfrac{s_1^2}{n_1} + \dfrac{s_2^2}{n_2}} = \sqrt{\dfrac{(.35)^2}{25} + \dfrac{(.45)^2}{25}} = .11401754$

Using the formula for degrees of freedom given on page 471 of the text, we obtain: $df = 45$

a. Area in each tail of the t distribution curve $= .5 - (.90/2) = .05$

The t value for $df = 45$ and $.05$ area in the right tail is 1.679.

The 90% confidence interval for $\mu_1 - \mu_2$ is:

$$(\bar{x}_1 - \bar{x}_2) \pm t s_{\bar{x}_1 - \bar{x}_2} = (9.43 - 9.94) \pm 1.679(.11401754) = -.51 \pm .19 = -\$.70 \text{ to } -\$.32$$

b. H_0: $\mu_1 - \mu_2 = 0$; $\quad$ H_1: $\mu_1 - \mu_2 \neq 0$.

$df = 45$

Area in each tail of the t distribution curve $= .05/2 = .025$

The critical values of t are -2.014 and 2.014.

$$t = \frac{(\bar{x}_1 - \bar{x}_2) - (\mu_1 - \mu_2)}{s_{\bar{x}_1 - \bar{x}_2}} = \frac{(9.43 - 9.94) - 0}{.11401754} = -4.473$$

Reject H_0. The mean prices of pizza are different for the two cities.

10.89 $\quad s_{\bar{x}_1 - \bar{x}_2} = \sqrt{\dfrac{s_1^2}{n_1} + \dfrac{s_2^2}{n_2}} = \sqrt{\dfrac{(14)^2}{25} + \dfrac{(12)^2}{20}} = 3.87814389$

Using the formula for degrees of freedom given on page 471 of the text, we obtain: $df = 42$

a. Area in each tail of the t distribution curve $= .5 - (.99/2) = .005$

The t value for $df = 42$ and $.005$ area in the right tail is 2.698.

The 99% confidence interval for $\mu_1 - \mu_2$ is:

$$(\bar{x}_1 - \bar{x}_2) \pm t s_{\bar{x}_1 - \bar{x}_2} = (97 - 89) \pm 2.698\,(3.87814389) = 8 \pm 10.46 = -\$2.46 \text{ to } \$18.46$$

b. $H_0: \mu_1 - \mu_2 = 0;$ $H_1: \mu_1 - \mu_2 > 0.$

$df = 42$ and   Area in right tail of the t distribution curve = .01

The critical value of t is 2.418.

$$t = \frac{(\bar{x}_1 - \bar{x}_2) - (\mu_1 - \mu_2)}{s_{\bar{x}_1 - \bar{x}_2}} = \frac{(97 - 89) - 0}{3.87814389} = 2.063$$

Do not reject H_0. The mean monthly auto insurance premium for company A is not higher than that of company B.

10.90 $s_{\bar{x}_1 - \bar{x}_2} = \sqrt{\dfrac{s_1^2}{n_1} + \dfrac{s_2^2}{n_2}} = \sqrt{\dfrac{(1.1)^2}{15} + \dfrac{(1)^2}{17}} = .37348386$

Using the formula for degrees of freedom given on page 471 of the text, we obtain: $df = 28$

a. Area in each tail of the t distribution curve = .5 − (.99/2) = .005

The t value for $df = 28$ and .005 area in the right tail is 2.673.

The 99% confidence interval for $\mu_1 - \mu_2$ is:

$(\bar{x}_1 - \bar{x}_2) \pm t s_{\bar{x}_1 - \bar{x}_2} = (5.1 - 4.2) \pm 2.763 (.37348386) = .9 \pm 1.03 = -.13$ to 1.93 minutes

b. $H_0: \mu_1 - \mu_2 = 0;$ $H_1: \mu_1 - \mu_2 > 0;$

$df = 28;$ and   Area in right tail of the t distribution curve = .05

The critical value of t is 1.701

$$t = \frac{(\bar{x}_1 - \bar{x}_2) - (\mu_1 - \mu_2)}{s_{\bar{x}_1 - \bar{x}_2}} = \frac{(5.1 - 4.2) - 0}{.37348386} = 2.410$$

Reject H_0. The mean evacuation time is faster at closing than during fire drills.

10.91

Guest	A	B	C	D	E	F	G	H	
Brand X	12	23	18	36	8	27	22	32	
Brand Y	9	20	21	27	6	18	15	25	
$d = X - Y$	3	3	-3	9	2	9	7	7	$\Sigma d = 37$
d^2	9	9	9	81	4	81	49	49	$\Sigma d^2 = 291$

$\bar{d} = \sum d / n = 37 / 8 = 4.625$

$$s_d = \sqrt{\frac{\sum d^2 - \frac{(\sum d)^2}{n}}{n-1}} = \sqrt{\frac{291 - \frac{(37)^2}{8}}{8-1}} = 4.138236339$$

$s_{\bar{d}} = s_d / \sqrt{n} = 4.13823634 / \sqrt{8} = 1.46308749$

a. $df = n - 1 = 8 - 1 = 7$ and   Area in each tail of the t distribution curve $= .5 - (.95/2) = .025$

The t value for $df = 7$ and $.025$ area in the right tail is 2.365.

The 95% confidence interval for μ_d is:

$$\overline{d} \pm t s_{\overline{d}} = 4.625 \pm 2.365(1.46308749) = 4.625 \pm 3.46 = 1.17 \text{ to } 8.09 \text{ bites}$$

b. $H_0: \mu_d = 0;$ $H_1: \mu_d \neq 0$

$df = n - 1 = 8 - 1 = 7$ and   Area in each tail of the t distribution curve $= .025$

The critical values of t are -2.365 and 2.365

$$t = \frac{\overline{d} - \mu_d}{s_{\overline{d}}} = \frac{4.63 - 0}{1.46308749} = 3.165$$

Reject H_0. There is a difference in the average number of bites between Brand X and Brand Y.

10.92

Before (B)	43	57	48	65	81	49	38	69	58	
After (A)	49	56	55	77	89	57	36	64	69	
$d = B - A$	–6	1	–7	–12	–8	–8	2	5	–11	$\Sigma d = -44$
d_2	36	1	49	144	64	64	4	25	121	$\Sigma d^2 = 508$

$$\overline{d} = \Sigma d / n = -44/9 = -4.89$$

$$s_d = \sqrt{\frac{\Sigma d^2 - \frac{(\Sigma d)^2}{n}}{n - 1}} = \sqrt{\frac{508 - \frac{(-44)^2}{9}}{9 - 1}} = 6.05071162$$

$$s_{\overline{d}} = s_d / \sqrt{n} = 6.05071162 / \sqrt{9} = 2.01690387$$

a. $df = n - 1 = 9 - 1 = 8$ and   Area in each tail of the t distribution curve $= .5 - (.95/2) = .025$

The t value for $df = 8$ and $.025$ area in the right tail is 2.306.

The 95% confidence interval for μ_d is:

$$\overline{d} \pm t s_{\overline{d}} = -4.89 \pm 2.306(2.01690387) = -4.89 \pm 4.65 = -9.54 \text{ to } -.24$$

b. $H_0: \mu_d = 0;$ $H_1: \mu_d < 0.$

$df = n - 1 = 9 - 1 = 8$ and   Area in left tail of the t distribution curve $= .01$

The critical value of t is -2.896

$$t = \frac{\overline{d} - \mu_d}{s_{\overline{d}}} = \frac{-4.89 - 0}{2.01690387} = -2.425$$

Do not reject the null hypothesis. The course does not make any improvement in the memory of students.

10.93 a. $s_{\hat{p}} = \sqrt{\dfrac{\hat{p}_1 \hat{q}_1}{n_1} + \dfrac{\hat{p}_2 \hat{q}_2}{n_2}} = \sqrt{\dfrac{(.24)(.76)}{800} + \dfrac{(.18)(.82)}{850}} = .02004113$

The z value for the 95% confidence level is 1.96

The 95% confidence interval for $p_1 - p_2$ is:

$(\hat{p}_1 - \hat{p}_2) \pm z s_{\hat{p}_1 - \hat{p}_2} = (.24 - .18) \pm 1.96(.02004113) = .06 \pm .039 = .021$ to $.099$

b. $H_0: p_1 - p_2 = 0$; $H_1: p_1 - p_2 \neq 0$.

For $\alpha = .02$, the critical values of z are -2.33 and 2.33.

$\bar{p} = \dfrac{n_1 \hat{p}_1 + n_2 \hat{p}_2}{n_1 + n_2} = \dfrac{800(.24) + 850(.18)}{800 + 850} = .209$

$s_{\hat{p}_1 - \hat{p}_2} = \sqrt{\bar{p}\bar{q}\left(\dfrac{1}{n_1} + \dfrac{1}{n_2}\right)} = \sqrt{(.209)(.791)\left(\dfrac{1}{800} + \dfrac{1}{850}\right)} = .02002852$

$z = \dfrac{(\hat{p}_1 - \hat{p}_2) - (p_1 - p_2)}{s_{\hat{p}_1 - \hat{p}_2}} = \dfrac{(.24 - .18) - 0}{.02002852} = 3.00$

Reject H_0. The proportions of all men and all women in this age group living with parent(s) are different.

10.94 a. $s_{\hat{p}_1 - \hat{p}_2} = \sqrt{\dfrac{\hat{p}_1 \hat{q}_1}{n_1} + \dfrac{\hat{p}_2 \hat{q}_2}{n_2}} = \sqrt{\dfrac{(.38)(.62)}{800} + \dfrac{(.28)(.72)}{900}} = .02277060$

The z value for the 99% confidence level is 2.58

The 99% confidence interval for $p_1 - p_2$ is:

$(\hat{p}_1 - \hat{p}_2) \pm z s_{\hat{p}_1 - \hat{p}_2} = (.38 - .28) \pm 2.58(.02277060) = .10 \pm .059 = .041$ to $.159$

b. $H_0: p_1 - p_2 = 0$; $H_1: p_1 - p_2 > 0$.

For $\alpha = .01$, the critical value of z is 2.33.

$\bar{p} = \dfrac{n_1 \hat{p}_1 + n_2 \hat{p}_2}{n_1 + n_2} = \dfrac{800(.38) + 900(.28)}{800 + 900} = .327$

$s_{\hat{p}_1 - \hat{p}_2} = \sqrt{\bar{p}\bar{q}\left(\dfrac{1}{n_1} + \dfrac{1}{n_2}\right)} = \sqrt{(.327)(.673)\left(\dfrac{1}{800} + \dfrac{1}{900}\right)} = .02279500$

$z = \dfrac{(\hat{p}_1 - \hat{p}_2) - (p_1 - p_2)}{s_{\hat{p}_1 - \hat{p}_2}} = \dfrac{(.38 - .28) - 0}{.02279500} = 4.39$

Reject H_0. Conclude that the proportion of boys who favor internet access exceeds that of girls.

10.95 a. $s_{\hat{p}_1-\hat{p}_2} = \sqrt{\dfrac{\hat{p}_1\hat{q}_1}{n_1} + \dfrac{\hat{p}_2\hat{q}_2}{n_2}} = \sqrt{\dfrac{(.65)(.35)}{900} + \dfrac{(.52)(.48)}{900}} = .02302414$

The z value for the 98% confidence level is 2.33

The 98% confidence interval for $p_1 - p_2$ is:

$(\hat{p}_1 - \hat{p}_2) \pm z s_{\hat{p}_1-\hat{p}_2} = (.65 - .52) \pm 2.33(.02302414) = .13 \pm .054 = .076 \text{ to} .184$

b. $H_0: p_1 - p_2 = 0;\qquad H_1: p_1 - p_2 \neq 0.$

For $\alpha = .02$, the critical values of z are -2.33 and 2.33.

$\bar{p} = \dfrac{n_1\hat{p}_1 + n_2\hat{p}_2}{n_1 + n_2} = \dfrac{900(.65) + 900(.52)}{900 + 900} = .585$

$s_{\hat{p}_1-\hat{p}_2} = \sqrt{\bar{p}\bar{q}\left(\dfrac{1}{n_1} + \dfrac{1}{n_2}\right)} = \sqrt{(.585)(.415)\left(\dfrac{1}{900} + \dfrac{1}{900}\right)} = .02322714$

$z = \dfrac{(\hat{p}_1 - \hat{p}_2) - (p_1 - p_2)}{s_{\hat{p}_1-\hat{p}_2}} = \dfrac{(.65 - .52) - 0}{.02322714} = 5.60$

Reject H_0. Conclude that the percentage of people favoring putting some of their social security funds in IRAs is different in 2000 than in 2002.

10.96 a. $s_{\hat{p}_1-\hat{p}_2} = \sqrt{\dfrac{\hat{p}_1\hat{q}_1}{n_1} + \dfrac{\hat{p}_2\hat{q}_2}{n_2}} = \sqrt{\dfrac{(.392)(.608)}{502} + \dfrac{(.313)(.687)}{495}} = .03015259$

The z value for the 95% confidence level is 1.96.

The 95% confidence interval for $p_1 - p_2$ is:

$(\hat{p}_1 - \hat{p}_2) \pm z s_{\hat{p}_1-\hat{p}_2} = (.392 - .313) \pm 2.33(.03025259) = .079 \pm .059 = .020 \text{ to } .138$

b. $H_0: p_1 - p_2 = 0;\qquad H_1: p_1 - p_2 > 0$

For $\alpha = .025$, the critical value of z is 1.96.

$\bar{p} = \dfrac{n_1\hat{p}_1 + n_2\hat{p}_2}{n_1 + n_2} = \dfrac{502(.392) + 495(.313)}{502 + 495} = .353$

$s_{\hat{p}_1-\hat{p}_2} = \sqrt{\bar{p}\,\bar{q}\left(\dfrac{1}{n_1} + \dfrac{1}{n_2}\right)} = \sqrt{(.353)(.647)\left(\dfrac{1}{502} + \dfrac{1}{495}\right)} = .03027141$

$z = \dfrac{(\hat{p}_1 - \hat{p}_2) - (p_1 - p_2)}{s_{\hat{p}_1-\hat{p}_2}} = \dfrac{(.392 - .313) - 0}{.03027141} = 2.610$

Conclude that the percentage of tenth graders who watch three of more hours of TV on school days is greater than the corresponding proportion of twelfth graders.

10.97   We are assuming that the populations are normally distributed and $\sigma_1 = \sigma_2 = 2$, so we can use the normal distribution for the following computation.

a.   The z value for a 90% confidence level is 1.65

The maximum error of estimate is: $z\sigma_{\bar{x}_1-\bar{x}_2} = 1.5$, so $\sigma_{\bar{x}_1-\bar{x}_2} = 1.5 / z = 1.5 / 1.65$

Also, $\sigma_{\bar{x}_1-\bar{x}_2} = \sqrt{\dfrac{\sigma_1^2}{n_1} + \dfrac{\sigma_2^2}{n_2}} = \sqrt{\dfrac{\sigma_1^2 + \sigma_2^2}{n}}$, where $n = n_1 = n_2$

Hence, $n = \dfrac{\sigma_1^2 + \sigma_2^2}{(\sigma_{\bar{x}_1-\bar{x}_2})^2} = \dfrac{2^2 + 2^2}{\left(\dfrac{1.5}{1.65}\right)} = 9.68 \approx 10$ For this estimate 10 cars are required.

b.   The mean of the sampling distribution of $\bar{x}_1 - \bar{x}_2 = \mu_1 - \mu_2 = 33{-}30 = 3$ mpg and the standard deviation of $\bar{x}_1 - \bar{x}_2$ is $\sigma_{\bar{x}_1-\bar{x}_2} = \sqrt{\dfrac{\sigma_1^2}{n_1} + \dfrac{\sigma_2^2}{n_2}} = \sqrt{\dfrac{2^2}{5} + \dfrac{2^2}{5}} = 1.26491106$ mpg.

$z = \dfrac{(\bar{x}_1 - \bar{x}_2) - (\mu_1 - \mu_2)}{\sigma_{\bar{x}_1-\bar{x}_2}} = \dfrac{0-3}{1.26491106} = -2.37$

For $\bar{x}_1 - \bar{x}_2 = 0$, $P(\bar{x}_1 - \bar{x}_2) \geq 0 = P(z \geq -2.37) = .4911 + .5 = .9911$

10.98 $\mu_1 = 200$, $\sigma_1 = 10$, $n_1 = 1$

$\mu_2 = 210$, $\sigma_2 = 12$, $n_2 = 1$

Since the distances are normally distributed with the above means and standard deviations, we can use the normal distribution despite the fact that the samples are small.

$\sigma_{\bar{x}_1-\bar{x}_2} = \sqrt{\dfrac{\sigma_1^2}{n_1} + \dfrac{\sigma_2^2}{n_2}} = \sqrt{\dfrac{10^2}{1} + \dfrac{12^2}{1}} = 15.62049935$

$\mu_{\bar{x}_1-\bar{x}_2} = 200{-}210 = -10$

$z = \dfrac{(\bar{x}_1 - \bar{x}_2) - (\mu_1 - \mu_2)}{\sigma_{\bar{x}_1-\bar{x}_2}} = \dfrac{0-(-10)}{15.62049935} = .64$

$P(\bar{x}_1 - \bar{x}_2) \geq 0 = P(z \geq .64) = .5 + .2389 = .2611$

The probability that Maria's throw is longer than Ellen's is 26.11%.

10.99 $H_0: \mu_1 - \mu_2 = 0$; $H_1: \mu_1 - \mu_2 > 0$

a. $s_{\bar{x}_1-\bar{x}_2} = \sqrt{\dfrac{s_1^2}{n_1} + \dfrac{s_2^2}{n_2}} = \sqrt{\dfrac{(3.5)^2}{1000} + \dfrac{(3.0)^2}{1000}} = .14577380$

$$z = \frac{(\bar{x}_1 - \bar{x}_2) - (\mu_1 - \mu_2)}{s_{\bar{x}_1 - \bar{x}_2}} = \frac{(15.4 - 15.0) - 0}{.14577380} = 2.74$$

The area to the right of $\bar{x}_1 - \bar{x}_2 = .4$ is equal to the area under the standard normal curve to the right of $z = 2.74$, which is .0031. That is, p–value = .0031

b. Since p–value = .0031 < .025, reject H_0. Thus, the result is statistically significant.

c. The z value for a 95% confidence level is 1.96.

The 95% confidence interval for $\mu_1 - \mu_2$ is:

$$(\bar{x}_1 - \bar{x}_2) \pm z s_{\bar{x}_1 - \bar{x}_2} = (15.4 - 15.0) \pm 1.96(.14577380) = .11 \text{ to } .69 \text{ minute}$$

d. Part c implies that in 95% of all cases the insomniacs fall asleep less than a minute earlier with the new pill than with the old pill. The result is not of great practical significance.

10.100 Confidence level is 90%.

Area in each tail of the t distribution curve = .10/2 = .05

Maximum error of estimate = $t s_{\bar{d}} \leq 2$ mpg, so $t \leq \frac{2mpg}{s_d} = \frac{2}{3/\sqrt{n}} = \frac{2}{3}\sqrt{n}$

Thus, we look for the smallest value of n for which $t \leq \frac{2}{3}\sqrt{n}$.

n	2	3	4	5	6	7	8	9
critical values of t	6.314	2.920	2.353	2.132	2.015	1.943	1.895	1.860
$\frac{2}{3}\sqrt{n}$	0.9428	1.1547	1.3333	1.4907	1.6330	1.7638	1.8856	2.000

Nine cars should be tested.

10.101

Salesperson	Difference in number of contacts (Before – After)	Difference in gas mileage (Before – After)
A	+1	−1
B	+3	−3
C	−5	+1
D	+4	−2
E	−6	−5
F	+5	−4
G	+8	−3

Number of contacts: $\sum d = +10; \sum d^2 = 176, n = 7$

$$\bar{d} = \frac{\sum d}{n} = \frac{+10}{7} = 1.43$$

$$s_d = \sqrt{\frac{\sum d^2 - \frac{(\sum d)^2}{n}}{n-1}} = \sqrt{\frac{176 - \frac{(10)^2}{7}}{6}} = 5.19156826;$$

$$s_{\bar{d}} = s_d / \sqrt{n} = 5.19156826 / \sqrt{7} = 1.96222836$$

Confidence level is 90%. Area in each tail of the t distribution is .05. The value for $df = 6$ and .05 area in the right tail is 1.943. The 90% confidence interval for μ_d is:

$$1.43 \pm 1.943\,(1.96222836) = -2.38 \text{ to } 5.24 \text{ contacts}$$

$H_0: \mu_d = 0;$ $\qquad\qquad$ $H_1: \mu_d > 0$

Area in right tail of the t distribution = .05

The critical value of t is 1.943

$$t = \frac{\bar{d} - \mu_d}{s_{\bar{d}}} = \frac{1.43 - 0}{1.96222836} = .729$$

Do not reject H_0. The mean number of contacts is the same.

Gas mileage: $\sum d = -17; \sum d^2 = 65, n = 7$

$$\bar{d} = -2.43\,;\ s_d = \sqrt{\frac{65 - \frac{(-17)^2}{7}}{6}} = 1.98805959;\ s_{\bar{d}} = s_d / \sqrt{n} = 1.98805959 / \sqrt{7} = .75141590$$

90% confidence interval: $\qquad\qquad\qquad$ $\bar{d} \pm t s_{\bar{d}} = -2.43 \pm 1.943\,(.75141590) = -3.89 \text{ to } -.97 \text{mpg}$

$H_0: \mu_d = 0;$ $\quad$ $H_1: \mu_d < 0$

The critical value of t is -1.943

$$t = \frac{\bar{d} - \mu_d}{s_{\bar{d}}} = \frac{-2.43 - 0}{.75141590} = -3.234$$

Reject H_0. The gas mileage (in mpg) is higher after the installation of the governors.
The 90% confidence intervals suggest that the number of contacts is slightly decreasing and that gas mileage (in mpg) is increasing. However, hypothesis tests with .05 significance levels could confirm only the latter. So the conclusion is that installing the governors does not (significantly) affect the number of contacts, but improves fuel consumption.

10.102 a. The z value for the 90% confidence interval is 1.65. The 90% confidence interval for $p_1 - p_2$ is given by $(\hat{p}_1 - \hat{p}_2) \pm z\sigma_{\hat{p}_1 - \hat{p}_2}$ with maximum error of estimate $z\sigma_{\hat{p}_1 - \hat{p}_2}$.

So, $1.65\sigma_{\hat{p}_1-\hat{p}_2} = .05$; $\sigma_{\hat{p}_1-\hat{p}_2} = \dfrac{.05}{1.65}$

We also have $\sigma_{\hat{p}_1-\hat{p}_2} = \sqrt{\dfrac{p_1 q_1}{n_1} + \dfrac{p_2 q_2}{n_2}} = \sqrt{\dfrac{(.5)(.5)}{n} + \dfrac{(.5)(.5)}{n}} = \sqrt{\dfrac{2 \cdot (.5)^2}{n}}$ where $n_1 = n_2 = n$.

So $n = \dfrac{2 \cdot (.5)^2}{(\sigma_{\hat{p}_1-\hat{p}_2})^2} = \dfrac{2(.5)^2}{(.05/1.65)^2} = 544.5 \approx 545$

b. $p_1 = .30$, $p_2 = .23$, $n_1 = n_2 = 100$

$\mu_{\hat{p}_1-\hat{p}_2} = p_1 - p_2 = .30 - .23 = .07$

$\sigma_{\hat{p}_1-\hat{p}_2} = \sqrt{\dfrac{(.30)(.70)}{100} + \dfrac{(.23)(.77)}{100}} = .06221736$

For $\hat{p}_1 - \hat{p}_2 = 0$ we get $z = \dfrac{0 - .07}{.06221736} = -1.13$

$P(\hat{p}_1 \geq \hat{p}_2) = P(z \geq -1.13) = .5 + .3708 = .8708$

10.103 a. No, a paired sample test would not be appropriate in this case. Different groups of cars are used in the two samples, so the samples are independent.

b. We must use the test for small and independent samples described in Section 10.3.

$H_0: \mu_1 - \mu_2 = 0$; $\qquad\qquad$ $H_1: \mu_1 - \mu_2 < 0$

From sample without additive: $n_1 = 6, \sum x = 140.4, \sum x^2 = 3432.36, \bar{x}_1 = \dfrac{\sum x}{n_1} = \dfrac{140.4}{6} = 23.40$

$s_1 = \sqrt{\dfrac{\sum x^2 - \dfrac{(\sum x)^2}{n_1}}{n_1 - 1}} = \sqrt{\dfrac{3432.36 - \dfrac{(140.4)^2}{6}}{6-1}} = 5.42217668$

From sample with additive: $n_2 = 6, \sum x = 150.7, \sum x^2 = 3957.61, \bar{x}_2 = \sum x / n_2 = 150.7/6 = 25.12$

$s_2 = \sqrt{\dfrac{\sum x^2 - \dfrac{(\sum x)^2}{n_2}}{n_2 - 1}} = \sqrt{\dfrac{3957.61 - \dfrac{(150.7)^2}{6}}{6-1}} = 5.87415242$

$s_p = \sqrt{\dfrac{(n_1-1)s_1^2 + (n_2-1)s_2^2}{n_1 + n_2 - 2}} = \sqrt{\dfrac{(6-1)(5.42217668)^2 + (6-1)(5.87415242)^2}{6+6-2}} = 5.65268373$

$s_{\bar{x}_1-\bar{x}_2} = s_p \sqrt{\dfrac{1}{n_1} + \dfrac{1}{n_2}} = 5.65268373\sqrt{\dfrac{1}{6} + \dfrac{1}{6}} = 3.26357847$

$df = n_1 + n_2 - 2 = 6+6-2 = 10$

For $df = 10$ and area in the left tail of the t curve = .025, the critical value of t is -2.228.

$$t = \frac{(\bar{x}_1 - \bar{x}_2) - (\mu_1 - \mu_2)}{s_{\bar{x}_1 - \bar{x}_2}} = \frac{(23.40 - 25.12) - 0}{3.26357847} = -.527$$

Do not reject H_0. Do not conclude that the mean gas mileage is lower without the additive.

c. The conclusions are different. In exercise 10.59, H_0 was rejected. In part b of this exercise, H_0 was not rejected.

10.104 We need to obtain sample information about the incidence of brain tumor in people who do not use cellular phones and the incidence of brain tumor in people who do use cellular phones. Since the occurrence of brain tumor is a fairly rare event, large samples of healthy people from each group must be used (say tens of thousands). For each group we would count the number that develop brain tumor over some period of time (say a year or two). The hypothesis would be:

H_0: $p_1 - p_2 = 0$; H_1: $p_1 - p_2 < 0$

where p_1 = P(brain tumor if do not use cellular phones)

p_2 = P(brain tumor if use cellular phones)

If the concern is about product liability, the worst mistake would be to conclude cellular phones do not increase the risk of brain tumor when they really do. This would be a Type II error and, hence, a higher α level would be appropriate. The test is based on the standard normal distribution with the following test statistic.

$$z = \frac{(\hat{p}_1 - \hat{p}_2) - (p_1 - p_2)}{s_{\hat{p}_1 - \hat{p}_2}} \quad \text{where,} \quad s_{\hat{p}_1 - \hat{p}_2} = \sqrt{\bar{p}\,\bar{q}\left(\frac{1}{n_1} + \frac{1}{n_2}\right)}$$

10.105 a. We require $z\sigma_{\bar{x}_1 - \bar{x}_2} \leq 5$ where $\sigma_{\bar{x}_1 - \bar{x}_2} = \sqrt{\dfrac{\sigma_1^2}{n_1} + \dfrac{\sigma_2^2}{n_2}}$

Let $n = n_1 = n_2$, so $\sigma_{\bar{x}_1 - \bar{x}_2} = \sqrt{\dfrac{\sigma_1^2}{n_1} + \dfrac{\sigma_2^2}{n_2}} = \sqrt{\dfrac{(15)^2}{n} + \dfrac{(10)^2}{n}} = \sqrt{\dfrac{325}{n}}$

For the 90% confidence level, $z = 1.65$.

Thus, $1.65\sqrt{\dfrac{325}{n}} \leq 5$ or $\sqrt{n} \geq \dfrac{1.65\sqrt{325}}{5}$ or $n \geq \dfrac{(1.65)^2(325)}{(5)^2} = 35.39 \approx 36$

Thus, use a sample size of 36 for each class.

b. H_0: $\mu_1 - \mu_2 = 0$; H_1: $\mu_1 - \mu_2 \neq 0$.

For $\alpha = .05$, the critical values of z are -1.96 and 1.96.

Using $n_1 = n_2 = 36$ (from part a), we obtain:

$$\sigma_{\bar{x}_1 - \bar{x}_2} = \sqrt{\frac{\sigma_1^2}{n_1} + \frac{\sigma_2^2}{n_2}} = \sqrt{\frac{(15)^2}{36} + \frac{(10)^2}{36}} = 3.00462606$$

To reject H_0, we require that z is either less than -1.96 or greater than 1.96, which can be written

as: $\quad |z| > 1.96$

or $\quad \dfrac{|(\bar{x}_1 - \bar{x}_2) - (\mu_1 - \mu_2)|}{\sigma_{x_1 - x_2}} > 1.96$

or $\quad \dfrac{|\bar{x}_1 - \bar{x}_2| - 0}{3.00462606} > 1.96$

or $\quad |\bar{x}_1 - \bar{x}_2| > 5.89$

Thus, the sample means must differ by at least 5.89 to conclude that the two population means are different.

10.106 $\quad \sigma_{\bar{x}_1 - \bar{x}_2} = \sqrt{\dfrac{\sigma_1^2}{n_1} + \dfrac{\sigma_2^2}{n_2}} = \sqrt{\dfrac{(.4)^2}{25} + \dfrac{(.7)^2}{36}} = .14146063$

a. For $\bar{x}_1 - \bar{x}_2 = .15$: $z = \dfrac{(\bar{x}_1 - \bar{x}_2) - (\mu_1 - \mu_2)}{\sigma_{\bar{x}_1 - \bar{x}_2}} = \dfrac{.15 - (1.3 - 1.5)}{.14146063} = 2.47$

Hence, $P\{(\bar{x}_1 - \bar{x}_2) < .15\} = P(z < 2.47) = .5 + .4932 = .9932$

b. For $\bar{x}_1 - \bar{x}_2 = 0$: $z = \dfrac{(\bar{x}_1 - x_2) - (\mu_1 - \mu_2)}{\sigma_{x_1 - x_2}} = \dfrac{0 - (1.3 - 1.5)}{.14146063} = 1.41$

Hence, $P(\bar{x}_1 > \bar{x}_2) = P\{(\bar{x}_1 - \bar{x}_2) > 0\} = P(z > 1.41) = .5 - .4207 = .0793$

c. $P\{(\bar{x}_1 - \bar{x}_2) < .15 \mid \bar{x}_1 = 2.0\} = P\{(2.0 - \bar{x}_2) < .15\} = P(\bar{x}_2 > 1.85)$

$\sigma_{\bar{x}_2} = \sigma_2 / \sqrt{n_2} = .7 / \sqrt{36} = .11666667$

For $x = 1.85$: $z = \dfrac{\bar{x}_2 - \mu_2}{\sigma_{\bar{x}_2}} = \dfrac{1.85 - 1.5}{.11666667} = 3.00$

Hence, $P(\bar{x}_2 > 1.85) = P(z > 3.00) = .5 - .4987 = .0013$

10.107 Let: x_1 = the number of male voters in 100 who favor the candidate

$\quad\quad x_2$ = the number of female voters in 100 who favor the candidate

Let $\hat{p}_1$ and $\hat{p}_2$ be the corresponding sample proportions.

$$\sigma_{\hat{p}_1-\hat{p}_2} = \sqrt{\frac{p_1 q_1}{n_1} + \frac{p_2 q_2}{n_2}} = \sqrt{\frac{(.65)(.35)}{100} + \frac{(.40)(.60)}{100}} = .06837397$$

$$P(x_1 > x_2 + 10) = P\{(x_1 - x_2) \geq 10\} = P\left(\frac{x_1 - x_2}{100} \geq \frac{10}{100}\right) = P\left(\frac{x_1}{100} - \frac{x_2}{100} \geq .10\right) = P\{(\hat{p}_1 - \hat{p}_2) \geq .10\}$$

For $\hat{p}_1 - \hat{p}_2 = .10$: $z = \dfrac{(\hat{p}_1 - \hat{p}_2) - (p_1 - p_2)}{\sigma_{\hat{p}_1-\hat{p}_2}} = \dfrac{.10 - (.65 - .40)}{.06837397} = -2.19$

Hence, $P(\hat{p}_1 - \hat{p}_2) \geq .10 = P(z \geq -2.19) = .4857 + .5 = .9857$

Self-Review Test for Chapter 10

1. a

2. See the solution to Exercise 10.1.

3. $s_{\bar{x}_1-\bar{x}_2} = \sqrt{\dfrac{s_1^2}{n_1} + \dfrac{s_2^2}{n_2}} = \sqrt{\dfrac{(.8)^2}{40} + \dfrac{(1.3)^2}{50}} = .22315914$

   a. The z value for a 99% confidence level is 2.58.

   The 99% confidence interval for $\mu_1 - \mu_2$ is:

   $(\bar{x}_1 - \bar{x}_2) \pm z s_{\bar{x}_1-\bar{x}_2} = (7.6 - 5.4) \pm 2.58(.22315914) = 2.2 \pm .58 = 1.62$ to 2.78

   b. H_0: $\mu_1 - \mu_2 = 0$; H_1: $\mu_1 - \mu_2 > 0$

   For $\alpha = .025$, the critical value of z is 1.96.

   $z = \dfrac{(\bar{x}_1 - \bar{x}_2) - (\mu_1 - \mu_2)}{s_{x_1-x_2}} = \dfrac{(7.6 - 5.4) - 0}{.22315914} = 9.86$

   Reject H_0. The mean stress score of all executives is higher than that of all professors.

4. Using the formulas given in Section 10.2 of the text, we calculate s_p and $s_{\bar{x}_1-\bar{x}_2}$ as follows.

   $s_p = \sqrt{\dfrac{(n_1-1)s_1^2 + (n_2-1)s_2^2}{n_1+n_2-2}} = \sqrt{\dfrac{(20-1)(.54)^2 + (25-1)(.8)^2}{20+25-2}} = .69717703$

   $s_{\bar{x}_1-\bar{x}_2} = s_p\sqrt{\dfrac{1}{n_1} + \dfrac{1}{n_2}} = .69717703 = \sqrt{\dfrac{1}{20} + \dfrac{1}{25}} = .20915311$

   a. $df = n_1 + n_2 - 2 = 20 + 25 - 2 = 43$

   Area in each tail of the t distribution curve = $.5 - (.95/2) = .025$

   The t value for $df = 43$ and .025 area in the right tail is 2.017.

   The 95% confidence interval for $\mu_1 - \mu_2$ is:

   $(\bar{x}_1 - \bar{x}_2) \pm t s_{\bar{x}_1-\bar{x}_2} = (2.3 - 4.6) \pm 2.017(.20915311) = -2.72$ to -1.88 hours

b. $H_0: \mu_1 - \mu_2 = 0; \quad H_1: \mu_1 - \mu_2 < 0$

$df = 20 + 25 - 2 = 43$ and Area in left tail of the t distribution curve = .01

The critical value of t is -2.416.

$$t = \frac{(\bar{x}_1 - \bar{x}_2) - (\mu_1 - \mu_2)}{s_{\bar{x}_1 - \bar{x}_2}} = \frac{(2.3 - 4.6) - 0}{.20915311} = -10.997$$

Reject the null hypothesis. The mean time spent per week playing with their children by all fathers who are alcoholic is less than that of fathers who are nonalcoholic.

5. $\quad s_{\bar{x}_1 - \bar{x}_2} = \sqrt{\dfrac{(.54)^2}{20} + \dfrac{(.8)^2}{25}} = .20044950$

Using the formula for degrees of freedom on page 471 of the text, we obtain: $df = 41$

a. Area in each tail of the t distribution curve = $.5 - (.95/2) = .025$

The t value for $df = 41$ and .025 area in the right tail is 2.020.

The 95% confidence interval for $\mu_1 - \mu_2$ is:

$(\bar{x}_1 - \bar{x}_2) \pm t s_{\bar{x}_1 - \bar{x}_2} = (2.3 - 4.6) \pm 2.020(.20044950) = -2.3 \pm .40 = -2.70$ to -1.90 hours

b. $H_0: \mu_1 - \mu_2 = 0; \quad H_1: \mu_1 - \mu_2 < 0$

$df = 41$ and Area in left tail of the t distribution curve = .01

The critical value of t is -2.421

$$t = \frac{(\bar{x}_1 - \bar{x}_2) - (\mu_1 - \mu_2)}{s_{\bar{x}_1 - \bar{x}_2}} = \frac{(2.3 - 4.6) - 0}{.20044950} = -11.474$$

Reject H_0. The mean time spent per week playing with their children by all alcoholic fathers is less than that of all nonalcoholic fathers.

6.

Zeke's	1058	544	1349	1296	676	998	1698	
Elmer's	995	540	1175	1350	605	970	1520	
d = Zeke–Elmer	63	4	174	-54	71	28	178	$\Sigma d = 464$
d^2	3969	16	30,276	2916	5041	784	31,684	$\Sigma d^2 = 74{,}686$

$$\bar{d} = \frac{\sum d}{n} = \frac{465}{7} = \$66.29$$

$$s_d = \sqrt{\frac{\sum d^2 - \dfrac{(\sum d)^2}{n}}{n-1}} = \sqrt{\frac{74{,}686 - \dfrac{(464)^2}{7}}{7-1}} = 85.56618157$$

$$s_{\overline{d}} = s_d / \sqrt{n} = 85.56618157 / \sqrt{7} = 32.34097672$$

a. $df = n - 1 = 7 - 1 = 6$ and Area in each tail of the t distribution curve $= .5 - (.99/2) = .005$

The t value for $df = 6$ and .005 area in the right tail is 3.707.

The 99% confidence interval for μ_d is:

$$\overline{d} \pm ts_{\overline{d}} = 66.29 \pm 3.707(32.34097672) = 66.29 \pm 119.89 = -\$53.60 \text{ to } \$186.18$$

b. $H_0: \mu_d = 0$; $H_1: \mu_d \neq 0$

$df = n - 1 = 7 - 1 = 6$ and Area in each tail of the t distribution curve $= .05/2 = .025$

The critical values of t are -2.447 and 2.447.

$$t = \frac{\overline{d} - \mu_d}{s_{\overline{d}}} = \frac{66.29 - 0}{32.34097672} = 2.050$$

Do not reject the null hypothesis. Do not conclude the mean μ_d of the population paired differences is different from zero.

7. a. $s_{\hat{p}_1 - \hat{p}_2} = \sqrt{\dfrac{\hat{p}_1 \hat{q}_1}{n_1} + \dfrac{\hat{p}_2 \hat{q}_2}{n_2}} = \sqrt{\dfrac{(.57)(.43)}{500} + \dfrac{(.55)(.45)}{400}} = .03330090$

The z value for a 97% confidence level is 2.17.

The 97% confidence interval for $p_1 - p_2$ is:

$$\left(\hat{p}_1 - \hat{p}_2\right) \pm zs_{\hat{p}_1 - \hat{p}_2} = (.57 - .55) \pm 2.17(.03330090) = .02 \pm .072 = -.052 \text{ to } .092$$

b. $H_0: p_1 - p_2 = 0$; $H_1: p_1 - p_2 \neq 0$.

For $\alpha = .01$, the critical values of z are -2.58 and 2.58.

$$\overline{p} = \frac{n_1 \hat{p}_1 + n_2 \hat{p}_2}{n_1 + n_2} = \frac{500(.57) + 400(.55)}{500 + 400} = .561 \qquad \text{and} \qquad \overline{q} = 1 - \overline{p} = 1 - .561 = .439$$

$$s_{\hat{p}_1 - \hat{p}_2} = \sqrt{\overline{p}\,\overline{q}\left(\frac{1}{n_1} + \frac{1}{n_2}\right)} = \sqrt{(.561)(.439)\left(\frac{1}{500} + \frac{1}{400}\right)} = .03329047$$

$$z = \frac{(\hat{p}_1 - \hat{p}_2) - (p_1 - p_2)}{s_{\hat{p}_1 - \hat{p}_2}} = \frac{(.57 - .55) - 0}{.03329047} = .60$$

Do not reject H_0. The proportion of all male voters who voted in the last presidential election is not different from that of all female voters who voted in the same election.

Chapter Eleven

11.1 See box on page 512 of the text.

11.2 For $df = 12$ and .025 area in the right tail: $\chi^2 = 23.337$

11.3 For $df = 28$ and .05 area in the right tail: $\chi^2 = 41.337$

11.4 For an area of .10 in the left tail, the area in the right tail is $1 - .10 = .90$. Hence, the value of chi–square for $df = 14$ and .10 area in the left tail is the same as for $df = 14$ and .90 area in the right tail, which is: $\chi^2 = 7.790$

11.5 For an area of .990 in the left tail, the area in the right tail is $1 - .990 = .01$. Hence, the value of chi–square for $df = 23$ and .990 area in the left tail is the same as for $df = 23$ and .01 area in the right tail, which is: $\chi^2 = 41.638$

11.6 a. For $df = 4$ and .005 area in the right tail: $\chi^2 = 14.860$

   b. For an area of .05 in the left tail, the area in the right tail is $1 - .05 = .95$. Hence, the value of chi–square for $df = 4$ and .05 area in the left tail is the same as for $df = 4$ and .95 area in the right tail, which is: $\chi^2 = .711$

11.7 a. For an area of .025 in the left tail, the area in the right tail is $1 - .025 = .975$. Hence, the value of chi–square for $df = 13$ and .025 area in the left tail is the same as for $df = 13$ and .975 area in the right tail, which is: $\chi^2 = 5.009$

   b. For $df = 13$ and .995 area in the right tail, $\chi^2 = 3.565$.

11.8 Following are the four characteristics of a multinomial experiment.
   1. It consists of n identical trials.
   2. Each trial results in one of k possible outcomes (or categories) where $k > 2$.
   3. The trials are independent.
   4. The probabilities of various outcomes remain constant for each trial.

11.9   A goodness–of–fit test compares the observed frequencies from a multinomial experiment with expected frequencies derived from a certain pattern or theoretical distribution. The test evaluates how well the observed frequencies fit the expected frequencies.

11.10   The observed frequencies are actual results obtained from a multinomial experiment. The expected frequencies are calculated from a theoretical distribution. These are the frequencies that we expect to obtain if the null hypothesis is true.

11.11   The expected frequency of a category is given by $E = np$ where n is the sample size and p is the probability that an element belongs to that category if the null hypothesis is true. The degrees of freedom for a goodness–of–fit test are $k - 1$, where k is the number of possible outcomes (or categories) for the experiment.

11.12   The minimum expected frequency for each category should be 5. If this condition is not satisfied, we may increase the sample size or combine two or more categories to make each expected frequency at least 5.

11.13   Step 1: H_0: The die is fair; H_1: The die is not fair.

   Step 2:   Because there are six categories (a die has six outcomes), it is a multinomial experiment. Consequently, we use chi–square distribution to conduct the test.

   Step 3:   The significance level is .05. Because a goodness – of – fit test is always a right – tailed test, area in the right tail = α = .05

   k = number of categories = 6, so $df = k - 1 = 6 - 1 = 5$

   From the chi–square distribution table, the critical value of χ^2 for $df = 5$ and .05 area in the right tail is 11.070.

   Step 4:   Note that the die will be fair if the probability of each of the six outcomes is the same, which is 1/6.

Outcome	O	p	$E = np$	$O - E$	$(O - E)^2$	$(O - E)^2/E$
1	7	1/6	10	-3	9	.900
2	12	1/6	10	2	4	.400
3	8	1/6	10	-2	4	.400
4	15	1/6	10	5	25	2.500
5	11	1/6	10	1	1	.100
6	7	1/6	10	-3	9	.900
	$n = 60$					Sum = 5.200

   The value of the test statistic is: $\chi^2 = \Sigma (O - E)^2 / E = 5.200$

   Step 5:   Do not reject the null hypothesis. The die is fair.

11.14 H_0: The current distribution of responses of all such workers is the same as that of 2002.

   H_1: The current distribution of responses of all such workers differs from that of 2002.

The significance level is .05. Because a goodness – of – fit test is always right – tailed, area in the right tail is $\alpha = .05$.

$k =$ number of categories $= 4$; so $\quad df = k - 1 = 4 - 1 = 3$;

From the chi–square distribution table, the critical value of χ^2 for $df = 3$ and .05 area in the right tail is 7.815.

Category	O	p	$E = np$	$O - E$	$(O - E)^2$	$(O - E)^2/E$
A	48	.08	40	8	64	1.600
B	242	.46	230	12	144	.626
C	120	.26	130	-10	100	.769
D	90	.20	100	-10	100	1.000
	$n = 500$					Sum $= 3.995$

Thus, the value of the test statistic is: $\chi^2 = \Sigma (O - E)^2 / E = 3.995$. Since $3.995 < 7.815$, do not reject H_0. The current distribution of worker responses is not different from that of 2002.

11.15 H_0: The current distribution of sources of stress is the same as that of the earlier survey.

H_1: The current distribution of sources of stress differs from that of the earlier survey.

$df = k - 1 = 5 - 1 = 4$

For $\alpha = .05$ and $df = 4$, the critical value of χ^2 is 13.277.

Response	O	p	$E = np$	$O - E$	$(O - E)^2$	$(O - E)^2/E$
Demands	404	.54	432	-28	784	1.815
Coworkers	183	.20	160	23	529	3.306
Boss	94	.10	80	14	196	2.450
Layoffs	80	.08	64	16	256	4.000
Others	39	.08	64	-25	625	9.766
	$n = 800$					Sum $= 21.337$

The value of the test statistic is: $\chi^2 = \Sigma (O - E)^2 / E = 21.337$

Since $21.337 > 13.277$, reject H_0. Conclude that the current distribution of sources of stress differs from the distribution in the earlier survey.

11.16 H_0: The distribution of office worker opinions concerning email responses has not changed since 2002.

H_1: The distribution of office worker opinions concerning email responses has changed since 2002.

$df = k - 1 = 4 - 1 = 3$ and   Area in the right tail $= \alpha = .01$

The critical value of $\chi^2 = 11.345$.

Response	O	p	$E = np$	$O - E$	$(O - E)^2$	$(O - E)^2/E$
A	128	.36	144	-16	256	1.778
B	142	.35	140	2	4	.029
C	116	.24	96	20	400	4.167
D	14	.05	20	-6	36	1.800
	$n = 400$					Sum $= 7.774$

The value of the test statistic is: $\chi^2 = \Sigma (O - E)^2 / E = 7.774$

Since $7.774 < 11.345$, do not reject H_0. Thus, we conclude that the distribution office worker opinions concerning email responses has not changed since 2002.

11.17 H_0: The orders are evenly distributed over all days of the week.

H_1: The orders are not evenly distributed over all days of the week.

$df = k - 1 = 5 - 1 = 4$

For $\alpha = .05$ and $df = 4$, the critical value of χ^2 is 9.488.

Day	O	p	$E = np$	$O - E$	$(O - E)^2$	$(O - E)^2 / E$
Monday	92	.20	80	12	144	1.800
Tuesday	71	.20	80	-9	81	1.013
Wednesday	65	.20	80	-15	225	2.813
Thursday	83	.20	80	3	9	.113
Friday	89	.20	80	9	81	1.013
	$n = 400$					Sum $= 6.752$

The value of the test statistic is: $\chi^2 = \Sigma (O - E)^2 / E = 6.752$

Since $6.752 < 9.488$ do not reject H_0. Thus, we conclude that orders are evenly distributed over all days of the week.

11.18 H_0: The distribution of payment methods has not changed.

H_1: The distribution of payment methods has changed.

$df = k - 1 = 4 - 1 = 3$

From the chi–square distribution table, the critical value of χ^2 for $df = 3$ and .01 area in the right tail is 11.345.

Day	O	p	$E = np$	$O - E$	$(O - E)^2$	$(O - E)^2 / E$
C	240	.41	205	35	1225	5.976
CK	104	.24	120	-16	256	2.133
D	111	.26	130	-19	361	2.777
N	45	.09	45	0	0	0
	$n = 500$					Sum $= 10.886$

Thus, the value of the test statistic is: $\chi^2 = \Sigma (O - E)^2 / E = 10.886$

Since $10.886 < 11.345$ do not reject H_0. Do not conclude that the incentive program has changed payment methods.

11.19 H_0: The number of cars sold is the same for each month.

H_1: The number of cars sold is not the same for each month.

$df = k - 1 = 12 - 1 = 11$

From the chi–square distribution table, the critical value of χ^2 for $df = 11$ and .10 area in the right tail is 17.275. The number of cars sold will be the same for each month if 1/12th of the cars sold during the whole year are sold each month.

Month	O	p	$E = np$	$O - E$	$(O - E)^2$	$(O - E)^2/E$
January	23	1/12	18.67	4.33	18.749	1.004
February	17	1/12	18.67	-1.67	2.789	.149
March	15	1/12	18.67	-3.67	13.469	.721
April	10	1/12	18.67	-8.67	75.169	4.026
May	14	1/12	18.67	-4.67	21.809	1.168
June	12	1/12	18.67	-6.67	44.489	2.383
July	13	1/12	18.67	-5.67	32.149	1.722
August	15	1/12	18.67	-3.67	13.469	0.721
September	23	1/12	18.67	4.33	18.749	1.004
October	26	1/12	18.67	7.33	53.729	2.878
November	27	1/12	18.67	8.33	69.389	3.717
December	29	1/12	18.67	10.33	106.709	5.716
	$\eta = 224$					Sum = 25.209

Thus the value of the test statistic is: $\chi^2 = \Sigma (O - E)^2 / E = 25.209$

Reject the null hypothesis. The number of cars sold is not the same for each month.

11.20 H_0: The sample is random.

H_1: The sample is not random.

$df = k - 1 = 4 - 1 = 3$

From the chi–square distribution table, the critical value of χ^2 for $df = 3$ and .10 area in the right tail is 6.251.

Category	O	p	$E = np$	$O - E$	$(O - E)^2$	$(O - E)^2/E$
Seniors	50	.19	38	12	144	3.789
Juniors	46	.23	46	0	0	.000
Sophomores	55	.27	54	1	1	.019
Freshmen	49	.31	62	-13	169	2.726
	$n = 200$					Sum = 6.534

Thus, value of the test statistic is: $\chi^2 = \Sigma (O - E)^2 / E = 6.534$

Since $6.534 > 6.251$ reject H_0. The sample is not random.

11.21 H_0: The percentage distribution of users' opinions is unchanged since the product was redesigned.

H_1: The percentage distribution of users' opinions has changed since the product was redesigned.

$df = k - 1 = 4 - 1 = 3$

For $\alpha = .025$ and $df = 3$, the critical value of χ^2 is 9.348

Opinion	O	p	$E = np$	$O - E$	$(O-E)^2$	$(O-E)^2/E$
Excellent	495	.53	424	71	5041	11.889
Satisfactory	255	.31	248	7	49	.198
Unsatisfactory	35	.07	56	−21	441	7.875
No opinion	15	.09	72	−57	3249	45.125
	$n = 800$					Sum = 65.087

The value of the test statistic is $\chi^2 = \Sigma (O - E)^2 / E = 65.087$

Since $65.087 > 9.348$, reject H_0. Conclude that the percentage distribution of users' opinions has changed since the product was redesigned.

11.22 H_0: The distribution of defects is the same as when the process is working properly and the process does not need an adjustment.

H_1: The distribution of defects is not the same as when the process is working properly and the process needs an adjustment.

$df = k - 1 = 3 - 1 = 2$

For $\alpha = .01$ and $df = 2$, the critical value of χ^2 is 9.210.

Defects	O	p	$E = np$	$O - E$	$(O-E)^2$	$(O-E)^2/E$
0	262	.92	276	−14	196	0.710
1	24	.05	15	9	81	5.400
2 or more	14	.03	9	5	25	2.778
	$n = 300$					Sum = 8.888

The value of the test statistic is: $\chi^2 = \Sigma (O - E)^2 / E = 8.888$. Since $8.888 < 9.210$, do not reject H_0. Conclude that the distribution of defects is the same as when the process is working properly. Thus, the process does not need an adjustment.

11.23   In a test of independence, we test the null hypothesis that two characteristics in a given population are independent against the alternative hypothesis that they are related. For example, we may want to test whether political party affiliation and opinion of voters on abortion are related.

In a test of homogeneity, we test whether two or more populations are the same with respect to the distribution of certain characteristics. For example, we may want to test whether the preferences of several different ethnic groups are similar for three television programs.

11.24   The expected frequency for a cell in a contingency table is given by $E = $ (Row total)(Column total)$/n$ where "Row total" represents the total of the observations in the row of that cell, "Column total" is the total of the observed values in the column of that cell, and n is the sample size. The degrees of freedom are given by $df = (R - 1)(C - 1)$, where R and C are the number of rows and columns, respectively, in the table.

11.25 The minimum expected frequency for each cell should be 5. If this condition is not satisfied, we may increase the sample size or combine some categories.

11.26 a. H_0: Columns and rows are independent. H_1: Columns and rows are dependent.

b. The expected frequencies are given in parentheses below the observed frequencies in the table below.

	Column 1	Column 2	Column 3	Total
Row 1	137	64	105	306
	(125.85)	(77.67)	(102.48)	
Row 2	98	71	65	234
	(96.24)	(59.39)	(78.37)	
Row 3	115	81	115	311
	(127.91)	(78.94)	(104.15)	
Total	350	216	285	851

c. $df = (R-1)(C-1) = (3-1)(3-1) = 4$

For $\alpha = .01$ and $df = 4$, the critical value of χ^2 is 13.277. The rejection region lies to the right of $\chi^2 = 13.277$. The nonrejection region lies to the left of $\chi^2 = 13.277$.

d. $\chi^2 = \Sigma (O-E)^2 / E$

$$= \frac{(137-125.85)^2}{125.85} + \frac{(64-77.67)^2}{77.67} + \frac{(105-102.48)^2}{102.48} + \frac{(98-96.24)^2}{96.24} + \frac{(71-59.39)^2}{59.39}$$
$$+ \frac{(65-78.37)^2}{78.37} + \frac{(115-127.91)^2}{127.91} + \frac{(81-78.94)^2}{78.94} + \frac{(115-104.15)^2}{104.15}$$

$= .988 + 2.406 + .062 + .032 + 2.270 + 2.281 + 1.303 + .054 + 1.130 = 10.526$

e. Since $10.526 < 13.277$, do not reject H_0. Columns and rows are independent.

11.27 a. H_0: The proportion in each row is the same for all four populations.

H_1: The proportion in each row is not the same for all four populations.

b. The expected frequencies are given in parentheses below the observed frequencies in the table below.

	Column 1	Column 2	Column 3	Column 4	Total
Row 1	24	81	60	121	286
	(31.62)	(63.95)	(89.95)	(100.49)	
Row 2	46	64	91	72	273
	(30.18)	(61.04)	(85.86)	(95.92)	
Row 3	20	37	105	93	255
	(28.19)	(57.01)	(80.20)	(89.59)	
Total	90	182	256	286	814

c. $df = (R-1)(C-1) = (3-1)(4-1) = 6$

For $\alpha = .025$ and $df = 6$, the critical value of χ^2 is 14.449. The rejection region lies to the right of $\chi^2 = 14.449$.

d. $\chi^2 = \Sigma (O-E)^2 / E = 1.836 + 4.546 + 9.972 + 4.186 + 8.293 + .144 + .308 + 5.965 + 2.379 + 7.023 + 7.669 + .130 = 52.451$

e. Since 52.451 > 14.449, reject H_0.

11.28 H_0: Gender and brand preference are independent.

H_1: Gender and brand preference are dependent.

$df = (R-1)(C-1) = (2-1)(2-1) = 1$

For $\alpha = .01$ and $df = 1$, the critical value of χ^2 is 6.635.

	National Brand	Store Brand	Total
Men	170	145	315
	(158.4)	(156.6)	
Women	182	203	385
	(193.6)	(191.4)	
Total	352	348	700

The test statistic is: $\chi^2 = \Sigma (O-E)^2 / E = .849 + .859 + .695 + .703 = 3.106$

Since $3.106 < 6.635$, do not reject H_0. Gender and brand preference are independent.

11.29 H_0: Gender and wearing or not wearing of seat belt are not related.

H_1: Gender and wearing or not wearing of seat belt are related.

$df = (R-1)(C-1) = (2-1)(2-1) = 1$

From the chi–square distribution table, the critical value of χ^2 for $df = 1$ and .025 area in the right tail is 5.024.

	Wearing a seat belt	Not wearing seat belt	Total
Men	34	21	55
	(36.3)	(18.7)	
Women	32	13	45
	(29.7)	(15.3)	
Total	66	34	100

The value of the test statistic is: $\chi^2 = \Sigma (O-E)^2 / E = .146 + .283 + .178 + .346 = .953$

Since $.953 < 5.024$, do not reject the null hypothesis. Being a male or a female and wearing or not wearing a seat belt are not related.

11.30 H_0: Gender and responses are independent. H_1: Gender and responses are dependent.

$df = (R-1)(C-1) = (2-1)(3-1) = 2$

From the chi–square distribution table the critical value of χ^2 for $df = 2$ and .01 area in the right tail is 9.210.

	Yes	No	Uncertain	Total
Women	125	59	21	205
	(115.54)	(70.55)	(18.92)	
Men	101	79	16	196
	(110.46)	(67.45)	(18.08)	
Total	226	138	37	401

The test statistic is: $\chi^2 = \Sigma (O - E)^2 / E = .775 + 1.891 + .229 + .810 + 1.978 + .239 = 5.922$

Do not reject H_0. Do not conclude gender and responses are dependent.

11.31 H_0: Use of a financial advisor is unrelated to stock ownership.

H_1: Use of a financial advisor is related to stock ownership.

$df = (R - 1)(C - 1) = (2 - 1)(2 - 1) = 1$

From the chi–square distribution table, the critical value of χ^2 for $df = 1$ and .05 area in the right tail is 3.841.

Use Financial Adviser	Own	Do not own	Total
Yes	165	135	300
	(156)	(144)	
No	43	57	100
	(52)	(48)	
Total	208	192	400

The value of the test statistic is: $\chi^2 = \Sigma (O - E)^2 / E = .519 + .563 + 1.558 + 1.688 = 4.328$

Reject H_0. Conclude that using a financial advisor is related to stock ownership.

11.32 H_0: Smoking and visits to the physician are not related.

H_1: Smoking and visits to the physician are related.

$df = (R - 1)(C - 1) = (2 - 1)(3 - 1) = 2$

From the chi–square distribution table, the critical value of χ^2 for $df = 2$ and .05 area in the right tail is 5.991.

	Visits to the Physician			Total
	$0 - 1$	$2 - 4$	≥ 5	
Smoker	25	60	75	160
	(54)	(60)	(46)	
Nonsmoker	110	90	40	240
	(81)	(90)	(69)	
Total	135	150	115	400

The value of the test statistic is:

$\chi^2 = \Sigma(O - E)^2 / E = 15.574 + .000 + 18.283 + 10.383 + .000 + 12.188 = 56.428$

Since $56.428 > 5.991$, reject the null hypothesis. Smoking and visits to the physician are related.

11.33 H_0: Causes of fires and region are unrelated.

H_1: Causes of fires and region are related.

$df = (R - 1)(C - 1) = (4 - 1)(2 - 1) = 3$

From the chi–square distribution table, the critical value of χ^2 for $df = 3$ and .05 area in the right tail is 7.815.

	Arson	Accident	Lightning	Unknown	Total
A	6	9	6	10	31
	(5.30)	(9.38)	(8.57)	(7.75)	
B	7	14	15	9	45
	(7.70)	(13.62)	(12.43)	(11.25)	
Total	13	23	21	19	76

The value of the test statistic is:

$\chi^2 = \sum (O - E)^2 / E = .092 + .015 + .771 + .653 + .064 + .011 + .531 + .450 = 2.587$

Since $2.587 < 7.815$ do not reject H_0. Do not conclude that causes of fires and region are related.

11.34 H_0: The distribution of good and defective parts is the same for both subsidiaries.

H_1: The distribution of good and defective parts is not the same for both subsidiaries.

$df = (R - 1)(C - 1) = (2 - 1)(2 - 1) = 1$ For $\alpha = .05$ and $df = 1$, the critical value of χ^2 is 3.841.

	Subsidiary A	Subsidiary B	Total
Good	284	381	665
	(285)	(380)	
Defective	16	19	35
	(15)	(20)	
Total	300	400	700

The value of the test statistic is: $\chi^2 = \sum (O - E)^2 / E = .004 + .003 + .067 + .050 = .124$

Since $.124 < 3.841$, do not reject H_0. The distribution of good and defective parts is the same for both subsidiaries.

11.35 H_0: The two drugs are similar in curing the patients.

H_1: The two drugs are not similar in curing the patients.

$df = (R - 1)(C - 1) = (2 - 1)(2 - 1) = 1$

From the chi–square distribution table, the critical value of χ^2 for $df = 1$ and .01 area in the right tail is 6.635.

	Cured	Not Cured	Total
Drug I	44	16	60
	(37.20)	(22.80)	
Drug II	18	22	40
	(24.80)	(15.20)	
Total	62	38	100

The value of the test statistic is: $\chi^2 = \sum (O - E)^2 / E = 1.243 + 2.028 + 1.865 + 3.042 = 8.178$

Reject the null hypothesis. The two drugs are not similar in curing the patients.

11.36 H_0: The distributions of opinions of users of this product are homogeneous for all four regions.

H_1: The distributions of opinions of users of this product are not homogeneous for all four regions.

$df = (R - 1)(C - 1) = (2 - 1)(4 - 1) = 3$ For $\alpha = .01$ and $df = 3$, the critical value of χ^2 is 11.345.

	East	South	Midwest	West	Total
Like	274	206	291	254	1025
	(256.25)	(256.25)	(256.25)	(256.25)	
Do not like	126	194	109	146	575
	(143.75)	(143.75)	(143.75)	(143.75)	
Total	400	400	400	400	1600

The value of the test statistic is:

$\chi^2 = \Sigma (O - E)^2 / E = 1.230 + 9.854 + 4.712 + .020 + 2.192 + 17.566 + 8.400 + .035 = 44.009$

Since $44.009 > 11.345$, reject H_0. The distributions of opinions of users of this product are not homogeneous for all four regions.

11.37 H_0: The distribution of media preference is the same for boys and girls.

H_1: The distribution of media preference is not the same for boys and girls.

$df = (R - 1)(C - 1) = (2 - 1)(5 - 1) = 3$ For $\alpha = .01$ and $df = 4$, the critical value of χ^2 is 13.277.

	Internet	TV	Phone	Radio	Other	Total
Boys	190	170	60	60	20	500
	(165)	(127.5)	(107.5)	(72.5)	(27.5)	
Girls	140	85	155	85	35	500
	(165)	(127.5)	(107.5)	(72.5)	(27.5)	
Total	330	255	215	145	55	1000

The value of the test statistic is: $\chi^2 = \Sigma (O - E)^2 / E = 3.788 + 14.167 + 20.988 + 2.155 + 2.045 + 3.788$

$+ 14.167 + 20.988 + 2.155 + 2.045 = 86.286$

Since $86.286 > 13.277$, reject H_0. The distribution of media preference is not the same for boys and girls.

11.38 H_0: The grade distributions are homogeneous for the three professors.

H_1: The grade distributions are not homogeneous for the three professors.

$df = (R - 1)(C - 1) = (4 - 1)(3 - 1) = 6$

From the chi–square distribution table, the critical value of χ^2 for $df = 6$ and .025 area in the right tail is 14.449.

		Professor			
		Miller	Smith	Moore	Total
	A	18	36	20	74
		(24.67)	(28.07)	(21.26)	
	B	25	44	15	84
Grade		(28.00)	(31.86)	(24.14)	
	C	85	73	82	240
		(80.00)	(91.03)	(68.97)	
	D & F	17	12	8	37
		(12.33)	(14.03)	(10.63)	
	Total	145	165	125	435

The value of the test statistic is: $\chi^2 = \Sigma\,(O - E)^2 / E = 1.803 + 2.240 + .075 + .321 + 4.626 + 3.461 +$

$.313 + 3.571 + 2.462 + 1.769 + .294 + .651 = 21.586$

Since $21.586 > 14.449$, reject the null hypothesis. The grade distributions are not homogeneous for the three professors.

11.39 H_0: The distributions of opinions are homogeneous for the two groups of workers.

H_1: The distributions of opinions are not homogeneous for the two groups of workers.

$df = (R - 1)(C - 1) = (2 - 1)(3 - 1) = 2$

From the chi–square distribution table, the critical value of χ^2 for $df = 2$ and .025 area in the right tail is 7.378.

	Opinion			
	Favor	Oppose	Uncertain	Total
Blue collar	44	39	12	95
Workers	(42.59)	(42.59)	(9.83)	
White collar	21	26	3	50
Workers	(22.41)	(22.41)	(5.17)	
Total	65	65	15	145

The value of the test statistic is: $\chi^2 = \Sigma\,(O - E)^2 / E = .047 + .303 + .479 + .089 + .575 + .911 = 2.404$

Do not reject the null hypothesis. The distributions of opinions are homogeneous for the two groups of workers.

11.40 $\alpha/2 = 5 - (.95/2) = .025$ and $1 - \alpha/2 = 1 - .025 = .975$

a. $df = n - 1 = 12 - 1 = 11$

χ^2 for 11 df and .025 area in the right tail = 21.920

χ^2 for 11 df and .975 area in the right tail = 3.816

The 95% confidence interval for σ^2 is:

$$\frac{(n-1)s^2}{\chi^2_{\alpha/2}} \text{ to } \frac{(n-1)s^2}{\chi^2_{1-\alpha/2}} = \frac{(12-1)(46)}{21.920} \text{ to } \frac{(12-1)(46)}{3.816} = 23.0839 \text{ to } 132.5996$$

b. $df = n - 1 = 16 - 1 = 15$

   χ^2 for 15 df and .025 area in the right tail = 27.488

   χ^2 for 15 df and .975 area in the right tail = 6.262

   The 95% confidence interval for σ^2 is:

$$\frac{(n-1)s^2}{\chi^2_{\alpha/2}} \text{ to } \frac{(n-1)s^2}{\chi^2_{1-\alpha/2}} = \frac{(16-1)(46)}{27.488} \text{ to } \frac{(16-1)(46)}{6.262} = 25.1019 \text{ to } 110.1884$$

c. $df = n - 1 = 25 - 1 = 24$

   χ^2 for 24 df and .025 area in the right tail = 39.364

   χ^2 for 24 df and .975 area in the right tail = 12.401

   The 95% confidence interval for σ^2 is:

$$\frac{(n-1)s^2}{\chi^2_{\alpha/2}} \text{ to } \frac{(n-1)s^2}{\chi^2_{1-\alpha/2}} = \frac{(25-1)(46)}{39.364} \text{ to } \frac{(25-1)(46)}{12.401} = 28.0459 \text{ to } 89.0251$$

11.41 $df = n - 1 = 25 - 1 = 24$

a. $\alpha/2 = .5 - (.99/2) = .005$ and $1 - \alpha/2 = 1 - .005 = .995$

   χ^2 for 24 df and .005 area in the right tail = 45.559

   χ^2 for 24 df and .995 area in the right tail = 9.886

   The 99% confidence interval for σ^2 is:

$$\frac{(n-1)s^2}{\chi^2_{\alpha/2}} \text{ to } \frac{(n-1)s^2}{\chi^2_{1-\alpha/2}} = \frac{(25-1)(35)}{45.559} \text{ to } \frac{(25-1)(35)}{9.886} = 18.4376 \text{ to } 84.9686$$

b. $\alpha/2 = .5 - (.95/2) = .025$ and $1 - \alpha/2 = 1 - .025 = .975$

   χ^2 for 24 df and .025 area in the right tail = 39.364

   χ^2 for 24 df and .975 area in the right tail = 12.401

   The 95% confidence interval for σ^2 is:

$$\frac{(n-1)s^2}{\chi^2_{\alpha/2}} \text{ to } \frac{(n-1)s^2}{\chi^2_{1-\alpha/2}} = \frac{(25-1)(35)}{39.364} \text{ to } \frac{(25-1)(35)}{12.401} = 21.3393 \text{ to } 67.7365$$

c. $\alpha/2 = .5 - (.90/2) = .005$ and $1 - \alpha/2 = 1 - .05 = .95$

   χ^2 for 24 df and .05 area in the right tail = 36.415

   χ^2 for 24 df and .95 area in the right tail = 13.848

   The 90% confidence interval for σ^2 is:

$$\frac{(n-1)s^2}{\chi^2_{\alpha/2}} \text{ to } \frac{(n-1)s^2}{\chi^2_{1-\alpha/2}} = \frac{(25-1)(35)}{36.415} \text{ to } \frac{(25-1)(35)}{13.848} = 23.0674 \text{ to } 60.6586$$

As the confidence level decreases, the confidence interval for σ^2 decreases in width.

11.42   a. H_0: $\sigma^2 = 14$; H_1: $\sigma^2 \neq 14$

b. $\alpha/2 = .05/2 = .025$ and $1 - \alpha/2 = 1 - .025 = .975$; $df = n - 1 = 22 - 1 = 21$

χ^2 for 21 df and .025 area in the right tail = 35.479.

χ^2 for 21 df and .975 area in the right tail = 10.283.

The rejection region lies to the left of $\chi^2 = 10.283$ and to the right of $\chi^2 = 35.479$

The nonrejection region lies between $\chi^2 = 10.283$ and $\chi^2 = 35.479$.

c.   The value of the test statistic is: $\chi^2 = (n-1)s^2/\sigma^2 = (22-1)(18)/14 = 27.000$

d.   Do not reject H_0.

11.43   a. H_0: $\sigma^2 = .80$; H_1: $\sigma^2 > .80$

b. $df = n - 1 = 16 - 1 = 15$

χ^2 for 15 df and .01 area in the right tail = 30.578.

The rejection region lies to the right of $\chi^2 = 30.578$.

The nonrejection region lies to the left of $\chi^2 = 30.578$.

c.   The value of the test statistic is: $\chi^2 = (n-1)s^2/\sigma^2 = (16-1)(1.10)/.80 = 20.625$

d.   Do not reject H_0.

11.44   a. H_0: $\sigma^2 = 1.25$; H_1: $\sigma^2 < 1.25$

b. $df = n - 1 = 25 - 1 = 24$

χ^2 for 24 df and .025 area in the left tail is the same as χ^2 for 24 df and .975 area in the right tail which gives $\chi^2 = 12.401$.

The rejection region lies to the left of $\chi^2 = 12.401$.

The nonrejection region lies to the right of $\chi^2 = 12.401$

c.   The value of the test statistic is: $\chi^2 = (n-1)s^2/\sigma^2 = (25-1)(.70)/1.25 = 13.440$

d.   Do not reject H_0.

11.45   a. H_0: $\sigma^2 = 2.2$; H_1: $\sigma^2 \neq 2.2$

b. $\alpha/2 = .05/2 = .025$ and $1 - \alpha/2 = 1 - .025 = .975$;

$df = n - 1 = 18 - 1 = 17$

χ^2 for 17 df and .025 area in the right tail = 30.191

χ^2 for 17 df and .975 area in the right tail = 7.564

The rejection region lies to the left of $\chi^2 = 7.564$ and to the right of $\chi^2 = 30.191$.

The nonrejection region lies between $\chi^2 = 7.564$ and $\chi^2 = 30.191$.

c. The value of the test statistic is: $\chi^2 = (n-1)\, s^2 / \sigma^2 = (18-1)(4.6)^2 / 2.2^2 = 35.545$

d. Reject H_0.

11.46 a. $\alpha/2 = .01/2 = .005$ and $1 - \alpha/2 = 1 - .005 = .995$

$df = n - 1 = 20 - 1 = 19$

χ^2 for 19 df and .005 area in the right tail $= 38.582$

χ^2 for 19 df and .995 area in the right tail $= 6.844$

The 99% confidence interval for the population variance σ^2 is:

$$\frac{(n-1)s^2}{\chi^2_{\alpha/2}} \text{ to } \frac{(n-1)s^2}{\chi^2_{1-\alpha/2}} = \frac{(20-1)(.014)}{38.582} \text{ to } \frac{(20-1)(.014)}{6.844} = .0069 \text{ to } .0389$$

The 99% confidence interval for σ is: $\sqrt{.0069}$ to $\sqrt{.0389}$ $= .083$ to $.197$.

b. $H_0: \sigma^2 = .01$; $H_1: \sigma^2 > .01$.

Area in the right tail $= \alpha = .025$ and $df = n - 1 = 20 - 1 = 19$

χ^2 for 19 df and .025 area in the right tail $= 32.852$

The value of the test statistic is: $\chi^2 = (n-1)\, s^2 / \sigma^2 = (20-1)(.014) / .01 = 26.600$

Do not reject the null hypothesis. The population variance is not greater than .01 square ounces.

11.47 a. $\alpha/2 = .02/2 = .01$ and $1 - \alpha/2 = 1 - .01 = .99$

$df = n - 1 = 23 - 1 = 22$

χ^2 for 22 df and .01 area in the right tail $= 40.289$

χ^2 for 22 df and .99 area in the right tail $= 9.542$

The 99% confidence interval for the population variance σ^2 is:

$$\frac{(n-1)s^2}{\chi^2_{\alpha/2}} \text{ to } \frac{(n-1)s^2}{\chi^2_{1-\alpha/2}} = \frac{(23-1)(2.7)}{40.289} \text{ to } \frac{(23-1)(2.7)}{9.542} = 1.4743 \text{ to } 6.2251$$

The 99% confidence interval for σ is: $\sqrt{1.4743}$ to $\sqrt{6.2251}$ $= 1.214$ to 2.495

b. $H_0: \sigma^2 \leq 2$; $H_1: \sigma^2 > 2$.

Area in the right tail $= \alpha = .01$ and $df = n - 1 = 23 - 1 = 22$

χ^2 for 22 df and .01 area in the right tail $= 40.289$

The value of the test statistic is: $\chi^2 = (n-1)\, s^2 / \sigma^2 = (23-1)(2.7) / 2 = 29.700$

Do not reject the null hypothesis. The population variance is not greater than 2

11.48 a. $df = n - 1 = 22 - 1 = 21$

$\alpha/2 = .5 - (.95/2) = .025$ and $1 - \alpha / 2 = 1 - .025 = .975$

χ^2 for 21 *df* and .025 area in the right tail = 35.479

χ^2 for 21 *df* and .975 area in the right tail = 10.283

The 95% confidence interval for the population variance σ^2 is:

$$\frac{(n-1)s^2}{\chi^2_{\alpha/2}} \text{ to } \frac{(n-1)s^2}{\chi^2_{1-\alpha/2}} = \frac{(22-1)(.62)}{35.479} \text{ to } \frac{(22-1)(.62)}{10.283} = .3670 \text{ to } 1.2662$$

The 95% confidence interval for σ is: $\sqrt{.3670}$ to $\sqrt{1.2662}$ = .606 to 1.125

b. $H_0: \sigma^2 = .30;$ $H_1: \sigma^2 \neq .30$

$\alpha/2 = .01/2 = .005$ and $1 - \alpha/2 = 1 - .005 = .995$

χ^2 for 21 *df* and .005 area in the right tail = 41.401

χ^2 for 21 *df* and .995 area in the right tail = 8.034

The value of the test statistic is: $\chi^2 = (n-1)\,s^2 \,/\, \sigma^2 = (22-1)(.62)\,/.30 = 43.400$

Since 43.400 > 41.401, reject H_o. Conclude that σ^2 is different from .30.

11.49   a. $df = n - 1 = 25 - 1 = 24$

$\alpha/2 = .5 - (.99/2) = .005$ and $1 - \alpha/2 = 1 - .005 = .995$

χ^2 for 24 *df* and .005 area in the right tail = 45.559

χ^2 for 24 *df* and .995 area in the right tail = 9.886

The 99% confidence interval for σ^2 is:

$$\frac{(n-1)s^2}{\chi^2_{\alpha/2}} \text{ to } \frac{(n-1)s^2}{\chi^2_{1-\alpha/2}} = \frac{(25-1)(5200)}{45.559} \text{ to } \frac{(25-1)(5200)}{9.886} = 2739.3051 \text{ to } 12,623.9126$$

The 99% confidence interval for σ is: $\sqrt{2739.3051}$ to $\sqrt{12,623.9126}$ = 52.338 to 112.356

b. $H_0: \sigma^2 = 4200;$ $H_1: \sigma^2 \neq 4200$

$\alpha/2 = (.05/2) = .025$ and $1 - \alpha/2 = 1 - .025 = .975$

χ^2 for 24 *df* and .025 area in the right tail = 39.364

χ^2 for 24 *df* and .975 area in the right tail = 12.401

The value of the test statistic is: $\chi^2 = (n-1)\,s^2 \,/\, \sigma^2 = (25-1)\,(5200)\,/\,4200 = 29.714$

Since 29.714 is between 12.401 and 39.364, do not reject H_0. Conclude that σ^2 is equal to 4200 square hours.

11.50 H_0: The current distribution of factors for judging a company's reputation is the same as the 2002 survey.

H_1: The current distribution of factors for judging a company's reputation differs from the 2002 survey.

$df = k - 1 = 6 - 1 = 5$

For $\alpha = .01$ and $df = 5$, the critical value of χ^2 is 15.086.

Factor	O	p	$E = np$	$O - E$	$(O - E)^2$	$(O - E)^2/E$
Quality	165	.29	174	-9	81	.466
Financial Status	138	.19	114	24	576	5.053
Accessibility	102	.18	108	-6	36	.333
Treatment	111	.16	96	15	225	2.344
Acknowledgement	72	.12	72	0	0	0
Other	12	.06	36	-24	576	16.000
	$n = 600$					Sum = 24.196

The value of the test statistic is: $\chi^2 = \Sigma (O - E)^2 / E = 24.196$

Since $24.196 > 15.086$, reject H_0. The current distribution of factors for judging a company's reputation is different from that of the 2002 survey.

11.51 H_0: The percentage of people who consume All – Bran Cereal is the same for all four brands.

      H_1: The percentage of people who consume All – Bran Cereal is not the same for all four brands.

      $df = k - 1 = 4 - 1 = 3$

      For $\alpha = .05$ and $df = 3$, the critical value of χ^2 is 7.815.

Brand	O	p	$E = np$	$O - E$	$(O - E)^2$	$(O - E)^2/E$
A	212	.25	250	-38	1444	5.776
B	284	.25	250	34	1156	4.624
C	254	.25	250	4	16	.064
D	250	.25	250	0	0	.000
	$n = 1000$					Sum = 10.464

The value of the test statistic is: $\chi^2 = \Sigma (O - E)^2 / E = 10.464$

Since $10.464 > 7.185$, reject H_0. The percentage of people who consume All – Bran Cereal is not the same for all four brands.

11.52 H_0: The current distribution of important rights and freedoms is the same as in 2002 survey.

      H_1: The current distribution of important rights and freedoms differs from the 2002 survey.

      $df = k - 1 = 7 - 1 = 6$

      For $\alpha = .025$ and $df = 6$, the critical value of χ^2 is 14.449.

Response	O	p	$E = np$	$O - E$	$(O - E)^2$	$(O - E)^2/E$
Speech	124	.26	130	-6	36	.277
Affordable Healthcare	108	.20	100	8	64	.640
Religion	101	.19	95	6	36	.379
Economic Advancement	70	.12	60	10	100	1.667
Education	64	.12	60	4	16	.267
Press	18	.03	15	3	9	19.455
Don't know	15	.08	40	-25	625	15.625
	$n = 500$					Sum = 19.445

The value of the test statistic is: $\chi^2 = \Sigma (O - E)^2 / E = 19.455$. Since $19.455 > 14.449$, reject H_0.

The current distribution of reasons given for important rights and freedoms is not the same as in 2002.

11.53 H_0: The present distribution of weight changes for adults is the same as that of the 2002 survey.

H_1: The present distribution of weight changes for adults differs from that of the 2002 survey.

$df = k - 1 = 5 - 1 = 4$ For $\alpha = .05$ and $df = 4$, the critical value of χ^2 is 9.488.

Category	O	p	$E = np$	$O - E$	$(O - E)^2$	$(O - E)^2/E$
Same	330	.35	350	-20	400	1.143
Gained a little	370	.34	340	30	900	2.647
Lost a little	100	.13	130	-30	900	6.923
Gained a lot	80	.10	100	-20	400	4.000
Lost a lot	120	.08	80	40	1600	20.000
	$n = 1000$					Sum = 34.713

The value of the test statistic is: $\chi^2 = \Sigma (O - E)^2 / E = 34.713$

Since $34.713 > 9.488$, reject H_0. Conclude that the present distribution of weight changes for adults differs from the 2002 survey.

11.54 H_0: The proportions of investors favoring stocks, bonds, real estate, and cash are the same.

H_1: The proportions of investors favoring stocks, bonds, real estate, and cash differs.

$df = k - 1 = 4 - 1 = 3$ For $\alpha = .025$ and $df = 3$, the critical value of χ^2 is 9.348.

Assets	O	p	$E = np$	$O - E$	$(O - E)^2$	$(O - E)^2/E$
Stock	46	.25	35	11	121	3.457
Bonds	41	.25	35	6	36	1.029
Real Estate	32	.25	35	-3	9	.257
Cash	21	.25	35	-14	196	5.600
	$n = 140$					Sum = 10.343

The value of the test statistic is: $\chi^2 = \Sigma (O - E)^2 / E = 10.343$

Since $10.343 > 9.348$, reject H_0. The proportion of investors favoring stocks, bonds, real estate, and cash differs.

11.55 H_0: The proportions of all allergic persons are equally distributed over the four seasons.

H_1: The proportions of all allergic persons are not equally distributed over the four seasons.

$df = k - 1 = 4 - 1 = 3$

From the chi–square distribution table, the value of χ^2 for $df = 3$ and .01 area in the right tail is 11.345.

Season	O	p	$E = np$	$O - E$	$(O - E)^2$	$(O - E)^2/E$
Fall	18	.25	25	-7	49	1.960
Winter	13	.25	25	-12	144	5.760
Spring	31	.25	25	6	36	1.440
Summer	38	.25	25	13	169	6.760
	$n = 100$					Sum = 15.920

The value of the test statistic is: $\chi^2 = \Sigma (O - E)^2 / E = 15.920$. Since $15.920 > 11.345$, reject the null hypothesis. The proportions of all allergic persons are not equally distributed over the four seasons.

11.56 H_0: Sentencing does not depend on which judge you get.

H_1: Sentencing depends on which judge you get.

$df = (R - 1)(C - 1) = (2 - 1)(2 - 1) = 1$

From the chi–square distribution table, the critical value of χ^2 for $df = 1$ and $\alpha = .05$ is 3.841.

	Jail	Other	Total
Stark	27	65	92
	(29.64)	(62.36)	
Rivera	31	57	88
	(28.36)	(59.64)	
Total	58	122	180

The value of the test statistic is: $\chi^2 = \Sigma (O - E)^2 / E = .235 + .112 + .246 + .117 = .710$

Do not reject the null hypothesis. Do not conclude that sentencing depends on which judge you get.

11.57 H_0: Marital status and time available to relax are unrelated for people with children.

H_1: Marital status and time available to relax are related for people with children.

$df = (R - 1)(C - 1) = (2 - 1)(3 - 1) = 2$

From the chi–square distribution table, the critical value of χ^2 for $df = 2$ and $\alpha = .05$ is 5.991.

	Time to Relax			
	Little	Some	Much	Total
Married with Children	158	90	40	288
	(165.35)	(82.67)	(39.98)	
Unmarried with Children	86	32	19	137
	(78.65)	(39.33)	(19.02)	
Total	244	122	59	425

The value of the test statistic is: $\chi^2 = \Sigma (O - E)^2 / E = .327 + .650 + .000 + .687 + 1.366 + .000 = 3.030$

Do not reject the null hypothesis. Conclude that marital status and time available to relax are unrelated for people with children.

11.58 H_0: The two attributes for all jurors are independent.

H_1: The two attributes for all jurors are dependent.

$df = (R - 1)(C - 1) = (2 - 1)(3 - 1) = 2$

From the chi–square distribution table, the critical value of χ^2 for $df = 2$ and $\alpha = .05$ is 5.991.

	Strict	Fair	Lenient	Total
Have been a victim	20	8	3	31
	(13.02)	(12.71)	(5.27)	
Have never been a victim	22	33	14	69
	(28.98)	(28.29)	(11.73)	
Total	42	41	17	100

The value of the test statistic is: $\chi^2 = \Sigma (O - E)^2 / E = 3.742 + 1.745 + .978 + 1.681 + .784 + .439 = 9.369$

Reject the null hypothesis. The two attributes for all jurors are dependent.

11.59 H_0: Gender and marital status are not related for all persons who hold more than one job.

H_1: Gender and marital status are related for all persons who hold more than one job.

$df = (R - 1)(C - 1) = (2 - 1)(3 - 1) = 2$

From the chi–square distribution table, the critical value of χ^2 for $df = 2$ and .10 area in the right tail is 4.605.

	Single	Married	Other	Total
Male	72	209	39	320
	(67.20)	(199.04)	(53.76)	
Female	33	102	45	180
	(37.80)	(111.96)	(30.24)	
Total	105	311	84	500

The value of the test statistic is: $\chi^2 = \Sigma (O - E)^2 / E = .343 + .498 + 4.052 + .610 + .886 + 7.204 = 13.593$

Reject the null hypothesis. Gender and marital status are related for all persons who hold more than one job.

11.60 H_0: The political distributions of veterans and non–veterans are the same.

H_1: The political distributions of veterans and non–veterans are different.

$df = (R - 1)(C - 1) = (2 - 1)(3 - 1) = 2$

From the chi–square distribution table, the critical value of χ^2 for $df = 2$ and .025 area in the right tail is 7.378.

	Conservative	Moderate	Liberal	Total
Veteran	84	90	26	200
	(92)	(76)	(32)	
Non–veteran	100	62	38	200
	(92)	(76)	(32)	
Total	184	152	64	400

The value of the test statistic is: $\chi^2 = \Sigma (O - E)^2 / E = .696 + 2.579 + 1.125 + .696 + 2.579 + 1.125 = 8.800$

Reject H_0. The political distributions of veterans and non–veterans are different.

11.61 H_0: The percentages of people with different opinions are similar for all four regions.

H_1: The percentages of people with different opinions are not similar for all four regions.

$df = (R - 1)(C - 1) = (4 - 1)(3 - 1) = 6$

From the chi–square distribution table, the critical value of χ^2 for $df = 6$ and .01 area in the right tail is 16.812.

	Favor	Oppose	Uncertain	Total
Northeast	56	33	11	100
	(63.75)	(29.75)	(6.50)	
Midwest	73	23	4	100
	(63.75)	(29.75)	(6.50)	
South	67	28	5	100
	(63.75)	(29.75)	(6.50)	
West	59	35	6	100
	(63.75)	(29.75)	(6.50)	
Total	255	119	26	400

The value of the test statistic is: $\chi^2 = \Sigma (O - E)^2 / E = .942 + .355 + 3.115 + 1.342 + 1.532 + .962 +$ $.166 + .103 + .346 + .354 + .926 + .038 = 10.181$

Do not reject the null hypothesis. The percentages of people with different opinions are similar for all four regions.

11.62   a. $\alpha = .1 - .98 = .02$

$\alpha/2 = .02/2 = .01$ and $1 - \alpha/2 = 1 - .01 = .99$; $df = n - 1 = 21 - 1 = 20$

χ^2 for 20 df and .01 area in the right tail = 37.566

χ^2 for 20 df and .99 area in the right tail = 8.260

The 98% confidence interval for the population variance σ^2 is:

$$\frac{(n-1)s^2}{\chi^2_{\alpha/2}} \text{ to } \frac{(n-1)s^2}{\chi^2_{1-\alpha/2}} = \frac{(21-1)(9.2)}{37.566} \text{ to } \frac{(21-1)(9.2)}{8.260} = 4.8980 \text{ to } 22.2760$$

The 98% confidence interval for σ is: $\sqrt{4.8980}$ to $\sqrt{22.2760}$ = 2.213 to 4.720

b. $\alpha = 1 - .98 = .02$, $\alpha/2 = .02/2 = .01$,   and $1 - \alpha/2 = 1 - .01 = .99$

$df = n - 1 = 17 - 1 = 16$

χ^2 for 16 df and .01 area in the right tail = 32.000

χ^2 for 16 df and .99 area in the right tail = 5.812

The 98% confidence interval for σ^2 is:

$$\frac{(n-1)s^2}{\chi^2_{\alpha/2}} \text{ to } \frac{(n-1)s^2}{\chi^2_{1-\alpha/2}} = \frac{(17-1)(1.7)}{32.000} \text{ to } \frac{(17-1)(1.7)}{5.812} = .8500 \text{ to } 4.6800$$

The 98% confidence interval for σ is:. $\sqrt{.8500}$ to $\sqrt{4.6800}$ = .922 to 2.163

11.63   a. $\alpha = 1 - .95 = .05,$ $\alpha/2 = .05/2 = .025,$ and $1 - \alpha/2 = 1 - .025 = .975$

$df = n - 1 = 10 - 1 = 9$

χ^2 for 9 df and .025 area in the right tail = 19.023

χ^2 for 9 df and .975 area in the right tail = 2.700

The 95% confidence interval for the population variance σ^2 is:

$$\frac{(n-1)s^2}{\chi^2_{\alpha/2}} \text{ to } \frac{(n-1)s^2}{\chi^2_{1-\alpha/2}} = \frac{(10-1)(7.2)}{19.023} \text{ to } \frac{(10-1)(7.2)}{2.700} = 3.4064 \text{ to } 24.0000$$

The 95% confidence interval for σ is: $\sqrt{3.4064}$ to $\sqrt{24.0000}$ = 1.846 to 4.899

b. $df = n - 1 = 18 - 1 = 17$

χ^2 for 17 df and .025 area in the right tail = 30.191

χ^2 for 17 df and .975 area in the right tail = 7.564

The 95% confidence interval for σ^2 is:

$$\frac{(n-1)s^2}{\chi^2_{\alpha/2}} \text{ to } \frac{(n-1)s^2}{\chi^2_{1-\alpha/2}} = \frac{(18-1)(14.8)}{30.191} \text{ to } \frac{(18-1)(14.8)}{7.564} = 8.3336 \text{ to } 33.2628$$

The 95% confidence interval for σ is: $\sqrt{8.3336}$ to $\sqrt{33.2628}$ = 2.887 to 5.767

11.64 $H_0: \sigma^2 = 6.5;$ $H_1: \sigma^2 \neq 6.5$

$\alpha/2 = .05/2 = .025;$ $1 - \alpha/2 = 1 - .025 = .975,$ and $df = n - 1 = 21 - 1 = 20$

χ^2 for 20 df and .975 area in the right tail = 9.591

χ^2 for 20 df and .025 area in the right tail = 34.170

The value of the test statistic is: $\chi^2 = (n-1) s^2 / \sigma^2 = (21 - 1)(9.2)/6.5 = 28.308$

Do not reject the null hypothesis. The population variance is not different from 6.5.

11.65 $H_0: \sigma^2 = 1.1;$ $H_1: \sigma^2 > 1.1$

Area in the right tail = $\alpha = .025.$ and $df = n - 1 = 17 - 1 = 16$

χ^2 for 16 df and .025 area in the right tail = 28.845

The value of the test statistic is: $\chi^2 = (n-1) s^2 / \sigma^2 = (17 - 1)(1.7)/1.1 = 24.727$

Do not reject the null hypothesis. The population variance is not greater than 1.1.

11.66 $H_0: \sigma^2 = 4.2;$ $H_1: \sigma^2 > 4.2$

Area in the right tail = $\alpha = .01.$ and $df = n - 1 = 10 - 1 = 9$

χ^2 for 9 df and .01 area in the right tail = 21.666

The value of the test statistic is: $\chi^2 = (n-1) s^2 / \sigma^2 = (10 - 1)(7.2)/4.2 = 15.429$

Do not reject the null hypothesis. The population variance is not greater than 4.2.

11.67 H_0: $\sigma^2 = 10.4$; H_1: $\sigma^2 \neq 10.4$

$\alpha/2 = .05/2 = .025, 1 - \alpha/2 = 1 - .025 = .975$, and $df = n - 1 = 18 - 1 = 17$

χ^2 for 17 *df* and .975 area in the right tail = 7.564

χ^2 for 17 *df* and .025 area in the right tail = 30.191

The value of the test statistic is: $\chi^2 = (n-1) s^2 / \sigma^2 = (18 - 1)(14.8) / 10.4 = 24.192$

Do not reject the null hypothesis. The population variance is not different from 10.4.

11.68 a. H_0: $\sigma^2 = 4.0$; H_1: $\sigma^2 > 4.0$

$df = n - 1 = 25 - 1 = 24$ and Area in the right tail = $\alpha = .01$

χ^2 for 24 *df* and .01 area in the right tail = 42.980

The value of the test statistic is: $\chi^2 = (n-1) s^2 / \sigma^2 = (25 - 1)(8.3) / 4.0 = 49.800$

Reject the null hypothesis. The population variance is greater than 4.0 square minutes.

b. $\alpha = .01$, $\alpha/2 = .01/2 = .005$, and $1 - \alpha/2 = 1 - .005 = .995$

$df = n - 1 = 25 - 1 = 24$

χ^2 for 24 *df* and .005 area in the right tail = 45.559

χ^2 for 24 *df* and .995 area in the right tail = 9.886

The 99% confidence interval for the population variance σ^2 is:

$$\frac{(n-1)s^2}{\chi^2_{\alpha/2}} \text{ to } \frac{(n-1)s^2}{\chi^2_{1-\alpha/2}} = \frac{(25-1)(8.3)}{45.559} \text{ to } \frac{(25-1)(8.3)}{9.886} = 4.3724 \text{ to } 20.1497$$

11.69 a. H_0: $\sigma^2 = 5000$; H_1: $\sigma^2 < 5000$

Area in the left tail = $\alpha = .025$ and Area in the right tail = $1 - \alpha = 1 - .025 = .975$

$df = n - 1 = 20 - 1 = 19$

χ^2 for 19 *df* and .975 area in the right tail = 8.907

The value of the test statistic is: $\chi^2 = (n-1) s^2 / \sigma^2 = (20 - 1)(3175) / 5000 = 12.065$

Do not reject the null hypothesis. The population variance of the SAT scores for students from the given school is not lower than 5000.

b. $\alpha = 1 - .98. = .02$, $\alpha/2 = .02/2 = .01$, and $1 - \alpha/2 = 1 - .01 = .99$

$df = n - 1 = 20 - 1 = 19$

χ^2 for 19 *df* and .01 area in the right tail = 36.191

χ^2 for 19 *df* and .99 area in the right tail = 7.633

The 98% confidence interval for σ^2 is:

$$\frac{(n-1)s^2}{\chi^2_{\alpha/2}} \text{ to } \frac{(n-1)s^2}{\chi^2_{1-\alpha/2}} = \frac{(20-1)(3175)}{36.191} \text{ to } \frac{(20-1)(3175)}{7.633} = 1666.8509 \text{ to } 7903.1835$$

The 98% confidence interval for the population standard deviation σ is: $\sqrt{1666.8509}$

to $\sqrt{7903.1835}$ = 40.827 to 88.900

11.70 a. $H_0: \sigma^2 \le .025;$ $H_1: \sigma^2 > .025;$ $df = n - 1 = 23 - 1 = 22$

χ^2 for 22 df and .05 area in the right tail = 33.924

The value of the test statistic is: $\chi^2 = (n-1) s^2 / \sigma^2 = (23 - 1) (.034) / .025 = 29.920$

Do not reject the null hypothesis. The population variance is not greater than .025 square

millimeters. Thus the machine does not need an adjustment.

b. $\alpha/2 = .05/2 = .025$, and $1 - \alpha/2 = 1 - .025 = .975$

$df = n - 1 = 23 - 1 = 22$

χ^2 for 22 df and .025 area in the right tail = 36.781

χ^2 for 22 df and .975 area in the right tail = 10.982

The 95% confidence interval for the population variance σ^2 is:

$$\frac{(n-1)s^2}{\chi^2_{\alpha/2}} \text{ to } \frac{(n-1)s^2}{\chi^2_{1-\alpha/2}} = \frac{(23-1)(.034)}{36.781} \text{ to } \frac{(23-1)(.034)}{10.982} = .0203 \text{ to } .0681$$

11.71 a. $\alpha/2 = .5 - (.99/2) = .005$ and $1 - \alpha/2 = 1 - .005 = .995$

$df = n - 1 = 25 - 1 = 24$

χ^2 for 24 df and .005 area in the right tail = 45.559

χ^2 for 24 df and .995 area in the right tail = 9.886

The 99% confidence interval for σ^2 is:

$$\frac{(n-1)s^2}{\chi^2_{\alpha/2}} \text{ to } \frac{(n-1)s^2}{\chi^2_{1-\alpha/2}} = \frac{(25-1)(.19)}{45.559} \text{ to } \frac{(25-1)(.19)}{9.886} = .1001 \text{ to } .4613$$

The 99% confidence interval for σ is: $\sqrt{.1001}$ to $\sqrt{.4613}$ = .316 to .679

b. $H_0: \sigma^2 = .13;$ $H_1: \sigma^2 \ne .13$

$\alpha/2 = .01/2 = .005$ and $1 - \alpha/2 = 1 - .005 = .995$ $df = n - 1 = 25 - 1 = 24$

χ^2 for 24 df and .005 area in the right tail = 45.559

χ^2 for 24 df and .995 area in the right tail = 9.886

The value of the test statistic is: $\chi^2 = (n-1) s^2 / \sigma^2 = (25 - 1)(.19) / .13 = 35.077$

Since 35.077 is between 9.886 and 45.559, do not reject H_0. The variance of the GPA's is still equal

to .13.

11.72 a. From the given data: $n = 7$, $\Sigma x = 140$, and $\Sigma x^2 = 3000$

$$s^2 = \frac{\Sigma x^2 - (\Sigma x)^2 / n}{n-1} = \frac{3000 - (140)^2 / 7}{7-1} = 33.3333$$

b. $\alpha/2 = .5 - (.98/2) = .01$ and $1 - \alpha/2 = 1 - .01 = .99$

$df = n - 1 = 7 - 1 = 6$

χ^2 for 6 df and .01 area in the right tail = 16.812

χ^2 for 6 df and .99 area in the right tail = .872

The 98% confidence interval for σ^2 is:

$$\frac{(n-1)s^2}{\chi^2_{\alpha/2}} \text{ to } \frac{(n-1)s^2}{\chi^2_{1-\alpha/2}} = \frac{(7-1)(33.3333)}{16.812} \text{ to } \frac{(7-1)(33.3333)}{.872} = 11.8963 \text{ to } 229.3576$$

The 98% confidence interval for σ is: $\sqrt{11.8963}$ to $\sqrt{229.3576}$ = 3.449 to 15.145

c. $H_0: \sigma^2 = 150;$ $H_1: \sigma^2 < 150$

χ^2 for 6 df and .99 area in the right tail = .872

The value of the test statistic is: $\chi^2 = (n-1)s^2/\sigma^2 = ((7-1)(33.3333))/150 = 1.333$

Since 1.333 is greater than .872, do not reject H_0. The population variance is not less than 150 square pounds.

11.73   a.   From the given data: $n = 8$, $\Sigma x = 6293$, and $\Sigma x^2 = 4,957,983$

$$s^2 = \frac{\Sigma x^2 - (\Sigma x)^2/n}{n-1} = \frac{4,957,983 - (6293)^2/8}{8-1} = 1107.4107$$

b. $\alpha/2 = .5 - (.95/2) = .025$ and $1 - \alpha/2 = 1 - .025 = .975;$

$df = n - 1 = 8 - 1 = 7$

χ^2 for 7 df and .025 area in the right tail = 16.013

χ^2 for 7 df and .975 area in the right tail = 1.690

The 95% confidence interval for σ^2 is:

$$\frac{(n-1)s^2}{\chi^2_{\alpha/2}} \text{ to } \frac{(n-1)s^2}{\chi^2_{1-\alpha/2}} = \frac{(8-1)(1107.4107)}{16.013} \text{ to } \frac{(8-1)(1107.4107)}{1.690} = 484.0989 \text{ to } 4586.9082$$

The 95% confidence interval for σ is: $\sqrt{484.0989}$ to $\sqrt{4586.9082}$ = \$22.002 to \$67.727

c. $H_0: \sigma^2 = 500;$ $H_1: \sigma^2 \neq 500$

$\alpha/2 = .05/2 = .025$ and $1 - \alpha/2 = 1 - .025 = .975$

χ^2 for 7 df and .025 area in the right tail = 16.013

χ^2 for 7 df and .975 area in the right tail = 1.690

The value of the test statistic is: $\chi^2 = (n-1)s^2/\sigma^2 = (8-1)(1107.4107)^2/500^2 = 15.504$

Since 15.504 is between 1.690 and 16.013, do not reject H_0. Conclude that σ^2 is not different from 500 square dollars.

11.74 H_0: The same proportion of students in each grade watches at least 3 hours of TV per school day.

H_1: The proportion of students in each grade watching at least 3 hours of TV per school day differs.

Assuming that all 1200 students answered either "yes" or "no", we obtain the following two – way table.

$df = (R – 1)(C – 1) = (4 – 1)(2 – 1) = 3$

For $\alpha = .01$ and $df = 3$, the critical value of χ^2 is 11.345.

Grade	Yes	No	Total
9	135	165	300
	(112.5)	(187.5)	
10	117	183	300
	(112.5)	(187.5)	
11	105	195	300
	(112.5)	(187.5	
12	93	207	300
	(112.5)	(187.5	
	450	750	1200

The test statistic is: $\chi^2 = \Sigma (O – E)^2 /E = 4.500 + 2.700 + .180 + .108 + .500 + .300 + 3.380 + 2.028 = 13.696$

Since $13.696 > 11.345$, reject H_0. The proportion of students in each grade watching at least 3 hours of TV per school day differs.

11.75 H_0: Opinions on disposal site are independent of gender.

H_1: Opinions on disposal site are dependent on gender.

$df = (R – 1)(C – 1) = (2 – 1)(3 – 1) = 2$

For $\alpha = .05$, and $df = 2$, the critical value of χ^2 is 5.991

From the given data we can calculate the following:

Total number opposed $= 200(.60) = 120$

Total number in favor $= 200(.32) = 64$

Total number undecided $= 200(.08) = 16$

Number of women opposed $= 120(.65) = 78$

Number of men opposed $= 120 – 78 = 42$

Number of men in favor $= 64(.625) = 40$

Number of women in favor $= 64 – 40 = 24$

Number of women undecided $= 110 – 78 – 24 = 8$

Number of men undecided $= 16 – 8 = 8$

Using these results, we may construct the following two–way table of observations and expected values.

| | | Opinion on Site | | |
	Opposed	In Favor	Undecided	Total
Women	78	24	8	110
	(66.0)	(35.2)	(8.8)	
Men	42	40	8	90
	(54.0)	(28.8)	(7.2)	
Total	120	64	16	200

The test statistic is: $\chi^2 = \Sigma (O - E)^2 / E = 2.182 + 3.564 + .073 + 2.667 + 4.356 + .089 = 12.931$

Since $12.931 > 5.991$, reject H_0. Conclude that opinions on the disposal site are dependent on gender.

11.76 a. H_0: $p_1 = p_2$; H_1: $p_1 \neq p_2$;

$\hat{p}_1 = 48/52 = .923$ and $\hat{p}_2 = 44/54 = .815$

$\bar{p} = \dfrac{48+44}{52+54} = .868$ and $\bar{q} = 1 - .868 = .132$

$s_{\hat{p}_1 - \hat{p}_2} = \sqrt{\bar{p}\bar{q}\left(\dfrac{1}{n_1} + \dfrac{1}{n_2}\right)}\sqrt{(.868)(.132)\left(\dfrac{1}{52} + \dfrac{1}{54}\right)} = .06576597$

$z = \dfrac{(\hat{p}_1 - \hat{p}_2) - (p_1 - p_2)}{s_{\hat{p}_1 - \hat{p}_2}} = \dfrac{(.923 - .815) - 0}{.06576597} = 1.64$

For $z = 1.64$, the area in the right tail of the normal distribution is $.5 - .4495 = .0505$. Hence,

$p - value = 2(.0505) = .1010$

b.

	Pass	Did Not Pass	Total
Instructor A	48	4	52
	(45.13)	(6.87)	
Instructor B	44	10	54
	(46.87)	(7.13)	
Total	92	14	106

For $df = 1$ and $\chi^2 = 2.713$, the area in the right tail is about $.1000$. Hence, $p - value = .1000$

c. $z^2 = (1.64)^2$ is approximately equal to $\chi^2 = 2.713$. The difference is due to rounding.

The $p - values$ are the same (the difference is due to rounding), which is always true since the tests in parts a and b are equivalent.

11.77 H_0: The proportions of red and green marbles are the same in all five boxes.

H_1: The proportions of red and green marbles are not the same in all five boxes.

$df = (R - 1)(C - 1) = (2 - 1)(5 - 1) = 4$

For $\alpha = .05$ and $df = 4$, the critical value of χ^2 is 9.488.

The following table lists the observed and expected frequencies.

The expected frequencies are given in parentheses below observed frequencies.

Box	1	2	3	4	5	Total
Red	20	14	23	30	18	105
	(21)	(21)	(21)	(21)	(21)	
Green	30	36	27	20	32	145
	(29)	(29)	(29)	(29)	(29)	
Total	50	50	50	50	50	250

The value of the test statistic is: $\chi^2 = \Sigma (O - E)^2 / E$

$$= \frac{(20-21)^2}{21} + \frac{(14-21)^2}{21} + \frac{(23-21)^2}{21} + \frac{(30-21)^2}{21} + \frac{(18-21)^2}{21} + \frac{(30-29)^2}{29} + \frac{(36-29)^2}{29} + \frac{(27-29)^2}{29}$$

$$+ \frac{(20-29)^2}{29} + \frac{(32-29)^2}{29} = .048 + 2.333 + .190 + 3.857 + .429 + .034 + 1.690 + .138 + 2.793 + .310 = 11.822$$

Since, $11.822 > 9.488$, reject H_0. The proportions of red and green marbles are not the same in all five boxes.

Self-Review Test for Chapter Eleven

1. b   **2.** a   **3.** c   **4.** a   **5.** b   **6.** b   **7.** c   **8.** b   **9.** a

10. H_0: The current distribution of opinions on this matter is the same as that of 2002.

H_1: The current distribution of opinions on this matter differs from that of 2002.

$df = k - 1 = 5 - 1 = 4$

For $\alpha = .01$ and $df = 4$, the critical value of χ^2 is 13.277.

Opinion	O	p	$E = np$	$O - E$	$(O - E)^2$	$(O - E)^2/E$
Merit	227	.39	234	−7	49	.209
Seniority	152	.26	156	−4	16	.103
Connections	118	.17	102	16	256	2.510
Luck	35	.06	36	−1	1	.028
Other	68	.12	72	−4	16	.222
	$n = 600$					Sum = 3.072

The value of the test statistic is: $\chi^2 = \Sigma (O - E)^2 / E = 3.072$

Since $3.072 < 13.277$, do not reject H_0.

Conclude that the current distribution of opinions on this matter is the same as that of 2002.

11. H_0: Educational level and ever being divorced are independent.

H_1: Educational level and ever being divorced are dependent.

$df = (R-1)(C-1) = (2-1)(4-1) = 3$

From the chi–square distribution table, the critical value of χ^2 for $df = 3$ and .01 area in the right tail is 11.345

	Educational Level				
	Less than High school	High school degree	Some college	College degree	Total
Divorced	173	158	95	53	479
	(160.47)	(136.04)	(98.20)	(84.30)	
Never Divorced	162	126	110	123	521
	(174.54)	(147.96)	(106.80)	(91.70)	
Total	335	284	205	176	1000

The value of the test statistic is:

$\chi^2 = \Sigma (O-E)^2 / E = .978 + 3.545 + .104 + 11.621 + .901 + 3.259 + .096 + 10.684 = 31.188$

Reject the null hypothesis. Educational level and ever being divorced are dependent.

12. H_0: The percentages of people who play the lottery often, sometimes, and never are the same for each income group.

H_1: The percentages of people who play the lottery often, sometimes, and never are not the same for each income group.

$df = (R-1)(C-1) = (3-1)(3-1) = 4$

For $\alpha = .05$ and $df = 4$, the critical value of χ^2 is 9.488

	Income Group			
	Low	Middle	High	Total
Play often	174	163	90	427
	(170.80)	(142.33)	(113.87)	
Play sometimes	286	217	120	623
	(249.20)	(207.67)	(166.13)	
Never Play	140	120	190	450
	(180.00)	(150.00)	(120.00)	
Total	600	500	400	1500

The value of the test statistic is: $\chi^2 = \Sigma (O-E)^2 / E = .060 + 3.002 + 5.004 + 5.434 + .419 + 12.809 + 8.889 + 6.000 + 40.833 = 82.450$

Since $82.450 > 9.488$, reject H_0. The percentages of people who play the lottery often, sometimes, and never are not the same for each income group.

13. a. $\alpha/2 = .01/2 = .005$ and $1 - \alpha/2 = 1 - .005 = .995$

$df = n - 1 = 20 - 1 = 19$

χ^2 for 19 df and .005 area in the right tail = 35.582

χ^2 for 19 df and .995 area in the right tail = 6.844

The 99% confidence interval for the population variance σ^2 is:

$$\frac{(n-1)s^2}{\chi^2_{\alpha/2}} \text{ to } \frac{(n-1)s^2}{\chi^2_{1-\alpha/2}} = \frac{(20-1)(.48)}{38.582} \text{ to } \frac{(20-1)(.48)}{6.844} = .2364 \text{ to } 1.3326$$

The 99% confidence interval for σ is: $\sqrt{.2364}$ to $\sqrt{1.3326} = .486$ to 1.154

b. $H_0: \sigma^2 = .25;$ $H_1: \sigma^2 > .25$

Area in the right tail = $\alpha = .01$

$df = n - 1 = 20 - 1 = 19$

χ^2 for 19 df and .01 area in the right tail = 36.191

The value of the test statistic is: $\chi^2 = (n-1)\, s^2 / \sigma^2 = (20 - 1)(.48) / .25 = 36.480$

Reject the null hypothesis. The population variance is greater than .25 square ounces.

Chapter Twelve

12.1   See description in the box on page 556 of the text.

12.2   a. From the F distribution table, the critical value of F for 6 df for the numerator, 12 df for the denominator, and .05 area in the right tail is: $F = 3.00$

b. From the F distribution table, the critical value of F for 4 df for the numerator, 18 df for the denominator, and .05 area in the right tail is: $F = 2.93$

c. From the F distribution table, the critical value of F for 12 df for the numerator, 6 df for the denominator, and .05 area in the right tail is: $F = 4.00$

12.3   a. From the F distribution table, the critical value of F for 6 df for the numerator, 12 df for the denominator, and .025 area in the right tail is: $F = 3.73$

b. From the F distribution table, the critical value of F for 4 df for the numerator, 18 df for the denominator, and .025 area in the right tail is: $F = 3.61$

c. From the F distribution table, the critical value of F for 12 df for the numerator, 6 df for the denominator, and .025 area in the right tail is: $F = 5.37$

12.4   a. From the F distribution table, the critical value of F for 7 df for the numerator, 11 df for the denominator, and .01 area in the right tail is: $F = 4.89$

b. From the F distribution table, the critical value of F for 5 df for the numerator, 12 df for the denominator, and .01 area in the right tail is: $F = 5.06$

c. From the F distribution table, the critical value of F for 15 df for the numerator, 6 df for the denominator, and .01 area in the right tail is: $F = 7.56$

12.5   a. From the F distribution table, the critical value of F for 4 df for the numerator, 14 df for the denominator, and .10 area in the right tail is: $F = 2.39$

b. From the F distribution table, the critical value of F for 9 df for the numerator, 11 df for the denominator, and .10 area in the right tail is: $F = 2.27$

c. From the F distribution table, the critical value of F for 11 df for the numerator, 5 df for the denominator, and .10 area in the right tail is: $F = 3.28$

12.6   a.   For $df = (3,12)$ and .05 area in the right tail, the critical value of F is 3.49.

      b.   For $df = (3,12)$ and .10 area in the right tail, the critical value of F is 2.61.

12.7   a.   For $df = (11, 5)$ and .01 area in the right tail, the critical value of F is 9.96.

      b.   For $df = (11, 5)$ and .025 area in the right tail, the critical value of F is 6.57.

12.8   a.   For $df = (4,11)$ and .025 area in the right tail, the critical value of F is 4.28.

      b.   For $df = (15,3)$ and .025 area in the right tail, the critical value of F is 14.25.

12.9   a.   For $df = (10, 10)$ and .01 area in the right tail, the critical value of F is 4.85.

      b.   For $df = (9, 25)$ and .01 area in the right tail, the critical value of F is 3.22.

12.10   One–way ANOVA is used to test the null hypothesis that the means of three or more populations are equal. Note that ANOVA can be used to test for the equality of the means of two populations, but procedures of Chapter 10 are easier to use in such cases.

12.11   See box on "Assumptions of one–way ANOVA" on page 559 of the text.

12.12   a.   For sample I: $n_1 = 6$, $\Sigma x = 170$, and $\Sigma x^2 = 4946$

$$\bar{x}_1 = \Sigma x / n_1 = 170 / 6 = 28.333$$

$$s_1 = \sqrt{\frac{\Sigma x^2 - (\Sigma x)^2 / n_1}{n_1 - 1}} = \sqrt{\frac{4946 - (170)^2 / 6}{6 - 1}} = 5.08592830$$

For sample II: $n_2 = 6$, $\Sigma x = 204$, and $\Sigma x^2 = 7048$

$$\bar{x}_2 = \Sigma x / n_2 = 204 / 6 = 34.000$$

$$s_2 = \sqrt{\frac{\Sigma x^2 - (\Sigma x)^2 / n_2}{n_2 - 1}} = \sqrt{\frac{7048 - (204)^2 / 6}{6 - 1}} = 4.73286383$$

      b. H_0: $\mu_1 - \mu_2 = 0$; $\qquad\qquad$ H_1: $\mu_1 - \mu_2 \neq 0$

$df = n_1 + n_2 - 2 = 6 + 6 - 2 = 10$ and $\qquad$ Area in each tail of the t–curve $= \alpha / 2 = .01 / 2 = .005$

For 10 df and .005 area in each tail, the critical values of t are -3.169 and 3.169.

$$s_p = \sqrt{\frac{(n_1 - 1)s_1^2 + (n_2 - 1)s_2^2}{n_1 - n_2 - 2}} = \sqrt{\frac{(6-1)(5.08592830)^2 + (6-1)(4.73286383)^2}{(6+6-2)}} = 4.91256892$$

$$s_{\bar{x}_1 - \bar{x}_2} = s_p \sqrt{\frac{1}{n_1} + \frac{1}{n_2}} = 4.91256892 \sqrt{\frac{1}{6} + \frac{1}{6}} = 2.83627299$$

The value of the test statistic is: $t = \dfrac{(\bar{x}_1 - \bar{x}_2) - (\mu_1 - \mu_2)}{s_{\bar{x}_1 - \bar{x}_2}} = \dfrac{(28.333 - 34.000) - 0}{2.83627299} = -1.998$

Do not reject H_0.

c. $H_0 : \mu_1 = \mu_2; \qquad H_1 : \mu_1 \neq \mu_2$

$k = 2, \; n_1 = 6, n_2 = 6, n = n_1 + n_2 = 6 + 6 = 12$

df for the numerator $= k - 1 = 2 - 1 = 1$

df for the denominator $= n - k = 12 - 2 = 10$

For $\alpha = .01$ and $df = (1, 10)$, the critical value of F is 10.04.

$T_1 = 170, T_2 = 204, \Sigma x = T_1 + T_2 = 170 + 204 = 374; \text{ and } \Sigma x^2 = 11{,}994$

$$\text{SSB} = \left[\frac{T_1^2}{n_1} + \frac{T_2^2}{n_2} \right] - \frac{(\Sigma x)^2}{n} = \left[\frac{(170)^2}{6} + \frac{(204)^2}{6} \right] - \frac{(374)^2}{12}$$

$$= 11{,}752.6667 - 11{,}656.3333 = 96.3334$$

$$\text{SSW} = \Sigma x^2 - \left[\frac{T_1^2}{n_1} + \frac{T_2^2}{n_2} \right] = 11{,}994 - \left[\frac{(170)^2}{6} + \frac{(204)^2}{6} \right] = 11{,}994 - 11{,}752.6667 = 241.3333$$

$\text{MSB} = \text{SSB} / (k - 1) = 96.3334 / (2 - 1) = 96.3334$

$\text{MSW} = \text{SSW} / (n - k) = 241.3333 / (12 - 2) = 24.1333$

The value of the test statistic is: $F = \text{MSB} / \text{MSW} = 96.3334 / 24.1333 = 3.99$

Since $3.99 < 10.04$, do not reject H_0. The population means are equal.

d.   The conclusion of parts b and c are the same: do not reject H_0.

12.13   a.   For sample I: $n_1 = 7, \Sigma x = 105, \text{ and } \Sigma x^2 = 1697$

$\bar{x}_1 = \Sigma x / n_1 = 105 / 7 = 15.000$

$$s_1 = \sqrt{\frac{\Sigma x^2 - (\Sigma x)^2 / n_1}{n_1 - 1}} = \sqrt{\frac{1697 - (105)^2 / 7}{7 - 1}} = 4.50924975$$

For sample II: $n_2 = 7, \Sigma x = 77, \text{ and } \Sigma x^2 = 963$

$\bar{x}_2 = \Sigma x / n_2 = 77 / 7 = 11.000$

$$s_2 = \sqrt{\frac{\Sigma x^2 - (\Sigma x)^2 / n_2}{n_2 - 1}} = \sqrt{\frac{963 - (77)^2 / 7}{7 - 1}} = 4.39696865$$

b. $H_0: \mu_1 - \mu_2 = 0; \qquad\qquad H_1: \mu_1 - \mu_2 \neq 0$

$df = n_1 + n_2 - 2 = 7 + 7 - 2 = 12;$ Area in each tail of the t curve $= \alpha / 2 = .05 / 2 = .025$

For 12 *df* and .025 area in each tail, the critical values of t are -2.179 and 2.179.

$$s_p = \sqrt{\frac{(n_1 - 1)s_1^2 + (n_2 - 1)s_2^2}{n_1 - n_2 - 2}} = \sqrt{\frac{(7-1)(4.50924975)^2 + (7-1)(4.39696865)^2}{(7+7-2)}} = 4.45346307$$

$$s_{\bar{x}_1 - \bar{x}_2} = s_p \sqrt{\frac{1}{n_1} + \frac{1}{n_2}} = 4.45346307 \sqrt{\frac{1}{7} + \frac{1}{7}} = 2.38047614$$

The value of the test statistic is: $t = \dfrac{(\bar{x}_1 - \bar{x}_2) - (\mu_1 - \mu_2)}{s_{\bar{x}_1 - \bar{x}_2}} = \dfrac{(15.000 - 11.000) - 0}{2.38047614} = 1.680$

Do not reject H_0.

c $H_0: \mu_1 = \mu_2$; $H_1: \mu_1 \neq \mu_2$;

   $k = 2$, $n_1 = 7$, $n_2 = 7$, $n = n_1 + n_2 = 7 + 7 = 14$

   df for the numerator $= k - 1 = 2 - 1 = 1$ and df for the denominator $= n - k = 14 - 2 = 12$

   For $\alpha = .05$ and $df = (1, 12)$, the critical value of F is 4.75.

   $T_1 = 105$, $T_2 = 77$, $\Sigma x = T_1 + T_2 = 105 + 77 = 182$ and $\Sigma x^2 = 2660$

$$SSB = \left[\frac{T_1^2}{n_1} + \frac{T_2^2}{n_2} \right] - \frac{(\Sigma x)^2}{n} = \left[\frac{(105)^2}{7} + \frac{(77)^2}{7} \right] - \frac{(182)^2}{14} = 2422 - 2366 = 56$$

$$SSW = \Sigma x^2 - \left[\frac{T_1^2}{n_1} + \frac{T_2^2}{n_2} \right] = 2660 - \left[\frac{(105)^2}{7} + \frac{(77)^2}{7} \right] = 2660 - 2422 = 238$$

   $MSB = SSB / (k - 1) = 56 / (2 - 1) = 56.0000$

   $MSW = SSW / (n - k) = 238 / (14 - 2) = 19.8333$

   The value of the test statistic is: $F = MSB / MSW = 56.0000 / 19.8333 = 2.82$

    Since $2.82 < 4.75$, do not reject H_0.

d. The conclusions of parts b and c are the same: do not reject H_0.

12.14 a.   ANOVA TABLE

Source of Variation	Degrees of Freedom	Sum of Squares	Mean Square	Value of the Test Statistic
Between	2	38.5626	19.2813	$F = \dfrac{19.2813}{8.9368} = 2.16$
Within	10	89.3677	8.9368	
Total	12	127.9303		

b. $H_0: \mu_1 = \mu_2 = \mu_3$;   and H_1: The means of the three populations are not all equal

    For $\alpha = .01$ and $df = (2, 10)$, the critical value of F is 7.56.

    Since $2.16 < 7.56$, do not reject H_0. The means of the three populations are equal.

12.15 a.

ANOVA TABLE

Source of Variation	Degrees of Freedom	Sum of Squares	Mean Square	Value of the Test Statistic
Between	3	112.5201	37.5067	$F = \dfrac{37.5067}{9.2154} = 4.07$
Within	15	138.2310	9.2154	
Total	18	250.7511		

b. $H_0: \mu_1 = \mu_2 = \mu_3 = \mu_4$;   and H_1: The means of the four populations are not all equal.

For $\alpha = .05$ and $df = (3,15)$, the critical value of F is 3.29.

Since $4.07 > 3.29$, reject H_0. The means of the four populations are not all equal.

12.16 a. $H_0: \mu_1 = \mu_2 = \mu_3$;   and H_1: All three populations means are not equal.

b. $k = 3$, and $n = n_1 + n_2 + n_3 = 7 + 7 + 7 = 21$

df for the numerator $= k - 1 = 3 - 1 = 2$ and df for the denominator $= n - k = 21 - 3 = 18$

For $\alpha = .025$ and $df = (2, 18)$, the critical value of F is 4.56.

The rejection region lies to the right of $F = 4.56$, and the nonrejection region lies to the left of $F = 4.56$.

c. $T_1 = 93$, $T_2 = 115$, and $T_3 = 80$

$\Sigma x = T_1 + T_2 + T_3 = 93 + 115 + 80 = 288$

$\Sigma x^2 = (15)^2 + (8)^2 + (17)^2 + (7)^2 + (26)^2 + (12)^2 + (8)^2 + (11)^2 + (16)^2 + (9)^2 + (16)^2 + (24)^2 + (20)^2$
$+ (19)^2 + (9)^2 + (17)^2 + (11)^2 + (8)^2 + (15)^2 + (6)^2 + (14)^2 = 4574$

$$\text{SSB} = \left[\frac{T_1^2}{n_1} + \frac{T_2^2}{n_2} + \frac{T_3^2}{n_2} \right] - \frac{(\Sigma x)^2}{n} = \left[\frac{(93)^2}{7} + \frac{(115)^2}{7} + \frac{(80)^2}{7} \right] - \frac{(288)^2}{21}$$

$= 4039.1429 - 3949.7143 = 89.4286$

$$\text{SSW} = \Sigma x^2 - \left[\frac{T_1^2}{n_1} + \frac{T_2^2}{n_2} + \frac{T_3^2}{n_3} \right] = 4574 - \left[\frac{(93)^2}{7} + \frac{(115)^2}{7} + \frac{(80)^2}{7} \right]$$

$= 4574 - 4039.1429 = 534.8571$

SST = SSB + SSW = 89.4286 + 534.8571 = 624.2857

d. df for the numerator $= k - 1 = 3 - 1 = 2$ and df for the denominator $= n - k = 21 - 3 = 18$

e. MSB = SSB / $(k - 1)$ = 89.4286 / $(3 - 1)$ = 44.7143

MSW = SSW / $(n - k)$ = 534.8571 / $(21 - 3)$ = 29.7143

f. For $\alpha = .025$ and $df = (2, 18)$, the critical value of F is 4.56

g. The value of the test statistic is: $F = MSB / MSW = 44.7143 / 29.7143 = 1.50$

h.   ANOVA TABLE

Sources of Variation	Degrees of Freedom	Sum of Squares	Mean Square	Value of the Test Statistic
Between	2	89.4286	44.7143	$F = \dfrac{44.7143}{29.7143} = 1.50$
Within	18	534.8571	29.7143	
Total	20	624.2857		

i. Since $1.50 < 4.56$, do not reject H_0. The mean weight loss for each diet is the same.

12.17   a. H_0: $\mu_1 = \mu_2 = \mu_3$;   and H_1: All three populations means are not equal.

b. $k = 3$ and $n = n_1 + n_2 + n_3 = 10 + 9 + 6 = 25$

   df for the numerator $= k - 1 = 3 - 1 = 2$ and df for the denominator $= n - k = 25 - 3 = 22$

c. $T_1 = 140$, $T_2 = 86$, and $T_3 = 37$; $\Sigma x = T_1 + T_2 + T_3 = 140 + 86 + 37 = 263$

   $\Sigma x^2 = (19)^2 + (13)^2 + (25)^2 + (10)^2 + (19)^2 + (4)^2 + (15)^2 + (10)^2 + (16)^2 + (9)^2 + (9)^2 + (6)^2 + (11)^2 +$
   $(14)^2 + (5)^2 + (9)^2 + (3)^2 + (11)^2 + (18)^2 + (5)^2 + (8)^2 + (2)^2 + (3)^2 + (10)^2 + (9)^2 = 3571$

   $$SSB = \left[\frac{T_1^2}{n_1} + \frac{T_2^2}{n_2} + \frac{T_3^2}{n_2}\right] - \frac{(\Sigma x)^2}{n} = \left[\frac{(140)^2}{10} + \frac{(86)^2}{9} + \frac{(37)^2}{6}\right] - \frac{(263)^2}{25}$$

   $= 3009.9444 - 2766.7600 = 243.1844$

   $$SSW = \Sigma x^2 - \left[\frac{T_1^2}{n_1} + \frac{T_2^2}{n_2} + \frac{T_3^2}{n_3}\right] = 3571 - \left[\frac{(140)^2}{10} + \frac{(86)^2}{9} + \frac{(37)^2}{6}\right] = 3571 - 3009.9444 = 561.0556$$

   $SST = SSB + SSW = 243.1844 + 561.0556 = 804.2400$

d. For $\alpha = .01$ and $df = (2,22)$, the critical value of F is 5.72.

   The rejection region lies to the right of $F = 5.72$.

   The nonrejection region lies to the left of $F = 5.72$.

e. $MSB = SSB / (k - 1) = 243.1844 / (3 - 1) = 121.5922$

   $MSW = SSW / (n - k) = 561.0556 / (25 - 3) = 25.5025$

f. For $\alpha = .01$ and $df = (2, 22)$, the critical value of F is 5.72

g. The value of the test statistic is: F = MSB / MSW = 121.5922 / 25.5025 = 4.77

h.   ANOVA TABLE

Source of Variation	Degrees of Freedom	Sum of Squares	Mean Square	Value of the Test Statistic
Between	2	243.1844	121.5922	$F = \dfrac{121.5922}{25.5025} = 4.77$
Within	22	561.0556	25.5025	
Total	24	804.2400		

i. Since 4.77 < 5.72, do not reject H_0. The mean number of classes missed by all three age groups is the same.

12.18 H_0: $\mu_1 = \mu_2 = \mu_3 = \mu_4$;   and H_1: All four population means are not equal

$n = n_1 + n_2 + n_3 + n_4 = 4 + 3 + 5 + 4 = 16$ and $k = 4$

df for the numerator = $k - 1 = 4 - 1 = 3$ and df for the denominator = $n - k = 16 - 4 = 12$

From the F distribution table, the critical value of F for 3 df for the numerator, 12 df for the denominator, and .01 area in the right tail is 5.95.

$T_1 = 278$, $T_2 = 234$, $T_3 = 395$, $T_4 = 246$; $\Sigma x = 1153$, $\Sigma x^2 = 86,355$

SSB = 818.9375, SSW = 2448.0000, MSB = 272.9792, MSW = 204.0000

The value of the test statistic is: F = MSB / MSW = 272.9792 / 204.0000 = 1.34

Do not reject the null hypothesis. The mean auto insurance premium paid per month by all drivers insured by each of the four companies is the same.

12.19 H_0: $\mu_1 = \mu_2 = \mu_3$;   and H_1: All three population means are not equal

$n = n_1 + n_2 + n_3 = 8 + 7 + 6 = 21$ and $k = 3$

df for the numerator = $k - 1 = 3 - 1 = 2$ and df for the denominator = $n - k = 21 - 3 = 18$

From the F distribution table, the critical value of F for 2 df for the numerator, 21 df for the denominator, and .05 area in the right tail is 3.55.

$T_1 = 408$, $T_2 = 232$, and $T_3 = 298$, $\Sigma x = 938$, $\Sigma x^2 = 49,322$,   SSB = 1400.4762,   SSW = 6024.1905

MSB = SSB / $(k - 1)$ = 1400.4762 / $(3 - 1)$ = 700.2381

MSW = SSW / $(n - k)$ = 6024.1905 / $(21 - 3)$ = 334.6773

The value of the test statistic is: F = MSB / MSW = 700.2381 / 334.6773 = 2.09

Do not reject the null hypothesis. Do not conclude all three means are not the same.

12.20 H_0: $\mu_1 = \mu_2 = \mu_3$;   and H_1: All three population means are not equal

$n = n_1 + n_2 + n_3 = 6 + 4 + 5 = 15$ and $k = 3$

df for the numerator $= k - 1 = 3 - 1 = 2$ and df for the denominator $n - k = 15 - 3 = 12$

From the F distribution table, the critical value of F for 2 df for the numerator, 12 df for the denominator, and .025 area in the right tail is $F = 5.10$.

$T_1 = 269$, $T_2 = 80$, and $T_3 = 268$; $\Sigma x = 617$, $\Sigma x^2 = 29,689$

SSB $= 2645.7000$, SSW $= 1664.0333$, MSB $= 1322.8500$, MSW $= 138.6694$

The value of the test statistic is: $F = \text{MSB} / \text{MSW} = 1322.8500 / 138.6694 = 9.54$

Reject the null hypothesis. The mean time taken to provide relief from a headache is not the same for all three drugs.

12.21 H_0: $\mu_1 = \mu_2 = \mu_3 = \mu_4$;   and H_1: The means of the four populations are not equal

$k = 4$ and $n = n_1 + n_2 + n_3 + n_4 = 7 + 7 + 7 + 7 = 28$

df for the numerator $= k - 1 = 4 - 1 = 3$ and df for the denominator $= n - k = 28 - 4 = 24$

For $\alpha = .025$ and $df = (3, 24)$, the critical value for F is 3.72.

$T_1 = 162$, $T_2 = 145$, $T_3 = 172$, $T_4 = 180$, $\Sigma x = 659$, $\Sigma x^2 = 15,751$.

SSB $= 97.5357$,   SSW $= 143.4286$,   MSB $= 32.5119$,   and MSW $= 5.9762$.

The value of the test statistic is: $F = \text{MSB} / \text{MSW} = 32.5119 / 5.9762 = 5.44$

Reject H_0. The mean life of bulbs for each of these four brands is not the same.

12.22 H_0: $\mu_1 = \mu_2 = \mu_3$;   and H_1: All three population means are not equal

$n = n_1 + n_2 + n_3 = 7 + 6 + 6 = 19$ and $k = 3$

df for the numerator $= k - 1 = 3 - 1 = 2$ and df for the denominator $= n - k = 19 - 3 = 16$

From the F distribution table, the critical value of F for 2 df for the numerator, 16 df for the denominator, and .05 area in the right tail is 3.63.

$T_1 = 58$, $T_2 = 33$, $T_3 = 56$, $\Sigma x = 147$, and $\Sigma x^2 = 1427$.

SSB $= 47.4223$, SSW $= 242.2619$, MSB $= 23.7112$, and MSW $= 15.1414$.

The value of the test statistic is: $F = \text{MSB} / \text{MSW} = 23.7112 / 15.1414 = 1.57$

Do not reject the null hypothesis. The mean number of violent crimes reported per day is the same for each of the three cities.

12.23   a. H_0: $\mu_1 = \mu_2 = \mu_3 = \mu_4$;   and H_1: All four population means are not equal

$n = n_1 + n_2 + n_3 + n_4 = 5 + 4 + 5 + 4 = 18$ and $k = 4$

df for the numerator $= k - 1 = 4 - 1 = 3$ and df for the denominator $= n - k = 18 - 4 = 14$

From the F distribution table, the critical value of F for 3 df for the numerator, 14 df for the denominator, and .05 area in the right tail is 3.34.

$T_1 = 324$, $T_2 = 210$, $T_3 = 354$, $T_4 = 199$, $\Sigma x = 1087$, and $\Sigma x^2 = 68,013$.

SSB $= 1340.9278$,   SSW $= 1029.3500$,   MSB $= 446.9759$,   and MSW $= 73.5250$.

The value of the test statistic is: $F = \text{MSB} / \text{MSW} = 446.9759 / 73.5250 = 6.08$

Reject the null hypothesis. The mean life of each of the four brands of batteries is not the same.

b. The Type I error would be to conclude that all four population means are not equal when actually they are equal. P(Type I error) $= \alpha = .05$.

12.24   a. H_0: $\mu_1 = \mu_2 = \mu_3$;   and H_1: The means of the three populations are not all equal.

$n = n_1 + n_2 + n_3 = 6 + 6 + 6 = 18$ and $k = 3$

df for the numerator $= k - 1 = 3 - 1 = 2$ and df for the denominator $= n - k = 18 - 3 = 15$

For $\alpha = .01$ and $df = (2, 15)$, the critical value of F is 6.36.

$T_1 = 5537$, $T_2 = 4840$, $T_3 = 4722$, $\Sigma x = 15,099$, and $\Sigma x^2 = 12,822,957$

SSB $= 64,664.3333$,   SSW $= 92,748.1667$,   MSB $= 32,332.1667$, and MSW $= 6183.2111$

The value of the test statistic is: $F =$ MSB / MSW $= 32,332.1667 / 6183.2111 = 5.23$

Do not reject H_0. The three population means are equal.

b.   The Type I error would be to conclude that the means of the three populations are not all equal when in fact they are equal. P(Type I error) $= \alpha = .01$

12.25   a. H_0: $\mu_1 = \mu_2 = \mu_3$;   and H_1: All three population means are not equal

$n = n_1 + n_2 + n_3 = 5 + 5 + 5 = 15$ and $k = 3$

df for the numerator $= k - 1 = 3 - 1 = 2$ and df for the denominator $= n - k = 15 - 3 = 12$

From the F distribution table, the critical value of F for 2 df for the numerator, 12 df for the denominator, and .01 area in the right tail is $F = 6.93$.

$T_1 = 5.3$, $T_2 = 3.6$, $T_3 = 5.1$, $\Sigma x = 14$, and $\Sigma x^2 = 15.08$

SSB $= .3453$,   SSW $= 1.6680$,   MSB $= .1727$,   and   MSW $= .1390$

The value of the test statistic is: $F =$ MSB / MSW $= .1727 / .1390 = 1.24$

Do not reject the null hypothesis. The mean weight gained by all chickens is the same for each of the three diets.

b.   By not rejecting H_0, we may have made a Type II error by concluding that the three population means are equal when in fact they are not.

12.26   a. H_0: $\mu_1 = \mu_2 = \mu_3 = \mu_4$;   and H_1: All four population means are not equal.

$n = n_1 + n_2 + n_3 + n_4 = 5 + 4 + 5 + 5 = 19$ and $k = 4$

df for the numerator $= k - 1 = 4 - 1 = 3$ and df for the denominator $= n - k = 19 - 4 = 15$

From the F distribution table, the critical value of F for 3 df for the numerator, 15 df for the denominator, and .05 area in the right tail is $F = 3.29$.

$T_1 = 466$, $T_2 = 367$, $T_3 = 450$, $T_4 = 483$, $\Sigma x = 1766$,   and $\Sigma x^2 = 166474$

SSB = 116.1974,   SSW = 2212.75,   MSB = 38.7325, and   MSW = 147.5167

The value of the test statistic is: $F = MSB / MSW = 38.7325 / 147.5167 = .2626$

Do not reject the null hypothesis. Do not conclude the mean prices of college textbooks in the four disciplines are not all equal.

b. If we do not reject the null hypothesis, we could be making a Type II error (concluding that the mean prices of college textbooks where the same in all four disciplines when actually they were different).

12.27 a. $H_0: \mu_1 = \mu_2 = \mu_3;$ and H_1: All three population means are not equal

$n = n_1 + n_2 + n_3 = 5 + 5 + 5 = 15$ and $k = 3$

df for the numerator $= k - 1 = 3 - 1 = 2$ and df for the denominator $= n - k = 15 - 3 = 12$

From the F distribution table, the critical value of F for 2 df for the numerator, 12 df for the denominator, and .05 area in the right tail is 3.89.

$T_1 = 521,$ $T_2 = 412,$ $T_3 = 464,$ $\Sigma x = 1397,$ and $\Sigma x^2 = 132,755$

SSB = 1188.9333,   SSW = 1458.8000,   MSB = 594.4667,   and MSW = 121.5667

The value of the test statistic is: $F = MSB / MSW = 594.4667 / 121.5667 = 4.89$

Reject the null hypothesis. Conclude the mean tips for the three restaurants are not equal.

b. If $\alpha = 0$, there is no rejection region, so we cannot reject H_0. We must conclude that the mean tips are the same for all three restaurants.

12.28 $H_0: \mu_1 = \mu_2 = \mu_3;$ and H_1: All three population means are not equal

$T_1 = 57, T_2 = 53, T_3 = 27,$ $\Sigma x = 137,$ and $\Sigma x^2 = 1503$

$n = n_1 + n_2 + n_3 = 5 + 5 + 5 = 15$ and $k = 3$

df for the numerator $= k - 1 = 3 - 1 = 2$ and df for the denominator $n - k = 15 - 3 = 12$

For $\alpha = .05$ and $df = (2, 12)$, the critical value of F is 3.89.

SSB = 106.1333   SSW = 146.6000   MSB = 53.0667   MSW = 12.1333

The value of the test statistic is: $F = MSB / MSW = 53.06667 / 12.1333 = 4.37$

Reject the null hypothesis. Conclude that the mean time taken to find a parking space just before 9AM on Monday, Wednesday, and Friday is not the same for all students.

12.29 $H_0: \mu_1 = \mu_2 = \mu_3 = \mu_4;$ and H_1: All four population means are not equal

$n_1 = 5,$ $\bar{x}_1 = 295$ $T_1 = 5(295) = 1475$

$n_2 = 5,$ $\bar{x}_2 = 380,$ $T_2 = 5(380) = 1900$

$n_3 = 5,$ $\bar{x}_3 = 405,$ $T_3 = 5(405) = 2025$

$n_4 = 5,$ $\bar{x}_4 = 345,$ $T_4 = 5(345) = 1725$

$\Sigma x = T_1 + T_2 + T_3 + T_4 = 7125,$ and $\Sigma x^2 = 2{,}890{,}000$

$n = n_1 + n_2 + n_3 + n_4 = 20$

df for the numerator $= k - 1 = 4 - 1 = 3$ and   *df* for the denominator $= n - k = 20 - 4 = 16$

For $\alpha = .01$ and *df* $= (3,16)$, the critical value of *F* is 5.29.

SSB = 34,093.75   and   SSW = 317,625

MSB = SSB / $(k - 1)$ = 34,093.75 / $(4 - 1)$ = 11,364.5833

MSW = SSW / $(n - k)$ = 317,625 / $(20 - 4)$ = 19,851.5625

The value of the test statistic is: $F =$ MSB / MSW = 11,364.5833 / 19851.5625 = .57

Do not reject H_0, since .57 < 5.29. Do not conclude that the means are not the same.

12.30   a.   Take random samples of carpenters, plumbers, electricians, and masons, respectively, from your city and collect information on each worker's hourly wage. Note that random samples will be representative samples irrespective of how many workers are union members and how many are not union members. Next, calculate the mean hourly wage for each sample and compare them in the article without using any statistical inferential procedures.

   b.   Using the ANOVA procedure and the data collected in part a for four samples, test the null hypothesis that the mean hourly wage rates for all four trades are equal against the alternative hypothesis that all four means are not equal. Select your own significance level. Then, write a report explaining and interpreting these results.

12.31   a.   Lay out a route for a test run in a typical city. Let each driver drive each car several times on this route. Make sure that the length of the route is selected in such a way that the total of all tests for each car does not exceed 500 miles. Then, calculate the gas mileage for each of these test runs. Next, compute the mean gas mileage for each of the three cars and compare them in the article.

   b.   Using the ANOVA procedure and the data collected in part a for three cars, test the null hypothesis that the mean gas mileage for all three cars is equal against the alternative hypothesis that all three means are not equal. Select your own significance level. Then, write a report explaining and interpreting these results.

12.32   a. $H_0: \mu_1 = \mu_2$;   and H_1 $\mu_1 \neq \mu_2$.

$\bar{x}_1 = 47.675$; $\bar{x}_2 = 38.250$; $s_1 = 11.57160514$; $s_2 = 4.70609261$

$$s_p = \sqrt{\frac{(n_1 - 1)s_1^2 + (n_2 - 1)s_2^2}{n_1 - n_2 - 2}} = \sqrt{\frac{(12 - 1)(11.57160514)^2 + (14 - 1)(4.70609261)^2}{(12 + 14 - 2)}} = 8.56552562$$

$$s_{\bar{x}_1-\bar{x}_2} = s_p\sqrt{\frac{1}{n_1}+\frac{1}{n_2}} = 8.56552562\sqrt{\frac{1}{12}+\frac{1}{14}} = 3.36965976$$

$$t = (\bar{x}_1 - \bar{x}_2)/s_{\bar{x}_1-\bar{x}_2} = (47.675 - 38.250)/3.36965976 = 2.797$$

From the t distribution table, the area in the right tail for $df = 24$ and $t = 2.797$ is .005

Hence, p–value = 2(.005) = .01

b.

Source	df	SS	MS	Test Statistic
Between	1	573.9825	573.9825	
Within	24	1760.8375	73.3682	F = 7.82
Total	25	2334.8200		

For $df = 1$ for the numerator, $df = 24$ for the denominator, and $F = 7.82$, the area in the right tail of the F distribution is .01.

Hence, p–value = $P(F > 7.82)$ = .01

c. For parts a and b, $(2.797)^2 = 7.82$. That is, $t^2 = F$. Actually for $df = 1$ for the numerator and the same area in the right tail, $t^2 = F$ for the df for t and for df for the denominator for F. The p–values are the same, which will always happen since these are equivalent tests.

Self-Review Test for Chapter Twelve

1. a   **2.** b   **3.** c   **4.** a   **5.** a   **6.** a   **7.** b   **8.** a

9. See box on "Assumptions of one–way ANOVA" on page 559 of the text.

10. a. H_0: $\mu_1 = \mu_2 = \mu_3 = \mu_4$;   and H_1: All four population means are not equal

$n = n_1 + n_2 + n_3 + n_4 = 6 + 6 + 6 + 6 = 24$ and $k = 4$

df for the numerator $= k - 1 = 4 - 1 = 3$ and df for the denominator $= n - k = 24 - 4 = 20$

From the F distribution table, the critical value of F for 3 df for the numerator, 20 df for the denominator, and .05 area in the right tail is 3.10.

$T_1 = 124.1$, $T_2 = 152.1$ $T_3 = 156.2$ $T_4 = 152$, $\Sigma x = 584.4$, $\Sigma x^2 = 14{,}503.26$

SSB = 109.4700,   SSW = 163.6500,   MSB = 36.4900,   MSW = 8.1825

The value of the test statistic is: $F = $ MSB / MSW $= 36.4900 / 8.1825 = 4.46$

Reject the null hypothesis. The mean price for all four pizza parlors are not the same.

b. By rejecting H_0, we may have committed a Type I error.

Chapter Thirteen

13.1 A regression model that includes only two variables, one independent and one dependent, is called a simple regression model. The dependent variable is the one being explained and the independent variable is the one used to explain the variation in the dependent variable. A (simple) regression model that gives a straight–line relationship between two variables is called a (simple) linear regression model.

13.2 The dependent variable is the variable to be predicted or explained. The independent variable is included in the regression model to help explain variation in the dependent variable.

13.3 In an exact relationship, the value of the dependent variable y is determined exactly by the independent variable x, that is, for a given value of x there is a unique value of y. In a nonexact relationship, there are many (perhaps infinitely many) values of y for a given value of x.

13.4 The graph of a linear relationship between two variables is a straight line. The graph of a nonlinear relationship is not a straight line.

13.5 A simple regression model has only one independent variable, while a multiple regression model has more than one independent variable. Both models have just one dependent variable.

13.6 In a deterministic model, the relationship between the dependent and independent variables is exact. In a probabilistic model, the independent variable does not determine the dependent variable exactly.

13.7 The random error term ε is included in a regression model to represent the following two phenomena:
1. Missing or omitted variables:
   Usually a dependent variable y is determined by a number of variables. However it is almost impossible to include all of these variables in the regression model. The random error term ε is included to capture the effect of all the missing or omitted variables which have not been included in the model.

2. Random Variation:

Human behavior is unpredictable. Even for the same value of x, the value of y may vary from element to element just because of random behavior. The random error term is included in a regression model to represent this random variation.

13.8 The least squares method fits a regression line through a scatter diagram by minimizing the error sum of squares. The least squares regression line is the line fitted to the data by the least squares method.

13.9 SSE denotes the error sum of squares, which is the sum of squared differences between the actual and predicted values of y, that is, SSE $= \Sigma(y - \hat{y})^2$. SSE represents the portion of the variation in y that is not explained by the regression model.

13.10 Whereas y is the actual value of the dependent variable for a given value of x, $\hat{y}$ is the predicted value of y obtained from the model $\hat{y} = a + bx$ using the same value of x.

13.11 When x and y have a positive linear relationship, y increases as x increases.

13.12 When x and y have a negative linear relationship, y decreases as x increases.

13.13 a. A regression line obtained by using the population data is called the population regression line. It gives the true values of A and B and is written as: $\mu_{y|x} = A + Bx$

b. A sample regression line is obtained from sample data. It uses estimated values, a and b, and is written as: $\hat{y} = a + bx$

Here, a is an estimate of A and b is an estimate of B.

c. The true values of A and B are the values obtained from the population regression line. They are the population parameters.

d. The estimated values of A and B are the values obtained from a regression model that is obtained by using the sample data. Such an estimated model is written as: $\hat{y} = a + bx$

13.14 See Section 13.2.4, Pages 590–593 of the text.

13.15   a.

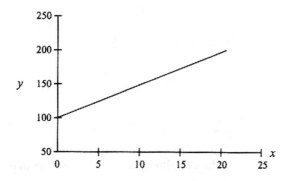

Here, the *y*–intercept is 100, which is the point where the line meets the *y*–axis. The slope is 5, which means that for a 1–unit increase in *x*, there will be a 5–unit increase in *y*. Since the slope has the positive value of 5, there is a positive relationship between *x* and *y*. (Note that the vertical axis in the graph is truncated as it starts at 50. This will be true of almost all graphs in this chapter.)

    b.

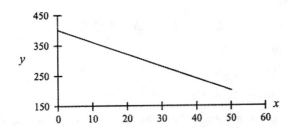

Here, the *y*–intercept is 400, which is the point where the line meets the *y*–axis. The slope is –4, which means that for a 1–unit increase in *x*, there will be a 4–unit decrease in *y*. Since the slope has the negative value of –4, there is a negative relationship between *x* and *y*.

13.16   a.

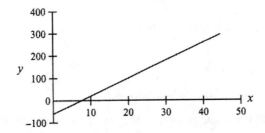

Here, the *y* intercept is –60, which is the point where the line meets the *y*–axis. The slope is 8, which means that for a 1–unit increase in *x*, there will be an 8–unit increase in *y*. Since the slope has the positive value of 8, there is a positive relationship between *x* and *y*.

b.

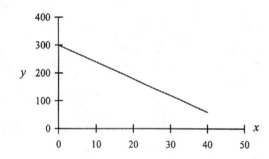

Here, the y–intercept is 300, which is the point where the line meets the y–axis. The slope is –6, which means that for a 1–unit increase in x, there will be a 6–unit decrease in y. Since the slope has the negative value of –6, there is a negative relationship between x and y.

13.17 Using the given information, we obtain:

$$SS_{xy} = \sum xy - \frac{(\Sigma x)(\Sigma y)}{N} = 85,080 - \frac{(9880)(1456)}{250} = 27,538.8800$$

$$SS_{xx} = \Sigma x^2 - \frac{(\Sigma x)^2}{N} = 485,870 - \frac{(9880)^2}{250} = 95,412.4000$$

$\mu_x = \Sigma x / N = 9880 / 250 = 39.5200$ $\qquad$ $\mu_y = \Sigma y/N = 1456/250 = 5.8240$

$B = SS_{xy}/SS_{xx} = 27,538.8800/95,412.4000 = .2886$

$A = \mu_y - B \mu_x = 5.8240 - (.2886)(39.5200) = -5.5815$

Thus, the population regression line is: $\mu_{y|x} = -5.5815 + .2886 x$

Note that because the given data are population data, we have used μ_x and μ_y to denote the means of the variables x and y, respectively.

13.18 $SS_{xy} = \sum xy - \frac{(\Sigma x)(\Sigma y)}{N} = 26,570 - (3920)(2650)/460 = 3987.3913$

$$SS_{xx} = \Sigma x^2 - \frac{(\Sigma x)^2}{N} = 48,530 - (3920)^2/460 = 15,124.7826$$

$\mu_x = \Sigma x / N = 3920/460 = 8.5217$ $\mu_y = \Sigma y / N = 2650/460 = 5.7609$

$B = SS_{xy}/SS_{xx} = 3987.3913/15,124.7826 = .2636$

$A = \mu_y - B \mu_x = 5.7609 - (.2636)(8.5217) = 3.5146$

Thus, the population regression line is: $\qquad$ $\mu_{y|x} = 3.5146 + .2636x$

Note that because the given data are population data, we have used μ_x and μ_y to denote the means of the variables x and y, respectively.

13.19 $SS_{xy} = \sum xy - \frac{(\Sigma x)(\Sigma y)}{n} = 3680 - (100)(220)/10 = 1480$

$$SS_{xx} = \Sigma x^2 - \frac{(\Sigma x)^2}{n} = = 1140 - (100)^2 / 10 = 140$$

$\bar{x} = \Sigma x / n = 100/10 = 10$ and $\bar{y} = \Sigma y / n = 220/10 = 22$

$b = SS_{xy}/SS_{xx} = 1480/140 = 10.5714$

$a = \bar{y} - b\bar{x} = 22 - (10.5714)(10) = -83.7140$

Thus, the estimated regression line is: $\hat{y} = -83.7140 + 10.5714x$

13.20 $SS_{xy} = \Sigma xy - \frac{(\Sigma x)(\Sigma y)}{n} = 2244 - (66)(588)/12 = -990$

$$SS_{xx} = \Sigma x^2 - \frac{(\Sigma x)^2}{n} = = 396 - (66)^2 / 12 = 33$$

$\bar{x} = \Sigma x / n = 66 / 12 = 5.5$ and $\bar{y} = \Sigma y / n = 588 / 12 = 49.0$

$b = SS_{xy}/SS_{xx} = -990/ 33 = -30$

$a = \bar{y} - b\bar{x} = 49 - (-30)(5.5) = 214$

Thus, the estimated regression line is: $\hat{y} = 214 - 30x$

13.21 a. $x = 100$, so $y = 40 + .20(100) = \$60$

b. Every person who rents a car from this agency for one day and drives it 100 miles will pay the same amount, $60. This is due to the fact that for any value x, the equation $y = 40 + .20x$ yields a unique value of y.

c. The relationship is exact.

13.22 a. $x = 3$, so $y = 50 + 20(3) = \$110$

b. Every person whose pest removal takes two hours will pay the same amount, $110. This is due to the fact that for any value of x, the equation $y = 50 + 20x$ yields a unique value of y.

c. The relationship is exact.

13.23 a. Here, $x = 2$, so expected gross sales for 1999 are: $y = 3.6 + 11.75(2) = \$27.1$ million

b. The four companies that spent $2 million each on advertising would not have the same actual gross sales for 1999. The $27.1 million obtained in part a is merely the mean gross sales for companies spending $2 million on advertising. The actual gross sales would differ due to the influence of variables not included in the model.

c. The relationship is nonexact.

13.24   a. Here, $x = 24$, so the expected average profits of all U.S. insurance companies are: $y = 342.6 - 2.10(24) = \$292.2$ million

b. The average profits for each of the three years would be different due to differences in the economic impact of the calamities and due to variables not included in the model. The $292.2 million obtained in part a is merely an average figure for years having 24 major calamities.

c. The relationship is nonexact.

13.25   Let: x = age of a car (in years)

y = price of a car (in hundreds of dollars)

a. & d.

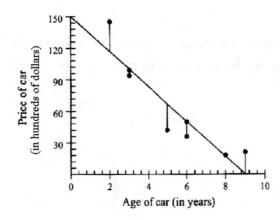

The scatter diagram exhibits a linear relationship between ages and prices of cars.

b.

x	y	xy	x^2
8	18	144	64
3	94	282	9
6	50	300	36
9	21	189	81
2	145	290	4
5	42	210	25
6	36	216	36
3	99	297	9
$\Sigma x = 42$	$\Sigma y = 505$	$\Sigma xy = 1928$	$\Sigma x^2 = 264$

$\bar{x} = \Sigma x / n = 42/8 = 5.250, \ \bar{y} = \Sigma y/n = 505/8 = 63.125$

$SS_{xx} = 43.5000$ and $SS_{xy} = -723.2500$

$b = SS_{xy}/SS_{xx} = -723.2500/43.5000 = -16.6264$

$a = \bar{y} - b\bar{x} = 63.125 - (-16.6264)(5.250) = 150.4136$

Thus the estimated regression model is: $\hat{y} = 150.4136 - 16.6264\, x$

c. The value of $a = 150.4136$ is the value of y for $x = 0$, which in this case represents the price of a new car (in hundreds of dollars). Thus, the price of a new car is expected to be (about) $15,041. The value of $b = -16.6264$ means that, on average, for every one year increase in the age of a car, its price decreases by $1663.

e. For $x = 7$: $y = 150.4136 - 16.6264(7) = 34.0288$ Thus, the price of a 7–year old car is $3403.

f. For $x = 18$: $y = 150.4136 - 16.6264(18) = -148.8616$

The negative price makes no sense. The regression line is based on data for cars from 2 to 8 years in age. Since $x = 18$ is outside this range, the estimate is invalid.

13.26 Let: $x =$ lowest temperature and $y =$ number of calls

a. & d.

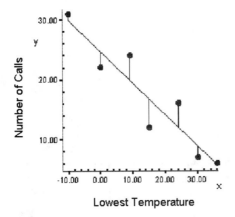

Lowest Temperature

The scatter diagram exhibts a linear relationship between lowest temperature and number of calls.

b. $n = 7, \Sigma x = 104, \Sigma y = 118, \Sigma x^2 = 3178, \Sigma xy = 896, \bar{x} = 14.8571, \bar{y} = 16.8571,$

$SS_{xx} = 1632.8571$, and $SS_{xy} = -857.1429$

$b = SS_{xy} / SS_{xx} = -857.1429 / 1632.8571 = -.5249$

$a = \bar{y} - b\bar{x} = 16.8571 - (-.5249)(14.8571) = 24.6556$

Thus, the estimated regression model is: $\hat{y} = 24.6556 - .5249\, x$

c. The value of $a = 24.6556$ is the value of y for $x = 0$. In this exercise it represents the number of calls when the temperature is at zero degrees.

The value of $b = -.5249$ means that on average the number of calls decreases by .5249 for every 1–degree increase in the temperature.

e. For $x = 20$: $\hat{y} = 24.6556 - .5249(20) = 14.1576$. Thus, the number of calls when the temperature is 20 degrees is about 14.

f. For $x = -20$: $\hat{y} = 24.6556 - .5249(-20) = 35.1536$. The regression model is based on data for temperatures ranging from –10 to 36 degrees. Since –20 is outside this range, little confidence should be placed in this estimate.

13.27 Let: x = annual income (in thousands of dollars)

y = amount of life insurance policy (in thousands of dollars)

a. & d.

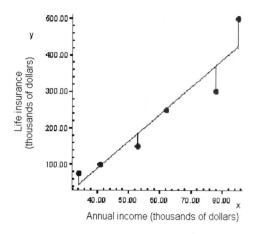

The scatter diagram shows a linear relationship between the annual incomes and amounts of life insurance.

b. $n = 6$, $\Sigma x = 353$, $\Sigma y = 1375$, $\Sigma x^2 = 22,799$, $\Sigma xy = 96,000$

$\bar{x} = 58.8333$, $\bar{y} = 229.1667$, $SS_{xx} = 2030.8333$, $SS_{xy} = 15,104.1667$

$b = SS_{xy} / SS_{xx} = 15,104.1667 / 2030.8333 = 7.4374$

$a = \bar{y} - b\bar{x} = 229.1667 - 7.4374 (58.8333) = -208.4001$

Thus, the estimated regression model is: $\hat{y} = -208.4001 + 7.4374x$

c. The value of $a = -208.4001$ is the value of y for $x = 0$. In this exercise it represents the amount of life insurance for a person with a zero income.

The value of $b = 7.4374$ means that, on average, the amount of life insurance increases by $7437 for every $1000 increase in the annual income of a person.

e. $x = 55$: $\hat{y} = -208.4001 + 7.4374 (55) = 200.6569$ or $200,656.90 Thus, the estimated value of life insurance for a person with an annual income of $55,000 is $200,656.90.

f. For $x = 78$: $\hat{y} = -208.4001 + 7.4374(78) = 371.7171$ or $371,717.10;

$e = y - \hat{y} = 300,000 - 371,717.10 = -\$71,717.10$

13.28 Let x = size of a house (in hundreds of square feet); y = monthly rent (in dollars)

a. & d.

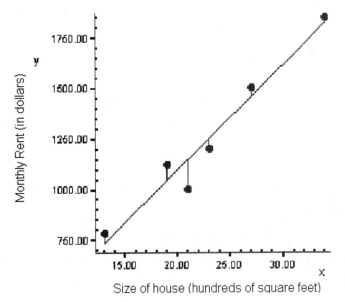

Size of house (hundreds of square feet)

The diagram exhibits a linear relationship between the sizes of houses and monthly rents.

b. $n = 6$, $\Sigma x = 137$, $\Sigma y = 7450$, $\Sigma x^2 = 3385$, $\Sigma xy = 183,420$ $\bar{x} = 22.8333$, $\bar{y} = 1241.6667$,

$SS_{xx} = 256.8333$, $SS_{xy} = 13,311.6667$

$b = SS_{xy} / SS_{xx} = 13,311.6667/256.8333 = 51.8300$

$a = \bar{y} - b\bar{x} = 1241.6667 - (51.8300)(22.8333) = 58.2168$

Thus, the estimated regression model is: $\hat{y} = 58.2168 + 51.8300\, x$

c. The value of $a = 58.2168$ is the value of y for $x = 0$. In this exercise it represents the rent for a house with an area of zero square feet. The value of $b = 51.8300$ means that, on average, the rent of a house increases by about \$51.83 for every 100 square feet increase in the size of the house.

e. For $x = 25$: $\hat{y} = 58.2168 + 51.8300\,(25) = \$1353.97$

Thus, the average monthly rent for a house with an area of 2500 square feet is about \$1353.97. Note that we have substituted 25 for x to calculate this rent. The reason is that x represents the size of a house in hundreds of square feet.

13.29   Let: x = total payroll (in millions of dollars); y = percentage of games won

a. $N = 16$, $\Sigma x = 1052$, $\Sigma y = 802.1$, $\Sigma x^2 = 76{,}630$, $\Sigma xy = 54{,}373.7$ $\mu_x = 65.75$, $\mu_y = 50.1313$, $SS_{xx} = 7461$, $SS_{xy} = 1635.625$. Note that because the given data are population data we have used μ_x and μ_y to denote the means of the variables x and y, respectively.

$B = SS_{xy}/SS_{xx} = 1635.625\,/\,7461 = .2192$

$A = \mu_y - B\,\mu_x = 50.1313 - (.2192)\,(65.75) = 35.7173$

Thus, the estimated regression model is: $\mu_{y|x} = 35.7173 + .2192x$

b. The regression line obtained in part a is the population regression line because the data are on all National League baseball teams. The values of the y–intercept and slope obtained above are those of A and B.

c. The value of $A = 35.7173$ is the value of $\mu_{y|x}$ for $x = 0$. In this exercise it represents the percentage of games won by a team with a total payroll of zero dollars.

The value of $B = .2192$ means that, on average, the percentage of games won increases by 21% for every \$1 million increase in payroll of a National League baseball team.

d. For $x = 55$: $\mu_{y|x} = 35.7173 + .2192(55) = 47.7733$. Thus, a team with a total payroll of \$55 million is expected to win about 47.77% of its games.

13.30   Let x = total payroll (in millions of dollars)

             y = percentage of games won

a. $N = 14$, $\Sigma x = 971$, $\Sigma y = 698.2$, $\Sigma x^2 = 77{,}629$, $\Sigma xy = 49{,}721.7$; $\mu_x = 69.3571$, $\mu_y = 49.8714$, $SS_{xx} = 10{,}283.2143$, $SS_{xy} = 1296.5429$. Note that because the given data are population data we have used μ_x and μ_y to denote the means of the variables x and y, respectively.

$B = SS_{xy}/SS_{xx} = 1296.5429 / 10,283.2143 = .1261$

$A = \mu_y - B\mu_x = 49.8714 - (.1261)(69.3571) = 41.1255$

Thus, the estimated regression model is: $\mu_{y|x} = 41.1255 + .1261\,x$

b. The regression line obtained in part a is the population regression line because the data are on all American League baseball teams. The values of the y–intercept and slope obtained above are those of A and B.

c. The value of $A = 41.1255$ is the value of $\mu_{y|x}$ for $x = 0$. In this exercise it represents the percentage of games won by a team with a total payroll of zero dollars.

The value of $B = .1261$ means that, on average, the percentage of games won increases by 13% for every \$1 million increase in payroll of an American League baseball team.

d. For $x = 65$: $\mu_{y|x} = = 41.1255 + .1261(65) = 49.3220$. Thus, a team with a total payroll of \$65 million is expected to win about 49% of its games.

13.31 For a simple linear regression model, $df = n - 2$.

13.32 The coefficient of determination represents the proportion of the total sum of squares (SST) that is explained by the regression model.

13.33 SST is the sum of squared differences between the actual y values and $\bar{y}$, that is, $SST = \Sigma\,(y - \bar{y})^2$. SSR is the portion of SST that is explained by the regression model.

13.34 $SS_{xx} = 95,412.4000$, $SS_{yy} = 127,195.2560$, and $SS_{xy} = 27,538.8800$

$B = SS_{xy}/SS_{xx} = 27,538.8800/95,412.4000 = .2886$

$$\sigma_\varepsilon = \sqrt{\frac{SS_{yy} - B(SS_{xy})}{N}} = \sqrt{\frac{127,195.2560 - (.2886)(27,538.8800)}{250}} = 21.8401$$

$\rho^2 = B(SS_{xy})/SS_{yy} = (.2886)(27,538.8800) / 127,195.2560 = .06$

13.35 $SS_{xx} = 15,124.7826$, $SS_{yy} = 24,080.6957$, and $SS_{xy} = 3987.3913$

$B = SS_{xy}/SS_{xx} = 3987.3913/15,124.7826 = .2636$

$$\sigma_\varepsilon = \sqrt{\frac{SS_{yy} - B(SS_{xy})}{N}} = \sqrt{\frac{24,080.6957 - (.2636)(3987.3913)}{460}} = 7.0756$$

$\rho^2 = B(SS_{xy})/SS_{yy} = (.2636)(3987.3913) / 24,080.6957 = .04$

13.36 $n = 10, SS_{xx} = 140, SS_{yy} = 20,432, \ SS_{xy} = 1480, \ b = SS_{xy}/SS_{xx} = 1480/140 = 10.5714$

$$s_e = \sqrt{\frac{SS_{yy} - b(SS_{xy})}{n-2}} = \sqrt{\frac{20,432 - (10.5714)(1480)}{10-2}} = 24.4600$$

$r^2 = b \ SS_{xy} / SS_{YY} = (10.5714)(1480) / 20,432 = .77$

13.37 $n = 12, SS_{xx,} = 33, SS_{yy} = 29,922, SS_{xy} = -990; \ b = SS_{xy}/SS_{xx} = -990 / 33 = -30$

$$s_e = \sqrt{\frac{SS_{yy} - b(SS_{xy})}{n-2}} = \sqrt{\frac{29,922 - (-30)(-990)}{12-2}} = 4.7117$$

$r^2 = b \ SS_{xy} / SS_{yy} = (-30)(-990)/29,922 = .99$

13.38 Let: x = hours worked per week, and y = GPA

a. $n = 7, \Sigma x = 86, \Sigma y = 22.7, \Sigma x^2 = 1238, \Sigma y^2 = 76.15, \Sigma xy = 260.4, \ \bar{x} = 12.2857,$ |
   $\bar{y} = 3.2429$

   $SS_{xx} = 181.4286, \qquad SS_{yy} = 2.5371, \qquad$ and $SS_{xy} = -18.4857$

b. $b = SS_{xy} / SS_{xx} = -18.4857 / 181.4286 = -.1019$

$$s_e = \sqrt{\frac{SS_{yy} - b(SS_{xy})}{n-2}} = \sqrt{\frac{2.5371 - (-.1019)(-18.4857)}{7-2}} = .3615$$

c. $a = \bar{y} - b\bar{x} = 3.2429 - (-.1019)12.2857 = 4.4948$

   The regression line is: $\hat{y} = 4.4948 - .1019x$

x	y	$\hat{y} = 4.4946 - .1019x$	$e = y - \hat{y}$	e^2
10	3.5	3.4758	.0242	.0006
8	3.7	3.6796	.0204	.0004
20	3.0	2.4568	.5432	.2951
15	2.8	2.9663	−.1663	.0277
18	2.1	2.6606	−.5606	.3143
5	4.0	3.9853	.0147	.0002
10	3.6	3.4758	.1242	.0154
				$\Sigma e^2 = .6537$

   $SST = SS_{yy} = 2.5371 \quad$ and $\qquad SSE = \Sigma e^2 = .6537$

   $SSR = SST - SSE = 2.5371 - .6537 = 1.8834$

d. $r^2 = bSS_{xy} / SS_{yy} = (-.1019)(-18.4857) / 2.5371 = .74$

13.39 Let: x = fat consumption (in grams) per day

   y = cholesterol level (in milligrams per hundred milliliters)

a. $n = 8$; $\Sigma x = 421$; $\Sigma y = 1514$; $\Sigma x^2 = 23{,}743$; $\Sigma y^2 = 292{,}116$ $\Sigma xy = 82{,}517$;

$\bar{x} = 52.625$; $\bar{y} = 189.25$, $SS_{xx} = 1587.8750$, $SS_{yy} = 5591.5000$, and $SS_{xy} = 2842.7500$

b. $b = SS_{xy}/SS_{xx} = 2842.7500/1587.8750 = 1.7903$

$$s_e = \sqrt{\frac{SS_{yy} - b(SS_{xy})}{n-2}} = \sqrt{\frac{5591.5000 - (1.7903)(2842.7500)}{8-2}} = 9.1481$$

c. $a = \bar{y} = b\bar{x} = 189.25 - 1.7903(52.625) = 95.0355$

The regression line is: $y = 95.0355 + 1.7903x$

$SST = SS_{yy} = 5591.5000$ $\qquad$ and $\qquad$ $SSE = \Sigma e^2 = 502.1652$

$SSR = SST - SSE = 5591.5000 - 502.1652 = 5089.3348$

d. $r^2 = b\, SS_{xy} / SS_{yy} = (1.7903)(2842.7500)/5591.5000 = .91$

13.40 Let: $x = $ age $\qquad$ and $\qquad$ $y = $ price

$SS_{yy} = 14{,}108.8750$, $SS_{xy} = -723.2500$, $b = -16.6264$, $n = 8$

a. $s_e = \sqrt{\dfrac{SS_{yy} - b(SS_{xy})}{n-2}} = \sqrt{\dfrac{14{,}108.8750 - (-16.6264)(-723.2500)}{8-2}} = 18.6361$

b. $r^2 = b\, SS_{xy}/ SS_{yy} = (-16.6264)(-723.2500)/14{,}108.8750 = .85$

Thus, 85%o of the total squared errors (SST) are explained by the regression model.

13.41 Let: $x = $ lowest temperature and $y = $ number of calls

$n = 7$, $SS_{yy} = 516.8571$; $SS_{xy} = -857.1429$; $b = -.5249$

a. $s_e = \sqrt{\dfrac{SS_{yy} - b(SS_{xy})}{n-2}} = \sqrt{\dfrac{516.8571 - (-.5249)(-857.1429)}{7-2}} = 3.6590$

b. $r^2 = b\, SS_{xy} / SS_{yy} = (-.5249)(-857.1429) /516.8571 = .87$

Thus, 87% of the total squared errors (SST) are explained by our regression model with lowest temperature as the independent variable and number of calls as the dependent variable.

13.42 Let: $x = $ annual income (in thousands of dollars)

$y = $ amount of life insurance policy (in thousands of dollars)

$n = 6$, $\Sigma y^2 = 440{,}625$, $SS_{yy} = 125{,}520.8333$

Referring to the calculations for Exercise 13.27: $SS_{xy} = 15{,}104.1667$, and $b = 7.4374$

a. $s_e = \sqrt{\dfrac{SS_{yy} - b(SS_{xy})}{n-2}} = \sqrt{\dfrac{125,520.8333 - (7.4374)(15,104.1667)}{6-2}} = 57.4132$

b. $r^2 = b\,SS_{xy} / SS_{yy} = (7.4374)(15,104.1667) / 125,520.8333 = .89$

Thus, 89% of the variation in life insurance amounts is explained by the annual incomes, and 11% is not explained.

13.43   Let: x = size of a house (in hundreds of square feet), y = monthly rent (in dollars)

$n = 6$, $SS_{yy} = 724,883.3333$, $SS_{xy} = 13,311.6667$, $b = 51.8300$

a. $s_e = \sqrt{\dfrac{SS_{yy} - b(SS_{xy})}{n-2}} = \sqrt{\dfrac{724,883.3333 - (51.8300)(13,311.6667)}{6-2}} = 93.4611$

b. $r^2 = = b\,SS_{xy} / SS_{yy} = (51.8300)(13,311.6667) / 724,883.3333 = .95$

Thus, 95% of the total squared errors (SST) are explained by the regression model with size of the house as the independent variable and monthly rent as the dependent variable, and 5% are not explained.

13.44   Let: x = total payroll, y = percentage of games won

a. $SS_{yy} = 978.2944$, $SS_{xy} = 1635.625$, $B = .2192$, $N = 16$

$\sigma_\varepsilon = \sqrt{\dfrac{SS_{yy} - B(SS_{xy})}{N}} = \sqrt{\dfrac{978.2944 - (.2192)(1635.625)}{16}} = 6.2238$

b. $\rho^2 = B(SS_{xy})/SS_{yy} = .2192(1635.625) / 978.2944 = .37$

13.45   Let: x = total payroll (in millions of dollars); y = percentage of games won

a. $SS_{yy} = 1447.6086$; $SS_{xy} = 1296.5429$; $B = .1261$; $N = 14$

$\sigma_\varepsilon = \sqrt{\dfrac{SS_{yy} - B(SS_{xy})}{N}} = \sqrt{\dfrac{1447.6086 - (.1261)(1296.5429)}{14}} = 9.5771$

b. $\rho^2 = B(SS_{xy})/SS_{yy} = .1261(1296.5429) / 1447.6086 = .11$

13.46   Under the assumption of normally distributed random errors, the sampling distribution of b is normal.

The mean of b is B and its standard deviation is $\sigma_\varepsilon / \sqrt{SS_{xx}}$.

13.47   a. $b = 6.32$ and $s_b = s_e / \sqrt{SS_{xx}} = 1.951 / \sqrt{340.700} = .1057$

$df = n - 2 = 16 - 2 = 14$

For the 99% confidence level, $\alpha/2 = .5 - (.99/2) = .005$

For 14 df and .005 area in the right tail of the t curve, $t = 2.977$.

The 99% confidence interval for B is: $b \pm t s_b = 6.32 \pm (2.977)(.1057) = 6.01$ to 6.63

b. $H_0: B = 0$; $H_1: B > 0$

For 14 df and .025 area in the right tail of the t curve, the critical value of t is 2.145.

The value of the test statistic is: $t = (b - B) / s_b = (6.32 - 0) / .1057 = 59.792$

Reject H_0. Conclude that B is positive.

c. $H_0: B = 0$; $H_1: B \neq 0$

For 14 df and .005 area in each tail of the t curve, the critical values of t are -2.977 and 2.977. The value of the test statistic is $t = 59.792$ from part b.

Reject H_0. Conclude that B is different from zero.

d. $H_0: B = 4.50$; $H_1: B \neq 4.50$

For 14 df and .01 area in each tail of the t curve, the critical values of t are -2.624 and 2.624. The value of the test statistic is: $t = (b - B) / s_b = (6.32 - 4.50) / .1057 = 17.219$

Reject H_0. Conclude that B is different from 4.50.

13.48   a. $b = -3.77$ and $s_b = s_e / \sqrt{SS_{xx}} = .932 / \sqrt{274.600} = .0562$

$df = n - 2 = 25 - 2 = 23$

For the 95% confidence level, $\alpha/2 = .5 - (.95/2) = .025$

For 23 df and .025 area in the right tail of the t curve, $t = 2.069$.

The 95% confidence interval for B is: $b \pm t s_b = -3.77 \pm (2.069)(.0562) = -3.89$ to -3.65

b. $H_0: B = 0$; $H_1: B < 0$

For 23 df and .01 area in the left tail of the t curve, the critical value of t is -2.500. The value of the test statistic is: $t = (b - B) / s_b = (-3.77 - 0) / .0562 = -67.082$

Reject H_0. Conclude that B is negative.

c. $H_0: B = 0$; $H_1: B \neq 0$

For 23 df and .025 area in each tail of the t curve, the critical values of t are -2.069 and 2.069. The value of the test statistic is $t = -67.082$ from part b.

Reject H_0. Conclude that B is different from zero.

d. $H_0: B = -5.20$; $H_1: B \neq -5.20$

For 23 df and .005 area in each tail of the t curve, the critical values of t are -2.807 and 2.807. The value of the test statistic is: $t = (b - B) / s_b = [-3.77 - (-5.20)] / .0562 = 25.445$

Reject H_0. Conclude that B is different from -5.20.

13.49   a. $b = 2.50$ and $s_b = s_e / \sqrt{SS_{xx}} = 1.464 / \sqrt{524.884} = .0639$

For the 98% confidence level, $z = 2.33$

The 98% confidence interval for B is: $b \pm z s_b = 2.50 \pm (2.33)(.0639) = 2.35$ to 2.65

b. $H_0: B = 0$; $H_1: B > 0$

For $\alpha = .02$, the critical value of z is 2.05.

The value of the test statistic is: $z = (b - B) / s_b = (2.50 - 0) / .0639 = 39.12$

Reject H_0. Conclude that B is positive.

c. $H_0: B = 0$; $H_1: B \neq 0$

For $\alpha = .01$, the critical values of z are -2.58 and 2.58.

The value of the test statistic is $z = 39.12$ from part b.

Reject H_0. Conclude that B is different from zero.

d. $H_0: B = 1.75$; $H_1 : B > 1.75$

For $\alpha = .01$, the critical value of z is 2.33.

The value of the test statistic is $z = (b - B) / s_b = (2.50 - 1.75) / .0639 = 11.74$

Reject H_0. Conclude that B is greater than 1.75.

13.50   a. $b = -2.70$ and $s_b = s_e / \sqrt{SS_{xx}} = .961 / \sqrt{380.592} = .0493$

For the 97% confidence level, $z = 2.17$

The 97% confidence interval for B is: $b \pm z s_b = -2.70 \pm (2.17)(.0493) = -2.81$ to -2.59

b. $H_0: B = 0$; $H_1: B < 0$

For $\alpha = .01$, the critical value of z is -2.33.

The value of the test statistic is: $z = (b - B) / s_b = (-2.70 - 0) / .0493 = -54.77$

Reject H_0. Conclude that B is negative.

c. $H_0: B = 0$; $H_1: B \neq 0$

For $\alpha = .01$, the critical values of z are -2.58 and 2.58.

The value of the test statistic is: $z = (b - B) / s_b = -54.77$ from part b.

Reject H_0. Conclude that B is different from zero.

d. $H_0: B = -1.25;$ $H_1: B < -1.25$

For $\alpha = .02$, the critical value of z is -2.05.

The value of the test statistic is: $z = (b - B) / s_b = [-2.70 - (-1.25)] / .0493 = -29.41$

Reject H_0. Conclude that B is less than -1.25.

13.51   Let: $x = $ age, $y = $ price

From the solutions to Exercises 13.25 and 13.40:

$SS_{xx} = 43.5000$, $SS_{yy} = 14,108.8750$, $SS_{xy} = -723.2500$, $b = -16.6264$, and $n = 8$

$$s_e = \sqrt{\frac{SS_{yy} - b(SS_{xy})}{n - 2}} = \sqrt{\frac{14,108.8750 - (-16.6264)(-723.2500)}{8 - 2}} = 18.6361$$

$s_b = s_e / \sqrt{SS_{xx}} = 18.6361 / \sqrt{43.5000} = 2.8256$

a. $df = n - 2 = 8 - 2 = 6$

For 6 df and the 95% confidence level, $t = 2.447$

The 95% confidence interval for B is $b \pm t s_b = -16.6264 \pm (2.447)(2.8256) = -23.5406$ to -9.7122

b. $H_0: B = 0;$ $H_1: B < 0$

For 6 df and .05 area in the left tail of the t distribution, the critical value of t is -1.943.

The value of the test statistic is: $t = (b - B) / s_b = (-16.6264 - 0) / 2.8256 = -5.884$

Reject H_0. Conclude that B is negative.

13.52   Let: $x = $ midterm score, $y = $ instructor score

$n = 10$, $\Sigma x = 809$; $\Sigma y = 28$; $\Sigma x^2 = 67,819$; $\Sigma y^2 = 88$; $\Sigma xy = 2376$, $\bar{x} = 80.90$; $\bar{y} = 2.80$

$SS_{xx} = 2370.9000$, $SS_{yy} = 9.6000$, and $SS_{xy} = 110.8000$; $b = SS_{xy} / SS_{xx} = 110.8000/2370.9000 = .0467$

$a = \bar{y} - b\bar{x} = 2.80 - .0467(80.90) = -.9780$

$$s_e = \sqrt{\frac{SS_{yy} - b(SS_{xy})}{n - 2}} = \sqrt{\frac{9.6000 - (.0467)(110.8000)}{10 - 2}} = .7438$$

$s_b = s_e / \sqrt{SS_{xx}} = .7438 / \sqrt{2370.9000} = .0153$

a. The regression line is: $\hat{y} = -.9780 + .0467 x$

b. $df = n - 2 = 10 - 2 = 8$

Area in each tail of the t curve $= \alpha / 2 = .5 - (.99/2) = .005$

From the t distribution table, the value of t for $df = 8$ and .005 area in the right tail is 3.355.

The 99% confidence interval for B is: $b \pm t s_b = .0467 \pm 3.355(.0153) = -.005$ to $.098$

c.. H_0: $B = 0$; H_1:$B > 0$

Area in the right tail of the t curve $= .01$ and $df = n - 2 = 10 - 2 = 8$

The critical value of t is 2.896.

The value of the test statistic is: $t = (b - B) / s_b = (.0467 - 0) / .0153 = 3.052$

Reject the null hypothesis. Hence, B is positive.

13.53   Let: $x =$ years of experience, $y =$ monthly salary

$n = 9$; $\Sigma x = 80$, $\Sigma y = 318$, $\Sigma x^2 = 968$, $\Sigma y^2 = 11{,}710$, $\Sigma xy = 3162$, $\bar{x} = 8.8889$, $\bar{y} = 35.3333$

$SS_{xx} = 256.8889$, $SS_{yy} = 474.0000$, and $SS_{xy} = 335.3333$

a. $b = SS_{xy}/SS_{xx} = 335.3333/256.8889 = 1.3054$

   $a = \bar{y} - b\bar{x} = 35.3333 - (1.3054)(8.8889) = 23.7297$

   The regression line is: $\hat{y} = 23.7297 + 1.3054\, x$

b. $s_e = \sqrt{\dfrac{SS_{yy} - b(SS_{xy})}{n-2}} = \sqrt{\dfrac{474.0000 - (1.3054)(335.3333)}{9-2}} = 2.2758$

   $s_b = s_e / \sqrt{SS_{xx}} = 2.2758 / \sqrt{256.8889} = .1420$

   $df = n - 2 = 9 - 2 = 7$

   For 7 df and .01 area in the right tail of the t curve, $t = 2.998$

   The 98% confidence interval for B is: $b \pm t s_b = 1.3054 \pm (2.998)(.1420) = .88$ to 1.73

c. H_0: $B = 0$; H_1: $B > 0$

   For 7 df and .025 area in the right tail of the t curve, the critical value of t is 2.365.

   The value of the test statistic is: $t = (b - B) / s_b = (1.3054 - 0) / .1420 = 9.193$

   Reject H_0. Conclude that B is greater than zero.

13.54   Let: $x =$ lowest temperature, $y =$ number of calls

From the solutions to Exercises 13.26 and 13.41:

$n = 7$, $SS_{xx} = 1632.8571$, $a = 24.6556$, $b = -.5249$ and $s_e = 3.6590$

$s_b = s_e / \sqrt{SS_{xx}} = 3.6590 / \sqrt{1632.8571} = .0905$

The regression line is: $\hat{y} = 24.6556 - .5249\, x$

a. $df = n - 2 = 7 - 2 = 5$

For 5 *df* and .025 area in the right tail of the *t* curve, $t = 2.571$

The 95% confidence interval for *B* is: $b \pm ts_b = -.5249 \pm 2.571(.0905) = -.758$ to $-.292$

b. $H_0: B = 0;$ $H_1: B < 0$

For 5 *df* and .025 area in the left tail of the *t* curve, $t = -2.571$.

The value of the test statistic is: $t = (b - B) / s_b = (-.5249) / .0905 = -5.8$

Reject H_0. *B* is negative.

13.55 Let: *x* = annual income,   *y* = amount of life insurance

From Exercises 13.27 and 13.42:

$n = 6$, $SS_{xx} = 2030.8333$, $s_e = 57.4132$, and $b = 7.4374$

$s_b = s_e / \sqrt{SS_{xx}} = 57.4132 / \sqrt{2030.833} = 1.2740$

a. $df = n - 2 = 6 - 2 = 4$

For 4 *df* and .005 area in the right tail of the *t* curve, $t = 4.604$

The 99% confidence interval for *B* is: $b \pm ts_b = 7.4374 \pm 4.604(1.2740) = 1.57$ to 13.30

b. $H_0: B = 0;$ $H_1: B \neq 0$

For 4 *df* and .005 area in each tail, the critical values of *t* are -4.604 and 4.604.

The value of the test statistic is: $t = (b - B) / s_b = (7.4374 - 0) / 1.2740 = 5.838$

Reject H_0. Conclude that B is different from zero.

13.56 Let: *x* = size of a house,   *y* = monthly rent

From Exercises 13.28 and 13.43:

$n = 6$, $SS_{xx} = 256.8333$, $s_e = 93.4611$, and $b = 51.8300$

$s_b = s_e / \sqrt{SS_{xx}} = 93.4611 / \sqrt{256.83333} = 5.8318$

a. $df = n - 2 = 6 - 2 = 4$

For 4 *df* and .01 area in the right tail of the *t* curve, $t = 3.747$

The 98% confidence interval for *B* is: $b \pm ts_b = 51.8300 \pm 3.747(5.8318) = 29.98$ to 73.68

b. $H_0: B = 0;$ $H_1: B \neq 0$

For 4 *df* and .025 area in each tail, the critical values of *t* are -2.776 and 2.776.

The value of the test statistic is: $t = (b - B) / s_b = (51.83 - 0) / 5.8318 = 8.887$

Reject H_0. Conclude that *B* is different from zero.

13.57   Let: x = hours worked, y = GPA

From the given data and the solution to Exercise 13.38:

$n = 7$, $SS_{xx} = 181.4286$, $b = -.1019$, $s_e = .3615$

a.   The regression line is: $\hat{y} = 4.4948 - .1019\,x$

b. $s_b = s_e / \sqrt{SS_{xx}}\ = .3615 / \sqrt{181.4286}\ = .0268$

     $df = n - 2 = 7 - 2 = 5$ and $\alpha/2 = .5 - (.95/2) = .025$

     For 5 df and .025 area in the right tail of the t distribution, $t = 2.571$.

     The 95% confidence interval for B is: $b \pm ts_b = -.1019 \pm 2.571(.0268) = -.171$ to $-.033$

c. H_0: $B = .04$; H_1: $B < -.04$.

     For 8 df and .05 area in the right tail of the t distribution, the critical value of t is -2.015

     The value of the test statistic is: $t = (b - B) / s_b = (-.1019 - (-.04)) / .0268 = -2.310$

     Reject H_0. Conclude that B is less than $-.04$.

13.58   From the solution to Exercise 13.39: $a = 95.0355$, $b = 1.7903$, $s_e = 9.1481$, $SS_{xx} = 1587.8750$

$s_b = s_e / \sqrt{SS_{xx}}\ = 9.1481 / \sqrt{1587.8750}\ = .2296$

a.   The regression line is: $\hat{y} = 95.0355 + 1.7903\,x$

b. $df = n - 2 = 8 - 2 = 6$

     Area in each tail of the t curve $= \alpha/2 = .5 - (.90/2) = .05$

     From the t distribution table, the value of t for $df = 6$ and .05 area in the right tail is 1.943.

     The 90% confidence interval for B is: $b \pm ts_b = 1.7903 \pm 1.943(.2296) = 1.34$ to 2.24

c. H_0: $B = 1.75$; H_1:$B \neq 1.75$

     Area in each tail of the t curve $= .05/2 = .025$

     $df = n - 2 = 8 - 2 = 6$

     The critical values of t are -2.447 and 2.447.

     The value of the test statistic is: $t = (b - B) / s_b = (1.7903 - 1.75) / .2296 = .176$

     Do not reject H_0. Hence, B is not different from 1.75.

13.59   The linear correlation coefficient measures the strength of the linear association between two variables. Its value always lies in the range -1 to 1.

13.60   While ρ is the correlation coefficient for an entire population, r is calculated from a sample.

13.61   a. Perfect positive linear correlation occurs when all the points in the scatter diagram lie on a straight line with positive slope. In this case, $r = 1$.

   b. Perfect negative linear correlation occurs when all the points in the scatter diagram lie on a straight line with negative slope. In this case, $r = -1$.

   c. If the correlation between two variables is positive and close to 1, they are said to have a strong positive correlation.

   d. If the correlation between two variables is negative and close to -1, they are said to have a strong negative correlation.

   e. If the correlation between two variables is positive and close to zero, they are said to have a weak positive correlation.

   f. If the correlation between two variables is negative and close to zero, they are said to have a weak negative correlation.

   g. If the data points are scattered all over the diagram (hence r is close to zero) there is no linear correlation between the variables.

13.62 B and ρ must have the same sign because both are obtained by dividing SS_{xy} by a positive quantity. Thus, both B and ρ have the same sign as SS_{xy}.

13.63   The answer is a, because r and b always have the same sign for a given sample.

13.64   The answer is b, because r and b always have the same sign for a given sample.

13.65   The linear correlation coefficient r measures only linear relationships. Thus, r may be zero and the variables might still have a nonlinear relationship.

13.66   a. We will expect a positive correlation between the grade of a student and the hours spent studying because, on average, an increase in the number of hours spent studying is expected to increase the grade of a student and a decrease in the number of hours spent studying is expected to decrease the grade of a student.

b. We will expect a positive correlation between the income and entertainment expenditure of a household because, on average, an increase in the income of a household is expected to increase the entertainment expenditure of that household and a decrease in the income of a household is expected to decrease the entertainment expenditure of that household.

c. We will expect a positive correlation between the age of a woman and the makeup expenses per month because, on average, with an increase in age a woman is expected to spend more on makeup.

d. The correlation between the price of a computer and the consumption of Coke is expected to be zero because these two variables are not related.

e. We will expect a negative correlation between the price and consumption of wine because, on average, an increase in the price of wine is expected to decrease its consumption (or demand) and a decrease in the price of wine is expected to increase its consumption.

13.67   a. We will expect a positive correlation between the SAT score and the GPA of a student because, on average, a student with a high SAT score is expected to have a high GPA.

b. We will expect a positive correlation between the stress level and blood pressure of a person because, on average, a person with a high stress level is expected to have high blood pressure.

c. We will expect a positive correlation between the amount of fertilizer used and the yield of corn per acre because, on average, an increase in the amount of fertilizer used will increase the yield of corn and a decrease in the amount of fertilizer used will decrease the yield of corn.

d. We will expect a negative correlation between the age and price of a house because, on average, as a house becomes older its price declines.

e. The correlation between the height of a husband and his wife's income is expected to be zero because these two variables are not related.

13.68 $SS_{xx} = 95,412.4000$, $SS_{yy} = 127,195.2560$, and $SS_{xy} = 27,538.8800$

$$\rho = \frac{SS_{xy}}{\sqrt{SS_{xx}SS_{yy}}} = \frac{27,538.8800}{\sqrt{(95,412.4000)(127,195.2560)}} = .25$$

13.69 $SS_{xx} = 15,124.7826$, $SS_{yy} = 24,080.6957$, and $SS_{xy} = 3987.3913$

$$\rho = \frac{SS_{xy}}{\sqrt{SS_{xx}SS_{yy}}} = \frac{3987.3913}{\sqrt{(15,124.7826)(24,080.6957)}} = .21$$

13.70   a. $SS_{xx} = 140$; $SS_{yy} = 20{,}432$; and $SS_{xy} = 1480$

$$r = \frac{SS_{xy}}{\sqrt{SS_{xx}SS_{yy}}} = \frac{1480}{\sqrt{(140)(20{,}432)}} = .88$$

b. $H_0: \rho = 0$; $H_1: \rho \neq 0$;

Area in each tail of the *t* curve = .02/2 = .01 and $df = n - 2 = 10 - 2 = 8$

The critical values of *t* are −2.896 and 2.896.

The value of the test statistic is: $t = r\sqrt{\dfrac{n-2}{1-r^2}} = .88\sqrt{\dfrac{10-2}{1-(.88)^2}} = 5.240$

Reject H_0. Conclude that ρ is different from zero.

13.71   a. $SS_{xx} = 33$; $SS_{yy} = 29{,}922$; and $SS_{xy} = -990$

$$r = \frac{SS_{xy}}{\sqrt{SS_{xx}SS_{yy}}} = \frac{-990}{\sqrt{(33)(29{,}922)}} = -.996$$

b. $H_0: \rho = 0$; $H_1: \rho < 0$

Area in the left tail of the *t* curve = .01;   and $df = n - 2 = 12 - 2 = 10$

The critical value of *t* is −2.764.

The value of the test statistic is: $t = r\sqrt{\dfrac{n-2}{1-r^2}} = -.996\sqrt{\dfrac{12-2}{1-(-.996)^2}} = -35.249$

Reject H_0. Hence ρ is negative.

13.72   a.   We expect the ages and prices of cars to be negatively related because, on average, the older a car
is, the less prospective buyers are willing to pay.

b.   From the solutions to Exercises 13.25 and 13.40:

$SS_{xx} = 43.5000$; $SS_{yy} = 14{,}108.8750$; and $SS_{xy} = -723.2500$

$$r = \frac{SS_{xy}}{\sqrt{SS_{xx}SS_{yy}}} = \frac{-723.2500}{\sqrt{(43.5000)(14{,}108.8750)}} = -.92$$

c. $H_0: \rho = 0$; $H_1: \rho < 0$

Area in the left tail of the *t* curve = .025   and $df = n - 2 = 8 - 2 = 6$

The critical value of *t* is −2.447.

The value of the test statistic is: $t = r\sqrt{\dfrac{n-2}{1-r^2}} = -.92\sqrt{\dfrac{8-2}{1-(-.92)^2}} = -5.750$

Reject H_0. Hence ρ is negative.

13.73   Let: x = years of experience, y = monthly salary

a.   We expect experience and monthly salaries to be positively related because, on average, more experienced secretaries command higher salaries.

b.   From the solution to Exercise 13.53: SS_{xx} = 256.8889, SS_{yy} = 474.0000, and SS_{xy} = 335.3333

$$r = \frac{SS_{xy}}{\sqrt{SS_{xx}SS_{yy}}} = \frac{335.3333}{\sqrt{(256.8889)(474.0000)}} = .96$$

c. $H_0: \rho = 0;$ $H_1: \rho > 0$

Area in the right tail of the t curve = .05 and $df = n - 2 = 9 - 2 = 7$
The critical value of t is 1.895.

The value of the test statistic is: $t = r\sqrt{\dfrac{n-2}{1-r^2}} = .96\sqrt{\dfrac{9-2}{1-(.96)^2}} = 9.071$

Reject H_0. Hence, ρ is positive.

13.74   a.   We expect the midterm scores and final examination scores to be positively correlated because, on average, a student with a high midterm score will also have a high final examination score.

b.   Let: x = midterm score, y = final exam score

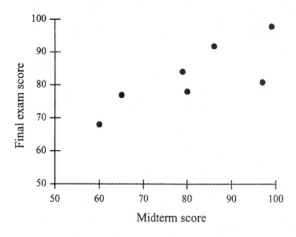

We expect the correlation coefficient to be close to 1 because the points in the scatter diagram show a very strong positive correlation.

c. $n = 7, \Sigma x = 561, \Sigma y = 581, \Sigma x^2 = 46,069, \Sigma y^2 = 48,875, \Sigma xy = 47,291,$

$SS_{xx} = 1108.8571, SS_{yy} = 652.0000, SS_{xy} = 728.0000$

$$r = \frac{SS_{xy}}{\sqrt{SS_{xx}SS_{yy}}} = \frac{728.0000}{\sqrt{(1108.8571)(652.0000)}} = .86$$

This value of r is consistent with what we expected in parts a and b.

    d. H_0: $\rho = 0$; H_1: $\rho > 0$

        Area in the right tail of the t curve = .01; and $df = n - 2 = 7 - 2 = 5$

        The critical value of t is 3.365.

        The value of the test statistic is: $t = r\sqrt{\dfrac{n-2}{1-r^2}} = .86\sqrt{\dfrac{7-2}{1-(.86)^2}} = 3.768$

        Reject H_0. Hence ρ is positive.

13.75   a. We expect the ages of husbands and wives to be positively correlated because, on average, a younger husband will have a younger wife and an older husband will have an older wife.

     b. Let: x = husband's age, y = wife's age

        We expect the correlation coefficient to be close to 1 because the points in the scatter diagram show a very strong positive correlation.

     c. $n = 6$, $\Sigma x = 221$, $\Sigma y = 211$, $\Sigma x^2 = 8989$, $\Sigma y^2 = 7927$, $\Sigma xy = 8411$,

        $SS_{xx} = 848.8333$, $SS_{yy} = 506.8333$, $SS_{xy} = 639.1667$

        $r = \dfrac{SS_{xy}}{\sqrt{SS_{xx}SS_{yy}}} = \dfrac{639.1667}{\sqrt{(848.8333)(506.8333)}} = .97$

        This value of r is consistent with what we expected in parts a and b.

     d. H_0: $\rho = 0$; H_1: $\rho \neq 0$

        Area in each tail of the t curve = .05/2 = .025   and $df = n - 2 = 6 - 2 = 4$

        The critical values of t are –2.776 and 2.776.

        The value of the test statistic is: $t = r\sqrt{\dfrac{n-2}{1-r^2}} = .97\sqrt{\dfrac{6-2}{1-(.97)^2}} = 7.980$

        Reject H_0. Hence the correlation coefficient is different from zero.

13.76   a.   Let: x = lowest temperature; y = number of calls

$n = 7$, $\Sigma x = 104$, $\Sigma y = 118$, $\Sigma x^2 = 3178$, $\Sigma y^2 = 2506$, $\Sigma xy = 896$,

$SS_{xx} = 1632.8571$, $SS_{yy} = 516.8571$, $SS_{xy} = -857.1429$

$$r = \frac{SS_{xy}}{\sqrt{SS_{xx}SS_{yy}}} = \frac{-857.1429}{\sqrt{(1632.8571)(516.8571)}} = -.93$$

The sign of b calculated in Exercise 13.26 is also negative.

   b. H_0: $\rho = 0$; H_1: $\rho < 0$

Area in the left tail of the t curve = .025; and $df = n - 2 = 7 - 2 = 5$

The critical value of t is -2.571.

The value of the test statistic is: $t = r\sqrt{\dfrac{n-2}{1-r^2}} = -.93\sqrt{\dfrac{7-2}{1-(-.93)^2}} = -5.658$

Reject H_0. Conclude that the linear correlation coefficient is negative. Yes, the decision is the same as in the test of B in Exercise 13.54, "Reject H_0".

13.77   Let: x = fat consumption (in grams) per day

       y = cholesterol level (in milligrams per hundred milliliters)

   a.   From the solutions to Exercises 13.39 and 13.58:

$n = 8$, $\Sigma x = 421$, $\Sigma y = 1514$, $\Sigma x^2 = 23,743$, $\Sigma y^2 = 292,116$, $\Sigma xy = 82,517$,

$SS_{xx} = 1587.8750$, $SS_{yy} = 5591.5000$, $SS_{xy} = 2842.7500$

$$r = \frac{SS_{xy}}{\sqrt{SS_{xx}SS_{yy}}} = \frac{2842.7500}{\sqrt{(1587.8750)(5591.5000)}} = .95$$

The sign of b calculated in Exercise 13.58 is also positive.

   b. H_0: $\rho = 0$; H_1: $\rho \neq 0$;

Area in each tail of the t curve = .01/2 = .005;   and $df = n - 2 = 8 - 2 = 6$

The critical values of t are -3.707 and 3.707

The value of the test statistic is: $t = r\sqrt{\dfrac{n-2}{1-r^2}} = .95\sqrt{\dfrac{8-2}{1-(.95)^2}} = 7.452$

Reject H_0. Conclude that ρ is different from zero.

13.78   Let: x = total payroll (in millions of dollars), y = percentage of games won

From the solutions to Exercises 13.29 and 13.44:

$SS_{xx} = 7461$, $SS_{yy} = 978.2944$, and $SS_{xy} = 1635.625$

$$\rho = \frac{SS_{xy}}{\sqrt{SS_{xx}SS_{yy}}} = \frac{1635.625}{\sqrt{(7461)(978.2944)}} = .61$$

13.79   Let: x = total payroll (in millions of dollars) y = percentage of games won

From the solutions to Exercises 13.30 and 13.45:

SS_{xx} = 10,283.2143, SS_{yy} = 1447.6086; and SS_{xy} = 1296.5429

$$\rho = \frac{SS_{xy}}{\sqrt{SS_{xx}SS_{yy}}} = \frac{1296.5429}{\sqrt{(10,283.2143)(1447.6086)}} = .34$$

13.80   a.   The pairs of gloves produced depend on temperature. When employees are more comfortable they work harder and produce more gloves. Hence, the relationship between the two variables is expected to be negative.

b.   Let: x = temperature, y = pairs of gloves produced

$n = 8$, $\Sigma x = 598$, $\Sigma y = 283$, $\Sigma x^2 = 44,824$, $\Sigma y^2 = 10,049$ $\Sigma xy = 21,091$, $\bar{x} = 74.75$, $\bar{y} = 35.375$

SS_{xx} = 123.500, SS_{yy} = 37.875, SS_{xy} = –63.250

c. $b = SS_{xy}/SS_{xx} = -63.250/123.500 = -.5121$

$a = \bar{y} - b\bar{x} = 35.375 - (-.5121)(74.75) = 73.6545$

The regression line is: $\hat{y} = 73.6545 + -.5122\,x$

d.   The value of $a = 73.6545$ is the value of y for $x = 0$. In this exercise it represents the number of pairs of gloves when the temperature is zero; the value makes no sense. This is because the temperatures in the sample range from 68 to 81, but 0 is far outside this range.

The value of $b = -.5121$ means that, on average, the pairs of gloves produced increase by .5121 for every 1 degree drop in temperature.

e.

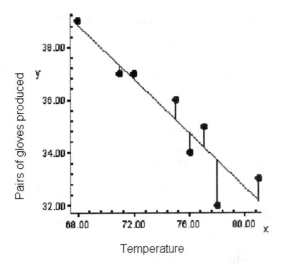

Temperature

f. $r = \dfrac{SS_{xy}}{\sqrt{SS_{xx}\,SS_{yy}}} = \dfrac{-63.250}{\sqrt{(123.500)(37.875)}} = -.92$

$r^2 = bSS_{xy}/SS_{yy} = (-.5121)(-63.250)/37.875 = .86$

The value of $r = -.92$ indicates that the two variables have a strongly negative correlation. The value of $r^2 = .86$ means that 86% of the total squared errors (SST) are explained by our regression model.

g. $s_e = \sqrt{\dfrac{SS_{yy} - b(SS_{xy})}{n-2}} = \sqrt{\dfrac{37.875 - (-.5121)(-63.25)}{8-2}} = .9561$

h. For $x = 74$: $y = 73.6545 - .5121(74) = 35.7591$

Thus, when the temperature is set to 74 degrees, approximately 36 pairs of gloves are made.

i. $s_b = s_e / \sqrt{SS_{xx}} = .9561 / \sqrt{123.500} = .0860$

$df = n - 2 = 8 - 2 = 6$ and   Area in each tail of the t curve $= \alpha/2 = .5 - (.99/2) = .005$
From the t distribution table, the value of t for $df = 6$ and .005 area in the right tail is 3.707.
The 99% confidence interval for B is: $b \pm t s_b = -.5121 \pm 3.707(.0860) = -.83$ to $-.19$

j. H_0: $B = 0$; H_1: $B < 0$

Area in the left tail of the t curve $= .05$;   and $df = n - 2 = 8 - 2 = 6$
The critical value of t is -1.943.

$t = (b - B) / s_b = (-.5122 - 0) / .0860 = -5.956$

Reject the null hypothesis. Hence, B is negative.

k. H_0: $\rho = 0$; H_1: $\rho < 0$

Area in the right tail of the t curve $= .01$;   and $df = n - 2 = 8 - 2 = 6$
The critical value of t is -3.143.

The value of the test statistic is: $t = r\sqrt{\dfrac{n-2}{1-r^2}} = -.92\sqrt{\dfrac{8-2}{1-(.92)^2}} = -5.750$

Reject H_0. Hence, ρ is negative.

13.81   a.   Let: x = age of man, y = cholesterol level

$n = 10$, $\Sigma x = 512$, $\Sigma y = 1896$, $\Sigma x^2 = 28{,}110$, $\Sigma y^2 = 364{,}280$ $\Sigma xy = 98{,}307$, $\bar{x} = 51.20$, $\bar{y} = 189.60$,

$SS_{xx} = 1895.6000$, $SS_{yy} = 4798.4000$, $SS_{xy} = 1231.8000$

b. $b = SS_{xy} / SS_{xx} = 1231.8000/1895.6000 = .6498$

$a = \bar{y} - b\bar{x} = 189.60 - (.6498)(51.20) = 156.3302$

The regression line is: $\hat{y} = 156.3302 + .6498x$

c. The value of $a = 156.3302$ is the value of y for $x = 0$. In this exercise it represents the cholesterol level of a man with an age of zero years.

The value of $b = .6498$ means that, on average, the cholesterol level of a man increases by .6498 for every 1–year increase in age.

d. $r = \dfrac{SS_{xy}}{\sqrt{SS_{xx}SS_{yy}}} = \dfrac{1231.8000}{\sqrt{(1895.6000)(4798.4000)}} = .41$

$r^2 = b \, SS_{xy} / SS_{yy} = (.6498)(1231.8000)/4798.4000 = .17$

The value of $r = .41$ indicates that the two variables have a positive correlation but they are not strongly related. The value of $r^2 = .17$ means that only 17% of the total squared errors (SST) are explained by our regression model.

e.

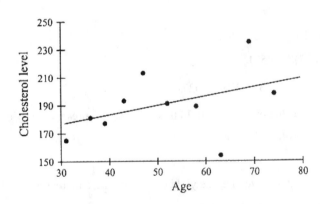

f. For $x = 60$: $\hat{y} = 156.3302 + .6498(60) = 195.3182$

Thus, a 60 year old man is expected to have a cholesterol level of about 195.

g. $s_e = \sqrt{\dfrac{SS_{yy} - b(SS_{xy})}{n-2}} = \sqrt{\dfrac{4798.4000 - (.6498)(1231.8000)}{10-2}} = 22.3550$

h. $s_b = s_e / \sqrt{SS_{xx}} = 22.3550 / \sqrt{1895.6000} = .5135$

$df = n - 2 = 10 - 2 = 8$ and   Area in each tail of the t curve $= \alpha/2 = .5 - (.95/2) = .025$

From the t distribution table, the value of t for $df = 8$ and .025 area in the right tail is 2.306.

The 95% confidence interval for B is: $b \pm ts_b = .6498 \pm 2.306(.5135) = -.53$ to 1.83

i. $H_0: B = 0;$ $H_1: B > 0$

Area in the right tail of the t curve = .05;   and $df = n - 2 = 10 - 2 = 8$

The critical value of t is 1.860.

$t = (b - B) / s_b = (.6498 - 0) / .5135 = 1.265$

Do not reject the null hypothesis. Hence, B is not positive.

j. $H_0: \rho = 0;$ $H_1: \rho > 0$

Area in the right tail of the t curve = .025;   and $df = n - 2 = 10 - 2 = 8$

The critical value of t is 2.306.

The value of the test statistic is: $t = r\sqrt{\dfrac{n-2}{1-r^2}} = .41\sqrt{\dfrac{10-2}{1-(.41)^2}} = 1.271$

Do not reject H_0. Hence, do not conclude that ρ is positive.

13.82 a. Let: x = amount of fertilizer used (in pounds)   and y = yield of corn (in bushels)

$n = 7, \Sigma x = 643, \Sigma y = 841, \Sigma x^2 = 61,169, \Sigma y^2 = 102,821, \Sigma xy = 79,152, \bar{x} = 91.8571,$

$\bar{y} = 120.1429, \ SS_{xx} = 2104.8571, SS_{yy} = 1780.8571, SS_{xy} = 1900.1429$

b. $b = SS_{xy} / SS_{xx} = 1900.1429 / 2104.8571 = .9027$

$a = \bar{y} - b\bar{x} = 120.1429 - (.9027)(91.8571) = 37.2235$

The regression line is: $\hat{y} = 37.2235 + .9027x$

c. The value of $a = 37.2235$ is the value of y for $x = 0$. In this exercise it represents the yield of corn (in bushels) per acre when no fertilizer is used.

The value of $b = .9027$ means that, on average, the yield of corn per acre increases by .9027 bushels for every 1 pound increase in fertilizer used.

d. $r = \dfrac{SS_{xy}}{\sqrt{SS_{xx}SS_{yy}}} = \dfrac{1900.1429}{\sqrt{(2104.8571)(1780.8571)}} = .98$

$r^2 = b\, SS_{xy} / SS_{yy} = (.9027)(1900.1429)/1780.8571 = .96$

The value of $r = .98$ indicates that the two variables have a very strong positive correlation. The value of $r^2 = .96$ means that 96% of the total squared errors (SST) are explained by our regression model.

e. $s_e = \sqrt{\dfrac{SS_{yy} - b(SS_{xy})}{n-2}} = \sqrt{\dfrac{1780.8571 - (.9027)(1900.1429)}{7-2}} = 3.6221$

f. For $x = 105$: $\hat{y} = 37.2235 + .9027 (105) = 132.0070$

Thus, if 105 pounds of fertilizer is used on an acre of land, the yield of corn on that acre is expected to be about 132 bushels.

g. $s_b = s_e / \sqrt{SS_{xx}} = 3.6221 / \sqrt{2104.8571} = .0798$

$df = n - 2 = 7 - 2 = 5$ and   Area in each tail of the t curve $= \alpha / 2 = .5 - (.98/2) = .01$

From the t distribution table, the value of t for $df = 5$ and .01 area in the right tail is 3.365.

The 98% confidence interval for B is: $b \pm t s_b = .9027 \pm 3.365(.0798) = .63$ to 1.17

h. $H_0: B = 0$; $H_1: B \neq 0$

$df = n - 2 = 7 - 2 = 5$, $\alpha / 2 = .05/2 = .025$

The critical values of t are -2.571 and 2.571.

$t = (b - B) / s_b = (.9027 - 0) / .0798 = 11.312$

Reject the null hypothesis. Hence, B is different from zero.

i. $H_0: \rho = 0$; $H_1: \rho \neq 0$

Area in the each tail of the t curve $= .05 / 2 = .025$ and $df = n - 2 = 7 - 2 = 5$

The critical values of t are -2.571 and 2.571.

The value of the test statistic is: $t = r \sqrt{\dfrac{n-2}{1-r^2}} = .98 \sqrt{\dfrac{7-2}{1-(.98)^2}} = 11.012$

Reject H_0. Conclude that ρ is different from zero.

13.83 Let: $x =$ income   and $y =$ charitable contributions

a. $n = 10$, $\Sigma x = 641$, $\Sigma y = 141$, $\Sigma x^2 = 45,349$, $\Sigma y^2 = 2927$, $\Sigma xy = 10934$, $\bar{x} = 64.10$, $\bar{y} = 14.10$,

$SS_{xx} = 4260.9000$, $SS_{yy} = 938.9000$, $SS_{xy} = 1895.9000$

b. $b = SS_{xy} / SS_{xx} = 1895.9000 / 4260.9000 = .4450$

$a = \bar{y} - b\bar{x} = 14.10 - (.4450)(64.10) = -14.4245$

The least squares regression line is: $\hat{y} = -14.4245 + .4450x$

c. The value of $a = -14.4245$ is the value of y for $x = 0$. Although $a = -14.4245$ represents the charitable contributions of a household with no income, the negative value makes no sense. This is because incomes in the sample varied from \$36,000 to \$102,000, but 0 is far outside that range. The value of $b = .4450$ means that, on average, charitable contributions increase by \$44.50 for every \$1000 increase in a household's income.

d. $r = \dfrac{SS_{xy}}{\sqrt{SS_{xx}SS_{yy}}} = \dfrac{1895.9000}{\sqrt{(4260.9000)(938.9000)}} = .95$

$r^2 = b\, SS_{xy} / SS_{yy} = (.4450)(1895.9000)/938.9000 = .90$

The value of $r = .95$ indicates that the two variables have a very strong positive linear correlation. The value of $r^2 = .90$ means that 90% of the total squared errors (SST) are explained by the regression model.

e. $s_e = \sqrt{\dfrac{SS_{yy} - b(SS_{xy})}{n-2}} = \sqrt{\dfrac{938.9000 - (.4950)(1895.9000)}{10-2}} = 3.4501$

f. $s_b = s_e / \sqrt{SS_{xx}} = 3.4501 / \sqrt{4260.9000} = .0529$

For $df = 8$ and .005 area in the right tail of the t curve, $t = 3.355$.

The 99% confidence interval for B is: $b \pm t s_b = .4450 \pm 3.355(.0529) = .27$ to $.62$

g. $H_0: B = 0;$ $\qquad\qquad$ $H_1: B > 0$

For $df = 8$ and .01 area in the right tail of the t curve, $t = 2.896$.

The value of the test statistic is: $t = (b - B) / s_b = (.4450 - 0) / .0529 = 8.412$

Reject the null hypothesis. Hence, B is positive.

h. $H_0: \rho = 0;$ $\qquad\qquad$ $H_1: \rho \neq 0$

Area in the each tail of the t curve $= .01 / 2 = .005$ $\qquad$ and $\qquad$ $df = n - 2 = 10 - 2 = 8$

The critical values of t are -3.355 and 3.355.

The value of the test statistic is: $\qquad\qquad t = r\sqrt{\dfrac{n-2}{1-r^2}} = .95\sqrt{\dfrac{10-2}{1-(.95)^2}} = 8.605$

Reject H_0. Conclude that the correlation coefficient is different from zero.

13.84 a. Let: $x =$ ticket price (in dollars), $\qquad$ and $\qquad$ $y =$ average attendance (in thousands)

$n = 6, \Sigma x = 170, \Sigma y = 358, \Sigma x^2 = 5157, \Sigma y^2 = 21952, \Sigma xy = 10{,}086.5, \bar{x} = 28.3333, \bar{y} = 59.6667,$

$SS_{xx} = 340.3333, SS_{yy} = 591.3333, SS_{xy} = -56.8333$

b. $b = SS_{xy} / SS_{xx} = -56.8333 / 340.3333 = -.1670x$

$a = \bar{y} - b\bar{x} = 59.6667 - (-.1670)(28.3333) = 64.3984$

The least squares regression line is: $\qquad\qquad\qquad \hat{y} = 64.3984 - .1670\, x$

c. The value of $a = 64.3984$ is the value of y for $x = 0$. In this exercise it represents the average attendance (64,398) if the ticket price is zero. The value of $b = -.1670$ means that, on average, the attendance will decrease by 167 for every \$1 increase in ticket price. Note that the units of y are in thousands.

d. $r = \dfrac{SS_{xy}}{\sqrt{SS_{xx}SS_{yy}}} = \dfrac{-56.8333}{\sqrt{(340.3333)(591.3333)}} = -.13$

$r^2 = b\, SS_{xy} / SS_{yy} = (-.1670)(-56.8333) / 591.3333 = .016$

The value of $r = -.13$ indicates that the two variables have a weak negative correlation. The value of $r^2 = .016$ means that 1.6% of the total squared errors (SST) are explained by our regression model.

e. $s_e = \sqrt{\dfrac{SS_{yy} - b(SS_{xy})}{n-2}} = \sqrt{\dfrac{591.3333 - (-.1670)(-56.8333)}{6-2}} = 12.0607$

f. $s_b = s_e / \sqrt{SS_{xx}} = 12.0607 / \sqrt{340.3333} = .6538$

$df = n - 2 = 6 - 2 = 4$ and   Area in each tail of the t curve $= \alpha/2 = .5 - (.90/2) = .05$.
From the t distribution table, the value of t for $df = 4$ and .05 area in the right tail is 2.132.
The 90% confidence interval for B is:

$\qquad b \pm t s_b = -.1670 \pm 2.132(.6538) = -.1670 \pm 1.39 = -1.56$ to 1.22

g. $H_0: B = 0;$ $\qquad\qquad H_1: B < 0$

$df = 8$ $\qquad$ and $\qquad$ .025 area in the right tail of the t curve, $t = -2.776$.
The value of the test statistic is: $t = (b - B) / s_b = (-.1670 - 0) / .6538 = .255$
Do not reject the null hypothesis. Hence, B is not negative.

h. $H_0: \rho = 0;$ $\qquad\qquad H_1 : \rho < 0$

Area in the left tail of the t curve $= .05/2 = .025;$ $\quad$ and $\quad$ $df = n - 2 = 6 - 2 = 4$
The critical value of t is -2.776.

The value of the test statistic is: $t = r\sqrt{\dfrac{n-2}{1-r^2}} = -.13\sqrt{\dfrac{6-2}{1-(-.13)^2}} = -.262$

Do not reject H_0. Conclude that the correlation coefficient is not negative.

13.85 a. Let: $x = $ GPA(grade point average), $\qquad y = $ starting salary (in thousands of dollars)

$n = 7$, $\Sigma x = 20.57$, $\Sigma y = 277.00$, $\Sigma x^2 = 63.8111$, $\Sigma y^2 = 11{,}247$, $\Sigma xy = 843.18$, $\bar{x} = 2.9386$,

$\bar{y} = 39.5714$, $SS_{xx} = 3.3647$, $SS_{yy} = 285.7143$, $SS_{xy} = 29.1957$

b. $b = SS_{xy}/SS_{xx} = 29.1957/3.3647 = 8.6771$

    $a = \bar{y} - b\bar{x} = 39.5714 - (8.6771)(2.9386) = 14.0729$

    The regression line is: $\hat{y} = 14.0729 + 8.6771x$

c.   The value of $a = 14.0729$ is the value of y for $x = 0$. In this exercise, it represents the starting salary (about \$14,073) for a college graduate with a GPA of zero. The value of $b = 8.6771$ means that, on average, the starting salary of a college graduate increases by \$8677 for every 1–point increase in GPA.

d. $r = \dfrac{SS_{xy}}{\sqrt{SS_{xx} SS_{yy}}} = \dfrac{29.1957}{\sqrt{(3.3647)(285.7143)}} = .94$

    $r^2 = b\,SS_{xy}/SS_{yy} = (8.6771)(29.1957)/285.7143 = .89$

The value of $r = .94$ indicates that the two variables have a very strong positive linear correlation. The value of $r^2 = .89$ means that 89% of the total squared errors (SST) are explained by the regression model.

e. $s_e = \sqrt{\dfrac{SS_{yy} - b(SS_{xy})}{n-2}} = \sqrt{\dfrac{285.7143 - (8.6771)(29.1957)}{7-2}} = 2.5448$

f. $s_b = s_e / \sqrt{SS_{xx}} = 2.5448 / \sqrt{3.3647} = 1.3873$

    $df = n - 2 = 7 - 2 = 5$ and   Area in each tail of the t curve $= \alpha/2 = .5 - (.95/2) = .025$.

    From the t distribution table, the value of t for $df = 5$ and .025 area in the right tail is 2.571.

    The 95% confidence interval for B is: $b \pm ts_b = 8.6771 \pm 2.571(1.3873) = 5.11$ to 12.24

g. $H_0: B = 0$; $H_1: B \neq 0$

    $df = n - 2 = 7 - 2 = 5$ and $\alpha/2 = .005$.

    The critical values of t are -4.032 and 4.032.

    The value of the test statistic is: $t = (b - B)/s_b = (8.6771 - 0)/1.3873 = 6.255$

    Reject the null hypothesis. Hence, B is different from zero.

h. $H_0: \rho = 0$; $H_1: \rho > 0$

    Area in the right tail of the t curve $= .01$ and $df = n - 2 = 7 - 2 = 5$

The critical value of t is 3.365.

The value of the test statistic is: $t = r\sqrt{\dfrac{n-2}{1-r^2}} = .94\sqrt{\dfrac{7-2}{1-(.94)^2}} = 6.161$

Reject H_0. Conclude that ρ is positive.

13.86   When we estimate $\mu_{y|x}$ for a given value of x, it is called estimating the mean value of y. For example, if we estimate the mean score in statistics for all students who spend exactly 5 hours studying statistics per week, we will be estimating $\mu_{y|x}$ for $x = 5$.

On the other hand, when we estimate y for a single element for a given value of x, it is called predicting the value of y, which is denoted by y_p. For example, if we estimate the statistics score for a given student who spends exactly 5 hours studying statistics per week, we will be predicting y_p for $x = 5$.

13.87   a.   For $x = 15 : \hat{y} = 3.25 + .80(15) = 15.25$

   $df = n - 2 = 10 - 2 = 8$ and   Area in each tail of the t curve $= \alpha/2 = 5 - (.99/2) = .005$

   From the t distribution table, the value of t for $df = 8$ and .005 area in the right tail is 3.355.

   The standard deviation of $\hat{y}$ for estimating the mean value of y for $x = 15$ is:

   $$s_{\hat{y}_m} = s_e \sqrt{\frac{1}{n} + \frac{(x_0 - \bar{x})^2}{SS_{xx}}} = (.954)\sqrt{\frac{1}{10} + \frac{(15 - 18.52)^2}{144.65}} = .4111$$

   The 99% confidence interval for $\mu_{y|15}$ is: $\hat{y} \pm ts_{\hat{y}_m} = 15.25 \pm 3.355(.4111) = 13.8708$ to 16.6292

   The standard deviation of $\hat{y}$ for predicting y for $x = 15$ is:

   $$s_{\hat{y}_p} = s_e \sqrt{1 + \frac{1}{n} + \frac{(x_0 - \bar{x})^2}{SS_{xx}}} = (.954)\sqrt{1 + \frac{1}{10} + \frac{(15 - 18.52)^2}{144.65}} = 1.0388$$

   The 99% prediction interval for y_p for $x = 15$ is:

   $\hat{y} \pm ts_{\hat{y}_p} = 15.25 \pm 3.355(1.0388) = 11.7648$ to 18.7352

   b.   For $x = 12 : \hat{y} = -27 + 7.67(12) = 65.04$

   $df = n - 2 = 10 - 2 = 8$ and   Area in each tail of the t curve $= \alpha/2 = .5 - (.99/2) = .005$

   From the t distribution table, the value of t for $df = 8$ and .005 area in the right tail is 3.355.

   The standard deviation of $\hat{y}$ for estimating the mean value of y for $x = 12$ is:

   $$s_{\hat{y}_m} = s_e \sqrt{\frac{1}{n} + \frac{(x_0 - \bar{x})^2}{SS_{xx}}} = (2.46)\sqrt{\frac{1}{10} + \frac{(12 - 13.43)^2}{369.77}} = .7991$$

The 99% confidence interval for $\mu_{y|12}$ is: $\hat{y} \pm ts_{\hat{y}_m} = 65.04 \pm 3.355(.7991) = 62.3590$ to 67.7210

The standard deviation of $\hat{y}$ for predicting y for $x = 12$ is:

$$s_{\hat{y}_p} = s_e \sqrt{1 + \frac{1}{n} + \frac{(x_0 - \bar{x})^2}{SS_{xx}}} = (2.46)\sqrt{1 + \frac{1}{10} + \frac{(12 - 13.43)^2}{369.77}} = 2.5865$$

The 99% prediction interval for y_p for $x = 12$ is:

$$\hat{y} \pm ts_{\hat{y}_p} = 65.04 \pm 3.355(2.5865) = 56.3623 \text{ to } 73.7177$$

13.88 a. For $x = 8: \hat{y} = 13.40 + 2.58(8) = 34.04$

$df = n - 2 = 12 - 2 = 10$ and   Area in each tail of the t curve $= \alpha/2 = .5 - (.95/2) = .025$

From the t distribution table, the value of t for $df = 10$ and $.025$ area in the right tail is 2.228.

The standard deviation of $\hat{y}$ for estimating the mean value of y for $x = 8$ is:

$$s_{\hat{y}_m} = s_e \sqrt{\frac{1}{n} + \frac{(x_0 - \bar{x})^2}{SS_{xx}}} = (1.29)\sqrt{\frac{1}{12} + \frac{(8 - 11.30)^2}{210.45}} = .4741$$

The 95% confidence interval for $\mu_{y|8}$ is: $\hat{y} \pm ts_{\hat{y}_m} = 34.04 \pm 2.228(.4741) = 32.9837$ to 35.0963

The standard deviation of $\hat{y}$ for predicting y for $x = 8$ is:

$$s_{\hat{y}_p} = s_e \sqrt{1 + \frac{1}{n} + \frac{(x_0 - \bar{x})^2}{SS_{xx}}} = (1.29)\sqrt{1 + \frac{1}{12} + \frac{(8 - 11.30)^2}{210.45}} = 1.3744$$

The 95% prediction interval for y_p for $x = 8$ is:

$$\hat{y} \pm ts_{\hat{y}_p} = 34.04 \pm 2.228(1.3744) = 30.9778 \text{ to } 37.1022$$

b. For $x = 24: \hat{y} = -8.6 + 3.72(24) = 80.68$

$df = n - 2 = 10 - 2 = 8$ and   Area in each tail of the t curve $= \alpha/2 = .5 - (.95/2) = .025$

From the t distribution table, the value of t for $df = 8$ and $.025$ area in the right tail is 2.306.

The standard deviation of $\hat{y}$ for estimating the mean value of y for $x = 24$ is:

$$s_{\hat{y}_m} = s_e \sqrt{\frac{1}{n} + \frac{(x_0 - \bar{x})^2}{SS_{xx}}} = (1.89)\sqrt{\frac{1}{10} + \frac{(24 - 19.70)^2}{315.40}} = .7527$$

The 95% confidence interval for $\mu_{y|24}$ is: $\hat{y} \pm ts_{\hat{y}_m} = 80.68 \pm 2.306(.7527) = 78.9443$ to 82.4157

The standard deviation of $\hat{y}$ for predicting y for $x = 24$ is:

$$s_{\hat{y}_p} = s_e \sqrt{1 + \frac{1}{n} + \frac{(x_0 - \bar{x})^2}{SS_{xx}}} = (1.89)\sqrt{1 + \frac{1}{10} + \frac{(24 - 19.70)^2}{315.40}} = 2.0344$$

The 95% prediction interval for y_p for $x = 24$ is:

$$\hat{y} \pm t s_{\hat{y}_p} = 80.68 \pm 2.306(2.0344) = 75.9887 \text{ to } 85.3713$$

13.89 From the solution to Exercise 13.53:

$n = 9, \; \bar{x} = 8.8889, \; SS_{xx} = 256.8889, \; s_e = 2.2758$

The regression line is: $\hat{y} = 23.7297 + 1.3054x$

For $x = 10$: $\hat{y} = 23.7297 + 1.3054(10) = 36.7837$

$df = n - 2 = 9 - 2 = 7$ and Area in each tail of the t curve $= \alpha/2 = .5 - (.90/2) = .05$

From the t distribution table, the value of t for $df = 7$ and .05 area in the right tail is 1.895.

The standard deviation of $\hat{y}$ for estimating the mean value of y for $x = 10$ is:

$$s_{\hat{y}_m} = s_e \sqrt{\frac{1}{n} + \frac{(x_0 - \bar{x})^2}{SS_{xx}}} = (2.2758)\sqrt{\frac{1}{9} + \frac{(10 - 8.8889)^2}{256.8889}} = .7748$$

The 90% confidence interval for $\mu_{y|10}$ is:

$$\hat{y} \pm t s_{\hat{y}_m} = 36.7837 \pm 1.895(.7748) = 36.7837 \pm 1.4682 = 35.3155 \text{ to } 38.2519$$

The standard deviation of $\hat{y}$ for predicting y for $x = 10$ is:

$$s_{\hat{y}_p} = s_e \sqrt{1 + \frac{1}{n} + \frac{(x_0 - \bar{x})^2}{SS_{xx}}} = (2.2758)\sqrt{1 + \frac{1}{9} + \frac{(10 - 8.8889)^2}{256.8889}} = 2.4041$$

The 90% prediction interval for y_p for $x = 10$ is:

$$\hat{y} \pm t s_{\hat{y}_p} = 36.7837 \pm 1.895(2.4041) = 36.7837 \pm 4.5558 = 32.2279 \text{ to } 41.3395$$

13.90 From the solution to Exercise 13.80: $n = 8, \; \bar{x} = 74.75, \; SS_{xx} = 123.5000, \; s_e = .9561$

The regression line is: $\hat{y} = 73.6545 - .5122x$

For $x = 77$: $\hat{y} = 73.6545 - .5122(77) = 34.2228$

$df = n - 2 = 8 - 2 = 6$ and Area in each tail of the t curve $= \alpha/2 = .5 - (.99/2) = .005$

From the t distribution table, the value of t for $df = 6$ and .005 area in the right tail is 3.707.

The standard deviation of $\hat{y}$ for estimating the mean value of y for $x = 77$ is:

$$s_{\hat{y}_m} = s_e \sqrt{\frac{1}{n} + \frac{(x_0 - \bar{x})^2}{SS_{xx}}} = (.9561)\sqrt{\frac{1}{8} + \frac{(77 - 74.75)^2}{123.5000}} = .3895$$

The 99% confidence interval for $\mu_{y|77}$ is:

$$\hat{y} \pm t s_{\hat{y}_m} = 34.2228 \pm 3.707(.3895) = 34.2228 \pm 1.4439 = 32.7789 \text{ to } 35.6667$$

The standard deviation of $\hat{y}$ for predicting y for $x = 77$ is:

$$s_{\hat{y}_p} = s_e \sqrt{1 + \frac{1}{n} + \frac{(x_0 - \bar{x})^2}{SS_{xx}}} = (.9561)\sqrt{1 + \frac{1}{8} + \frac{(77 - 74.75)^2}{123.5000}} = 1.0324$$

The 99% prediction interval for y_p for $x = 77$ is:

$$\hat{y} \pm t s_{\hat{y}_p} = 34.2228 \pm 3.707(1.0324) = 34.2228 \pm 3.8271 = 30.3957 \text{ to } 38.0499$$

13.91 From the solution to Exercise 13.82: $\quad n = 7,\ \bar{x} = 91.8571,\ SS_{xx} = 2104.8571,\ s_e = 3.6221$

The regression line is: $\hat{y} = 37.2235 + .9027x$

For $x = 90$: $\hat{y} = 37.2235 + .9027(90) = 118.4665$

$df = n - 2 = 7 - 2 = 5 \quad$ and $\quad$ Area in each tail of the t curve $= \alpha/2 = .5 - (.99/2) = .005$

From the t distribution table, the value of t for $df = 5$ and .005 area in the right tail is 4.032.

The standard deviation of $\hat{y}$ for estimating the mean value of y for $x = 90$ is:

$$s_{\hat{y}_m} = s_e \sqrt{\frac{1}{n} + \frac{(x_0 - \bar{x})^2}{SS_{xx}}} = (3.6221)\sqrt{\frac{1}{7} + \frac{(90 - 91.8571)^2}{2104.8571}} = 1.3769$$

The 99% confidence interval for $\mu_{y|90}$ is:

$$\hat{y} \pm t s_{\hat{y}_m} = 118.4665 \pm 4.032(1.3769) = 118.4665 \pm 5.5517 = 112.9148 \text{ to } 124.0182$$

The standard deviation of $\hat{y}$ for predicting y for $x = 90$ is:

$$s_{\hat{y}_p} = s_e \sqrt{1 + \frac{1}{n} + \frac{(x_0 - \bar{x})^2}{SS_{xx}}} = (3.6221)\sqrt{1 + \frac{1}{7} + \frac{(90 - 91.8571)^2}{2104.8571}} = 3.8750$$

The 99% prediction interval for y_p for $x = 90$ is:

$$\hat{y} \pm t s_{\hat{y}_p} = 118.4665 \pm 4.032(3.8570) = 118.4665 \pm 15.6240 = 102.8425 \text{ to } 134.0905$$

13.92 From the solution to Exercise 13.81: $n = 10,\ \bar{x} = 51.20,\ SS_{xx} = 1895.6000,\ s_e = 22.3550$

The regression line is: $\hat{y} = 156.3302 + .6498x$

For $x = 53$: $\hat{y} = 156.3302 + .6498(53) = 190.7696$

$df = n - 2 = 10 - 2 = 8 \quad$ and $\quad$ Area in each tail of the t curve $= \alpha/2 = .5 - (.95/2) = .025$

From the t distribution table, the value of t for $df = 8$ and .025 area in the right tail is 2.306.

The standard deviation of $\hat{y}$ for estimating the mean value of y for $x = 53$ is:

$$s_{\hat{y}_m} = s_e \sqrt{\frac{1}{n} + \frac{(x_0 - \bar{x})^2}{SS_{xx}}} = (22.3550)\sqrt{\frac{1}{10} + \frac{(53 - 51.20)^2}{1895.6000}} = 7.1294$$

The 95% confidence interval for $\mu_{y|53}$ is:

$$\hat{y} \pm t s_{\hat{y}_m} = 190.7696 \pm 2.306(7.1294) = 190.7696 \pm 16.4404 = 174.3292 \text{ to } 207.2100$$

The standard deviation of $\hat{y}$ for predicting y for $x = 53$ is:

$$s_{\hat{y}_p} = s_e \sqrt{1 + \frac{1}{n} + \frac{(x_0 - \bar{x})^2}{SS_{xx}}} = (22.3550) \sqrt{1 + \frac{1}{10} + \frac{(53 - 51.20)^2}{1895.6000}} = 23.4643$$

The 95% prediction interval for y_p for $x = 53$ is:

$$\hat{y} \pm ts_{\hat{y}_p} = 190.7696 \pm 2.306(23.4643) = 190.7696 \pm 54.1087 = 136.6609 \text{ to } 244.8783$$

13.93 From the solution to Exercise 13.83: $n = 10$, $\bar{x} = 64.1$, $SS_{xx} = 4260.9000$, $s_e = 3.4501$

The regression line is: $\hat{y} = -14.4215 + .4450x$

For $x = 64$: $\hat{y} = -14.4245 + .4450(64) = 14.0555$

$df = n - 2 = 10 - 2 = 8$ and Area in each tail of the t curve $= \alpha/2 = .5 - (.95/2) = .025$

From the t distribution table, the value of t for $df = 8$ and .025 area in the right tail is 2.306.

The standard deviation of $\hat{y}$ for estimating the mean value of y for $x = 64$ is:

$$s_{\hat{y}_m} = s_e \sqrt{\frac{1}{n} + \frac{(x_0 - \bar{x})^2}{SS_{xx}}} = (3.4501) \sqrt{\frac{1}{10} + \frac{(64 - 64.1)^2}{4260.9000}} = 1.0910$$

The 95% confidence interval for $\mu_{y|64}$ is:

$$\hat{y} \pm ts_{\hat{y}_m} = 14.0555 \pm 2.306(1.0910) = 11.5397 \text{ to } 16.5713 \text{ or } \$1153.97 \text{ to } \$1657.13$$

The standard deviation of $\hat{y}$ for predicting y for $x = 64$ is:

$$s_{\hat{y}_p} = s_e \sqrt{1 + \frac{1}{n} + \frac{(x_0 - \bar{x})^2}{SS_{xx}}} = (3.4501) \sqrt{1 + \frac{1}{10} + \frac{(64 - 64.10)^2}{4260.9000}} = 3.6185$$

The 95% prediction interval for y_p for $x = 64$ is:

$$\hat{y} \pm ts_{\hat{y}_p} = 14.0555 \pm 2.306(3.6185) = 5.7112 \text{ to } 22.3998 \text{ or } \$571.12 \text{ to } \$2239.98$$

13.94 From the solution to Exercise 13.85: $n = 7$, $\bar{x} = 2.9386$, $SS_{xx} = 3.3647$, $s_e = 2.5448$

The regression line is: $\hat{y} = 4.0729 + 8.6771x$

For $x = 3.15$: $\hat{y} = 14.0729 + 8.6771(3.15) = 31.4058$

$df = n - 2 = 7 - 2 = 5$ and Area in each tail of the t curve $= \alpha/2 = .5 - (.98/2) = .01$

From the t distribution table, the value of t for $df = 5$ and .01 area in the right tail is 3.365.

The standard deviation of $\hat{y}$ for estimating the mean value of y for $x = 3.15$ is:

$$s_{\hat{y}_m} = s_e \sqrt{\frac{1}{n} + \frac{(x_0 - \bar{x})^2}{SS_{xx}}} = (2.5448) \sqrt{\frac{1}{7} + \frac{(3.15 - 2.9386)^2}{3.3647}} = 1.0056$$

The 98% confidence interval for $\mu_{y|3.15}$ is:

$$\hat{y} \pm ts_{\hat{y}_m} = 41.4058 \pm 3.365(1.0056) = 41.4058 \pm 3.3838 = 38.0220 \text{ to } 44.7896$$

The standard deviation of $\hat{y}$ for predicting y for $x = 3.15$ is:

$$s_{\hat{y}_p} = s_e \sqrt{1 + \frac{1}{n} + \frac{(x_0 - \bar{x})^2}{SS_{xx}}} = (2.5448)\sqrt{1 + \frac{1}{7} + \frac{(3.15 - 2.9386)^2}{3.3647}} = 2.7363$$

The 98% prediction interval for y_p for $x = 3.15$ is:

$$\hat{y} \pm ts_{\hat{y}_p} = 41.4058 \pm 3.365(2.7363) = 41.4058 \pm 9.2076 = 32.1982 \text{ to } 50.6134$$

13.95 Let: $x =$ age (in years) of a machine, $\quad y =$ the number of breakdowns

a. As the age of a machine increases (that is, the machine becomes older), the number of breakdowns is expected to increase. Hence, we expect a positive relationship between these two variables. Consequently, B is expected to be positive.

b. $n = 7$, $\sum x = 55$, $\sum y = 41$, $\sum x^2 = 527$, $\sum y^2 = 339$, $\sum xy = 416$

$\bar{x} = 7.8571$, $\bar{y} = 5.8571$, $SS_{xx} = 94.8571$, $SS_{yy} = 98.8571$, $SS_{xy} = 93.8571$

$b = SS_{xy}/SS_{xx} = 93.8571 / 94.8571 = .9895$

$a = \bar{y} - b\bar{x} = 5.8571 - (.9895)(7.8571) = -1.9175$

The regression line is: $\hat{y} = -1.9175 + .9895x$

The sign of $b = .9895$ is positive, which is consistent with what we expected.

c. The value of $a = -1.9175$ is the value of $\hat{y}$ for $x = 0$. In this exercise it represents the number of breakdowns per month for a new machine.

The value of $b = .9895$ means that the average number of breakdowns per month increases by about .99 for every one year increase in the age of such a machine.

d. $r = \dfrac{SS_{xy}}{\sqrt{SS_{xx}SS_{yy}}} = \dfrac{93.8571}{\sqrt{(94.8571)(98.8571)}} = .97$

$r^2 = bSS_{xy}/SS_{yy} = (.9895)(93.8571)/98.8571 = .94$

The value of $r = .97$ indicates that the two variables have a very strong positive correlation.

The value of $r^2 = .94$ means that 94% of the total squared errors (SST) are explained by our regression model.

e. $s_e = \sqrt{\dfrac{SS_{yy} - bSS_{xy}}{n-2}} = \sqrt{\dfrac{98.8571 - (.9895)(93.8571)}{7-2}} = 1.0941$

f. $s_b = s_e / \sqrt{SS_{xx}} = 1.0941 / \sqrt{94.8571} = .1123$

   $df = n - 2 = 7 - 2 = 5$ and   Area in each tail of the t curve $= \alpha / 2 = .5 - (.99 / 2) = .005$

   From the t distribution table, the value of t for $df = 5$ and .005 area in the right tail is 4.032.

   The 99% confidence interval for B is: $b \pm t s_b = .9895 \pm 4.032(.1123) = .54$ to 1.44

g. $H_0 : B = 0;$ $H_1 : B > 0$

   Area in the right tail of the t curve $= .025;$ and $df = n - 2 = 7 - 2 = 5$

   The critical value of t is 2.571.

   The value of the test statistic is: $t = \dfrac{b - B}{s_b} = \dfrac{.9895 - 0}{.1123} = 8.811$

   Reject H_0. Hence, B is positive.

h. $H_0 : \rho = 0;$ $H_1 : \rho > 0$

   Area in the right tail of the t curve $= .025;$ and $df = n - 2 = 7 - 2 = 5$

   The critical value of t is 2.571.

   The value of the test statistic is: $t = r \sqrt{\dfrac{n - 2}{1 - r^2}} = .97 \sqrt{\dfrac{7 - 2}{1 - (.97)^2}} = 8.922$

   Reject H_0. Conclude that ρ is positive.

   The conclusion is the same as that of part g (reject H_0).

13.96   Let: $x =$ air pollution index, $y =$ the number of emergency admissions

a. As the air pollution index rises, air pollution is getting worse. It gets harder to breathe for those with chronic breathing problems, so on average more emergency admissions occur. Hence, we expect a positive relationship between these two variables. Consequently, B is expected to be positive.

b. $n = 7, \ \sum x = 38.1, \ \sum y = 405, \ \sum x^2 = 224.75, \ \sum y^2 = 27{,}551, \ \sum xy = 2440.9$

   $\bar{x} = 5.4429, \ \bar{y} = 57.8571, \ SS_{xx} = 17.3771, \ SS_{yy} = 4118.8571, \ SS_{xy} = 236.5429$

   $b = SS_{xy} / SS_{xx} = 236.5429 / 17.3771 = 13.6123$

   $a = \bar{y} - b\bar{x} = 57.8571 - (13.6123)(5.4429) = -16.2333$

   The regression line is: $\hat{y} = -16.2333 + 13.6123x$

The sign of $b = 13.6123$ is positive, which is consistent with what we expected.

c. $r = \dfrac{SS_{xy}}{\sqrt{SS_{xx}SS_{yy}}} = \dfrac{236.5429}{\sqrt{(17.3771)(4118.8571)}} = .88$

$r^2 = b\,SS_{xy}\big/SS_{yy} = (13.6123)(236.5429)/4118.8571 = .78$

The value of $r = .88$ indicates that the two variables have a very strong positive correlation.
The value of $r^2 = .78$ means that 78% of the total squared errors (SST) are explained by our regression model.

d. $s_e = \sqrt{\dfrac{SS_{yy} - bSS_{xy}}{n-2}} = \sqrt{\dfrac{4118.8571 - (13.6123)(236.5429)}{7-2}} = 13.4087$

e. $s_b = s_e\big/\sqrt{SS_{xx}} = 13.4087/\sqrt{17.3771} = 3.2116$

$df = n-2 = 7-2 = 5$ and Area in each tail of the t curve $= \alpha/2 = .5-(.90/2) = .05$

From the t distribution table, the value of t for $df = 5$ and .05 area in the right tail is 2.015.
The 90% confidence interval for B is: $b \pm t s_b = 13.6123 \pm 2.015(3.2116) = 7.13$ to 20.09

f. $H_0 : B = 0; \quad H_1 : B > 0$

Area in the right tail of the t curve $= .05$; and $df = n-2 = 7-2 = 5$
The critical value of t is 2.015.
The value of the test statistic is: $t = (b-B)/s_b = (13.6123-0)/3.2166 = 4.232$
Reject H_0. Hence, B is positive.

g. $H_0 : \rho = 0; \quad H_1 : \rho > 0$

Area in the right tail of the t curve $= .05$; and $df = n-2 = 7-2 = 5$
The critical value of t is 2.015.

The value of the test statistic is: $t = r\sqrt{\dfrac{n-2}{1-r^2}} = .88\sqrt{\dfrac{7-2}{1-(.88)^2}} = 4.123$

Reject H_0. Conclude that ρ is positive. The conclusion is the same as that of part f (reject H_0).

13.97 Let: $x =$ number of promotions per day, $y =$ number of units (in hundreds) sold per day

a. We would expect an increase in the number of promotions to yield increased sales, implying a positive relationship between the two variables. Consequently, we expect B to be positive.

b. From the given data:

$n = 7$, $\sum x = 177$, $\sum y = 144$, $\sum x^2 = 5285$, $\sum y^2 = 3224$, $\sum xy = 4049$

$\bar{x} = 25.2857$, $\bar{y} = 20.5714$, $SS_{xx} = 809.4286$, $SS_{yy} = 261.7143$, $SS_{xy} = 407.8571$

$b = SS_{xy}/SS_{xx} = 407.8571/809.4286 = .5039$

$a = \bar{y} - b\bar{x} = 20.5714 - (.5039)(25.2857) = 7.8299$

The regression line is: $\hat{y} = 7.8299 + .5039x$

The sign of b is positive, agreeing with the prediction of part a.

c. The value of $a = 7.8299$ is the value of $\hat{y}$ for $x = 0$. In this exercise it represents the number of units (in hundreds) sold if there are no promotions.

The value of $b = .5039$ means that the sales are expected to increase by about 50 units per day for each additional promotion.

d. $r = \dfrac{SS_{xy}}{\sqrt{SS_{xx}SS_{yy}}} = \dfrac{407.8571}{\sqrt{(809.4286)(261.7143)}} = .89$

$r^2 = bSS_{xy}/SS_{yy} = (.5039)(407.8571)/261.7143 = .79$

The value of $r = .89$ indicates strong positive linear correlation between the two variables.

The value of $r^2 = .79$ means that 79% of the total squared errors (SST) are explained by the model.

e. For $x = 35$: $\hat{y} = 7.8299 + .5039(35) = 25.4664$

Thus, we expect sales of about 2547 units in a day with 35 promotions.

f. $s_e = \sqrt{\dfrac{SS_{yy} - bSS_{xy}}{n-2}} = \sqrt{\dfrac{261.7143 - (.5039)(407.8571)}{7-2}} = 3.3525$

g. $s_b = s_e/\sqrt{SS_{xx}} = 3.3525/\sqrt{809.4286} = .1178$

$df = n - 2 = 7 - 2 = 5$ and   For 5 df and .01 area in each tail of the t curve, $t = 3.365$.

The 98% confidence interval for B is: $b \pm ts_b = .5039 \pm 3.365(.1178) = .11$ to .90

h. $H_0 : B = 0$; $H_1 : B > 0$

For 5 df and .01 area in the right tail of the t curve, the critical value of t is 3.365.

The value of the test statistic is:
$$t = \frac{b-B}{s_b} = \frac{.5039-0}{.1178} = 4.278$$

Reject H_0. Conclude that B is positive.

i. $H_0 : \rho = 0;$ $H_1 : \rho \neq 0;$

Area in each tail of the t curve $= .02/2 = .01;$ and $df = n - 2 = 7 - 2 = 5$

The critical values of t are -3.365 and 3.365.

The value of the test statistic is: $t = r\sqrt{\dfrac{n-2}{1-r^2}} = .89\sqrt{\dfrac{7-2}{1-(.89)^2}} = 4.365$

Reject H_0. Conclude that the correlation coefficient is different from zero.

13.98   Let: $x =$ temperature   and $y =$ volume of ice cream (in pounds) sold

a. $n = 8$, $\sum x = 711$, $\sum y = 1488$, $\sum x^2 = 63{,}713$, $\sum y^2 = 297{,}428$, $\sum xy = 135{,}466$

   $\bar{x} = 88.875$, $\bar{y} = 186.000$, $SS_{xx} = 522.8750$, $SS_{yy} = 20{,}660.0000$, $SS_{xy} = 3220.0000$

   $b = SS_{xy}\big/SS_{xx} = 3220.0000 / 522.8750 = 6.1583$

   $a = \bar{y} - b\bar{x} = 186.000 - (6.1583)(88.875) = -361.3189$

   The regression line is: $\hat{y} = -361.3189 + 6.1583x$

b.   The value of $a = -361.3189$ is the value of $\hat{y}$ for $x = 0$. In this exercise it gives the volume of ice cream (in pounds) sold on a day with a zero temperature.

   The value of $b = 6.1583$ means that, on average, the amount of ice cream sold increases by about 6.16 pounds per day for every one degree increase in the temperature.

c. $r = \dfrac{SS_{xy}}{\sqrt{SS_{xx}SS_{yy}}} = \dfrac{3220.0000}{\sqrt{(522.8750)(20{,}660.0000)}} = .98$

   $r^2 = bSS_{xy}\big/SS_{yy} = (6.1583)(3220.0000)/20{,}660.0000 = .96$

   The value of $r = .98$ indicates that the two variables have a very strong positive correlation.

   The value of $r^2 = .96$ means that 96% of the total squared errors (SST) are explained by our regression model.

d.   For $x = 95$: $\hat{y} = -361.3189 + 6.1583(95) = 223.7196$

Thus, about 223.7 pounds of ice cream will be sold at the given ice cream parlor on a day with a temperature of 95 degrees.

e. $s_e = \sqrt{\dfrac{SS_{yy} - b\,SS_{xy}}{n-2}} = \sqrt{\dfrac{20{,}660.0000 - (6.1583)(3220.0000)}{8-2}} = 11.7635$

f. $s_b = s_e / \sqrt{SS_{xx}} = 11.7635 / \sqrt{522.8750} = .5144$

   $df = n-2 = 8-2 = 6$ and   Area in each tail of the t curve $= \alpha/2 = .5 - (.99/2) = .005$

   From the t distribution table, the value of t for $df = 6$ and .005 area in the right tail is 3.707.

   The 99% confidence interval for B is: $b \pm t s_b = 6.1583 \pm 3.707(.5144) = 4.25$ to 8.07

g. $H_0 : B = 0$; $H_1 : B \neq 0$

   Area in each tail of the t curve $= .01/2 = .005$;   and $df = n-2 = 8-2 = 6$

   The critical values of t are -3.707 and 3.707.

   The value of the test statistic is: $t = \dfrac{b-B}{s_b} = \dfrac{6.1583 - 0}{.5144} = 11.972$

   Reject H_0. Conclude that B is different from zero.

h. $H_0 : \rho = 0$; $H_1 : \rho \neq 0$

   Area in each tail of the t curve $= .01/2 = .005$;   and $df = n-2 = 8-2 = 6$

   The critical values of t are -3.707 and 3.707.

   The value of the test statistic is: $t = r\sqrt{\dfrac{n-2}{1-r^2}} = .98\sqrt{\dfrac{8-2}{1-(.98)^2}} = 12.063$

   Reject H_0. Conclude that the correlation coefficient is different from zero.

13.99   Let: $x =$ time, $y =$ average hotel room rate

a.

x	0	1	2	3	4	5	6	7	8	9
y	59.39	60.99	63.35	66.34	70.68	74.77	78.24	81.59	85.69	84.58

b. $n = 10$, $\sum x = 45$, $\sum y = 725.62$, $\sum x^2 = 285$, $\sum y^2 = 53{,}522.3638$, $\sum xy = 3530.59$

   $\bar{x} = 4.5$, $\bar{y} = 72.562$, $SS_{xx} = 82.5000$, $SS_{yy} = 869.9254$, $SS_{xy} = 265.3000$

c.

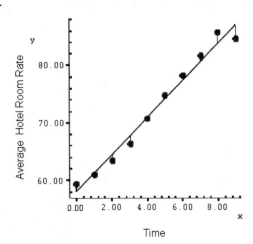

The scatter diagram exhibits a positive linear relationship between time and hotel room rates.

d. $b = SS_{xy}/SS_{xx} = 265.3000 / 82.5000 = 3.2158$

$a = \bar{y} - b\bar{x} = 72.562 - (3.2158)(4.5) = 58.0909$

The regression line is: $\hat{y} = 58.0909 + 3.2158x$

e. The value of $a = 58.0909$ is the value of $\hat{y}$ for $x = 0$. In this exercise it gives the average hotel room rate at time zero. The value of $b = 3.2158$ means that the linear relationship between time and the average hotel room rate shows an average increase of $3.22 per year in hotel room rates from 1992 to 2001.

f. $r = \dfrac{SS_{xy}}{\sqrt{SS_{xx}SS_{yy}}} = \dfrac{265.3000}{\sqrt{(82.5000)(869.9254)}} = .99$

g. For $x = 14$: $\hat{y} = 58.0909 + 3.2158 (14) = 103.1123$

Thus, the predicted average hotel room rate for year 15 (that is, 2006) is $103.11.

Note that this predicted average hotel room rate is based on the regression equation derived from data for 1992 through 2001. This prediction assumes that the same linear relationship will continue for 5 or more years into the future, a questionable assumption.

13.100 Let: $x =$ time and $y =$ tax returns filed electronically (in millions)

a.

x	0	1	2	3	4	5	6	7	8	9
y	11	12	14	12	15	19	25	29	35	40

b. $n = 10$, $\sum x = 45$, $\sum y = 212$, , $\sum x^2 = 285$, $\sum y^2 = 5482$, $\sum xy = 1224$

$\bar{x} = 4.5$, $\bar{y} = 21.2$, $SS_{xx} = 82.5000$, $SS_{yy} = 987.6000$, $SS_{xy} = 270$

c.

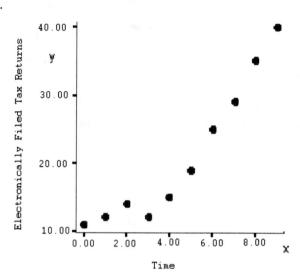

The scatter diagram exhibits a positive linear relationship between time and number of households with cable TV.

d. $b = SS_{xy} / SS_{xx} = 270.0000/82.5000 = 3.2727$

$a = \bar{y} - b\bar{x} = 21.2 - (3.2727)(4.5) = 6.4729$

The regression line is: $\hat{y} = 6.4729 + 3.2727x$

e. The value of $a = 6.4729$ is the value of $\hat{y}$ for $x = 0$. In this exercise it gives the number of electronically filed tax returns at time zero.

The value of $b = 3.2727$ means that the linear relationship between time and the number of electronically filed tax returns shows an average increase of 3.2727 millions of electronically filed tax returns per year from 1992 to 2001.

f. $r = \dfrac{SS_{xy}}{\sqrt{SS_{xx} SS_{yy}}} = \dfrac{270.0000}{\sqrt{(82.5000)(987.6000)}} = .95$

g. For $x = 15$: $\hat{y} = 6.4729 + 3.2727(15) = 55.5634$

Thus, the predicted number of electronically filed tax returns for year 15, that is, 2007, is about 55 million. Note that this prediction of the number of electronically filed tax returns in 2007 is based on the regression equation derived from data for 1992 to 2001. This prediction assumes that the same linear relationship will continue for 6 or more years into the future, a questionable assumption.

13.101 Let: $x =$ time, $y =$ students per computer

a.

x	0	1	2	3	4	5	6	7	8	9	10
y	20	18	16	14	10.5	10	7.8	6.1	5.7	5.4	5

b. $n = 11$, $\sum x = 55$, $\sum y = 118.5$, $\sum x^2 = 385$, $\sum y^2 = 1570.95$, $\sum xy = 417.7$

$\bar{x} = 5$, $\bar{y} = 10.7727$, $SS_{xx} = 110.0000$, $SS_{yy} = 294.3818$, $SS_{xy} = -174.8000$

c.

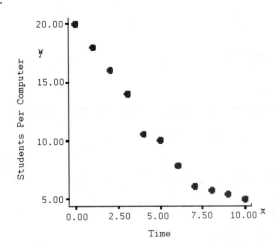

The scatter diagram exhibits a negative linear relationship between time and students per computer.

d. $b = SS_{xy}/SS_{xx} = -174.8000 / 110.0000 = -1.5891$

$a = \bar{y} - b\bar{x} = 10.7727 - (-1.5891)(5) = 18.7182$

The regression line is: $\hat{y} = 18.7182 - 1.5891x$

e. The value of $a = 18.7182$ is the value of $\hat{y}$ for $x = 0$. In this exercise, it gives the students per computer at time zero, which is 1990–91. The value of $b = -1.5891$ means that the linear relationship between time and students per computer shows an average decrease of 1.5891 students per computer per year between years 1990–91 and 2000–01.

f. $r = \dfrac{SS_{xy}}{\sqrt{SS_{xx}SS_{yy}}} = \dfrac{-174.8000}{\sqrt{(110.0000)(294.3818)}} = -.97$

g. For $x = 15$: $\hat{y} = 18.7182 - 1.5891 (15) = -5.1183$

Thus, the predicted students per computer in 2005–'06 is –5.1183.

Note that this prediction of students per computer in 2005–'06 is based on the regression equation derived from data for 1990–'91 through 2000–'01. This prediction assumes that the same linear relationship will continue for 5 or more years into the future, a questionable assumption.

13.102 From the solution to Exercise 13.95: $n = 7$, $\bar{x} = 7.8571$, $SS_{xx} = 94.8571$, $s_e = 1.0941$

The regression line is: $\hat{y} = -1.9175 + .9895x$

For $x = 8$: $\hat{y} = -1.9175 + .9895(8) = 5.9985$

$df = n - 2 = 7 - 2 = 5$ and   Area in each tail of the t curve $= \alpha/2 = .5 - (.99/2) = .005$

From the t distribution table, the value of t for $df = 5$ and .005 area in the right tail is 4.032.

The standard deviation of $\hat{y}$ for estimating the mean value of y for $x = 8$ is:

$$s_{\hat{y}_m} = s_e \sqrt{\frac{1}{n} + \frac{(x_0 - \bar{x})^2}{SS_{xx}}} = (1.0941)\sqrt{\frac{1}{7} + \frac{(8 - 7.8571)^2}{94.8571}} = .4138$$

The 99% confidence interval for $\mu_{y|8}$ is:

$\hat{y} \pm ts_{\hat{y}_m} = 5.9985 \pm (4.032)(.4138) = 5.9985 \pm 1.6684 = 4.3301$ to 7.6669

The standard deviation of $\hat{y}$ for predicting the value of y for $x = 8$ is:

$$s_{\hat{y}_p} = s_e \sqrt{1 + \frac{1}{n} + \frac{(x_0 - \bar{x})^2}{SS_{xx}}} = (1.0941)\sqrt{1 + \frac{1}{7} + \frac{(8 - 7.8571)^2}{94.8571}} = 1.1698$$

The 99% prediction interval for y_p for $x = 8$ is:

$\hat{y} \pm ts_{\hat{y}_p} = 5.9985 \pm (4.032)(1.1698) = 5.9985 \pm 4.7166 = 1.2819$ to 10.7151

13.103 From the solution to Exercise 13.96: $n = 7$, $\bar{x} = 5.4429$, $SS_{xx} = 17.3771$, $s_e = 13.4087$

The regression line is: $\hat{y} = -16.2333 + 13.6123x$

For $x = 7$: $\hat{y} = -16.2333 + 13.6123(7) = 79.0528$

$df = n - 2 = 7 - 2 = 5$ and   Area in each tail of the t curve $= \alpha/2 = .5 - (.95/2) = .025$

From the t distribution table, the value of t for $df = 5$ and .025 area in the right tail is 2.571.

The standard deviation of $\hat{y}$ for estimating $\mu_{y|x}$ for $x = 7$ is:

$$s_{\hat{y}_m} = s_e \sqrt{\frac{1}{n} + \frac{(x_0 - \bar{x})^2}{SS_{xx}}} = (13.4087)\sqrt{\frac{1}{7} + \frac{(7 - 5.4429)^2}{17.3771}} = 7.1253$$

The 95% confidence interval for $\mu_{y|7}$ is: $\hat{y} \pm ts_{\hat{y}_m} = 79.0528 \pm (2.571)(7.1253) = 60.7337$ to 97.3719

The standard deviation of $\hat{y}$ for predicting y_p for $x = 7$ is:

$$s_{\hat{y}_p} = s_e\sqrt{1 + \frac{1}{n} + \frac{(x_0 - \bar{x})^2}{SS_{xx}}} = (13.4087)\sqrt{1 + \frac{1}{7} + \frac{(7 - 5.4429)^2}{17.3771}} = 15.1843$$

The 95% prediction interval for y_p for $x = 7$ is:

$$\hat{y} \pm ts_{\hat{y}_p} = 79.0528 \pm (2.571)(15.1843) = 40.0140 \text{ to } 118.0916$$

13.104   From the solution to Exercise 13.97: $n = 7$, $\bar{x} = 25.2857$, $SS_{xx} = 809.4286$, $s_e = 3.3525$

The regression line is: $\hat{y} = 7.8299 + .5039x$

For $x = 35$: $\hat{y} = 7.8299 + .5039(35) = 25.4664$

$df = n - 2 = 7 - 2 = 5$ and   Area in each tail of the t curve $= \alpha/2 = .5 - (.90/2) = .05$

From the t distribution table, the value of t for $df = 5$ and .05 area in the right tail is 2.015.

The standard deviation of $\hat{y}$ for estimating $\mu_{y|x}$ for $x = 35$ is:

$$s_{\hat{y}_m} = s_e\sqrt{\frac{1}{n} + \frac{(x_0 - \bar{x})^2}{SS_{xx}}} = (3.3525)\sqrt{\frac{1}{7} + \frac{(35 - 25.2857)^2}{809.4286}} = 1.7076$$

The 90% confidence interval for $\mu_{y|35}$ is:

$$\hat{y} \pm ts_{\hat{y}_m} = 25.4664 \pm (2.015)(1.7076) = 25.4664 \pm 3.4408 = 22.0256 \text{ to } 28.9072$$

The standard deviation of $\hat{y}$ for predicting y_p for $x = 35$ is:

$$s_{\hat{y}_p} = s_e\sqrt{1 + \frac{1}{n} + \frac{(x_0 - \bar{x})^2}{SS_{xx}}} = (3.3525)\sqrt{1 + \frac{1}{7} + \frac{(35 - 25.2857)^2}{809.4286}} = 3.7623$$

The 90% prediction interval for y_p for $x = 35$ is:

$$\hat{y} \pm ts_{\hat{y}_p} = 25.4664 \pm (2.015)(3.7623) = 25.4664 \pm 7.5810 = 17.8854 \text{ to } 33.0474$$

13.105   From the solution to Exercise 13.98: $n = 8$, $\bar{x} = 88.8750$, $SS_{xx} = 522.8750$, $s_e = 11.7635$

The regression line is: $\hat{y} = -361.3189 + 6.1583x$

For $x = 95$: $\hat{y} = -61.3189 + 6.1583(95) = 223.7196$

$df = n - 2 = 8 - 2 = 6$ and   Area in each tail of the t curve $= \alpha/2 = .5 - (.98/2) = .01$

From the t distribution table, the value of t for $df = 6$ and .01 area in the right tail is 3.143.

The standard deviation of $\hat{y}$ for estimating $\mu_{y|x}$ for $x = 95$ is:

$$s_{\hat{y}_m} = s_e\sqrt{\frac{1}{n} + \frac{(x_0 - \bar{x})^2}{SS_{xx}}} = (11.7635)\sqrt{\frac{1}{8} + \frac{(95 - 88.8750)^2}{522.8750}} = 5.2179$$

The 98% confidence interval for $\mu_{y|95}$ is:

$\hat{y} \pm ts_{\hat{y}_m} = 223.7196 \pm (3.143)(5.2179) = 223.7196 \pm 16.3999 = 207.3197 \text{ to } 240.1195$

The standard deviation of $\hat{y}$ for predicting y for $x = 95$ is:

$$s_{\hat{y}_p} = s_e \sqrt{1 + \frac{1}{n} + \frac{(x_0 - \bar{x})^2}{SS_{xx}}} = (11.7635)\sqrt{1 + \frac{1}{8} + \frac{(95 - 88.8750)^2}{522.8750}} = 12.8688$$

The 98% prediction interval for y_p for $x = 95$ is:

$\hat{y} \pm ts_{\hat{y}_p} = 223.7196 \pm (3.143)(12.8688) = 223.7196 \pm 40.4466 = 183.2730 \text{ to } 264.1662$

13.106 $n = 6, \; \sum x = 210, \; \sum y = 122, \; \sum x^2 = 9100, \; \sum y^2 = 2696, \; \sum xy = 4880$

$\bar{x} = 35, \; \bar{y} = 20.3333, \; SS_{xx} = 1750, \; SS_{yy} = 215.3333, \; SS_{xy} = 610$

a. $b = SS_{xy} / SS_{xx} = 610/1750 = .3486$

   $a = \bar{y} - b\bar{x} = 20.3333 - .3486(35) = 8.1323$

   The regression line is: $\hat{y} = 8.1323 + .3486x$

   $$r = \frac{SS_{xy}}{\sqrt{SS_{xx}SS_{yy}}} = \frac{610}{\sqrt{(1750)(215.3333)}} = .99$$

b. r should not change, since the data points all move up the same amount. The regression line in part a shifted up by 5 units will fit the new data points just as well as the regression line in part a fit the original data points.

c. $n = 6, \; \sum x = 210, \; \sum y = 152, \; \sum x^2 = 9100, \; \sum y^2 = 4066, \; \sum xy = 5930$

   $\bar{x} = 35, \; \bar{y} = 25.3333, \; SS_{xx} = 1750, \; SS_{yy} = 215.3333, \; SS_{xy} = 610$

   $b = SS_{xy} / SS_{xx} = 610/1750 = .3486$

   $a = \bar{y} - b\bar{x} = 25.3333 - .3486(35) = 13.1323$

   The regression line is: $\hat{y} = 13.1323 + .3486x$

   $$r = \frac{SS_{xy}}{\sqrt{SS_{xx}SS_{yy}}} = \frac{610}{\sqrt{(1750)(215.3333)}} = .99, \text{ as we expected.}$$

13.107 a. $y = -432 + 7.7x, \; s_e = 28.17, \; SS_{xx} = 607, \; \bar{x} = 87.5, \; n = 20$

   $b = 7.7, \; s_b = s_e / \sqrt{SS_{xx}} = 28.17 / \sqrt{607} = 1.1434$

   $H_0 : B = 0; \quad H_1 : B > 0$

Area in the right tail of the t distribution curve = .05   and $df = n - 2 = 20 - 2 = 18$

The critical value of t is 1.734

$$t = \frac{b - B}{s_e} = \frac{7.7 - 0}{1.1434} = 6.734$$ Reject H_0. The maximum temperature and bowling activity

between twelve noon and 6:00pm have a positive association.

b. For $x = 90$: $\hat{y} = -432 + 7.7(90) = 261$

From the t distribution table, the value of t for $df = 20-2 = 18$ and $.05/2 = .025$ area in the right tail

is 2.101.

The standard deviation of $\hat{y}$ for estimating $\mu_{y|x}$ for $x = 90$ is:

$$s_{\hat{y}_m} = s_e \sqrt{\frac{1}{n} + \frac{(x_0 - \bar{x})^2}{SS_{xx}}} = (28.17)\sqrt{\frac{1}{20} + \frac{(90 - 87.5)^2}{607}} = 6.9172$$

The 95% confidence interval for $\mu_{y|90}$ is:

$$\hat{y} \pm t s_{\hat{y}_m} = 261 \pm (2.101)(6.9172) = 246.4670 \text{ to } 275.5330 \text{ lines.}$$

c. The standard deviation of $\hat{y}$ for predicting y for $x = 90$ is:

$$s_{\hat{y}_p} = s_e \sqrt{1 + \frac{1}{n} + \frac{(x_0 - \bar{x})^2}{SS_{xx}}} = (28.17)\sqrt{1 + \frac{1}{20} + \frac{(90 - 87.5)^2}{607}} = 29.0068$$

The 95% prediction interval for y_p for $x = 90$ is:

$$\hat{y} \pm t s_{\hat{y}_p} = 261 \pm (2.101)(29.0068) = 200.0567 \text{ to } 321.9433 \text{ lines.}$$

d. The mean value $\mu_{y|90}$ could be at either extreme of the interval in part b. Given a particular mean,

the individual data points for this mean will have a certain variation, hence the prediction interval

for y_p must be larger than the prediction interval for $\mu_{y|x}$.

e. $y = -432 + 7.7(100) = 338$ lines

Our regression line is only valid for the range of x–values in our sample ($77°$ to $95°$ Fahrenheit).

We should interpret this estimate very cautiously and not attach too much value to it.

13.108 The correlation coefficient suggests a moderate to strong positive relation between the two variables in

this example. However it does not reveal whether 30–year olds earn more than their fathers. In order

to determine that, we would need to know the mean income of each group (the fathers and the 30–

year–old children). All we know is that the higher the father's income, the higher his son or daughter's

income tends to be.

13.109 Burton's logic is faulty. The correlation coefficient merely describes the quantitative relationship between the two variables (frequency of mowing the lawn and size of corn ears). The high correlation does not prove that there is a cause–and–effect relation between the two variables. In this case, the correlation is due to the effect of other variables, such as amounts of sunshine and rain, and fertility of the soil. In years in which there are favorable amounts of sun and rain (and perhaps when Burton applies optimal amounts of fertilizers to both lawn and garden) the corn grows larger and the grass grows faster, thus requiring more frequent mowing. Thus, each of these other variables (amount of sunshine, amount of rain, and amount of fertilizer) is highly correlated with the size of the corn ears. Each of them is also highly correlated with the growth rate of the grass, (and therefore with the frequency of mowing). To obtain larger corn ears next year, Burton should be sure to plant the corn in a sunny part of his garden, water the corn during periods of dry weather, and apply fertilizer consistently.

13.110 a. Assuming a linear relationship of the form $y = A + Bx$, we must select reasonable values of A and B where:

$A = $ average GPA of students who do not work.

$B = $ average change in GPA for each additional hour worked.

Assume that students who do not work have, on average, a GPA of 3.0.

Thus, $A = 3.0$.

We expect B to be negative, since work reduces the time available for study.

However, $|B|$ should not be so large that a student working 40 hours per week should be expected to have a GPA that is zero or negative. Try $B = -.02$.   Thus, $\mu_{y|x} = 3.0 - .02x$

Note that different students will propose different values of A and B. Trying several values of x yields the following predictions of GPA.

x	10	15	20	25	30	35	40
y	2.8	2.7	2.6	2.5	2.4	2.3	2.2

The predicted GPA's seem to conform roughly to the GPA's in the data for comparable values of x.

b. $n = 10$, $\sum x = 183$, $\sum y = 27.9$, $\sum x^2 = 4923$, $\sum xy = 463.9$

$\bar{x} = 18.3$, $\bar{y} = 2.79$, $SS_{xx} = 1574.1$, $SS_{xy} = -46.67$

$b = SS_{xy} / SS_{xx} = -46.67/1574.1 = -.0296$

$a = \bar{y} - b\bar{x} = 2.79 - (-.0296)(18.3) = 3.3317$

The regression line is: $\hat{y} = 3.3317 - .0296x$

Thus, the estimate of A obtained from the data is about .33 higher than the value proposed in part a.

The estimate of B from the data is about .01 lower than the value from part a.

13.111 a. Let: $x =$ number of students living at each address, $\quad y =$ monthly phone bill

A linear relationship of the form $y = A + Bx$ seems reasonable where:

$A =$ the phone company's basic monthly charge.

$B =$ the average student's monthly accumulation of toll charges.

We might assume $A = \$15$ and $B = \$25$ per month. Thus $\mu_{y|x} = 15 + 25x$

Note that different students will propose different values of A and B. Trying several values of x yields the following predictions of y.

x	1	2	3	4	5
y	40	65	90	115	140

The predicted phone bills seem to be lower than most of the bills in the data for comparable values of x.

b. $n = 15$, $\sum x = 48$, $\sum y = 1683.88$, $\sum x^2 = 184$, $\sum xy = 6332.23$

$\bar{x} = 3.2$, $\bar{y} = 112.2587$, $SS_{xx} = 30.4$, $SS_{xy} = 943.814$

$b = SS_{xy}/SS_{xx} = 943.814/30.4 = 31.0465$

$a = \bar{y} - b\bar{x} = 112.2587 - 31.0465\,(3.2) = 12.9099$

The regression line is: $\hat{y} = 12.9099 + 31.0465x$

Thus, the estimate of A obtained from the data is about 2.09 lower than the value proposed in part a. The estimate of B from the data is about 6.05 higher than the value from part a.

13.112 Let: $x =$ executive's test score, $\quad y =$ executive's salary

a. We are given that $\bar{x} = 44$ and $\bar{y} = 200{,}000$

For U.S. executives, a loss of $\$16{,}836$ for every five points scored above average on the test is equivalent to a loss of $\$3367.20$ for each point scored above average. Thus, based on the given information, $b = -3367.20$.

$a = \bar{y} - b\bar{x} = 200{,}000 - (-3367.20)(44) = 348{,}156.80$

Thus, the regression equation is: $\hat{y} = 348{,}156.80 - 3367.20x$

b. Nothing is said about the salaries of U.S. executives who scored below average, so the equation may not be valid for values of x below 44. It is also given that the maximum possible score on the test is 60. Thus, the equation is valid for the values of x from 44 to 60.

Self -Review Test for Chapter Thirteen

1. d	**2.** a	**3.** b	**4.** a	**5.** b
6. b	**7.** True	**8.** True	**9.** a	**10.** b

11. See the solution to Exercise 13.7.

12. The values of A and B for a regression model are obtained by using the population data. On the other hand, if a regression model is estimated by using the sample data, then we obtain the values of a and b.

13. See section 13.2.4, Pages 590 – 593 of the text.

14. A regression line obtained by using the population data is called the population regression line. It gives values of A and B and is written as: $\mu_{y|x} = A + B\,x$

A regression line obtained by using the sample data is called the sample regression line. It gives the estimated values of A and B, which are denoted by a and b. The sample regression line is written as: $\hat{y} = a + bx$

15. a. The attendance depends on temperature. With a higher temperature more people attend the minor league baseball game. Hence, a higher temperature is expected to draw bigger crowds.

     b. As mentioned in part a, a higher temperature is expected to bring in more ticket buyers on average. Consequently, we expect B to be positive.

c.

The scatter diagram exhibits a linear relationship between temperature and the attendance at a minor league baseball game but this relationship does not seem to be strong.

d. Let: x = temperature (in degrees)   and y = attendance (in hundreds)

$n = 7,\ \sum x = 422,\ \sum y = 99,\ \sum x^2 = 26{,}084,\ \sum y^2 = 1513,\ \sum xy = 6143$

$\bar{x} = 60.2857,\ \bar{y} = 14.1429,\ SS_{xx} = 643.4286,\ SS_{yy} = 112.8571,$ and $SS_{xy} = 174.7143$

$b = SS_{xy}/SS_{xx} = 174.7143 / 643.4286 = .2715$

$a = \bar{y} - b\bar{x} = 14.1429 - .2715(60.2857) = -2.2247$

The regression line is: $\hat{y} = -2.2247 + .2715x$

The sign of b is consistent with what we expected in part b.

e. The value of $a = -2.2247$ is the value of $\hat{y}$ for $x = 0$. In this exercise it represents the number of people attending a minor league game when the temperature is zero.

The value of $b = .2715$ means that, on average, the people attending a minor league games increases by about .27 for every one degree increase in temperature.

f. $r = \dfrac{SS_{xy}}{\sqrt{SS_{xx}SS_{yy}}} = \dfrac{174.7143}{\sqrt{(643.4286)(112.8571)}} = .65$

$r^2 = bSS_{xy}/SS_{yy} = (.2715)(174.7143)/112.8571 = .42$

The value of $r = .65$ indicates that the two variables have a positive correlation, which is not very strong. The value of $r^2 = .42$ means that 42% of the total squared errors (SST) are explained by our regression model.

g. For $x = 60$: $\hat{y} = -2.2247 + .2715(60) = 14.0653$

Thus, with a sixty degree temperature the minor league game is expected to sell about 1407 tickets.

h. $s_e = \sqrt{\dfrac{SS_{yy} - b\,SS_{xy}}{n-2}} = \sqrt{\dfrac{112.8571 - .2715(174.7143)}{7-2}} = 3.6172$

i. $s_b = s_e / \sqrt{SS_{xx}} = 3.6172 / \sqrt{643.4286} = .1426$

$df = n - 2 = 7 - 2 = 5$ and Area in each tail of the t curve $= \alpha/2 = .5 - (.99/2) = .005$

From the t distribution table, the value of t for $df = 5$ and .005 area in the right tail is 4.032.

The 99% confidence interval for B is: $b \pm t s_b = .2715 \pm 4.032(.1426) = .2715 \pm .57 = -.30$ to $.84$

j. $H_0 : B = 0; H_1 : B > 0$;

Area in the right tail of the t curve $= .01$; and $df = n - 2 = 7 - 2 = 5$

The critical value of t is 3.365.

The value of the test statistic is: $t = (b - B)/s_e = (.2715 - 0)/.1426 = 1.904$

Do not reject the null hypothesis. Hence, B is not positive.

k. For $x = 60$: $\hat{y} = -2.2247 + .2715(60) = 14.0653$

$df = n - 2 = 7 - 2 = 5$ and Area in each tail of the t curve $= \alpha/2 = .5 - (.95/2) = .025$

From the t distribution table, the value of t for $df = 5$ and .025 area in the right tail is 2.571.

The standard deviation of $\hat{y}$ for estimating the mean value of y for $x = 60$ is:

$s_{\hat{y}_m} = s_e \sqrt{\dfrac{1}{n} + \dfrac{(x_0 - \bar{x})^2}{SS_{xx}}} = (3.6172)\sqrt{\dfrac{1}{7} + \dfrac{(60 - 60.2857)^2}{643.4286}} = 1.3678$

The 95% confidence interval for $\mu_{y|60}$ is:

$\hat{y} \pm t s_{\hat{y}_m} = 14.0653 \pm 2.571(1.3678) = 14.0653 \pm 3.5166 = 10.5487$ to 17.5819

l. The standard deviation of $\hat{y}$ for predicting y for $x = 60$ is:

$s_{\hat{y}_p} = s_e \sqrt{1 + \dfrac{1}{n} + \dfrac{(x_0 - \bar{x})^2}{SS_{xx}}} = (3.6172)\sqrt{1 + \dfrac{1}{7} + \dfrac{(60 - 60.2857)^2}{643.4286}} = 3.8672$

The 95% prediction interval for y_p for $x = 60$ is:

$$\hat{y} \pm ts_{\hat{y}_p} = 14.0653 \pm 2.571(3.8672) = 14.0653 \pm 9.9426 = 4.1227 \text{ to } 24.0079$$

m. $H_0 : \rho = 0; \quad H_1 : \rho > 0;$

Area in the right tail of the t curve $= .01;$ and $df = n - 2 = 7 - 2 = 5$

The critical value of t is 3.365.

The value of the test statistic is: $t = r\sqrt{\dfrac{n-2}{1-r^2}} = .65\sqrt{\dfrac{7-2}{1-(.65)^2}} = 1.913$

Do not reject H_0. Do not conclude that the linear correlation coefficient is positive.

Chapter Fourteen

14.1   Data that are divided into different categories for identification purposes are called categorical data. For example, dividing adults according to whether or not they smoke, or classifying registered voters by their party affiliation: Republican, Democrat, Independent, etc. results in categorical data.

14.2   To use the normal distribution for the rejection region of a sign test for categorical data requires a sample size $n > 25$.

14.3   When using the sign test for the median of a single population, Table XI must be used if the sample size $n \leq 25$.

14.4   When using the sign test for the difference between medians of two dependent populations, the large sample case requires that $n > 25$.

14.5   a. The rejection region is $X \geq 12$.

        b. The rejection region is $X \leq 3$ and $X \geq 17$.

        c. The rejection region lies to the left of $z = -1.65$.

14.6   a. Step 1: $H_0: p = .50$; $H_1: p > .50$;   A one–tailed test.

                Step 2: $n = 14$. Since $n < 25$, use the binomial distribution.

                Step 3:   For $\alpha = .05$, the rejection region is $X \geq 11$.

                Step 4:   The observed value of $X = 10$.

                Step 5:   Do not reject the null hypothesis.

        b. Step 1: $H_0: p = .50$; $H_1: p \neq .50$;   A two–tailed test.

                Step 2: $n = 10$. Since $n < 25$, use the binomial distribution.

                Step 3:   For $\alpha = .05$, the rejection region is $X \leq 1$ or $X \geq 9$.

                Step 4:   The observed value of $X = 1$.

                Step 5:   Reject the null hypothesis.

        c. Step 1: $H_0: p = .50$; $H_1: p < .50$;   A one–tailed test.

Step 2: $n = 30$. Since $n > 25$, use the normal distribution.

Step 3:   For $\alpha = .05$, the rejection region lies to the left of $z = -1.65$.

Step 4:   For $p = .50$ and $n = 30$

$$\mu = np = 30\,(.50) = 15 \qquad \text{and} \qquad \sigma = \sqrt{npq} = \sqrt{30(.50)(.50)} = 2.73861279$$

$$X = 12 \text{ and } \frac{n}{2} = \frac{30}{2} = 15 \qquad \text{Since } X < \frac{n}{2}, z = \frac{(X+.5)-\mu}{\sigma} = \frac{(12+.5)-15}{2.73861279} = -.91$$

Step 5:   Do not reject the null hypothesis.

d. Step 1: $H_0: p = .50$; $H_1: p > .50$;   A one–tailed test.

Step 2: $n = 27$. Since $n > 25$, use the normal distribution.

Step 3:   For $\alpha = .05$, the rejection region lies to the right of $z = 1.65$.

Step 4:   For $p = .50$ and $n = 27$

$$\mu = np = 27\,(.50) = 13.5 \qquad \text{and} \qquad \sigma = \sqrt{npq} = \sqrt{27(.50)(.50)} = 2.59807621$$

$$X = 20 \text{ and } \frac{n}{2} = \frac{27}{2} = 13.5 \qquad \text{Since } X > \frac{n}{2}, z = \frac{(X-.5)-\mu}{\sigma} = \frac{(20-.5)-13.5}{2.59807621} = 2.31$$

Step 5:   Reject the null hypothesis.

14.7   a. Step 1: H_0: Median $= 28$; H_1: Median > 28;   A one–tailed test.

Step 2: $n = 10$. Since $n < 25$, use the binomial distribution.

Step 3:   For $\alpha = .05$, the rejection region is $X \geq 9$.

Step 4:   The observed value of $X = 8$.

Step 5:   Do not reject the null hypothesis.

b. Step 1: H_0: Median $= 100$; H_1: Median < 100;   A one–tailed test.

Step 2: $n = 11$. Since $n < 25$, use the binomial distribution.

Step 3:   For $\alpha = .05$, the rejection region is $X \leq 2$.

Step 4:   The observed value of $X = 1$.

Step 5:   Reject the null hypothesis.

c. Step 1: H_0: Median $= 180$; H_1: Median $\neq 180$;   A two–tailed test.

Step 2: $n = 26$. Since $n > 25$, use the normal distribution.

Step 3:   For $\alpha = .05$, the rejection region lies to the left of $z = -1.96$ and to the right of $z = 1.96$.

Step 4:   For $n = 26$, $p = .50$, $q = 1 - p = .50$.

$$\mu = np = 26\,(.50) = 13 \qquad \text{and} \qquad \sigma = \sqrt{npq} = \sqrt{26(.50)(.50)} = 2.54950976$$

$$X = 3 \text{ and } \frac{n}{2} = \frac{26}{2} = 13 \qquad \text{Since } X < \frac{n}{2}, z = \frac{(X+.5)-\mu}{\sigma} = \frac{(3+.5)-13}{2.54950976} = -3.73$$

Step 5: Reject the null hypothesis.

d. Step 1: H_0: Median $= 55$; H_1: Median < 55;   A one–tailed test.

Step 2: $n = 30$. Since $n > 25$, use the normal distribution.

Step 3: For $\alpha = .05$, the rejection region lies to the left of $z = -1.65$.

Step 4: For $p = .50$ and $n = 30$

$$\mu = np = 30\,(.50) = 15 \qquad \text{and} \qquad \sigma = \sqrt{npq} = \sqrt{30(.50)(.50)} = 2.73861279$$

$$X = 6 \text{ and } \frac{n}{2} = \frac{30}{2} = 15 \quad \text{Since } X < \frac{n}{2}, z = \frac{(X+.5)-\mu}{\sigma} = \frac{(6+.5)-15}{2.73861279} = -3.10$$

Step 5: Reject the null hypothesis.

14.8 a. Step 1: H_0: $M = 0$; H_1: $M < 0$;   A one–tailed test.

Step 2: $n = 20$. Since $n < 25$, use the binomial distribution.

Step 3: For $\alpha = .01$, the rejection region is $X \leq 4$.

Step 4: The observed value of $X = 6$.

Step 5: Do not reject the null hypothesis.

b. Step 1: H_0: $M = 0$; H_1: $M > 0$;   A one–tailed test.

Step 2: $n = 8$. Since $n < 25$, use the binomial distribution.

Step 3: For $\alpha = .01$, the rejection region is $X \geq 8$.

Step 4: The observed value of $X = 8$.

Step 5: Reject H_0.

c. Step 1: H_0: $M = 0$; H_1: $M \neq 0$;   A two–tailed test.

Step 2: $n = 29$. Since $n > 25$, use the normal distribution.

Step 3: For $\alpha = .01$, the rejection region lies to the left of $z = -2.58$ and to the right of $z = 2.58$.

Step 4: $n = 29$, $p = .50$, and $q = 1 - p = 1 - .50 = .50$,

$$\mu = np = 29(.50) = 14.5 \qquad \text{and} \qquad \sigma = \sqrt{npq} = \sqrt{29(.50)(.50)} = 2.69258240$$

$$X = 4 \text{ and } \frac{n}{2} = \frac{29}{2} = 14.5, \quad \text{Since } X < \frac{n}{2}, z = \frac{(X+.5)-\mu}{\sigma} = \frac{(4+.5)-14.5}{2.69258240} = -3.71$$

Step 5: Reject the null hypothesis.

14.9 Let p be the proportion of residents who prefer bottled water, and let B represents bottled water and C represent city water.

Step 1: H_0: $p = .50$; H_1: $p \neq .50$;   A two–tailed test.

Step 2: $n = 12$. Since $n < 25$, use the binomial distribution.

Step 3: For a two–tailed test with $n = 12$ and $\alpha = .05$, the rejection region is $X \leq 2$ and $X \geq 10$.

Step 4:   Since p is the proportion of people preferring bottled water, we mark a plus sign for each person who prefers bottled water and a minus sign for each person preferring city water, which yields the following table.

Person	1	2	3	4	5	6	7	8	9	10	11	12
Water Source	B	C	B	C	C	B	C	C	C	C	B	C
Sign	+	–	+	–	–	+	–	–	–	–	+	–

There are 4 plus signs, indicating that 4 of the 12 residents prefer bottled water. Thus, the observed value of $X = 4$.

Step 5:   The observed value of X is not in the rejection region, so do not reject H_0. Do not conclude that the residents prefer either of these two water sources over the other.

14.10   Let p be the proportion of all such families who would prefer Brand A over Brand B.

Step 1: $H_0: p = .50$; $H_1: p > .50$;   A right–tailed test.

Step 2:   One family had no preference, so the true value of $n = 15 - 1 = 14$. Since $n < 25$, use the binomial distribution.

Step 3:   For $n = 14$ and $\alpha = .05$, the rejection region is $X \geq 11$.

Step 4:   Since p is the proportion of families who prefer Brand A, we mark a plus sign for each family that prefers Brand A, a minus sign for each family that prefers Brand B, and a zero for the family that had no preference. The result is summarized in the following table.

Family	1	2	3	4	5	6	7	8	9	10	11	12	13	14	15
Brand	A	A	A	B	A	A	B	A	A	A	0	A	A	B	A
Sign	+	+	+	–	+	+	–	+	+	+	0	+	+	–	+

There are 11 plus signs, indicating that 11 of the 15 families prefer Brand A. Thus, the observed value of $X = 11$.

Step 5:   The observed value of X is in the rejection region, so reject H_0. Conclude that families prefer Brand A over Brand B.

14.11   Let p be the proportion of all drinkers of JW's beer who can distinguish JW's from the rival brand.

$H_0: p = .50$; $H_1: p > .50$;   A right–tailed test.

$n = 20$. Since $n < 25$, use the binomial distribution.

For $n = 20$ and $\alpha = .025$, the rejection region is $X \geq 15$.

Thirteen drinkers in the sample correctly identified JW's. Thus, the observed value of $X = 13$.

Do not reject H_0. Do not conclude that drinkers of JW's are more likely to identify it than not.

14.12   Let p be the proportion of all such voters who favor the Republican candidate.

$H_0: p = .50$; $H_1: p \neq .50$;   A two–tailed test.

Since 20 of the 200 voters have no opinion, the true value of $n = 200 - 20 = 180$. Thus, $n > 25$, so use the normal distribution.

For $\alpha = .01$, the rejection region lies to the left of $z = -2.58$ and to the right of $z = 2.58$.

$n = 180$, $p = .50$ and $q = 1 - p = 1 - .50 = .50$

$\mu = np = 180(.50) = 90$ and $\sigma = \sqrt{npq} = \sqrt{180(.50)(.50)} = 6.70820393$

$X = 95$ and $\dfrac{n}{2} = \dfrac{180}{2} = 90$ Since $X > \dfrac{n}{2}$, $z = \dfrac{(X - .5) - \mu}{\sigma} = \dfrac{(95 - .5) - 90}{6.70820393} = .67$

Do not reject H_0. Do not conclude that voters prefer one candidate over the other.

14.13   Let p be the proportion of adult North Dakota residents who would prefer to stay in North Dakota.

H_0: $p = .50$; H_1: $p < .50$;   A left–tailed test.

Four of the 100 adults have no preference, so the true value of $n = 100 - 4 = 96$. Since $n > 25$, use the normal distribution.

For $\alpha = .025$, the rejection region lies to the left of $z = -1.96$.

$n = 96$ and $p = q = .50$

$\mu = np = 96(.50) = 48$ and $\sigma = \sqrt{npq} = \sqrt{96(.50)(.50)} = 4.89897949$

$X = 41$ and $\dfrac{n}{2} = \dfrac{96}{2} = 48$ Since $X < \dfrac{n}{2}$, $z = \dfrac{(X + .5) - \mu}{\sigma} = \dfrac{(41 + .5) - 48}{4.89897949} = 1.33$

Do not reject H_0. Do not conclude that less than half of all adult residents of North Dakota would prefer to stay.

14.14   Let p be the proportion of doctors who feel that diet is more important than exercise in weight control.

H_0: $p = .50$; H_1: $p > .50$;   A right–tailed test.

Since 21 of the 300 doctors thought that diet and exercise were equally important, the true value of $n = 300 - 21 = 279$. Since $n > 25$, use the normal distribution.

For $\alpha = .01$, the rejection region lies to the right of $z = 2.33$ $n = 279$, $p = q = .50$.

$\mu = np = 279(.50) = 139.5$; and $\sigma = \sqrt{npq} = \sqrt{279(.50)(.50)} = 8.35164654$

$X = 162$ and $\dfrac{n}{2} = \dfrac{279}{2} = 139.5$ Since $X > \dfrac{n}{2}$, $z = \dfrac{(X - .5) - \mu}{\sigma} = \dfrac{(162 - .5) - 139.5}{8.35164654} = 2.63$

Reject H_0. Conclude that for all doctors, the number who favor diet exceeds the number who favor exercise.

14.15   Let p be the proportion of adults that frequently experience stress.

H_0: $p = .50$; H_1: $p > .50$;   A right–tailed test.

$n = 700$. Since $n > 25$, use the normal distribution.

For $\alpha = .01$, the rejection region lies to the right of $z = 2.33$.

$n = 700$ and $p = q = .50$

$$\mu = np = 700(.50) = 350 \qquad \text{and} \qquad \sigma = \sqrt{npq} = \sqrt{700(.50)(.50)} = 13.22875656$$

$$X = 370 \text{ and } \frac{n}{2} = \frac{500}{2} = 350 \qquad \text{Since } X > \frac{n}{2}, z = \frac{(X - .5) - \mu}{\sigma} = \frac{(370 - .5) - 350}{13.22875656} = 1.47$$

Do not reject H_0. Do not conclude that over half of adults frequently experience stress in their daily lives.

14.16 Step 1: H_0: Median = 18 hours; H_1: Median > 18 hours; A right–tailed test.

Step 2: $n = 10$. Since $n < 25$, use the binomial distribution.

Step 3:   For $n = 10$ and $\alpha = .05$, the rejection region is $X \geq 9$.

Step 4:   We mark a plus sign for each time exceeding 18 hours, and a minus sign for each time that is less than 18, which yields the following table.

Adult	1	2	3	4	5	6	7	8	9	10
Time	14	25	22	38	16	26	19	23	41	33
Sign	−	+	+	+	−	+	+	+	+	+

There are eight plus signs, indicating that eight of the 10 times exceed 18 hours. Thus, the observed value of $X = 8$.

Step 5:   Since the observed value of X is not in the rejection region, do not reject H_0. Do not conclude that the median amount of time spent per week on leisure activities by all adults is greater than 18 hours.

14.17 Step 1: H_0: Median = 12 ounces; H_1: Median ≠ 12 ounces;   A two–tailed test.

Step 2:   Since one of the 10 bottles had exactly 12.00 ounces, the true sample size is $n = 10 - 1 = 9$. Since $n \leq 25$, use the binomial distribution.

Step 3:   For $n = 9$ and $\alpha = .05$, the rejection region is $X \leq 1$ and $X \geq 8$.

Step 4:   We mark a plus sign for each bottle that has more than 12 ounces, a minus sign for each bottle that has less than 12 ounces, and a zero for each bottle holding exactly 12 ounces. This yields the following table.

Bottle	1	2	3	4	5	6	7	8	9	10
Amount	12.10	11.95	12.00	12.01	12.02	12.05	12.02	12.03	12.04	12.06
Sign	+	−	0	+	+	+	+	+	+	+

There are eight plus signs, corresponding to the eight bottles which had more than 12 ounces of soda. Thus, the observed value of $X = 8$.

Step 5:   Since the observed value of X is in the rejection region, Reject H_0. Conclude that the median amount of soda in all such bottles differs from 12 ounces.

Note that we could have based the test on the number of minus signs which would have given $X = 1$. Since this value is in the lower part of the rejection region, we would reject H_0, as we did in Step 5 above.

14.18 Step 1: H_0: Median = \$589,133; H_1: Median > \$589,133;   A right–tailed test.

Step 2: $n = 10$. Since $n \leq 25$, we use the binomial distribution.

Step 3:   For $n = 10$ and $\alpha = .05$, the rejection region is $X \geq 9$.

Step 4:   Assigning a plus sign for each salary above \$589,133 and a minus sign to each salary below \$589,133 yields eight plus signs and two minus signs. Thus, the observed value of $X = 8$.

Step 5:   Do not reject H_0. Do not conclude that the median salary of all offensive linemen in the NFL now exceeds \$589,133.

14.19 H_0: Median = 4 minutes; H_1: Median > 4 minutes;   A right–tailed test.

Two of the 28 response times are exactly 4 minutes, so the true sample size is $n = 28 - 2 = 26$.

Since $n > 25$, use the normal distribution.

For a right–tailed test with $\alpha = .01$, the rejection region lies to the right of $z = 2.33$.

Let p be the proportion of response times that exceed 4 minutes. If H_0 is true, we would expect about half of the times to exceed 4 minutes.

Thus, $n = 26$ and $p = q = .50$, so

$$\mu = np = 26(.50) = 13 \quad \text{and} \quad \sigma = \sqrt{npq} = \sqrt{26(.50)(.50)} = 2.54950976$$

If we assign a plus sign to every response time above 4 minutes, a minus sign to every time below 4 minutes, and a zero to every time of exactly 4 minutes, there will be 21 plus signs, 5 minus signs, and 2 zeroes.

For a right–tailed test, we use the larger of the values (21 and 5) as the observed value of X, so $X =$

$$21. \quad \frac{n}{2} = \frac{26}{2} = 13 \qquad\qquad \text{Since } X > \frac{n}{2}, z = \frac{(X-.5)-\mu}{\sigma} = \frac{(21-.5)-13}{2.54950976} =$$

2.94

Reject H_0. Conclude that the median response time to all 911 calls in the inner city is greater than 4 minutes.

14.20 H_0: Median income = \$54,226; H_1: Median income $\neq$ \$54,226;   A two–tailed test.

$n = 400$. Since $n > 25$, use the normal distribution.

For $\alpha = .02$, the rejection region lies to the left of $z = -2.33$ and to the right of $z = 2.33$.

$n = 400$ and $p = q = .50$, so

$$\mu = np = 400(.50) = 200 \qquad\qquad \text{and} \qquad \sigma = \sqrt{npq} = \sqrt{400(.50)(.50)} = 10$$

If we assign a plus sign to every household income above \$54,226 and a minus sign to every income below \$54,226, we obtain 220 plus signs and 180 minus signs. If we use the larger value, $X = 220$.

$$\frac{n}{2} = \frac{400}{2} = 200 \qquad\qquad \text{Since } X > \frac{n}{2}, z = \frac{(X-.5)-\mu}{\sigma} = \frac{(220-.5)-200}{10} = 1.95$$

Do not reject H_0. Do not conclude that the median household income currently differs from \$54,226.

If we had used the smaller value (180 minus signs), we would have:

$$z = \frac{(180+.5)-200}{10} = -1.95 \qquad \text{and our conclusion would be the same (Do not reject } H_0).$$

14.21 H_0: Median = 42 months; H_1: Median < 42 months;   A left–tailed test.

$n = 35$. Since $n > 25$, use the normal distribution.

For $\alpha = .01$, the rejection region lies to the left of $z = -2.33$.

$n = 35$ and $p = q = .50$, so

$$\mu = np = 35(.50) = 17.5 \qquad \text{and} \qquad \sigma = \sqrt{npq} = \sqrt{35(.50)(.50)} = 2.95803989$$

If we mark a plus sign for every time that exceeds 42 months and a minus sign for every time that is less than 42 months, we obtain 10 plus signs and 25 minus signs. Since the test is left–tailed, we use the smaller value, $X = 10$.

$$\frac{n}{2} = \frac{35}{2} = 17.5 \qquad\qquad \text{Since } X < \frac{n}{2}, z = \frac{(X+.5)-\mu}{\sigma} = \frac{(10+.5)-17.5}{2.95803989} = -2.37$$

Reject H_0. Conclude that the median time served by all such prisoners is less than 42 months.

14.22   Let M denote the difference in median weight before and after the special diet. Thus:

Paired difference = Weight before – Weight after

Step 1: H_0: $M = 0$ (The diet does not increase the median weight)

H_1: $M < 0$ (The diet increases the median weight)

A left–tailed test.

Step 2:   One boy's weight remained the same, so the true sample size is $n = 12 - 1 = 11$.

Since $n < 25$, use the binomial distribution.

Step 3:   For $n = 11$ and $\alpha = .025$, the rejection region is $X \leq 1$.

Step 4:   We assign a plus sign to each boy who lost weight, a minus sign to each boy who gained weight, and a zero to any boy whose weight did not change

Before	65	63	71	60	66	72	78	74	58	59	77	65
After	70	68	75	60	69	70	81	81	66	56	79	71
Sign	−	−	−	0	−	+	−	−	−	+	−	−

There are two plus signs, nine minus signs and one zero. Since the test is left–tailed we use the lower number. Thus, the observed value of $X = 2$.

Step 5: X is not in the rejection region, so do not reject H_0. Do not conclude that the diet increases the median weight of all such boys.

14.23   For each employee: Paired difference = score before course − score after course

Let M denote the difference in median test scores before and after the course, where a plus sign indicates an employee's score was lower after the course. Thus, if the course improves test scores, we would expect $M < 0$.

Step 1: H_0: $M = 0$; H_1: $M < 0$;   A left–tailed test.

Step 2:   One of the employee's scores were the same before and after the course, so the true sample size is $n = 6$.

Step 3:   For $n = 6$ and $\alpha = .05$, the rejection region is $X = 0$.

Step 4:   We assign a plus sign to each employee whose score decreased, a minus sign to each employee whose score increased, and a zero to each employee whose score was the same before and after the course. The results are shown in the following table.

Before	8	5	4	9	6	9	5
After	10	8	5	11	6	7	9
Sign	−	−	−	−	0	+	−

We find one plus sign, five minus signs, and one zero, indicating that one employee's score decreased, five increased, and one remained the same after the course. Since the test is left–tailed, we use the smaller number of signs for X. Thus, the observed value of $X = 1$.

Step 5: $X = 1$ is not in the rejection region so do not reject H_0. Do not conclude that attending this course increases the median self–confidence test score of all employees.

14.24   For each day: Paired difference = (Day shift hours of sick leave) − (Night shift hours of sick leave)

H_0: $M = 0$; H_1: $M < 0$;   A left–tailed test.

$n = 12 \leq 25$, so use the binomial distribution.

For $n = 12$ and $\alpha = .05$, the rejection region is $X \leq 2$.

Assigning a plus sign to each day in which the day shift used more sick leave than the night shift and a minus sign to each day in which the day shift used fewer hours results in three plus signs and nine minus signs. Since the test is left–tailed we use the smaller number of signs. Thus:

The observed value of $X = 3$.

Since $X = 3$ is not in the rejection region do not reject H_0. Do not conclude that the median number of hours of sick leave used by workers per day is lower for the day shift than for the night shift.

14.25   For each employee: Paired difference = (Bikes assembled before)−(Bikes assembled after new system)

H_0: $M = 0$; H_1: $M \neq 0$;   A two–tailed test.

Since one worker assembled the same number of bikes under both systems, the true value of $n = 27 - 1 = 26$.

$n = 26 > 25$, so use the normal distribution.

For $\alpha = .02$, the rejection region lies to the left of $z = -2.33$ and to the right of $z = 2.33$.

$n = 26$ and $p = q = .50$, so,

$$\mu = np = 26(.50) = 13 \qquad \text{and} \qquad \sigma = \sqrt{npq} = \sqrt{26(.50)(.50)} = 2.54950976$$

Assigning a plus sign to each employee who produced more under the new payment system, and a minus sign to each employee who produced less, results in 19 plus signs and 7 minus signs with one zero, as one employee produced the same number of bikes . If we use the larger number of signs, $X = 19$.

$$\frac{n}{2} = \frac{26}{2} = 13 \qquad\qquad \text{Since } X > \frac{n}{2}, z = \frac{(X - .5) - \mu}{\sigma} = \frac{(19 - .5) - 13}{2.54950976} = 2.16$$

Do not reject H_0. Do not conclude that the median number of bikes assembled differs from the median before the new payment system was instituted.

If we had used the smaller number of signs (7 minus signs) we would have $z = \dfrac{(7 + .5) - 13}{2.54950976} = -2.16$

and our conclusion would be the same (Do not reject H_0).

14.26   For each blood sample: Paired difference = (Lab A's cholesterol level) – (Lab B's cholesterol level)

$H_0: M = 0$; $\qquad\qquad$ $H_1: M > 0$; $\qquad\qquad$ A right–tailed test.

Since the two labs' readings were the same on one sample, the true sample size is $n = 30 - 1 = 29$.

Since $n > 25$, use the normal distribution.

For $\alpha = .01$, the rejection region lies to the right of $z = 2.33$.

$n = 29$ and $p = q = .50$, so

$$\mu = np = 29(.50) = 14.5 \qquad \text{and} \qquad \sigma = \sqrt{npq} = \sqrt{29(.50)(.50)} = 2.69258240$$

Assigning a plus sign to each sample for which Lab A gives a higher cholesterol level than lab B, and a minus sign for each sample for which lab B gives a higher level, and zero for each sample for which the two labs give the same level results in 21 plus signs, 8 minus signs, and one zero. The test is right–tailed, so we use the larger number of signs. Thus, $X = 21$.

$$\frac{n}{2} = \frac{29}{2} = 14.5 \qquad\qquad \text{Since } X > \frac{n}{2}, z = \frac{(X - .5) - \mu}{\sigma} = \frac{(21 - .5) - 14.5}{2.69258240} = 2.23$$

Do not reject H_0. Do not conclude that the median cholesterol level for all such adults as determined by lab A is higher than that of lab B.

14.27   For each matched pair of cows:

Paired difference = Milk production of cow without hormone –Milk production of cow with hormone.

$H_0: M = 0$; $\qquad\qquad$ $H_1: M \neq 0$; $\qquad\qquad$ A two–tailed test.

Since milk production was the same for two pairs of cows, the true value of $n = 30 - 2 = 28$.

$n > 25$, so use the normal distribution.

For a two–tailed test with $\alpha = .05$, the rejection region lies to the left of $z = -1.96$ and to the right of $z = 1.96$.

$n = 28$ and $p = q = .50$, so

$$\mu = np = 28(.50) = 14 \qquad \text{and} \qquad \sigma = \sqrt{npq} = \sqrt{28(.50)(.50)} = 2.64575131$$

Assigning a plus sign to each pair of cows in which the cow taking the hormone produced less milk, a minus sign to each pair in which the cow taking the hormone produced more milk, and a zero to each pair in which both cows produced the same amount of milk results in 9 plus signs, 19 minus signs, and two zeroes.

Using the larger number of signs, $X = 19$.

$$\frac{n}{2} = \frac{28}{2} = 14 \qquad\qquad \text{Since } X > \frac{n}{2}, z = \frac{(X - .5) - \mu}{\sigma} = \frac{(19 - .5) - 14}{2.64575131} =$$

1.70

Do not reject H_0. Do not conclude that the hormone changes the milk production of such cows.

Note that if we had used the smaller number of signs ($X = 9$), then $z = \dfrac{(9 + .5) - 14}{2.64575131} = -1.70$ and the conclusion would be the same (do not reject H_0).

14.28   When $n < 30$ and the population of paired differences does not have a normal distribution, we would use the Wilcoxon signed–rank test procedure instead of the paired–sample test of Chapter 10.

14.29   The null hypothesis of the Wilcoxon signed–rank test usually states that the medians of the two population distributions are equal.

14.30   When two or more absolute differences have the same value in the Wilcoxon signed–rank test, they are assigned the average of their ranks.

14.31   a.   The rejection region is $T \le 11$.

b.   The rejection region is $T \le 7$.

c.   The rejection region lies to the left of $z = -1.96$.

d.   The rejection region lies to the right of $z = 2.33$.

14.32   a.   Step 1: $H_0: M_A = M_B$; $H_1: M_A < M_B$.

Step 2: $n = 8 < 15$. Use the Wilcoxon signed–rank test for the small–sample case.

Step 3:   For $n = 8$ and $\alpha = .05$, the rejection region is $T \le 6$.

Step 4:   The observed value of $T = 5$.

Step 5:   Reject H_0

b.   Step 1: $H_0: M_A = M_B$; $H_1: M_A > M_B$.

Step 2: $n = 15$. Use the Wilcoxon signed–rank test for the small–sample case.

Step 3:   For $n = 15$ and $\alpha = .01$, the rejection region is $T \leq 20$.

Step 4:   The observed value of $T = 20$.

Step 5:   Reject H_0

c. Step 1: H_0: $M_A = M_B$; H_1: $M_A \neq M_B$.

Step 2: $n = 25$. Since $n > 15$, use the Wilcoxon signed–rank test with the normal distribution approximation.

Step 3:   For a two–tailed test with $\alpha = .02$, the rejection region lies to the left of $z = -2.33$ and to the right of $z = 2.33$

Step 4: $\mu_T = \dfrac{n(n+1)}{4} = \dfrac{25(25+1)}{4} = 162.5$

$\sigma_T = \sqrt{\dfrac{n(n+1)(2n+1)}{24}} = \sqrt{\dfrac{25(25+1)(50+1)}{24}} = 37.16517187$

$z = \dfrac{T - \mu_T}{\sigma_T} = \dfrac{51 - 162.5}{37.16517187} = -3.00$

Step 5:   Reject H_0.

d. Step 1: H_0: $M_A = M_B$; H_1: $M_A < M_B$.

Step 2: $n = 36$. Since $n > 15$, use the Wilcoxon signed–rank test procedure with the normal distribution approximation.

Step 3:   For $\alpha = .01$ with a left–tailed test, the rejection region lies to the left of $z = -2.33$.

Step 4: $\mu_T = \dfrac{n(n+1)}{4} = \dfrac{36(36+1)}{4} = 333$

$\sigma_T = \sqrt{\dfrac{n(n+1)(2n+1)}{24}} = \sqrt{\dfrac{36(36+1)(72+1)}{24}} = 63.65139433$

$z = \dfrac{T - \mu_T}{\sigma_T} = \dfrac{238 - 333}{63.65139433} = -1.49$

Step 5:   Do not reject H_0.

14.33 a.   For each salesperson:

Paired difference = (Number of contacts before) − (Number of contacts after)

Let M_A and M_B be the median number of contacts by all such salespersons after and before the installation of governors, respectively.

Step 1: H_0: $M_A = M_B$; H_1: $M_A < M_B$;   A left–tailed test.

Step 2: $n = 7$. Since $n < 15$, use the Wilcoxon signed–rank test for the small–sample case.

Step 3:   For a one–tailed test with $n = 7$ and $\alpha = .05$, the rejection region is $T \leq 4$.

Step 4:   The signed ranks are calculated in the following table.

Before	After	Differences (Before–After)	Absolute Differences	Ranks of Differences	Signed Ranks
50	49	+1	1	1	+1
63	60	+3	3	2	+2
42	47	−5	5	4.5	−4.5
55	51	+4	4	3	+3
44	50	−6	6	6	−6
65	60	+5	5	4.5	+4.5
66	58	+8	8	7	+7

Sum of positive ranks = 1 + 2 + 3 + 4.5 + 7 = 17.5

Sum of absolute values of negative ranks = 4.5 + 6 = 10.5

For a left–tailed test, T is the sum of the absolute values of the negative ranks. Thus:

The observed value of $T = 10.5$

Step 5:   Since the observed value of T is not in the rejection region, do not reject H_0. Do not conclude that the use of governors tends to reduce the number of contacts made per week by the Gamma Corporation's salespersons.

b. The conclusion in part a of this exercise is the same as that of the corresponding test of Exercise 10.101 (do not reject H_0).

14.34   For each salesperson: Paired difference = Gas mileage before – Gas mileage after

Let M_A and M_B be the median gas mileage after and before the installation of governors for all such salespersons.

a. Step 1: $H_0: M_A = M_B$; $H_1: M_A > M_B$.   A right–tailed test.

Step 2: $n = 7$. Since $n \leq 15$, use the Wilcoxon signed–rank test for the small–sample case.

Step 3:   For $n = 7$, $\alpha = .05$, and a one–tailed test the rejection region is $T \leq 4$.

Step 4:   The signed ranks are found in the table below.

Before	After	Differences (Before–After)	Absolute Differences	Ranks of Differences	Signed Ranks
25	26	−1	1	1.5	−1.5
21	24	−3	3	4.5	−4.5
27	26	+1	1	1.5	+1.5
23	25	−2	2	3	−3
19	24	−5	5	7	−7
18	22	−4	4	6	−6
20	23	−3	3	4.5	−4.5

Sum of positive ranks = 1.5

Sum of absolute values of negative ranks = 1.5 + 4.5 + 3 + 7 + 6 + 4.5 = 26.5

For a right–tailed test, T is the sum of the positive ranks. Thus:

The observed value of $T = 1.5$

Step 5:   Since the observed value of T is in the rejection region, reject H_0. Conclude that the use of governors tends to increase the median gas mileage for the Gamma Corporation's salespersons.

b.   The conclusion in part a of this exercise is the same as that of the corresponding test in Exercise 10.101 (Reject H_0).

14.35   For each employee: Paired difference = score before course − score after course

Let M_A and M_B denote the median self–confidence test scores of all such employees after and before attending the course.

a. H_0: $M_A = M_B$; H_1: $M_A > M_B$;   A right–tailed test.

One of the employees had the same test score before and after attending the course, so the true sample size is $n = 7 - 1 = 6$. Since $n < 15$, use the Wilcoxon signed–rank test procedure for the small–sample case.

For $n = 6$, $\alpha = .05$, and a one–tailed test, the rejection region is $T \leq 2$.

Before	After	Differences (Before–After)	Absolute Differences	Ranks of Differences	Signed Ranks
8	10	−2	2	3	−3
5	8	−3	3	5	−5
4	5	−1	1	1	−1
9	11	−2	2	3	−3
6	6	−	−	−	−
9	7	+2	2	3	3
5	9	−4	4	6	−6

Sum of positive ranks = 3

Sum of absolute values of negative ranks = $3 + 5 + 1 + 3 + 6 = 18$

For a right–tailed test, T is the sum of the positive ranks. Thus, the observed value of $T = 3$

Do not reject H_0. Do not conclude that attending this course increases the median self–confidence test scores of employees.

b.   In both exercises H_0 is not rejected.

14.36   a.   For each employee:

Paired difference = Number under old payment system − Number under new payment system

Let M_A and M_B denote the median number of bicycles assembled under the new and old payment systems respectively.

H_0: $M_A = M_B$; H_1: $M_A \neq M_B$;   A two–tailed test.

Since one worker assembled the same number of bikes under both systems, the true value of $n = 27 - 1 = 26$.

$n = 26 > 15$, so use the Wilcoxon signed–rank test procedure with the normal distribution approximation.

For $\alpha = .02$ and a two–tailed test, the rejection region lies to the left of $z = -2.33$ and to the right of $z = 2.33$.

For $n = 27$, $\mu_T = \dfrac{n(n+1)}{4} = \dfrac{26(26+1)}{4} = 175.5$

$$\sigma_T = \sqrt{\dfrac{n(n+1)(2n+1)}{24}} = \sqrt{\dfrac{26(26+1)(54+1)}{24}} = 39.37321425$$

Calculating the signed ranks yields:

Sum of positive ranks = 61

Sum of absolute values of negative ranks = 290

Since the test is two–tailed, we may use either sum of ranks (61 or 290) as the observed value of T.

If we let $T = 61$: $z = \dfrac{T - \mu_T}{\sigma_T} = \dfrac{61 - 175.5}{39.37321425} = -2.91$

Reject H_0. Conclude that the median number of bicycles assembled differs under the new payment system. Note that if we had taken $T = 290$, then $z = 2.91$ and the conclusion would be the same (Reject H_0).

b. The two conclusions differ. Using the sign test in Exercise 14.25, H_0 was not rejected, but using the Wilcoxon signed–rank test in part a, H_0 was rejected.

14.37 For each adult: Paired difference = hiking time before course – hiking time after course

Let M_A and M_B denote the median time to complete the hike after and before the fitness course for all such adults.

H_0: $M_A = M_B$; H_1: $M_A < M_B$;   A left–tailed test.

$n = 20 > 15$, so use the Wilcoxon signed–rank test with the normal distribution approximation.

For $\alpha = .025$ and a left–tailed test, the rejection region lies to the left of $z = -1.96$.

For $n = 20$:

$$\mu_T = \dfrac{n(n+1)}{4} = \dfrac{20(20+1)}{4} = 105$$

$$\sigma_T = \sqrt{\dfrac{n(n+1)(2n+1)}{24}} = \sqrt{\dfrac{20(20+1)(40+1)}{24}} = 26.78619047$$

Calculating the differences, absolute differences, ranks, and signed ranks yields:

Sum of positive ranks = 182.5

Sum of absolute values of negative ranks = 27.5

Since the test is left–tailed, T is the sum of the absolute values of the negative ranks. Thus, the observed value of $T = 27.5$

$$z = \frac{T - \mu_T}{\sigma_T} = \frac{27.5 - 105}{26.78619047} = -2.89$$

Reject H_0. Conclude that the fitness course tends to reduce the median time required to complete the two–mile hike.

14.38 For each adult: Paired difference = Balance before program – Balance after program

Let M_A and M_B denote the median credit card debt of all such adults after and before the group therapy program.

H_0: $M_A = M_B$; $\quad$ H_1: $M_A < M_B$; $\quad\quad\quad$ A left–tailed test.

$n = 90 > 15$, so use the Wilcoxon signed–rank test procedure with the normal distribution.

For $\alpha = .05$ and a left–tailed test, the rejection region lies to the left of $z = -1.65$.

For $n = 90$:

$$\mu_T = \frac{n(n+1)}{4} = \frac{90(90+1)}{4} = 2047.5$$

$$\sigma_T = \sqrt{\frac{n(n+1)(2n+1)}{24}} = \sqrt{\frac{90(90+1)(180+1)}{24}} = 248.52816742$$

Since the test is left–tailed, we use the sum of the absolute values of the negative ranks.

Thus, the observed value of $T = 1588$

$$z = \frac{T - \mu_T}{\sigma_T} = \frac{1588 - 2047.5}{248.52816742} = -1.85$$

Reject H_0. Conclude that the group therapy program tends to reduce credit card debt for all such adults.

14.39 The Wilcoxon signed–rank test is used for paired samples, while the Wilcoxon rank–sum test is used for independent samples.

14.40 a. The rejection region is $T \geq 71$.

b. The rejection region is $T \leq 83$ and $T \geq 127$.

c. The rejection region lies to the left of $z = -1.65$.

d. The rejection region lies to the left of $z = -2.58$ and to the right of $z = 2.58$.

14.41 a. Step 1: H_0: The two population distributions are identical.

$\quad\quad\quad\quad\quad\quad$ H_1: The two population distributions are different.

$\quad\quad$ Step 2: Because $n_1 < 10$ and $n_2 < 10$, use the Wilcoxon rank–sum test for small samples.

$\quad\quad$ Step 3: The rejection region is $T \leq 28$ and $T \geq 56$.

$\quad\quad$ Step 4: The observed value of $T = 22$.

$\quad\quad$ Step 5: Reject H_0.

b. **Step 1:** H_0: The two population distributions are identical.

            H_1: The distribution of population 1 lies to the right of the distribution of population 2.

   Step 2: Because $n_2 = 12 > 10$, use the normal distribution.

   Step 3: The rejection region lies to the right of $z = 1.96$.

   Step 4: $\mu_T = \dfrac{n_1(n_1 + n_2 + 1)}{2} = \dfrac{10(10 + 12 + 1)}{2} = 115$

$$\sigma_T = \sqrt{\dfrac{n_1 n_2 (n_1 + n_2 + 1)}{12}} = \sqrt{\dfrac{10(12)(10 + 12 + 1)}{12}} = 15.16575089$$

$$z = \dfrac{T - \mu_T}{\sigma_T} = \dfrac{137 - 115}{15.16575089} = 1.45$$

   Step 5: Do not reject H_0.

c. **Step 1:** H_0: The two population distributions are identical.

            H_1: The distribution of population 1 lies to the left of the distribution of population 2.

   Step 2: Because $n_2 = 11 > 10$, use the normal distribution.

   Step 3: The rejection region lies to the left of $z = -1.65$.

   Step 4: $\mu_T = \dfrac{n_1(n_1 + n_2 + 1)}{2} = \dfrac{9(9 + 11 + 1)}{2} = 94.5$

$$\sigma_T = \sqrt{\dfrac{n_1 n_2 (n_1 + n_2 + 1)}{12}} = \sqrt{\dfrac{9(11)(9 + 11 + 1)}{12}} = 13.16244658$$

$$z = \dfrac{T - \mu_T}{\sigma_T} = \dfrac{68 - 94.5}{13.16244658} = -2.01$$

   Step 5: Reject H_0.

d. **Step 1:** H_0: The two population distributions are identical.

            H_1: The two population distributions are different.

   Step 2: Because n_1 and n_2 are greater than 10, use the normal distribution.

   Step 3: The rejection region lies to the left of $z = -2.58$ and to the right of $z = 2.58$.

   Step 4: $\mu_T = \dfrac{n_1(n_1 + n_2 + 1)}{2} = \dfrac{22(22 + 23 + 1)}{2} = 506$

$$\sigma_T = \sqrt{\dfrac{n_1 n_2 (n_1 + n_2 + 1)}{12}} = \sqrt{\dfrac{22(23)(22 + 23 + 1)}{12}} = 44.04164696$$

$$z = \dfrac{T - \mu_T}{\sigma_T} = \dfrac{638 - 506}{44.04164696} = 3.00$$

   Step 5: Reject H_0.

14.42   Step 1: H_0: The population distribution of caffeine contents are identical for the two brands of coffee.

                         H_1: The population distribution of caffeine contents are different for the two brands of coffee.

        Step 2:   Because n_1 and n_2 are less than 10, use the Wilcoxon rank–sum test for small samples.

        Step 3:   For $n_1 = n_2 = 8$ and $\alpha = .10$, the critical values are $T_L = 52$ and $T_U = 84$. Thus the rejection region is $T \leq 52$ and $T \geq 84$.

        Step 4:

Brand I		Brand II	
Amount of Caffeine	Rank	Amount of Caffeine	Rank
82	13.5	75	5
77	7	80	10
85	16	76	6
73	2.5	81	11.5
84	15	72	1
79	9	74	4
81	11.5	73	2.5
82	13.5	78	8
	Sum = 88		Sum = 48

              Since $n_1 = n_2$ and the test is two–tailed, the observed value of T is given by the sum of ranks for either sample.

        Step 5:   Using $T = 88$, $T > 84$, so reject H_0. Conclude that the two brands of coffee have different median caffeine contents per jar.

              Note that using $T = 48$ would lead to $T < 52$, so the conclusion would be the same (reject H_0).

14.43   Step 1: H_0: The population distributions of times in the 500–meter event are identical for the two types of skates.

                         H_1: The population distribution of times with the new skates lies to the left of the population distribution of times with the traditional skates.

        Step 2:   Because n_1 and n_2 are less than 10, use the Wilcoxon rank–sum test for small samples.

        Step 3:   For a left–tailed test with $n_1 = 7$, $n_2 = 8$ and $\alpha = .05$, $T_L = 41$. Thus, the rejection region is $T \leq 41$.

Step 4:

New Skates		Traditional Skates	
Time	Rank	Time	Rank
40.5	7.5	41.0	13
40.3	6	40.8	11
39.5	1	40.9	12
39.7	2	39.8	3
40.0	5	40.6	9
39.9	4	40.7	10
41.5	15	41.1	14
		40.5	7.5
	Sum = 40.5		Sum = 79.5

Since the test is one–tailed, the observed value of T is the sum of ranks for the smaller sample. Thus, the observed value of $T = 40.5$.

Step 5:   Reject H_0. Conclude that the new skates tend to produce faster times in this event.

14.44 H_0: The current median sale price of houses is the same in Phoenix and Las Vegas.

H_1: The current median sale price of houses is not the same in Phoenix and Las Vegas.

Because $n_1 \leq 10$ and $n_2 \leq 10$, use the Wilcoxon rank–sum test for small samples.

The test is two–tailed, so the rejection region is $T \leq 66$ and $T \geq 114$.

Phoenix		Las Vegas	
Price	Rank	Price	Rank
158	8	180	15
169	13	145	5
129	1	219	19
179	14	189	16
149	6	159	9
160	10	168	12
142	4	195	17
140	3	139	2
207	18	162	11
		150	7
	Sum = 77		Sum = 113

Since the sample sizes are unequal, the observed value of T is given by the sum of ranks for the smaller sample. Hence, the observed value of $T = 77$, T is between the critical values 66 and 114.

Thus, do not reject H_0. Do not conclude that the median sale prices of houses are different for Phoenix and Las Vegas.

14.45 H_0: The population distributions of numbers of good parts produced by the two groups are identical.

H_1: The population distribution of numbers of good parts produced by Group A lies to the right of the corresponding distribution for Group B.

Because n_1 and n_2 are greater than 10, use the normal distribution.

The test is right–tailed, so the rejection region is $z > 2.33$.

Calculating the ranks yields:

Sum of ranks for Group A = 174.5

Sum of ranks for Group B = 125.5

For a right–tailed test with equal sample sizes, the observed value of T is given by the sum of ranks for the first sample. Thus, the observed value of $T = 174.5$.

$$\mu_T = \frac{n_1(n_1 + n_2 + 1)}{2} = \frac{12(12+12+1)}{2} = 150$$

$$\sigma_T = \sqrt{\frac{n_1 n_2(n_1 + n_2 + 1)}{12}} = \sqrt{\frac{12(12)(12+12+1)}{12}} = 17.32050808$$

$$z = \frac{T - \mu_T}{\sigma_T} = \frac{174.5 - 150}{17.32050808} = 1.41$$

Do not reject H_0. Do not conclude that the median number of good parts produced by machinists who take a five–minute break every hour is higher than the corresponding median for machinists who do not take such breaks.

14.46 H_0: The population distributions of mileages are identical for the two brands of tires.

H_1: The population distribution of mileages for Brand X lies to the right of the distribution of mileages for Brand Y.

Because n_1 and n_2 are greater than 10, use the normal distribution.

The test is right–tailed, so the rejection region lies to the right of $z = 1.65$.

Calculating the ranks yields:

Sum of ranks for Brand X = 161.5   Sum of ranks for Brand Y = 114.5

The observed value of T is given by the sum of ranks for the smaller sample. Thus:

The observed value of $T = 161.5$

$$\mu_T = \frac{n_1(n_1 + n_2 + 1)}{2} = \frac{11(11+12+1)}{2} = 132$$

$$\sigma_T = \sqrt{\frac{n_1 n_2(n_1 + n_2 + 1)}{12}} = \sqrt{\frac{11(12)(11+12+1)}{12}} = 16.24807681$$

$$z = \frac{T - \mu_T}{\sigma_T} = \frac{161.5 - 132}{16.24807681} = 1.82$$

Reject H_0. Conclude that the median mileage for Brand X tires is greater than the median mileage for Brand Y tires.

14.47 H_0: The population distribution of travel times are identical for the planes and buses.

H_1: The population distribution of travel times for the plane lies to the right of the distribution of travel times for the bus.

Because n_1 and n_2 are greater than 10, use the normal distribution.

The test is right–tailed, so the rejection region lies to the right of $z = 1.65$.

The observed value of T is given by the sum of ranks for the smaller sample, that of the planes. Thus:

The observed value of $T = 295$.

$$\mu_T = \frac{n_1(n_1 + n_2 + 1)}{2} = \frac{15(15 + 17 + 1)}{2} = 247.5$$

$$\sigma_T = \sqrt{\frac{n_1 n_2 (n_1 + n_2 + 1)}{12}} = \sqrt{\frac{15(17)(15 + 17 + 1)}{12}} = 26.48112535$$

$$z = \frac{T - \mu_T}{\sigma_T} = \frac{295 - 247.5}{26.48112535} = 1.79$$

Reject H_0. Conclude that the median travel time for the plane trip is higher than for the bus trip.

14.48 The Kruskal–Wallis test is used to test whether or not the medians of three or more populations are equal.

14.49 In the ANOVA procedure of Chapter 12, the populations being compared are assumed to have normal distributions. This assumption is not required for the Kruskal–Wallis test.

14.50 In a Kruskal–Wallis test, the null hypothesis states that the population distributions are identical. The alternative hypothesis is that at least one of the population distributions differs from the others.

14.51 a. Step 1: H_0: The three population distributions are identical.

                H_1: The three population distributions are not identical.

      Step 2:   Use the χ^2 distribution.

      Step 3:   For $\alpha = .05$ and $df = k - 1 = 3 - 1 = 2$, the critical value of $\chi^2 = 5.991$, so the rejection region is $\chi^2 > 5.991$.

      Step 4: $n = n_1 + n_2 + n_3 = 9 + 8 + 5 = 22$

$$H = \frac{12}{n(n+1)}\left[\frac{R_1^2}{n_1} + \frac{R_2^2}{n_2} + \frac{R_3^2}{n_3}\right] - 3(n+1) = \frac{12}{22(22+1)}\left[\frac{(81)^2}{9} + \frac{(102)^2}{8} + \frac{(70)^2}{5}\right] - 3(22+1)$$

$$= 2.372$$

      Step 5:   Do not reject H_0.

    b. Step 1: H_0: The four population distributions are identical.

                H_1: The four population distributions are not identical.

      Step 2:   Use the x^2 distribution.

      Step 3:   For $\alpha = .05$ and $df = k - 1 = 4 - 1 = 3$, the critical value of $\chi^2 = 7.815$, so the rejection region is $\chi^2 > 7.815$.

Step 4: $n = n_1 + n_2 + n_3 + n_4 = 5 + 5 + 5 + 5 = 20$

$$H = \frac{12}{n(n+1)}\left[\frac{R_1^2}{n_1} + \frac{R_2^2}{n_2} + \frac{R_3^2}{n_3} + \frac{R_4^2}{n_4}\right] - 3(n+1)$$

$$= \frac{12}{20(20+1)}\left[\frac{(27)^2}{5} + \frac{(30)^2}{5} + \frac{(83)^2}{5} + \frac{(70)^2}{5}\right] - 3(20+1) = 13.674$$

Step 5:   Reject H_0.

c. Step 1: H_0: The three population distributions are identical.

    H_1: The three population distributions are not identical.

Step 2:   Use the χ^2 distribution.

Step 3:   For $\alpha = .05$ and $df = k - 1 = 3 - 1 = 2$, the critical value of $\chi^2 = 5.991$, so the rejection region is $\chi^2 > 5.991$.

Step 4: $n = n_1 + n_2 + n_3 = 6 + 10 + 6 = 22$

$$H = \frac{12}{n(n+1)}\left[\frac{R_1^2}{n_1} + \frac{R_2^2}{n_2} + \frac{R_3^2}{n_3}\right] - 3(n+1) = \frac{12}{22(22+1)}\left[\frac{(93)^2}{6} + \frac{(70)^2}{10} + \frac{(90)^2}{6}\right] - 3(22+1)$$

$$= 8.822$$

Step 5:   Reject H_0.

d. Step 1: H_0: The five population distributions are identical.

    H_1: The five population distributions are not identical.

Step 2:   Use the χ^2 distribution.

Step 3:   For $\alpha = .05$ and $df = k - 1 = 5 - 1 = 4$, the critical value of $x^2 = 9.488$, so the rejection region is $\chi^2 > 9.488$.

Step 4: $n = n_1 + n_2 + n_3 + n_4 + n_5 = 8 + 9 + 8 + 10 + 9 = 44$

$$H = \frac{12}{n(n+1)}\left[\frac{R_1^2}{n_1} + \frac{R_2^2}{n_2} + \frac{R_3^2}{n_3} + \frac{R_4^2}{n_4} + \frac{R_5^2}{n_5}\right] - 3(n+1)$$

$$= \frac{12}{44(44+1)}\left[\frac{(210)^2}{8} + \frac{(195)^2}{9} + \frac{(178)^2}{8} + \frac{(212)^2}{10} + \frac{(195)^2}{9}\right] - 3(44+1) = .863$$

Step 5: Do not reject H_0.

14.52   Step 1: H_0: The population distributions are identical.

    H_1: The population distributions are not identical.

Step 2:   Use the χ^2 distribution.

Step 3:   For $\alpha = .01$ and $df = k - 1 = 3 - 1 = 2$, the critical value of $\chi^2 = 9.210$, so the rejection region is $\chi^2 > 9.210$.

Step 4: $n = n_1 + n_2 + n_3 = 6 + 5 + 5 = 16$

$$R_1 = 3 + 1 + 10 + 7 + 9 + 6 = 36$$

$$R_2 = 14 + 11 + 16 + 15 + 12 = 68$$

$$R_3 = 2 + 4.5 + 13 + 4.5 + 8 = 32$$

$$H = \frac{12}{n(n+1)}\left[\frac{R_1^2}{n_1} + \frac{R_2^2}{n_2} + \frac{R_3^2}{n_3}\right] - 3(n+1) = \frac{12}{16(16+1)}\left[\frac{(36)^2}{6} + \frac{(68)^2}{5} + \frac{(32)^2}{5}\right] - 3(16+1)$$

$$= 8.365$$

Step 5:   Do not reject H_0.

14.53   a.   Step 1: H_0: The population distributions of test scores of fourth–grade students taught by the three methods are identical.

H_1: The population distributions of test scores of fourth–grade students taught by the three methods are not identical.

Step 2:   Use the χ^2 distribution.

Step 3:   For $\alpha = .01$ and $df = k - 1 = 3 - 1 = 2$, the critical value of $\chi^2 = 9.210$, so the rejection region is $\chi^2 > 9.210$.

Step 4:

Method I		Method II		Method III	
Score	Rank	Score	Rank	Score	Rank
48	1	55	3	84	11
73	9	85	12	68	6
51	2	70	8	95	15
65	4	69	7	74	10
87	13	90	14	67	5
$n_1 = 5$	$R_1 = 29$	$n_2 = 5$	$R_2 = 44$	$n_3 = 5$	$R_3 = 47$

Here, $n = n_1 + n_2 + n_3 = 5 + 5 + 5 = 15$

$$H = \frac{12}{n(n+1)}\left[\frac{R_1^2}{n_1} + \frac{R_2^2}{n_2} + \frac{R_3^2}{n_3}\right] - 3(n+1) = \frac{12}{15(15+1)}\left[\frac{(29)^2}{5} + \frac{(44)^2}{5} + \frac{(47)^2}{5}\right] - 3(15+1) = 1.860$$

Step 5:   Do not reject H_0. Do not conclude that the median test scores of all fourth–grade students taught by the three methods are different.

b.   The conclusion is the same (do not reject H_0).

14.54   Step 1: H_0: The population distributions of auto insurance premiums for the four companies are identical.

H_1: The population distributions of auto insurance premiums for the four companies are not identical.

Step 2:   Use the χ^2 distribution.

Step 3:   For $\alpha = .05$ and $df = k - 1 = 4 - 1 = 3$, the critical value of $\chi^2 = 7.815$, so the rejection region is $\chi^2 > 7.815$.

Step 4:

Company A		Company B		Company C		Company D	
Premium	Rank	Premium	Rank	Premium	Rank	Premium	Rank
65	11	48	4	57	8	62	10
73	16	69	12.5	61	9	53	6
54	7	88	19	89	20	45	3
43	1	75	17	77	18	51	5
70	14	72	15	69	12.5	44	2
$n_1 = 5$	$R_1 = 49$	$n_2 = 5$	$R_2 = 67.5$	$n_3 = 5$	$R_3 = 67.5$	$n_4 = 5$	$R_4 = 26$

Here, $n = n_1 + n_2 + n_3 + n_4 = 5 + 5 + 5 + 5 = 20$

$$H = \frac{12}{n(n+1)} \left[\frac{R_1^2}{n_1} + \frac{R_2^2}{n_2} + \frac{R_3^2}{n_3} + \frac{R_4^2}{n_4} \right] - 3(n+1)$$

$$= \frac{12}{20(20+1)} \left[\frac{(49)^2}{5} + \frac{(67.5)^2}{5} + \frac{(67.5)^2}{5} + \frac{(26)^2}{5} \right] - 3(20+1) = 6.654$$

Step 5: Do not reject H_0. Do not conclude that the distribution of auto insurance premiums paid per month by all such drivers are not identical for the four companies.

14.55 a. H_0: The median delivery time for all pizza parlors is the same.

H_1: The median delivery time for all pizza parlors is not the same.

For $\alpha = .05$ and $df = k - 1 = 4 - 1 = 3$, the critical value of $\chi^2 = 7.815$, so the rejection region is $\chi^2 > 7.815$.

Tony's		Luigi's		Angelo's		Kowalski's	
Time	Rank	Time	Rank	Time	Rank	Time	Rank
20.0	3	22.1	7	23.9	8	23.9	9
24.0	10.5	27.0	20	24.1	18.5	24.1	12
18.3	1	20.2	4	25.8	10.5	25.8	16.5
22.0	6	32.0	24	29.0	23	29.0	22
20.8	5	26.0	18.5	25.0	21	25.0	15
19.0	2	24.8	14	24.2	16.5	24.2	13
$n_1 = 6$	$R_1 = 27.5$	$n_2 = 6$	$R_2 = 87.5$	$n_3 = 6$	$R_3 = 97.5$	$n_4 = 6$	$R_4 = 87.5$

Here, $n = n_1 + n_2 + n_3 + n_4 = 6 + 6 + 6 + 6 = 24$

$$H = \frac{12}{n(n+1)} \left[\frac{R_1^2}{n_1} + \frac{R_2^2}{n_2} + \frac{R_3^2}{n_3} + \frac{R_4^2}{n_4} \right] - 3(n+1)$$

$$= \frac{12}{24(24+1)} \left[\frac{(27.5)^2}{6} + \frac{(87.5)^2}{6} + \frac{(97.5)^2}{6} + \frac{(87.5)^2}{6} \right] - 3(24+1) = 10.250$$

Reject H_0. Conclude that the median delivery time for all four pizza parlors are not the same.

b. The conclusion is the same in both exercises (Reject H_0).

14.56   a.   Step 1: H_0: The population distributions of Friday night tips in the 3 restaurants are identical.

   H_1: The population distributions of Friday night tips in the 3 restaurants are not identical.

   Step 2:   Use the χ^2 distribution.

   Step 3:   For $\alpha = .05$ and $df = k - 1 = 3 - 1 = 2$, the critical value of $\chi^2 = 5.991$, so the rejection region is $\chi^2 > 5.991$.

   Step 4:

Barzini's		Hwang's		Jack's	
Tips	Rank	Tips	Rank	Tips	Rank
97	10	67	1	93	9
114	14	85	4.5	102	12
105	13	92	8	98	11
85	4.5	78	2	80	3
120	15	90	6	91	7
$n_1 = 5$	$R_1 = 56.5$	$n_2 = 5$	$R_2 = 21.5$	$n_3 = 5$	$R_3 = 42$

   Here, $n = n_1 + n_2 + n_3 = 5 + 5 + 5 = 15$

   $$H = \frac{12}{n(n+1)}\left[\frac{R_1^2}{n_1} + \frac{R_2^2}{n_2} + \frac{R_3^2}{n_3}\right] - 3(n+1) = \frac{12}{15(15+1)}\left[\frac{(56.5)^2}{5} + \frac{(21.5)^2}{5} + \frac{(42)^2}{5}\right]$$

   $$- 3(15+1) = 6.185$$

   Step 5:   Reject H_0. Conclude that the population distributions of Friday night tips in the three restaurants are not identical.

   b.   The conclusion is the same in both exercises (Reject H_0).

   c.   If $\alpha = 0$, there is no rejection region, so we cannot reject H_0. We must conclude that the distributions of Friday night tips in the three restaurants are identical.

14.57 H_0: The median number of defective parts is the same for all three shifts.

   H_1: The median number of defective parts is not the same for all three shifts.

   For $\alpha = .05$ and $df = k - 1 = 3 - 1 = 2$, the critical value of $\chi^2 = 5.991$, so the rejection region is $\chi^2 > 5.991$.

First Shift		Second Shift		Third Shift	
Number	Rank	Number	Rank	Number	Rank
23	1	25	2	33	4
36	6	35	5	44	10
32	3	41	9	50	12.5
40	8	38	7	52	14
45	11	50	12.5	60	15
$n_1 = 5$	$R_1 = 29$	$n_2 = 5$	$R_2 = 35.5$	$n_3 = 5$	$R_3 = 55.5$

   Here, $n = n_1 + n_2 + n_3 = 5 + 5 + 5 = 15$

$$H = \frac{12}{n(n+1)}\left[\frac{R_1^2}{n_1}+\frac{R_2^2}{n_2}+\frac{R_3^2}{n_3}\right]-3(n+1) = \frac{12}{15(15+1)}\left[\frac{(29)^2}{5}+\frac{(35.5)^2}{5}+\frac{(55.5)^2}{5}\right]-3(15+1)=3.815$$

Do not reject H_0. Do not conclude that the numbers of defective parts are not the same for all three shifts.

14.58 H_0: The population distributions of service times at the three restaurants are identical.

H_1: The population distributions of service times at the three restaurants are not identical.

For $\alpha = .10$ and $df = k - 1 = 3 - 1 = 2$, the critical value of $\chi^2 = 4.605$, so the rejection region is $\chi^2 > 4.605$.

Al's		Eduardo's		Patel's	
Time	Rank	Time	Rank	Time	Rank
7.0	18	3.3	9	1.1	1
8.3	21	11.0	22	2.4	6
6.9	17	5.7	12	1.8	4
1.3	2	8.1	20	3.0	7
6.7	16	6.6	14.5	4.1	10
7.1	19	13.0	24	12.0	23
5.5	11	2.3	5	1.5	3
6.6	14.5	5.9	13	3.1	8
$n_1 = 8$	$R_1 = 118.5$	$n_2 = 8$	$R_2 = 119.5$	$n_3 = 8$	$R_3 = 62$

Here, $n = n_1 + n_2 + n_3 = 8 + 8 + 8 = 24$

$$H = \frac{12}{n(n+1)}\left[\frac{R_1^2}{n_1}+\frac{R_2^2}{n_2}+\frac{R_3^2}{n_3}\right]-3(n+1) = \frac{12}{24(24+1)}\left[\frac{(118.5)^2}{8}+\frac{(119.5)^2}{8}+\frac{(62)^2}{8}\right]-3(24+1)=5.416$$

Reject H_0. Conclude that the population distributions of service times at the three restaurants are not identical.

14.59 To make the test of hypothesis about ρ in Chapter 13, both variables (x and y) must be normally distributed. No such assumption is required for testing a hypothesis about the Spearman rho rank correlation coefficient.

14.60 In the following tables, $d = u - v$

a.

u	2	1	3	4	6	5	7	8	
v	8	6	7	4	5	2	1	3	
d	−6	−5	−4	0	1	3	6	5	
d^2	36	25	16	0	1	9	36	25	$\Sigma d^2 = 148$

$$r_s = 1 - \frac{6\sum d^2}{n(n^2-1)} = 1 - \frac{6(148)}{8(64-1)} = 1 - \frac{888}{504} = -.762$$

b.

u	1	2	3	4	5	6	7	

v	4	2	1	5	3	7	6	
d	−3	0	2	−1	2	−1	1	
d^2	9	0	4	1	4	1	1	$\Sigma d^2 = 20$

$$r_s = 1 - \frac{6\sum d^2}{n(n^2-1)} = 1 - \frac{6(20)}{7(49-1)} = 1 - \frac{120}{336} = .643$$

14.61   a.

x	5	10	15	20	25	30	
y	17	15	12	14	10	9	
u	1	2	3	4	5	6	
v	6	5	3	4	2	1	
d	−5	−3	0	0	3	5	
d^2	25	9	0	0	9	25	$\Sigma d^2 = 68$

$$r_s = 1 - \frac{6\sum d^2}{n(n^2-1)} = 1 - \frac{6(68)}{6(36-1)} = 1 - \frac{408}{210} = -.943$$

b.

x	27	15	32	21	16	40	8	
y	95	81	102	88	75	120	62	
u	5	2	6	4	3	7	1	
v	5	3	6	4	2	7	1	
d	0	−1	0	0	1	0	0	
d^2	0	1	0	0	1	0	0	$\Sigma d^2 = 2$

$$r_s = 1 - \frac{6\sum d^2}{n(n^2-1)} = 1 - \frac{6(2)}{7(49-1)} = 1 - \frac{12}{336} = .964$$

14.62   a.   For a right–tailed test with $n = 9$ and $\alpha = .025$, the critical value from Table XIV is .700, so the rejection region is $r_s \geq .700$.

Since $r_s = .575$, do not reject H_0.

b.   For a left–tailed test with $n = 15$ and $\alpha = .005$, the critical value from Table XIV is −.688, so the rejection region is $r_s \leq -.688$.

Since $r_s = -.575$, do not reject H_0.

c.   For a two–tailed test with $n = 20$ and $\alpha = .01$, the critical values from Table XIV are ±.591, so the rejection region is $r_s \leq -.591$ and $r_s \geq .591$.

Since $r_s = .554$, do not reject H_0.

d.   For a right–tailed test with $n = 20$ and $\alpha = .01$, the critical value from Table XIV is .534, so the rejection region is $r_s \geq .534$.

Since $r_s = .554$, reject H_0.

14.63 a. We would expect r_s to be positive, because as height increases, we would expect weight to increase.

b. In the table below, u and v denote the ranks for height and weight, respectively, and $d = u - v$.

u	9.5	3	4.5	4.5	9.5	1	7.5	6	7.5	2	
v	8	4	3	5	10	1	7	6	9	2	
d	1.5	−1	1.5	−.5	−.5	0	.5	0	−1.5	0	
d^2	2.25	1	2.25	.25	.25	0	.25	0	2.25	0	$\Sigma d^2 = 8.5$

$$r_s = 1 - \frac{6\sum d^2}{n(n^2 - 1)} = 1 - \frac{6(8.5)}{10(100 - 1)} = 1 - \frac{51}{990} = .948$$

The value of r_s is positive, which agrees with part a.

14.64 For a right–tailed test with $n = 10$ and $\alpha = .01$, the rejection region is $r_s \geq .745$.
From Exercise 14.63, $r_s = .948$.   Reject H_0.

14.65 a. The regression line has a positive slope, which indicates that as x increases, y tends to increase. Therefore, we would expect the Spearman rho rank correlation coefficient to be positive.

b. In the following table, u and v denote the ranks of x and y, respectively, and $d = u - v$.

u	5	7	2	6	1	4	3	
v	4.5	7	2	6	1	3	4.5	
d	.5	0	0	0	0	1	−1.5	
d^2	.25	0	0	0	0	1	2.25	$\Sigma d^2 = 3.50$

$$r_s = 1 - \frac{6\sum d^2}{n(n^2 - 1)} = 1 - \frac{6(3.50)}{7(49 - 1)} = 1 - \frac{21}{336} = .938$$

The value of r_s is positive, which agrees with part a.

14.66 a. The result of the hypothesis test for Example 13–7 suggests a positive linear relationship between income and food expenditures for households. Rejecting H_0: $\rho_s = 0$ in favor of the alternative H_1: $\rho_s > 0$ would also indicate a positive relationship between the two variables. Therefore, we would expect to reject H_0.

b. For a right–tailed test with $n = 7$ and $\alpha = .01$, the rejection region is $r_s \geq .893$. From Exercise 14.65, part b, $r_s = .938$.   Reject H_0.

14.67   a.   In the following table, u and v denote the ranks of x and y, respectively, and $d = u - v$.

u	6	2	4	7	9	1	8	3	5	
v	6	3	5	7	8	2	9	1	4	
d	0	−1	−1	0	1	−1	−1	2	1	
d^2	0	1	1	0	1	1	1	4	1	$\Sigma d^2 = 10$

$$r_s = 1 - \frac{6\sum d^2}{n(n^2 - 1)} = 1 - \frac{6(10)}{9(81-1)} = 1 - \frac{60}{720} = .917$$

   b.   For a right–tailed test with $n = 9$ and $\alpha = .05$, the rejection region is $r_s \geq .600$. Since $r_s = .917$, which is in the rejection region, reject H_0.

   c.   Since $r_s > 0$ and we reject H_0, the test indicates a positive relationship between the variables x and y.

14.68   a.   In the following table, u and v denote the ranks for aggressiveness and weekly time spent watching TV, respectively.

u	1	2	3	4	5	6	7	
v	2	4	6	1	5	7	3	
d	−1	−2	−3	3	0	−1	4	
d^2	1	4	9	9	0	1	16	$\Sigma d^2 = 40$

$$r_s = 1 - \frac{6\sum d^2}{n(n^2 - 1)} = 1 - \frac{6(40)}{7(49-1)} = 1 - \frac{240}{336} = .286$$

   b. H_0: $\rho_s = 0$; H_1: $\rho_s > 0$;   A right–tailed test.

For $n = 7$ and $\alpha = .05$, the rejection region is $r_s \geq .714$.

Since $r_s = .286$, which is not in the rejection region, do not reject H_0. Do not conclude that there is a positive relationship between aggressiveness and hours spent watching television.

14.69   In a string of occurrences in which there are only two possible outcomes, a run is a sequence of one or more consecutive occurrences of the same outcome. For example, in the sequence AABBAAABAA, there are three runs of outcome A and two runs of outcome B, for a total of five runs.

14.70   The usual form of the null hypothesis in a runs test for randomness is that the symbols occur in random order in the sequence.

14.71 Let n_1 be the number of times the first outcome occurs in a string of outcomes, and let n_2 be the number of times the second outcome occurs. If either $n_1 > 15$ or $n_2 > 15$, the normal approximation may be used.

14.72   a.   Step 1: H_0: The sequence is random. H_1: The sequence is not random.

Step 2:   Since $n_1 < 15$ and $n_2 < 15$, use the runs test with critical values from Table XV.

Step 3:   For $n_1 = 10$, $n_2 = 12$, and $\alpha = .05$, the critical values of R are 7 and 17. Thus, H_0 will be rejected if $R \leq 7$ or $R \geq 17$.

Step 4:   The observed value of $R = 17$.

Step 5:   Reject H_0.

b.   Step 1: H_0: The sequence is random. H_1: The sequence is not random.

Step 2:   Since $n_1 > 15$ and $n_2 > 15$, use the normal distribution.

Step 3:   For $\alpha = .01$, the rejection region lies to the left of $z = -2.58$ and to the right of $z = 2.58$.

Step 4: $\mu_R = \dfrac{2n_1 n_2}{n_1 + n_2} + 1 = \dfrac{2(20)(23)}{20 + 23} + 1 = 22.40$

$\sigma_R = \sqrt{\dfrac{2n_1 n_2 (2n_1 n_2 - n_1 - n_2)}{(n_1 + n_2)^2 (n_1 + n_2 - 1)}} = \sqrt{\dfrac{2(20)(23)(2 \cdot 20 \cdot 23 - 20 - 23)}{(20 + 23)^2 (20 + 23 - 1)}} = 3.22329912$

$z = \dfrac{R - \mu_R}{\sigma_R} = \dfrac{35 - 22.40}{3.22329912} = 3.91$

Step 5:   Reject H_0.

c.   Step 1: H_0: The sequence is random. H_1: The sequence is not random.

Step 2:   Since $n_2 > 15$, use the normal distribution.

Step 3:   For $\alpha = .05$, the rejection region lies to the left of $z = -1.96$ and to the right of $z = 1.96$.

Step 4: $\mu_R = \dfrac{2n_1 n_2}{n_1 + n_2} + 1 = \dfrac{2(15)(17)}{15 + 17} + 1 = 16.94$

$\sigma_R = \sqrt{\dfrac{2n_1 n_2 (2n_1 n_2 - n_1 - n_2)}{(n_1 + n_2)^2 (n_1 + n_2 - 1)}} = \sqrt{\dfrac{2(15)(17)(2 \cdot 15 \cdot 17 - 15 - 17)}{(15 + 17)^2 (15 + 17 - 1)}} = 2.77120217$

$z = \dfrac{R - \mu_R}{\sigma_R} = \dfrac{7 - 16.94}{2.77120217} = -3.59$

Step 5:   Reject H_0.

d.   Step 1: H_0: The sequence is random. H_1: The sequence is not random.

Step 2:   Since $n_1 < 15$ and $n_2 < 15$, use the runs test with critical values from Table XV.

Step 3: For $n_1 = 14$, $n_2 = 13$, and a = .05, the critical values of R are 9 and 20. Reject H_0 if $R \leq 9$ or $R \geq 20$.

Step 4: The observed value of $R = 21$.

Step 5: Reject H_0.

14.73 In Example 14–13, replacing "M" and "F" by "0" and "1", respectively would not affect the test. The values of n_1, n_2, and R would be unchanged, so the rejection region and the observed value of R would be the same. Thus, the conclusion would be the same.

14.74 a. Using n_1 for "X" and n_2 for "Y" yields: $n_1 = 8$, $n_2 = 7$, and $R = 10$.

b. Using n_1 for "F" and n_2 for "M" yields: $n_1 = 11$, $n_2 = 3$, and $R = 5$.

c. Using n_1 for "+" and n_2 for "–" yields: $n_1 = 13$, $n_2 = 9$, and $R = 9$.

d. Using n_1 for "1" and n_2 for "0" yields: $n_1 = 8$, $n_2 = 6$, and $R = 5$.

14.75 Step 1: H_0: The sequence of heads and tails is random.

H_1: The sequence is not random.

Step 2: Using n_1 for "H" and n_2 for "T" yields $n_1 = 11$ and $n_2 = 9$.

Since $n_1 < 15$ and $n_2 < 15$, use the runs test with critical values from Table XV.

Step 3: For $n_1 = 11$, $n_2 = 9$ and $\alpha = .05$, the critical values of R are 6 and 16. Thus, H_0 is rejected if $R \leq 6$ or $R \geq 16$.

Step 4: The observed value of $R = 13$.

Step 5: Since R is between 6 and 16, do not reject H_0. Do not conclude that the psychic's claim is true.

14.76 Step 1: H_0: The sequence of underfilled and overfilled bottles is random.

H_1: The sequence is not random.

Step 2: Using n_1, for "U" and n_2 for "0" yields $n_1 = 10$ and $n_2 = 8$.

Since $n_1 < 15$ and $n_2 < 15$, use the runs test with critical values from Table XV.

Step 3: For $n_1 = 10$, $n_2 = 8$ and $\alpha = .05$, the critical values of R are 5 and 15. Thus, H_0 is rejected if $R \leq 5$ or $R \geq 15$.

Step 4: The observed value of $R = 7$.

Step 5: Since R is between 5 and 15, do not reject H_0. Do not conclude that there is a nonrandom pattern of overfilling and underfilling such bottles.

14.77 H_0: Diseased and normal trees are randomly mixed in the row.

H_1: The trees are not randomly mixed.

Using n_1 for "N" and n_2 for "D" yields $n_1 = 13$ and $n_2 = 7$.

Since n_1 and n_2 are less than 15, use the runs test with critical values from Table XV.

For $n_1 = 13$, $n_2 = 7$, and $\alpha = .05$, the critical values of R are 5 and 15. Thus, reject H_0 if $R \leq 5$ or $R \geq 15$.

The observed value of $R = 5$.

Reject H_0. Conclude that there is a non–random pattern in the sequence.

14.78 H_0: The correct (and incorrect) responses are randomly distributed among the population of all 4th graders.

H_1: The correct (and incorrect) responses are randomly distributed among the population of all 4th graders. Using n_1 for correct answers and n_2 for incorrect answers yields $n_1 = 16$ and $n_2 = 6$.

Since $n_1 > 15$, use the normal distribution to make the runs test.

For $\alpha = .05$, the rejection region lies to the left of $z = -1.96$ and to the right of $z = 1.96$. The observed value of $R = 7$.

$$\mu_R = \frac{2n_1n_2}{n_1+n_2} + 1 = \frac{2(16)(6)}{16+6} + 1 = 9.73$$

$$\sigma_R = \sqrt{\frac{2n_1n_2(2n_1n_2 - n_1 - n_2)}{(n_1+n_2)^2(n_1+n_2-1)}} = \sqrt{\frac{2(16)(6)(2\cdot16\cdot6-16-6)}{(16+6)^2(16+6-1)}} = 7.493112948$$

$$z = \frac{R-\mu_R}{\sigma_R} = \frac{7-9.73}{7.493112948} = -.36 \qquad \text{Do not reject } H_0.$$

14.79 H_0: The hits occur randomly among all at–bats. H_1: The hits do not occur randomly.

Since $n_1 > 15$ and $n_2 > 15$, use the normal distribution.

For $\alpha = .01$, the rejection region lies to the left of $z = -2.58$ and to the right of $z = 2.58$.

$$\mu_R = \frac{2n_1n_2}{n_1+n_2} + 1 = \frac{2(22)(53)}{22+53} + 1 = 32.09$$

$$\sigma_R = \sqrt{\frac{2n_1n_2(2n_1n_2 - n_1 - n_2)}{(n_1+n_2)^2(n_1+n_2-1)}} = \sqrt{\frac{2(22)(53)(2\cdot22\cdot53-22-53)}{(22+53)^2(22+53-1)}} = 3.55592776$$

$$z = \frac{R-\mu_R}{\sigma_R} = \frac{37-32.09}{3.55592776} = 1.38 \quad \text{Do not reject } H_0. \text{ Conclude that hits occur randomly for this player.}$$

14.80 H_0: Movements in the stock market are random.

H_1: Movements in the stock market are not random.

Using n_1 for movements up and n_2 for movements down, $n_1 = 16$ and $n_2 = 24$.

Since $n_1 > 15$, use the normal distribution to make the runs test.

For $\alpha = .05$, the rejection region lies to the left of $z = -1.96$ and to the right of $z = 1.96$. The observed value of $R = 11$.

$$\mu_R = \frac{2n_1n_2}{n_1+n_2}+1 = \frac{2(16)(24)}{16+24}+1 = 20.2$$

$$\sigma_R = \sqrt{\frac{2n_1n_2(2n_1n_2-n_1-n_2)}{(n_1+n_2)^2(n_1+n_2-1)}} = \sqrt{\frac{2(16)(24)(2\cdot16\cdot24-16-24)}{(16+24)^2(16+24-1)}} = 2.99332591$$

$$z = \frac{R-\mu_R}{\sigma_R} = \frac{11-20.2}{2.99332591} = -3.07$$

Reject H_0. Conclude that movements in the stock market are not random.

14.81 H_0: The sequence of all the state's daily numbers is random.

H_1: The sequence of all the state's daily numbers is not random.

Since n_1 and n_2 are greater than 15, use the normal distribution.

For $\alpha = .025$, the rejection region lies to the left of $z = -2.24$ and to the right of $z = 2.24$. The observed value of $R = 11$.

$$\mu_R = \frac{2n_1n_2}{n_1+n_2}+1 = \frac{2(27)(23)}{27+23}+1 = 25.84$$

$$\sigma_R = \sqrt{\frac{2n_1n_2(2n_1n_2-n_1-n_2)}{(n_1+n_2)^2(n_1+n_2-1)}} = \sqrt{\frac{2(27)(23)(2\cdot27\cdot23-27-23)}{(27+23)^2(27+23-1)}} = 3.47640913$$

$$z = \frac{R-\mu_R}{\sigma_R} = \frac{11-25.84}{3.47640913} = -4.27$$

Reject H_0. Conclude that the sequence of all this state's daily numbers is not random.

14.82 Let p be the proportion of all cola drinkers who would prefer Brand A over Brand B.

H_0: $p = .50$; H_1: $p > .50$;   A right–tailed test.

$n = 15 < 25$, so use the binomial distribution.

For $n = 15$ and $\alpha = .05$, the rejection region is $X \geq 12$. The observed value of $X = 10$.

Do not reject H_0. Do not conclude that among all cola drinkers there is a preference for Brand A over Brand B.

14.83 Let p be the proportion of all people who would prefer Brand A over Brand B.

H_0: $p = .50$; H_1: $p < .50$;   A left–tailed test.

$n = 24 < 25$, so use the binomial distribution.

For $n = 24$ and $\alpha = .05$, the rejection region is $X \leq 7$. The observed value of $X = 7$.

Reject H_0. Conclude that among all people there is a preference for Brand B over Brand A.

14.84 Let p be the proportion of all fans that prefer professional to college football.

H_0: $p = .50$; H_1: $p \neq .50$;   A two–tailed test.

Since 12 of the 400 fans had no preference, the true value of $n = 400 - 12 = 388$.

Because $n > 25$, use the normal distribution.

For $\alpha = .02$, the rejection region lies to the left of $z = -2.33$ and to the right of $z = 2.33$.

$$\mu = np = 388(.50) = 194; \quad \sigma = \sqrt{npq} = \sqrt{388(.50)(.50)} = 9.84885780$$

$X = 220$ and $\dfrac{n}{2} = \dfrac{388}{2} = 194$ \qquad Since $X > \dfrac{n}{2}, z = \dfrac{(X - .5) - \mu}{\sigma} = \dfrac{(220 - .5) - 194}{9.84885780} = 2.59$

Reject H_0. Conclude that among all fans there is a preference for professional over college football.

14.85 Let p be the proportion of all such bank customers who prefer an ATM to a human teller.

$H_0: p = .50;$ \qquad $H_1: p > .50;$ \qquad A right–tailed test.

Since 12 of the 200 customers have no preference, the true value of n is $200 - 12 = 188$.

Because $n > 25$, use the normal distribution.

For $\alpha = .01$, the rejection region lies to the right of $z = 2.33$.

$$\mu = np = 188(.50) = 94; \quad \sigma = \sqrt{npq} = \sqrt{188(.50)(.50)} = 6.85565460$$

$X = 122$ and $\dfrac{n}{2} = \dfrac{188}{2} = 94$ \qquad Since $X > \dfrac{n}{2}, z = \dfrac{(X - .5) - \mu}{\sigma} = \dfrac{(122 - .5) - 94}{6.85565460} = 4.01$

Reject H_0. Conclude that more than half of all customers of this bank prefer an ATM.

14.86 $H_0:$ Median $= 5$ minutes; \qquad $H_1:$ Median > 5 minutes; \qquad A right–tailed test.

Since one of the 10 times was exactly five minutes, the true value of $n = 10 - 1 = 9$. $n = 9 < 25$. Use the binomial distribution.

For $n = 9$ and $\alpha = .05$, the rejection region is $X \geq 8$.

Assigning a plus sign to each time above 5 minutes and a minus sign to each time below 5 minutes yields eight plus signs and one minus sign. Thus: \qquad The observed value of $X = 8$.

Reject H_0. Conclude that the median time for the survey exceeds five minutes.

14.87 $H_0:$ Median $= 45$ years old; \qquad $H_1:$ Median > 45 years old; \qquad A right–tailed test.

Since two buyers were 45 years old, the true value of $n = 25 - 2 = 23$.

$n = 23 \leq 25$, so we use the binomial distribution.

For $n = 23$ and $\alpha = .05$, the rejection region is $X \geq 16$.

Since 16 recent purchasers were over 45 years old, the observed value of $X = 16$.

Reject H_0. Conclude that the median age exceeded 45 years.

14.88 a. $H_0:$ Median $= 25$ minutes; \qquad $H_1:$ Median > 25 minutes; \qquad A right–tailed test.

Since two of the 30 times were exactly 25 minutes, the true value of $n = 30 - 2 = 28$.

$n = 28 > 25$. Use the normal distribution.

For $\alpha = .05$, the rejection region lies to the right of $z = 1.65$.

$\mu = np = 28(.50) = 14; \sigma = \sqrt{npq} = \sqrt{28(.50)(.50)} = 2.64575131$

If we assign a plus sign to every time greater than 25 minutes and a minus sign to every time

less than 25 minutes, we obtain 19 plus signs and 9 minus signs.

Thus, the observed value of $X = 19$.

$\dfrac{n}{2} = \dfrac{28}{2} = 14$ Since $X > \dfrac{n}{2}, z = \dfrac{(X - .5) - \mu}{\sigma} = \dfrac{(19 - .5) - 14}{2.64575131} = 1.70$

Reject H_0. Conclude that the median waiting time at this site during these hours exceeds 25

minutes.

b. For $\alpha = .025$, the rejection region lies to the right of $z = 1.96$.

From part a, the observed value of $z = 1.70$ Do not reject H_0.

c. Note the opposite conclusions of parts a and b. The observed value of $z = 1.70$ is just marginally

inside the rejection region of part a. When the significance level was lowered from .05 to .025 in

part b, the rejection region moved to the right, leaving $z = 1.70$ in the nonrejection region.

14.89 H_0: Median = \$250; H_1: Median ≠ \$250;   A two–tailed test.

Since one of the 35 students spent exactly \$250, the true value of $n = 35 - 1 = 34$. $n = 34 > 25$. Use the

normal distribution.

For $\alpha = .05$, the rejection region lies to the left of $z = -1.96$ and to the right of $z = 1.96$

$\mu = np = 34(.50) = 17; \sigma = \sqrt{npq} = \sqrt{34(.50)(.50)} = 2.91547595$

If we assign a plus sign to every value above \$250 and a minus sign to every value below \$250, we

obtain 26 plus signs and 8 minus signs. Since the test is two–tailed we may use either value as the

observed value of X.

$\dfrac{n}{2} = \dfrac{34}{2} = 17$. If we use $X = 8, X < \dfrac{n}{2},$ $z = \dfrac{(X + .5) - \mu}{\sigma} = \dfrac{(8 + .5) - 17}{2.91547595} = -2.92$

Reject H_0. Conclude that the median expenditure on text books by all such students in 2002–2003 was

different from \$250.

14.90 Let M denote the difference in median morning commutes.

Then for each morning: Paired difference = Morris' commute time – Bob's commute time

H_0: $M = 0$; H_1: $M < 0$;   A left–tailed test.

Since $n = 21 < 25$, use the binomial distribution.

For $n = 21$ and $\alpha = .05$, the rejection region is $X \le 6$.

Assign a plus sign to each day Bob arrives first and a minus sign to each day Morris arrives first.

We obtain 16 plus signs and 5 minus signs. Since the test is left–tailed, use the lower number of signs

for X. Thus, the observed value of $X = 5$.

Reject H_0. Conclude that Bob's median morning commute time is less than Morris' commute time.

14.91 a. Let M denote the difference in median gas mileage before and after the installation of governors.

Then for each salesperson: Paired difference = Gas mileage before – Gas mileage after

H_0: $M = 0$; H_1: $M < 0$;   A left–tailed test.

Since $n = 7 < 25$, use the binomial distribution.

For $n = 7$ and $\alpha = .05$, the rejection region is $X = 0$.

Before	25	21	27	23	19	18	20
After	26	24	26	25	24	22	23
Sign (Before-After)	—	—	+	—	—	—	—

There are one plus sign and six minus signs.

For a left–tailed test the observed value of X is the smaller number of signs. Thus, the observed value of $X = 1$.

Do not reject H_0. Do not conclude that the use of governors tends to increase the median gas mileage for the Gamma Corporation's salespersons' cars.

b. The conclusion of part a (Do not reject H_0) is different from the conclusions of Exercises 14.34 and 10.101 (Reject H_0 in both cases).

c. The Wilcoxon signed–rank test of Exercise 14.34 and the test based on the t–distribution of Exercise 10.101 both take into account the magnitudes of the paired differences, but the sign test of part a uses only the signs of the paired differences. This makes the sign test less efficient, so it is more prone to Type II errors (failing to reject H_0 when H_0 is false).

14.92 Let M denote the difference in median fares quoted by Agency X and Agency Y. For each destination:

Paired difference = Quote by Agency X – Quote by Agency Y

H_0: $M = 0$; H_1: $M \neq 0$;   A two–tailed test.

For six of the 32 destinations, the fares quoted were the same, so the true value of $n = 32 - 6 = 26$. Since $n > 25$, so use the normal distribution.

For $\alpha = .02$, the rejection region lies to the left of $z = -2.33$ and to the right of $z = 2.33$.

$$\mu = np = 26(.50) = 13; \quad \sigma = \sqrt{npq} = \sqrt{26(.50)(.50)} = 2.54950976$$

Assigning a plus sign to each destination for which Agency X quotes a higher fare than Agency Y and a minus sign to each destination for which Agency X quotes a lower fare than Agency Y yields 8 plus signs and 18 minus signs. For a two–tailed test we may use either value for X. If we use the smaller number of signs: The observed value of $X = 8$

$$\frac{n}{2} = \frac{26}{2} = 13. \qquad\qquad \text{Since, } X < \frac{n}{2}, z = \frac{(X+.5)-\mu}{\sigma} = \frac{(8+.5)-13}{2.54950976} = -1.77$$

Do not reject H_0. Do not conclude that there is any difference in the median fares quoted by Agency X and Agency Y for all destinations they both offer.

14.93 Let M denote the difference in median blood pressures before and after taking the medication.

For each patient: Paired difference = Blood pressure before – Blood pressure after

H_0: $M = 0$; H_1: $M > 0$; A right–tailed test.

For three of the 35 patients there was no change in blood pressure, so the true value of $n = 35 - 3 = 32$.

Since $n > 25$, use the normal distribution.

For $\alpha = .025$, the rejection region lies to the right of $z = 1.96$.

$$\mu = np = 32(.50) = 16; \sigma = \sqrt{npq} = \sqrt{32(.50)(.50)} = 2.82842712$$

Assigning a plus sign to each patient whose blood pressure decreased and a minus sign to each patient whose blood pressure increased yields 25 plus signs and 7 minus signs. For a right–tailed test we use the larger number of signs. Thus: The observed value of $X = 25$

$$\frac{n}{2} = \frac{32}{2} = 16.$$ Since, $X > \frac{n}{2}, z = \frac{(X - .5) - \mu}{\sigma} = \frac{(25 - .5) - 16}{2.82842712} = 3.01$

Reject H_0. Conclude that the median blood pressure in all such patients is lower after the medication than before.

14.94 a. For each sales person: Paired difference = Sales before course – Sales after course

Let M_A and M_B denote the median weekly sales of all such salespersons after and before the course, respectively.

H_0: $M_A = M_B$; H_1: $M_A > M_B$; A right–tailed test.

$n = 6 < 15$. Use the Wilcoxon signed–rank test for the small–sample case.

For a one–tailed test with $n = 6$ and $\alpha = .05$, the rejection region is $T < 2$.

Before	After	Differences (Before–After)	Absolute Differences	Ranks of Differences	Signed Ranks	Sign
12	18	–6	6	5.5	–5.5	–
18	24	–6	6	5.5	–5.5	–
25	24	+1	1	1	+1	+
9	14	–5	5	3.5	–3.5	–
14	19	–5	5	3.5	–3.5	–
16	20	–4	4	2	–2	–

For a right–tailed test, T = sum of positive ranks Thus, the observed value of $T = 1$.

Reject H_0. Conclude that the weekly sales for all salespersons tend to increase as a result of attending this course.

b. Let M denote: (Median weekly sales before course) – (Median weekly sales after course)

H_0: $M = 0$; H_1: $M < 0$; Note that the sign test here is left–tailed.

$n = 6 < 25$. Use the binomial distribution.

For a left–tailed test with $n = 6$ and $\alpha = .05$, the rejection region is $X = 0$.

There are five minus signs and one plus sign in the seventh column of the table above. Since the test is left–tailed, we use the lower number. Thus, the observed value of $X = 1$. Do not reject H_0.

c. In part a, using the Wilcoxon signed–rank test, we rejected H_0. In part b, using the sign test with the same data and the same level of significance, we could not reject H_0.

14.95 For each skater: Paired difference = Judge A's score – Judge B's score

Let M_A and M_B denote the median scores for all such skaters given by judges A and B, respectively.

$H_0: M_A = M_B$; $\qquad\qquad H_1: M_A \neq M_B$; $\qquad\qquad$ A two–tailed test.

One skater was scored the same by both judges, so the true sample size is $n = 8 - 1 = 7$.

Since $n < 15$, use the Wilcoxon signed–rank test procedure for the small–sample case.

For $n = 7$, $\alpha = .05$ and a two–tailed test, the rejection region is $T \leq 2$.

| Scores | | Differences | Absolute | Ranks of | Signed |
Judge A	Judge B	(A–B)	Differences	Differences	Ranks
5.8	5.4	+.4	.4	4.5	+4.5
5.7	5.5	+.2	.2	2.5	+2.5
5.6	5.7	−.1	.1	1	−1
5.9	5.4	+.5	.5	6	+6
5.8	5.6	+.2	.2	2.5	+2.5
5.9	5.3	+.6	.6	7	+7
5.8	5.4	+.4	.4	4.5	+4.5
5.6	5.6	—	—	—	—

For a two–tailed test, T is the smaller sum of ranks. Here, T is the sum of the absolute values of the negative ranks. Thus, the observed value of $T = 1$.

Reject H_0. Conclude that one judge tends to give higher scores than the other.

14.96 a. For each blood sample: Paired difference = lab B's cholesterol level – lab A's level

Let M_A and M_B denote the median cholesterol levels for all such adults as determined by labs A and B, respectively.

$H_0: M_A = M_B$; $\qquad\qquad H_1: M_A > M_B$; $\qquad\qquad$ A right–tailed test.

Since the two labs' levels were the same for one adult, the true sample size is $n = 30 - 1 = 29$. Since $n > 15$, use the Wilcoxon signed–rank test with the normal distribution approximation. For a right–tailed test with $\alpha = .01$, the rejection region lies to the right of $z = 2.33$.

$$\mu_T = \frac{n(n+1)}{4} = \frac{29(29+1)}{4} = 217.5$$

$$\sigma_T = \sqrt{\frac{n(n+1)(2n+1)}{24}} = \sqrt{\frac{29(29+1)(2 \cdot 29+1)}{24}} = 46.24662150$$

Calculating the signed ranks yields:

Sum of positive ranks = 42

Sum of absolute values of negative ranks = 393

For a right–tailed test, $T =$ the sum of the absolute values of the negative ranks. Thus:

The observed value of $T = 393$

$$z = \frac{T - \mu_T}{\sigma_T} = \frac{393 - 217.5}{46.24662150} = 3.79$$

Reject H_0. Conclude that the median cholesterol level for all such adults as determined by lab A is higher than that determined by lab B.

b. In part a, using the Wilcoxon signed–rank test we rejected H_0. In Exercise 14.26 using the sign test with the same data and significance level, we could not reject H_0.

14.97 Let M_M and M_R denote the median gas mileages for all such drivers with the M car and the R car, respectively.

Since one of the drivers obtained the same mileage with both cars, the true value of $n = 18 - 1 = 17$.

H_0: $M_R = M_M$; H_1: $M_R > M_M$;   A right–tailed test.

Since $n > 15$, use the Wilcoxon signed–rank test with the normal distribution approximation.

For a right–tailed test with $\alpha = .025$, the rejection region lies to the right of $z = 1.96$.

$$\mu_T = \frac{n(n+1)}{4} = \frac{17(17+1)}{4} = 76.5$$

$$\sigma_T = \sqrt{\frac{n(n+1)(2n+1)}{24}} = \sqrt{\frac{17(17+1)(34+1)}{24}} = 21.12463017$$

For a right–tailed test, $T =$ the sum of the absolute values of the negative ranks = 122

$$z = \frac{T - \mu_T}{\sigma_T} = \frac{122 - 76.5}{21.12463017} = 2.15$$

Reject H_0. Conclude that the R car gets better gas mileage than the M car.

14.98 Let M_A and M_B denote the median cost of a week's supply of groceries for all such shoppers at Al's and Bart's, respectively.

H_0: $M_A = M_B$; H_1: $M_A \neq M_B$;   A two–tailed test.

Since $n = 50 > 15$, use the Wilcoxon signed–rank test procedure with the normal distribution approximation.

For a two–tailed test with $\alpha = .01$, the rejection region lies to the left of $z = -2.58$ and to the right of $z = 2.58$.

$$\mu_T = \frac{n(n+1)}{4} = \frac{50(50+1)}{4} = 637.5$$

$$\sigma_T = \sqrt{\frac{n(n+1)(2n+1)}{24}} = \sqrt{\frac{50(50+1)(100+1)}{24}} = 103.59174678$$

For a two–tailed test we may use either sum of ranks. Let $T =$ the sum of positive ranks = 527.

$$z = \frac{T - \mu_T}{\sigma_T} = \frac{527 - 637.5}{103.59174678} = -1.07$$

Do not reject H_0. Do not conclude that either store is less expensive than the other.

14.99 H_0: The population distributions of egg prices in the suburbs and cities are identical.

H_1: The population distribution of egg prices in the cities lies to the right of the population distribution of egg prices in the suburbs.

$n_1 = 6$ and $n_2 = 7$ are less than 10, so use the Wilcoxon rank–sum test for small samples. For a right–tailed test with $n_1 = 6$, $n_2 = 7$ and $\alpha = .05$, the rejection region is $T \geq 54$.

City		Suburb	
Price	Rank	Price	Rank
1.49	12	.99	1
1.29	6	1.09	3
1.35	8	1.39	9
1.58	13	1.28	5
1.33	7	1.16	4
1.47	11	1.44	10
		1.05	2
	Sum = 57		Sum = 34

Since the test is one–tailed, the observed value of T is the sum of ranks for the smaller sample. Thus:

The observed value of $T = 57$.

Reject H_0. Conclude that egg prices tend to be higher in the city.

14.100 H_0: The population distributions of learning times are identical for the two brands.

H_1: The population distribution of learning times for Brand X lies to the right of that of than Brand Y.

n_1 and n_2 are less than 10, so use the Wilcoxon rank–sum test for small samples.

For a right–tailed test with $n_1 = 6$, $n_2 = 7$ and $\alpha = .05$, the rejection region is $T \geq 54$.

Brand X		Brand Y	
Time	Rank	Time	Rank
32	8	33	9
36	10	18	2
28	7	21	3
43	12	25	5
98	13	24	4
39	11	27	6
		17	1
	Sum = 61		Sum = 30

Since the test is one–tailed, the observed value of T is the sum of ranks for the smaller sample. Thus: The observed value of $T = 61$.

Reject H_0. Conclude the learning times for programming Brand X's VCR is longer than for Brand Y's VCR.

14.101 a. H_0: The population distributions of times required for elementary statistics students to complete the assignment are identical for softwares A and B.

H_1: The population distribution of times required for elementary statistics students to complete the assignment using software A lies to the right of the corresponding distribution for software B.

n_1 and n_2 are greater than 10, so use the normal distribution.

The test is right–tailed, so the rejection region lies to the right of $z = 1.65$.

Group A		Group B	
Time	Rank	Time	Rank
123	20	65	2
101	12	115	18.5
112	16	95	8
85	4	100	11
87	5	94	7
133	22	72	3
129	21	60	1
114	17	110	14.5
150	23	99	10
110	14.5	102	13
180	24	88	6
115	18.5	97	9
	Sum = 197		Sum = 103

Since the test is one–tailed and $n_1 = n_2$, the observed value of T is the sum of ranks for the first sample. Thus, the observed value of $T = 197$.

$$\mu_T = \frac{n_1(n_1 + n_2 + 1)}{2} = \frac{12(12 + 12 + 1)}{2} = 150$$

$$\sigma_T = \sqrt{\frac{n_1 n_2 (n_1 + n_2 + 1)}{12}} = \sqrt{\frac{12(12)(12 + 12 + 1)}{12}} = 17.32050808$$

$$z = \frac{T - \mu_T}{\sigma_T} = \frac{197 - 150}{17.32050808} = 2.71$$

Reject H_0. Conclude that the median time required for all students taking elementary statistics at this university to complete this assignment is greater for software A than for software B.

b. A paired–sample sign test would not be appropriate here. Each of the 24 students uses software A only or software B only. Thus, the samples are not paired, but are independent.

14.102 H_0: The population distributions of scores of elementary statistics students on this assignment are identical for software A and software B.

H_1: The population distributions of scores of elementary statistics students on this assignment are different for software A and software B.

n_1 and n_2 are greater than 10, so use the normal distribution.

The test is two–tailed with $\alpha = .10$, so the rejection region lies to the left of $z = -1.65$ and to the right of $z = 1.65$.

Group A		Group B	
Score	Rank	Score	Rank
48	22	37	10
38	11	21	3
45	20	40	14
31	7	27	5
42	17	49	23
25	4	44	19
40	14	36	9
43	18	41	16
50	24	20	2
30	6	39	12
33	8	18	1
46	21	40	14
	Sum = 172		Sum = 128

For a two–tailed test with $n_1 = n_2$, either sum of ranks may be used as the observed value of T. Let $T = 172$

$$\mu_T = \frac{n_1(n_1 + n_2 + 1)}{2} = \frac{12(12 + 12 + 1)}{2} = 150$$

$$\sigma_T = \sqrt{\frac{n_1 n_2(n_1 + n_2 + 1)}{12}} = \sqrt{\frac{12(12)(12 + 12 + 1)}{12}} = 17.32050808$$

$$z = \frac{T - \mu_T}{\sigma_T} = \frac{172 - 150}{17.32050808} = 1.27$$

Do not reject H_0. Do not conclude that there is a difference in the median scores of all students using software A and all students using software B.

14.103 H_0: The population distributions of test scores for these three countries are identical.

H_1: The population distributions of test scores for these three countries are not identical

For $\alpha = .05$ and $df = k - 1 = 3 - 1 = 2$, the rejection region is $\chi^2 > 5.991$.

Country					
Sweden		Germany		Italy	
Number	Rank	Number	Rank	Number	Rank
35	4.5	40	8	35	4.5
49	12	44	11	42	9
55	15	29	1	29	7
37	6	33	3	54	14
43	10	53	13	30	2
$n_1 = 5$	$R_1 = 47.5$	$n_2 = 5$	$R_2 = 36$	$n_3 = 5$	$R_3 = 36.5$

Here, $n = n_1 + n_2 + n_3 = 5 + 5 + 5 = 15$

$$H = \frac{12}{n(n+1)}\left[\frac{R_1^2}{n_1} + \frac{R_2^2}{n_2} + \frac{R_3^2}{n_3}\right] - 3(n+1) = \frac{12}{15(15+1)}\left[\frac{(47.5)^2}{5} + \frac{(36)^2}{5} + \frac{(36.5)^2}{5}\right] - 3(15+1) = .845$$

Do not reject H_0. Do not conclude that there is a difference in median test scores.

14.104 H_0: The population distributions of starting salaries are identical for all three fields.

H_1: The population distributions of starting salaries are not identical for all three fields.

For $\alpha = .05$ and $df = k - 1 = 3 - 1 = 2$, the rejection region is $\chi^2 > 5.991$.

Here, $n_1 = 8$, $n_2 = 10$, $n_3 = 7$, $n = n_1 + n_2 + n_3 = 8 + 10 + 7 = 25$, $R_1 = 137$, $R_2 = 126$, and $R_3 = 62$

$$H = \frac{12}{n(n+1)} \left[\frac{R_1^2}{n_1} + \frac{R_2^2}{n_2} + \frac{R_3^2}{n_3} \right] - 3(n+1) = \frac{12}{25(25+1)} \left[\frac{(137)^2}{8} + \frac{(126)^2}{10} + \frac{(62)^2}{7} \right] - 3(25+1) = 4.761$$

Do not reject H_0. Do not reject the null hypothesis that the median salary is the same for all three major fields.

14.105 H_0: The population distributions of lengths of drives are the same for all three brands of golf balls.

H_1: The population distributions of lengths of drives are not the same for all three brands of golf balls.

For $\alpha = .05$ and $df = k - 1 = 3 - 1 = 2$, the rejection region is $\chi^2 > 5.991$.

Here, $n = n_1 + n_2 + n_3 = 6 + 6 + 6 = 18$

$$H = \frac{12}{n(n+1)} \left[\frac{R_1^2}{n_1} + \frac{R_2^2}{n_2} + \frac{R_3^2}{n_3} \right] - 3(n+1) = \frac{12}{18(18+1)} \left[\frac{(87)^2}{6} + \frac{(37)^2}{6} + \frac{(47)^2}{6} \right] - 3(18+1) = 8.187$$

Reject H_0. Conclude that the median lengths of drives by this golfer are not the same for all three brands of golf balls.

14.106 a. We are given $n = 30$.

The sum of ranks for all 30 observations will be $1 + 2 + 3 + \ldots + 30 = \frac{30(30+1)}{2} = 465$

We are given $R_1 = 134$ and $R_2 = 157$.

Hence, $R_1 + R_2 + R_3 = 134 + 157 + R_3 = 465$

So $R_3 = 465 - 134 - 157 = 174 =$ the rank sum for Psychology majors

b. H_0: The population distributions of textbook costs for such students in all three majors are identical.

H_1: The population distributions of textbook costs for such students are not identical in all three majors.

For $\alpha = .025$ and $df = k - 1 = 3 - 1 = 2$, the rejection region is $\chi^2 > 7.378$.

Here, $n_1 = 10$, $n_2 = 9$, and $n_3 = 11$

$$H = \frac{12}{n(n+1)} \left[\frac{R_1^2}{n_1} + \frac{R_2^2}{n_2} + \frac{R_3^2}{n_3} \right] - 3(n+1) = \frac{12}{30(30+1)} \left[\frac{(134)^2}{10} + \frac{(157)^2}{9} + \frac{(174)^2}{11} \right] - 3(30+1) = 1.022$$

Do not reject H_0. Do not reject the null hypothesis that the median textbook cost is the same for second–semester juniors in all three majors.

14.107 a. In states with a small percentage of students who take the SAT, those taking it are usually the better students and tend to score high. In other states the exam is taken by a broad spectrum of students, so the average scores would tend to be lower. Thus, we would expect ρ_s to be negative.

b. In the following table u and v are the ranks of Math SAT scores and percentage of graduates taking the SAT test, respectively.

u	4	2	9	6	7	3	1	8	5	10	
v	10	7	3.5	5	3.5	9	6	1	8	2	
d	−6	−5	5.5	1	3.5	−6	−5	7	−3	8	
d^2	36	25	30.25	1	12.25	36	25	49	9	64	$\Sigma d^2 = 287.5$

$$r_s = 1 - \frac{6\sum d^2}{n(n^2-1)} = 1 - \frac{6(287.5)}{10(100-1)} = 1 - \frac{1725}{990} = -.742$$

r_s is negative, which is consistent with the answer to part a.

c. For a two–tailed test with $n = 10$ and $\alpha = .05$, the critical values from Table XIV are $\pm.648$, so the rejection region is $r_s \le -.648$ and $r_s \ge .648$. From part b, $r_s = -.742$, so reject H_0.

14.108 a. In the following table u and v are the critic's ranks and the ranks of box office receipts, respectively.

u	7	3	10	1	4	5	2	6	8	9	
v	9	3	10	1	2	4	7	5	8	6	
d	−2	0	0	0	2	1	−5	1	0	3	
d^2	4	0	0	0	4	1	25	1	0	9	$\Sigma d^2 = 44$

$$r_s = 1 - \frac{6\sum d^2}{n(n^2-1)} = 1 - \frac{6(44)}{10(100-1)} = 1 - \frac{264}{990} = .733$$

b. For a right–tailed test with $n = 10$ and $\alpha = .05$, the rejection region is $r_s \ge .564$

From part a, $r_s = .733$, thus: Reject H_0.

c. In part b, H_0, was rejected at the .05 level of significance, so there is sufficient evidence that this critic can predict the box–office performance of films.

14.109 a. We would expect to reject H_0.

b. In the following table, u and v are the ranks of driving experience and insurance premiums, respectively.

u	2	1	5	4	6	3	8	7	
v	6	8	3	7	2	4	1	5	
d	−4	−7	2	−3	4	−1	7	2	
d^2	16	49	4	9	16	1	49	4	$\Sigma d^2 = 148$

$$r_s = 1 - \frac{6 \sum d^2}{n(n^2 - 1)} = 1 - \frac{6(148)}{8(64 - 1)} = 1 - \frac{888}{504} = -.762$$

For a left–tailed test with $n = 8$ and $\alpha = .05$, the rejection region is $r_s \leq -.643$. Reject *Ho* as we expected in part a.

14.110 H_0: The order of boarding the elevator by gender is random.

H_1: The order of boarding the elevator by gender is not random.

Let n_1 be the number of women and n_2 be the number of men in the sample of 20 people. Since $n_1 = n_2 = 10 < 15$, use the runs test with critical values from Table XV. For $\alpha = .05$, the critical values of R are 6 and 16.

Thus, reject H_0 if $R \leq 6$ or $R \geq 16$. The observed value of $R = 8$.

Do not reject H_0. Do not conclude that the order of boarding by gender is nonrandom.

14.111 H_0: The sequence of defective and good tools is random.

H_1: The sequence of defective and good tools is not random.

Let n_1 be the number of good tools and n_2 be the number of defective tools in the sample of 18 tools. Then $n_1 = 13$ and $n_2 = 5$.

Since n_1 and, n_2 are less than 15, use the runs test with critical values from Table XV. For $\alpha = .05$, the critical values of R are 4 and 12. Thus, reject H_0 if $R \leq 4$ or $R \geq 12$.

The given data has 7 runs, so $R = 7$.

Do not reject H_0. There is insufficient evidence of nonrandomness in this sequence.

14.112 H_0: The test results for this emissions test station are in random order.

H_1: The test results for this emissions test station are not in random order.

Since n_1 and n_2 are greater than 15, use the normal distribution.

For $\alpha = .01$, the rejection region lies to the left of $z = -2.58$ and to the right of $z = 2.58$.

$$\mu_R = \frac{2n_1 n_2}{n_1 + n_2} + 1 = \frac{2(157)(143)}{157 + 143} + 1 = 150.67$$

$$\sigma_R = \sqrt{\frac{2n_1 n_2 (2n_1 n_2 - n_1 - n_2)}{(n_1 + n_2)^2 (n_1 + n_2 - 1)}} = \sqrt{\frac{2(157)(143)(2 \cdot 157 \cdot 143 - 157 - 143)}{(157 + 143)^2 (157 + 143 - 1)}} = 8.62686815$$

$$z = \frac{R - \mu_R}{\sigma_R} = \frac{41 - 150.67}{8.62686815} = -12.71$$

Reject H_0. Conclude that the test results for this emissions test station are not in random order.

14.113 H_0: The wins and losses were in random order.

H_1: The wins and losses were not in random order.

n_1 = number of wins = 10 and n_2 = number of losses = 20. Since n_2 is greater than 15, use the normal distribution.

For α = .02, the rejection lies to the left of $z = -2.33$ and to the right of $z = 2.33$.

From the data, the observed value of $R = 17$.

$$\mu_R = \frac{2n_1n_2}{n_1+n_2}+1 = \frac{2(10)(20)}{10+20}+1 = 14.33$$

$$\sigma_R = \sqrt{\frac{2n_1n_2(2n_1n_2-n_1-n_2)}{(n_1+n_2)^2(n_1+n_2-1)}} = \sqrt{\frac{2(10)(20)(2\cdot10\cdot20-10-20)}{(10+20)^2(10+20-1)}} =$$

$$z = \frac{R-\mu_R}{\sigma_R} = \frac{17-14.33}{2.38128077} = 1.12$$

Do not reject H_0. Conclude that the wins and losses were in random order.

14.114 a. H_0: The population distributions of life spans of laboratory mice are identical for the two diets.

H_1: The population distribution of life spans for the standard diet lies to the left of the population distribution for the low–calorie diet.

Since $n_1 = n_2 = 10$, use the Wilcoxon rank–sum test for small samples.

For a left–tailed test with α = .025, the critical value of $T = T_L = 79$. Thus, the rejection region is $T \le 79$.

The mouse in Group B that survived the experiment lived longer than the other 19 mice. Therefore it is reasonable to assign its life span the rank of 20.

Group A		Group B	
Life Span	Rank	Life Span	Rank
900	8	1037	17
907	11	905	10
751	1	1023	15
833	3	988	13
920	12	1078	19
787	2	1011	14
850	5	*	20
877	6	1063	18
848	4	898	7
901	9	1033	16
	Sum = 61		Sum = 149

Since the test is one–tailed and $n_1 = n_2$ the observed value of T is the sum of ranks for sample 1. Thus, the observed value of $T = 61$

Reject H_0. Conclude that the low–calorie diet appears to lengthen the lives of laboratory mice.

b. A Wilcoxon signed–rank test would not be appropriate in this example. Each mouse is given either the standard diet or the low–calorie diet, but not both. Thus the samples are independent, not paired, as required for the Wilcoxon signed–rank test.

14.115 a. Lay out a route for a test run in a typical city. Let each driver drive each car several times on this route. Make sure that the length of the route is selected so that the total mileage of all test runs for each car does not exceed 500 miles. Then calculate the gas mileage for each test run. Next, compute the median gas mileage for each of the three cars and compare them in the article without using any inferential statistical procedures.

   b. Since the gas mileage cannot be assumed to be normally distributed, the ANOVA procedure of Chapter 12 would not be appropriate. Instead, use the Kruskal–Wallis test on the data collected in part a for the three cars. Test the null hypothesis that the population distributions of gas mileage for the three cars are identical against the alternative hypothesis that the three distributions are not identical. Select an appropriate significance level. Then, write a report explaining and interpreting these results, being sure to point out what the test results imply about the median gas mileage for the three cars.

14.116

Number of Contacts		Gas Mileage	
Difference (Before–After)	Signed Rank	Difference (Before–After)	Signed Rank
+1	+1	−1	−1.5
+3	+2	−3	−4.5
−5	−4.5	+1	+1.5
+4	+3	−2	−3
−6	−6	−5	−7
+5	+4.5	−4	−6
+8	+7	−3	−4.5

Number of Contacts

Let M_B and M_A denote the median number of contacts made by all such salespersons before and after the installation of governors, respectively.

H_0: $M_A = M_B$; H_1: $M_A < M_B$;   A left–tailed test.

Since we cannot assume that the differences in number of contacts are normally distributed and since $n < 15$, use the Wilcoxon small–sample signed–rank test.

For $n = 7$, $\alpha = .05$, with a left–tailed test, the rejection region is $T \leq 4$.

For a left–tailed test, T is the sum of the absolute values of the negative ranks, so $T = 4.5 + 6 = 10.5$.

Therefore, do not reject H_0. Conclude that the median number of contacts is unchanged by the installation of governors.

Gas Mileage

Let M_B and M_A denote the median gas mileages for all such salespersons' cars before and after the installation of governors, respectively.

H_0: $M_A = M_B$; H_1: $M_A > M_B$;   A right–tailed test.

Since we cannot assume that the differences in gas mileages are normally distributed, and since $n < 15$, use the Wilcoxon signed–rank test for small samples.

For a right–tailed test with $n = 7$ and $\alpha = .05$, the rejection region is $T \leq 4$.

The observed value of T is the sum of the positive ranks, so $T = 1.5$. Therefore, reject H_0. Conclude that the median gas mileage is higher after the installation of governors.

The hypothesis tests indicate that the governors improved fuel economy without reducing the number of contacts made by salespersons.

14.117 a. Take random samples of carpenters, plumbers, electricians, and masons from your city and record each worker's hourly wage. Note that random samples will be representative of the populations of the four trades and should have approximately the same proportions of union members, respectively, as the four populations. Next, find the median hourly wage for each sample and compare them in the article without using any inferential statistical procedures.

b. Since the hourly wages cannot be assumed to be normally distributed, the ANOVA procedure of Chapter 12 would not be appropriate. Instead, use the Kruskal–Wallis test on the data collected in part a for the four trades. Test the null hypothesis that the population distributions of hourly wages are identical against the alternative that the four distributions are not identical. Select an appropriate significance level. Then write a report explaining and interpreting these results, being sure to point out what the test results imply about the median hourly wages for the four trades.

14.118 a. Since each value of y increases by the same amount, the rank of each value of y would stay the same. Because r_s is based solely on the ranks, which would not change, the value of r_s would remain the same.

b. Using the original data, let u and v denote the ranks of x and y, respectively.

x	10	20	30	40	50	60	
y	12	15	19	21	25	30	
u	1	2	3	4	5	6	
v	1	2	3	4	5	6	
d	0	0	0	0	0	0	
d^2	0	0	0	0	0	0	$\Sigma d^2 = 0$

$$r_s = 1 - \frac{6\sum d^2}{n(n^2 - 1)} = 1 - \frac{6(0)}{6(36 - 1)} = 1$$

Next, adding 5 to each value of y, let u and w denote the ranks of x and y, respectively.

x	10	20	30	40	50	60	
$y+5$	17	20	24	26	30	35	
u	1	2	3	4	5	6	
w	1	2	3	4	5	6	
d	0	0	0	0	0	0	
d^2	0	0	0	0	0	0	$\Sigma d^2 = 0$

$$r_s = 1 - \frac{6\sum d^2}{n(n^2-1)} = 1 - \frac{6(0)}{6(36-1)} = 1 = \text{original value of } r_s$$

Note that the value of r_s is unchanged by adding 5 to each value of y, which agrees with the answer to part a.

14.119 a. Since the data are paired (two grades for each essay) the department head could use the sign test for paired data of Section 14.1.3 or the Wilcoxon signed–rank test of Section 14.2.

b. We will use the Wilcoxon signed–rank test. For each essay:

Paired difference = Professor A's grade – instructor's grade

Let M_A and M_B denote the median grades for all such essays by Professor A and the instructor, respectively.

$H_0: M_A = M_B$; $H_1: M_A \neq M_B$;   A two–tailed test.

Since $n = 10 < 15$, use the Wilcoxon signed–rank test procedure for the small–sample case. For $n = 10$, $\alpha = .05$, and a two–tailed test, the rejection region is $T \leq 8$.

Grades		Differences	Absolute	Ranks of	Signed
Professor A	Professor B	(Prof. A–Inst.)	Differences	Differences	Ranks
75	80	–5	5	6.5	–6.5
62	50	+12	12	10	+10
90	85	+5	5	6.5	+6.5
48	55	–7	7	9	–9
67	63	+4	4	3.5	+3.5
82	78	+4	4	3.5	+3.5
94	89	+5	5	6.5	+6.5
76	81	–5	5	6.5	–6.5
78	75	+3	3	2	+2
84	83	+1	1	1	+1

For a two–tailed test, T is the smaller sum of ranks. Here, T is the sum of the absolute values of the negative ranks. Thus, the observed value of $T = 6.5 + 9 + 6.5 = 22$.

Do not reject H_0. Do not conclude that the instructor tends to grade higher or lower than Professor A.

c. To determine whether or not the instructor and Professor A tend to rank the essays similarly, Spearman's rho rank correlation coefficient of Section 14.5 would be appropriate. A positive value of r_s would suggest some consistency.

Appropriate hypotheses would be: H_0: $\rho_s = 0$; H_1: $\rho_s > 0$

d. For a right–tailed test with $n = 10$ and $\alpha = .05$, the rejection region is $r_s \geq .564$.

In the following table, u and v are the ranks of Professor A's grades and the instructor's grades, respectively.

u	4	2	9	1	3	7	10	5	6	8	
v	6	1	9	2	3	5	10	7	4	8	
d	−2	1	0	−1	0	2	0	−2	2	0	
d^2	4	1	0	1	0	4	0	4	4	0	$\Sigma d^2 = 18$

$$r_s = 1 - \frac{6\sum d^2}{n(n^2 - 1)} = 1 - \frac{6(18)}{10(100 - 1)} = 1 - \frac{108}{990} = .891$$

Reject H_0. Conclude that the instructor and Professor A tend to rank such essays similarly.

14.120 a. The ANOVA procedure of Chapter 12 and the Kruskal–Wallis test of Section 14.4 might be appropriate.

b. The ANOVA procedure of Chapter 12 is based on means while the Kruskal–Wallis test of Section 14.4 concerns medians.

c. The ANOVA procedure from Chapter 12 requires the assumptions that the populations from which the samples are drawn are approximately normally distributed and have the same variance.

14.121 a. To base a test on the linear correlation coefficient of Chapter 13 requires that both variables (GPAs and SAT scores) be normally distributed.

b. If the assumptions of part a are not satisfied, the test based on Spearman's rho rank correlation coefficient may be used.

14.122 a. The following tests from Chapter 10 and 14 might be appropriate.

1. The paired–sample test of Section 10.4

2. The sign test for paired data of Section 14.1.3

3. The Wilcoxon signed–rank test of Section 14.2

b. The paired sample test of Section 10.4 involves the mean of paired differences. The sign test for paired data of Section 14.1.3 and the Wilcoxon signed–rank test of Section 14.2 both involve the median of paired differences.

c. If the paired differences are not normally distributed, the sign test for paired data and the Wilcoxon signed–rank test could be used.

14.123   Place B's and A's alternately, starting with a B until all 10 A's have been used, then finish with the remaining B's, as shown below.

BABABABABABABABABABBBBB

This sequence has 21 runs. Any sequence which begins with an A or has two or more A's in a row will have less than 21 runs.

Self-Review Test for Chapter Fourteen

1. b	**2.** a	**3.** a	**4.** c	**5.** b	**6.** c	**7.** a, b
8. b	**9.** c	**10.** c	**11.** a, b	**12.** c	**13.** b	

14.   Each time a juror is selected from this pool, let p be the probability that the person is a woman.

H_0: $p = .50$; H_1: $p \neq .50$;   A two–tailed test.

$n = 12 < 25$. So use the binomial distribution.

For a two–tailed test with $n = 12$ and $\alpha = .05$, the rejection region (from Table XI) is $X \leq 2$ and $x \geq 10$.

The observed value of $X = 2$.

Reject H_0. Conclude that the selection process is biased with respect to gender.

15.   Let p be the proportion of Americans in favor of putting Social Security money into personal retirement accounts.

H_0: $p = .50$; H_1: $p > .50$;   A right–tailed test.

$n = 1000 > 25$, so use the normal distribution.

For a right–tailed test with $\alpha = .025$, the rejection region lies to the right of $z = 1.96$.

$\mu = np = 1000(.50) = 500$; $\sigma = \sqrt{npq} = \sqrt{1000(.50)(.50)} = 15.81138830$

Since 520 respondents favored the proposal, $X = 520$.

$\dfrac{n}{2} = \dfrac{1000}{2} = 500$.   Since, $X > \dfrac{n}{2}$, $z = \dfrac{(X - .5) - \mu}{\sigma} = \dfrac{(520 - .5) - 500}{15.81138830} = 1.23$

Do not reject H_0. Do not conclude that over half of Americans favor this proposal.

16. H_0: Median = \$65; H_1: Median > \$65;   A right–tailed test.

$n = 12 < 25$, so we use the binomial distribution.

For a right–tailed test with $n = 12$ and $\alpha = .05$, the rejection region is $X \geq 10$.

Assigning a plus sign to each amount above \$65 and a minus sign to each amount below \$65 yields, 7 plus signs and 5 minus signs. Since the test is right–tailed, the observed value of X is given by the larger number of signs. Thus, the observed value of $X = 7$

Do not reject H_0. Do not conclude that the median amount spent by all customers at this store after the campaign exceeds $65.

17. H_0: Median = $20,264; H_1: Median $\neq$ $20,264;   A two–tailed test.

     $n = 400 > 25$, so use the normal distribution.

     For a two–tailed test with $\alpha = .01$, the rejection regions lies to the left of -2.58 and to the right of $z = 2.58$.

$$\mu = np = 400(.50) = 200; \quad \text{and} \quad \sigma = \sqrt{npq} = \sqrt{400(.50)(.50)} = 10$$

     If we assign a plus sign to every woman earning more than $20,264 and a minus sign to every woman earning less than $20,264, there are 171 plus signs and 229 minus signs. For a two–tailed test either value can be X, so let's use $X = 229$.

$$\frac{n}{2} = \frac{400}{2} = 200 \qquad \text{Since, } X > \frac{n}{2}, \ z = \frac{(X - .5) - \mu}{\sigma} = \frac{(229 - .5) - 200}{10} = 2.85$$

     Reject H_0. Conclude that the median income of women living alone is different from $20,264.

18.   a.   For each adult: Paired difference = Cholesterol level before diet – Cholesterol level after diet

Before	After	Differences (Before–After)	Absolute Differences	Ranks of Differences	Signed Ranks	Sign
210	193	+17	17	7	+7	+
180	186	−6	6	2	−2	—
195	186	+9	9	3.5	+3.5	+
220	223	−3	3	1	−1	—
231	220	+11	11	5	+5	+
199	183	+16	16	6	+6	+
224	233	−9	9	3.5	−3.5	—

     Let M denote Median cholesterol level before diet – Median cholesterol level after diet

     H_0: $M = 0$; H_1: $M \neq 0$;   A two–tailed test.

     $n = 7 < 25$. Use the binomial distribution.

     For a two–tailed test with $n = 7$ and $\alpha = .05$, the rejection region is $X = 0$ and $X = 7$.

     There are four plus signs and three minus signs in the seventh column of the preceding table.

     For a two–tailed test we may use either number of signs as the observed value of X.

     If we let $X = 3$, then:

     Do not reject H_0. Conclude that the median cholesterol level before the diet is the same as after the diet.

   b.   Let M_B and M_A be the median cholesterol levels before and after the diet, respectively.

     H_0: $M_A = M_B$; H_1: $M_A \neq M_B$;   A two–tailed test.

     $n = 7 < 15$. Use the Wilcoxon signed–rank test for the small–sample case.

     For a two–tailed test with $n = 7$ and $\alpha = .05$, the rejection region is $T \leq 2$.

For a two–tailed test the observed value of T is the smaller sum of ranks. Thus The observed value of T = Sum of absolute values of negative ranks = 2 +1 + 3.5 = 6.5, from the preceding table. Do not reject H_0. Conclude that the median cholesterol level before the diet is the same as after the diet.

c. The conclusions of parts a and b are the same; Do not reject H_0.

19. Let M = Median age of such artifacts dated by Method I minus Median age of such artifacts dated by Method II.

H_0: M = 0; H_1: M ≠ 0;   A two–tailed test.

For two of the 33 artifacts, there was no difference in ages for the two methods, so the true value of n = 33 – 2 = 31. Since $n > 25$, use the normal distribution.

For a two–tailed test with α = .02, the rejection region lies to the left of z = –2.33 and to the right of z = 2.33.

$$\mu = np = 31(.50) = 15.5; \sigma = \sqrt{npq} = \sqrt{31(.50)(.50)} = 2.78388218$$

The data yield 11 plus signs and 20 minus signs. For a two–tailed test we may use either number of signs as the observed value of X. If we let X = 20, then:

$$\frac{n}{2} = \frac{31}{2} = 15.5.$$ Since, $X > \frac{n}{2}$, $z = \frac{(X-.5)-\mu}{\sigma} = \frac{(20-.5)-15.5}{2.78388218} = 1.44$

Do not reject H_0. Do not conclude that the median estimated ages of such artifacts differ for the two methods.

20. For each student: Paired difference = Fall GPA – Spring GPA

Let M_F and M_S denote the median GPAs for all Sophomore Electrical Engineering majors in the Fall and Spring semesters, respectively.

H_0: $M_S = M_F$; H_1: $M_S < M_F$;   A left–tailed test.

One student's GPA was the same for both semesters, so the true value of n = 10 – 1 = 9. Since $n < 15$, use the Wilcoxon signed–rank test procedure for the small–sample case. For n = 9, α = .05, and a one–tailed test, the rejection region is $T \leq 8$.

GPAs		Differences	Absolute	Ranks of	Signed
Fall	Spring	(Fall–Spring)	Differences	Differences	Ranks
3.20	3.15	+.05	.05	2.5	+2.5
3.56	3.40	+.16	.16	6	+6
3.05	2.88	+.17	.17	7.5	+7.5
3.78	3.67	+.11	.11	4	+4
4.00	4.00	—	—	—	—
2.85	3.00	–.15	.15	5	–5
3.33	3.30	+.03	.03	1	+1
2.67	3.05	–.38	.38	9	–9
3.00	2.95	+.05	.05	2.5	+2.5
3.67	3.50	+.17	.17	7.5	+7.5

For a left–tailed test, the observed value of T is the sum of the absolute values of the negative ranks. Thus, the observed value of $T = 5 + 9 = 14$.

Do not reject H_0. Do not conclude that the median GPA for all Sophomore Electrical Engineering majors at this university is lower in the Spring semester than in the Fall.

21.   Let M_A and M_B denote the median memory test scores after and before the course, respectively.

H_0: $M_A = M_B$; H_1: $M_A > M_B$;   A right–tailed test.

Since three of the 30 students scored the same before and after the course, the true value of $n = 30 - 3 = 27$. $n > 15$. Use the Wilcoxon signed–rank test with the normal distribution approximation.

For a right–tailed test with $\alpha = .025$, the rejection region lies to the right of $z = 1.96$.

$$\mu_T = \frac{n(n+1)}{4} = \frac{27(27+1)}{4} = 189$$

$$\sigma_T = \sqrt{\frac{n(n+1)(2n+1)}{24}} = \sqrt{\frac{27(27+1)(54+1)}{24}} = 41.62331078$$

For a right–tailed test, the observed value of T is the sum of the absolute values of the negative ranks. Thus, $T = 276$.

$$z = \frac{T - \mu_T}{\sigma_T} = \frac{276 - 189}{41.62331078} = 2.09$$

Reject H_0. Conclude that the course tends to improve scores on memory tests.

22. H_0: The population distributions of commuting times are the same for both routes.

H_1: The population distributions of commuting times are different for the two routes.

$n_1 = n_2 = 8 < 10$, so use the Wilcoxon rank–sum test for small samples.

For a two–tailed test with $n_1 = n_2 = 8$ and $\alpha = .05$, the rejection region is $T \leq 49$ and $T \geq 87$.

Route I		Route II	
Time	Rank	Time	Rank
45	12	38	3.5
43	9.5	40	6
38	3.5	39	5
56	16	42	8
41	7	50	15
43	9.5	37	2
46	13.5	46	13.5
44	11	36	1
	Sum = 82		Sum = 54

Since the test is two–tailed and $n_1 = n_2$, the observed value of T is the sum of ranks for either sample. If we take $T = 82$, the conclusion is:

Do not reject H_0. Do not reject the hypothesis that the median commuting time is the same for both routes.

23. H_0: The population distributions of times to prepare such tax returns are the same for employees A and B.

H_1: The population distributions of times to prepare such tax returns are different for employees A and B.

Here, $n_1 = n_2 = 18 > 10$. Use the normal distribution. The test is two–tailed with $\alpha = .025$, so the rejection region lies to the left of $z = -2.24$ and to the right of $z = 2.24$.

We are given: Sum of ranks for $A = 298$ and sum of ranks for $B = 368$.

Since the test is two–tailed and $n_1 = n_2$, the observed value of T is given by either sum of ranks. Then, we may let $T = 298$.

$$\mu_T = \frac{n_1(n_1+n_2+1)}{2} = \frac{18(18+18+1)}{2} = 333$$

$$\sigma_T = \sqrt{\frac{n_1 n_2(n_1+n_2+1)}{12}} = \sqrt{\frac{18(18)(18+18+1)}{12}} = 31.60696126$$

$$z = \frac{T-\mu_T}{\sigma_T} = \frac{298-333}{31.60696126} = -1.11$$

Do not reject H_0. Do not conclude that there is a difference in the median times taken by A and B to prepare such tax returns.

24. a. H_0: The population distributions of reported cases of telemarketing fraud in the three cities are identical.

H_1: The population distributions of reported cases of telemarketing fraud in the three cities are not identical.

For $\alpha = .025$ and $df = k - 1 = 3 - 1 = 2$, the rejection region is $\chi^2 > 7.378$.

City A		City B		City C	
Number	Rank	Number	Rank	Number	Rank
53	11	29	1	75	16
46	6	35	4	49	8
59	12	44	5	62	14
33	3	31	2	68	15
60	13	50	9	52	10
		48	7		
$n_1 = 5$	$R_1 = 45$	$n_2 = 6$	$R_2 = 28$	$n_3 = 5$	$R_3 = 63$

Here, $n = n_1 + n_2 + n_3 = 5 + 6 + 5 = 16$

$$H = \frac{12}{n(n+1)}\left[\frac{R_1^2}{n_1} + \frac{R_2^2}{n_2} + \frac{R_3^2}{n_3}\right] - 3(n+1) = \frac{12}{16(16+1)}\left[\frac{(45)^2}{5} + \frac{(28)^2}{6} + \frac{(63)^2}{5}\right] - 3(16+1) = 7.653$$

Reject H_0. Reject the hypothesis that the distributions of numbers of such reported cases are identical for the three cities.

b. For $\alpha = .01$ and $df = 2$, the rejection region is $\chi^2 > 9.210$.
Do not reject H_0.

c. Parts a and b show that the sample does not support the alternative hypothesis very strongly because lowering the significance level from .025 to .01 reverses the conclusion.

25.   a. Since y tends to increase as x increases, we would expect the value of the Spearman rho rank correlation coefficient to be positive.

b. In the table below, u and v are the ranks of x and y, respectively.

u	7	5	9	1	6	2.5	10	8	4	2.5	
v	8	5	9	2	7	4	10	6	3	1	
d	−1	0	0	−1	−1	−1.5	0	2	1	1.5	
d^2	1	0	0	1	1	2.25	0	4	1	2.25	$\Sigma d^2 = 12.5$

$$r_s = 1 - \frac{6\sum d^2}{n(n^2 - 1)} = 1 - \frac{6(12.5)}{10(100 - 1)} = 1 - \frac{75}{990} = .924$$

r_s is positive, which agrees with the answer to part a.

c. For a right–tailed test with $n = 10$ and $\alpha = .025$, the rejection region is $r_s \geq .648$.
From part b, $r_s = .924$, so reject H_0.

26. H_0: Keepers occur randomly in the sequence of fish.
H_1: Keepers do not occur randomly in the sequence of fish.
Let $n_1 =$ number of keepers and $n_2 =$ number of short bass.
From the data, $n_1 = 7$ and $n_2 = 7$.
Since n_1 and n_2 are less than 15, use the runs test with critical values from Table XV.
For $\alpha = .05$, the critical values of R are 3 and 13. Thus, reject H_0 if $R \leq 3$ or $R \geq 13$.
From the data, the observed value of $R = 6$.
Do not reject H_0. This sequence does not support Ramon's theory.

27. H_0: The wins and losses occur randomly among the 54 games.
H_1: The wins and losses do not occur randomly among the 54 games.
Since $n_1 = 30$ and $n_2 = 24$ are greater than 15, use the normal distribution.
For $\alpha = .05$, the rejection region lies to the left of $z = -1.96$ and to the right of $z = 1.96$.

The observed value of $R = 15$.

$$\mu_R = \frac{2n_1 n_2}{n_1 + n_2} + 1 = \frac{2(30)(24)}{30 + 24} + 1 = 27.67$$

$$\sigma_R = \sqrt{\frac{2n_1 n_2 (2n_1 n_2 - n_1 - n_2)}{(n_1 + n_2)^2 (n_1 + n_2 - 1)}} = \sqrt{\frac{2(30)(24)(2 \cdot 30 \cdot 24 - 30 - 24)}{(30 + 24)^2 (30 + 24 - 1)}} = 3.59361185$$

$$z = \frac{R - \mu_R}{\sigma_R} = \frac{15 - 27.67}{3.59361185} = -3.53$$

Reject H_0. Conclude that the wins and losses do not occur randomly among the 54 games.

Appendix A

A.1   Data sources can be divided into three categories: (1) internal sources, (2) external sources, and (3) surveys and experiments.

1. INTERNAL SOURCES

   The data sources such as a company's own personal files or accounting records are called internal sources. For example, a company that wants to forecast the future sales of its product might use data from its own records for past time periods.

2. EXTERNAL SOURCES

   Many times we may obtain data from sources outside the company. Such sources of data are called external sources. Data obtained from external sources may be primary or secondary data. Primary data are the data obtained from the organization which originally collected them. Secondary data are data obtained from a source which did not originally collect them.

3. SURVEYS AND EXPERIMENTS

   Sometimes the data we need may not be available from internal or external sources. In such cases, we may obtain data by conducting our own surveys or experiments.

A.2   Internal sources of data are a company's own files or records; external sources are data obtained from other companies or institutions.

A.3   A census is a survey that includes all members of the population. A survey based on a portion of the population is called a sample survey. A sample survey is preferred over a census for the following reasons.

    i.   Conducting a census is very expensive because the size of the population is usually very large.

    ii.   Conducting a census is very time consuming.

    iii. In many cases it is almost impossible to identify every member of the target population.

A.4   In a survey, data are collected without exercising any control over the factors that may affect the characteristics of interest or the results of a survey. In an experiment, the researchers exercise control over some or all of these factors.

A.5   a. *Random sample*: A sample drawn in such a way that each element of the population has a chance of being included in the sample.

b. *Nonrandom sample*: A sample in which some members of the population may have no chance of being selected.

c. *Convenience sample*: A sample in which the most accessible members of the population are selected.

d. *Judgment sample*: A sample in which members of a population are selected according to an expert's judgment.

e. *Quota sample*: A sample selected in such a way that each group or subpopulation is represented in the sample in exactly the same proportion as in the target population.

A.6   Following are the four sampling techniques discussed in Appendix A.

a. SIMPLE RANDOM SAMPLING

A sample that assigns the same probability to each member of the population for being selected is called a simple random sample. One way to select a simple random sample is that of a lottery or drawing.

b. SYSTEMATIC RANDOM SAMPLING

In systematic random sampling, we first randomly select one member from the first k units. Then every k^{th} member, starting with the first selected member, is included in the sample. For example, if we need to select 100 students from a list of 4000 by using systematic random sampling, we will adopt the following procedure. We would first arrange all the 4000 students alphabetically (or based on some other characteristic). Since, the sample size is to be 100, the ratio of the population to sample size is 4000/100=40. Using this ratio, we will randomly select one student from the first 40 students in the arranged list either by using the lottery system or by using a table of random numbers. Suppose, by using either of these methods, we select the 18th student. We will then select every 18th student from every 40 students in the list. In other words, our sample will include the students with the numbers 18, 58, 98, 138,…, and so on. Note that systematic random sampling does not give a simple random sample because we cannot select two adjacent elements. Hence, every member of the population does not have the same probability of being selected.

c. STRATIFIED RANDOM SAMPLING

In a stratified random sample, we first divide the population into subpopulations which are called strata. Then, one sample is taken from each stratum. The collection of all samples from all strata gives the stratified random sample. We divide the population on the basis of any characteristic such as income, expenditure, sex, education, race, employment or family size.

d. CLUSTER SAMPLING

Sometimes the target population may be scattered over a much wider geographical area. Consequently, if a simple random sample is taken, it may be costly to contact each member of the sample. In such a case, cluster sampling may be the best alternative. In cluster sampling, the whole population is divided into (geographical) groups called clusters. Each cluster is representative of the population. Then a random sample of clusters is selected. Finally a random sample of elements from each of the selected clusters is selected.

A.7   Each member of the population has the same chance of being included in a sample under the simple random sampling technique.

A.8   a. This is a nonrandom sample since students in the university who were not in her statistics class had no chance of being included in the sample.

b. This is a convenience sample since students in her class were the most accessible members of the population.

c. This sample suffers from selection error. The population consists of all students at the university, but the sampling frame is limited to members of her statistics class.

A.9   a. This is a nonrandom sample. The students are not picked randomly but are chosen according to the professor's preferences.

b. This is a judgment sample since the professor uses his knowledge and expertise to determine which students to include in the sample.

c. This sample is subject to selection error since only those students the professor considers appropriate are included in the sample.

A.10   a. This is a random sample since it is selected randomly from a complete list of students at the university. Thus, each student in the population has a chance of being included in the sample.

         b. This is a simple random sample since the software package would give each student an equal chance of being selected.

         c. There should be no systematic error, since the sampling frame is the entire population and the use of the software would give each student an equal chance of being included in the sample.

A.11   a. This is a random sample since the sampling frame is the entire class.

         b. This is a simple random sample since the software package gives each student an equal chance of being chosen.

         c. There should be no systematic error, since the sampling frame is the entire population and the software package would give each student an equal chance of being selected.

A.12   This is a quota sample since it is composed of 58% males and 42% females, the same proportions found in the population of 1000 employees. It is also a nonrandom sample because men and women were selected by interviewers as they wished.

A.13   This is a non-random sample. Only those who have a computer and use monster.com were able to answer the survey. This sample is subject to voluntary response error, since only those who feel strongly enough about the issues to complete the questionnaire will respond. It also suffers from selection error since only the website's users are included in the sampling frame.

A.14   The survey is subject to voluntary response error since it receives responses from only those companies that are willing to take the trouble to complete the questionnaire and mail it in. These respondents may not be representative of all major companies. It also suffers from nonresponse error because many companies did not respond.

A.15   This survey is subject to response error since some parents may be reluctant to give honest answers to an interviewer's questions about sensitive family matters.

A.16   Since the sample includes only people from one borough of New York City, it is not representative of the entire city. Therefore the researcher is not justified in applying the result to New York City.

A.17   a. This is an observational study since the doctors did not control the assignment of volunteers to the treatment and control groups.

        b. The study is not double-blind because:

           i. There was no placebo, so the volunteers knew whether or not they were receiving the treatment, and

          ii. The doctors knew who was getting the treatment and who was not.

A.18   a. This is a designed experiment since the physicians controlled the assignment of people to the treatment and control groups.

        b. The experiment is not double-blind since the physicians knew who was given aspirin and who was given the placebo.

A.19   a. This is a designed experiment since the physicians controlled the assignment of people to the treatment and control groups.

        b. The study is double-blind since neither the patients nor the doctors knew who was given the aspirin and who was given the placebo.

A.20   This is an observational study since the researchers relied on volunteers to form the treatment and control groups.

A.21   This is a designed experiment since the researchers selected participants randomly from the entire population of families on welfare and then controlled which families received the treatment (job training) and which did not.

A.22   The conclusion is unjustified. The families volunteered; they were not randomly selected from the population of all families on welfare, thus they may not be representative of the entire population.

A.23   If the data showed that the percentage of families who got off welfare was higher in the group that received job training, the conclusion is justified. Since families were randomly assigned to treatment and control groups, the two groups should have been similar, and the difference in outcomes should be due to treatment (job training).

A.24   a. This is an observational study since each participant decided how much meat to consume. Thus, the treatment is not controlled by the experimenters.

         b. Because this is an observational study, no cause-and-effect relationship between meat consumption and cholesterol level may be inferred. The effect of meat consumption on cholesterol level may be confounded with other variables such as other dietary habits, amount of exercise, and other features of the participants' lifestyles.

A.25   a. Since the study relies on volunteers, it may not be representative of the entire population of people suffering from compulsive behavior. Furthermore, the doctors used their own judgment to form the treatment and control groups. Thus, subjective factors may have influenced them, and the two groups may not be comparable. As a result, the effect of the medicine on compulsive behavior may be confounded with other variables. Therefore, the conclusion is not justified.

         b. Although this study technically satisfies the criteria for a designed experiment (experimenters controlled the assignment of people to treatment groups), it suffers from the weaknesses of an observational study, as pointed out in part a.

         c. The study is not double-blind since the physicians knew who received the treatment.

A.26   a. Since the patients were randomly selected from the population of all people suffering from compulsive behavior and were randomly assigned to treatment and control groups, the two groups should be comparable and representative of the entire population. The patients did not know whether or not they were getting the treatment, so any improvement in their condition should be due to the medicine and not merely to the power of suggestion. Thus, the conclusion is justified.

         b. This is a designed experiment since the doctors controlled the assignment of patients to the treatment and control groups.

         c. The study is not double-blind since the doctors knew who received the medicine.

A.27   a. This is a designed experiment, since the doctors controlled the assignment of patients to the treatment and control groups.

         b. The study is double-blind since neither patients nor doctors know who was receiving the medicine.

A.28 a. This is a designed experiment since the doctors controlled the assignment of dieters to the lower sugar and control groups.

b. Yes, those who received as much as ten percent of their calories from sucrose were the control group.

c. No, this was not a double-blind experiment since both the doctors and dieters knew who was on the low sugar diet and who was not.

A.29 a. These 10 pigs represent a convenience sample since the first ten (easiest to catch) pigs comprise the sample. Convenience samples are nonrandom samples.

b. From part a we know these 10 pigs comprise a nonrandom sample. Therefore, they are not likely to be representative of the entire population. Faster pigs, for example, are not as likely to be included in the sample.

c. They form a convenience sample.

d. Answers will vary, but one better procedure is as follows. Assign numbers 1 through 40 to the pigs, and write the numbers 1 through 40 on separate pieces of paper, put them in a hat, mix them, and then draw 10 numbers. Pick the pigs whose numbers were drawn.

A.30 a. No, this method is not likely to produce a random sample.

b. The following types of biases are likely to be present:

Voluntary Response Error: Only readers that have a strong opinion and are willing to pay $1 to respond will do so.

Selection Error: Not all members of the population are included; only those who actually read that newspaper may participate.

Response Error: A group may have a financial interest in the casino and place many calls in order to influence the outcome of the poll.

A.31 Seventy-eight percent of members of Health Maintenance Organizations (HMOs) responding to a recent survey reported that they had experienced denial of claims by their HMOs. Of those who had suffered such denials, 25% had unable to resolve the problem to their satisfaction in at least one case. The survey was based on questionnaires sent to 5000 randomly chosen HMO members, of which 1200 actually completed their questionnaires and returned them.

The results of this survey should be interpreted with caution, because the percentage may not be representative of all HMO members. The most likely source of bias is *nonresponse error*, since only 1200 of the 5000 questionnaires were actually returned. Members who have had a claim denied may be angry and consequently have more motivation to return their questionnaires. Thus, there is likely to be a higher percentage of denied claims among the 1200 members who actually responded. Therefore, 78% may be an overestimate of the true percentage of HMO members who have experienced denied claims. Similarly, those who were unable to resolve their problem would be even more strongly motivated to respond to the survey, so 25% is likely to overestimate the corresponding percentage for the whole group.

A.32   a. Total student body = 4000+3200+2800+2000= 12,000

Proportion of freshman in student body = 4000/1200= 1/3

Thus, 1/3 of the sample should be freshman.

Hence, number of freshman in sample = 300(1/3)= 100

b. Number of sophomores in sample = 300(3200/12,000)= 80

Number of juniors in sample = 300(2800/12,000)= 70

Number of seniors in sample = 300(2000/12,000)=50

A.33   a. We would expect \$61,200 to be a biased estimate of the current mean annual income for all 5432 alumni because only 1240 of the 5432 alumni answered the income question. These 1240 are unlikely to be representative of the entire group of 5432.

b. The following types of bias are likely to be present.

Nonresponse error: Alumni with low incomes may be ashamed to respond. Thus, the 1240 who actually returned their questionnaires and answered the income question would tend to have higher than average incomes.

Response error: Some of those who answered the income question may give a value that is higher than their actual income in order to appear more successful.

c. We would expect the estimate of \$61,200 to be above the current mean annual income of all 5432 alumni, for the given reasons in part b.

A.34   a. Yes, the unvaccinated dogs are the control group.

        b. No, the experiment is not double-blind. The owners and the veterinarians who examined the dogs for Lyme Disease know which dogs were vaccinated.

        c. The following are potential sources of bias:

           Selection error: Dogs whose owners permitted vaccination may not be comparable to other dogs. Their owners may be more concerned about keeping them away from ticks.

           Not double-blind: Since owners of vaccinated dogs know their dogs were vaccinated, they may have a different degree of concern about keeping their dogs away from ticks than owners of unvaccinated dogs. Also, the veterinarians know which dogs were vaccinated, which may influence their diagnosis for a dog having symptoms resembling Lyme Disease.

        d. The experiment could be improved by making it a randomized double-blind study. Select 200 dogs at random from dogs whose owners will permit vaccination. Randomly assign 100 of these dogs to the treatment group to receive the vaccine. The other 100 dogs would form the control group and would be given a placebo. The veterinarians who examined the dogs later for Lyme Disease would not be told which dogs were vaccinated, to avoid bias in diagnosis.

Data Sets

**These data sets are available on the textbook's website.
See the Preface of the textbook.**

Data Set I: City Data

C1	C2	C3	C4	C5	C6	C7	C8	C9	C10	C11	C12	C13	C14
Auburn-Opelika AL	7.66	1.86	4.05	1.28	501.00	192000.00	108.72	26.00	1.30	446.00	29.39	6.02	5.84
Birmingham AL	7.64	1.96	5.21	1.29	615.00	202781.00	108.43	25.36	1.37	647.00	24.79	8.33	6.29
Cullman County AL	6.54	1.71	4.43	1.38	510.00	195750.00	106.16	25.05	1.37	505.00	22.60	6.53	5.95
Decatur-Hartselle AL	5.80	1.70	4.05	1.48	486.00	177029.00	97.87	25.41	1.39	501.00	29.00	7.52	5.74
Florence AL	6.35	1.58	4.13	1.19	505.00	185000.00	100.60	25.43	1.33	440.00	26.20	7.72	6.49
Huntsville AL	7.60	1.66	4.28	1.24	578.00	192930.00	91.63	27.05	1.39	547.50	29.40	7.29	6.28
Marshall County AL	6.33	1.57	4.35	1.18	542.00	197600.00	89.62	24.37	1.32	335.00	20.00	7.69	6.09
Mobile AL	6.55	1.86	4.05	1.12	546.00	199369.00	118.09	23.75	1.34	294.40	22.00	6.81	5.69
Montgomery AL	7.39	1.87	4.20	1.10	576.00	205000.00	120.63	22.28	1.34	680.00	31.20	6.52	6.40
Tuscaloosa AL	6.46	1.69	3.87	1.15	684.00	220000.00	111.29	21.95	1.35	595.00	35.00	8.92	6.49
Anchorage AK	7.89	2.42	4.38	1.72	941.00	294007.00	110.32	21.32	1.47	821.00	32.00	10.56	6.44
Fairbanks AK	7.09	2.27	4.67	1.85	862.00	289615.00	201.92	19.48	1.43	998.00	29.08	10.00	7.59
Juneau AK	6.89	2.37	5.03	1.57	1020.00	341613.00	182.77	18.26	1.58	1550.00	29.47	9.95	7.29
Flagstaff AZ	6.19	1.91	4.39	1.28	849.00	261300.00	124.53	24.47	1.47	845.00	24.33	8.64	5.15
Phoenix AZ	6.12	1.83	4.20	1.14	657.00	207739.00	114.79	19.69	1.31	738.15	29.00	8.96	5.28
Sierra Vista AZ	5.99	1.72	4.35	1.11	531.00	179257.00	139.15	22.28	1.30	614.00	18.78	9.27	5.41
Tucson AZ	6.39	1.62	4.82	1.25	689.00	195000.00	135.96	21.03	1.35	853.17	30.00	10.19	5.29
Yuma AZ	5.83	1.97	3.60	1.05	583.00	205000.00	149.21	23.70	1.33	625.00	29.20	7.73	5.79
Fayetteville AR	6.23	1.49	3.83	0.96	566.00	216250.00	105.51	22.97	1.34	425.00	26.40	6.40	5.86
Fort Smith AR	7.16	2.20	4.05	1.25	528.00	165600.00	133.76	22.30	1.30	300.00	28.40	6.30	6.35
Hot Springs AR	5.52	1.84	4.43	1.17	532.00	182333.00	124.15	23.84	1.37	636.50	26.67	5.70	6.02
Jonesboro AR	6.49	1.71	4.20	1.10	495.00	196170.00	103.46	24.39	1.31	337.50	22.00	7.50	6.75
Little Rock-N Little Rock AR	6.06	1.99	3.30	1.03	595.00	181654.00	112.80	28.56	1.38	362.40	24.00	7.50	5.39
Hemet-San Jacinto CA	7.28	2.09	4.13	0.80	659.00	199440.00	173.90	22.35	1.62	530.00	24.70	6.07	4.86
Lancaster/Palmdale CA	7.79	2.16	4.13	1.04	575.00	213600.00	125.64	19.08	1.58	775.00	31.67	6.53	4.17
Orange County CA	5.91	2.17	4.65	1.13	1116.00	428164.00	133.32	16.69	1.65	849.67	20.74	6.55	4.69
Palm Springs CA	7.77	2.36	4.05	1.51	875.00	243296.00	169.75	21.25	1.69	1834.11	38.80	8.35	5.01
Riverside City CA	7.61	2.10	4.50	1.06	871.00	223618.00	124.10	16.96	1.59	928.67	35.19	8.04	4.07
Sacramento CA	7.39	2.24	5.10	1.28	749.00	290450.00	165.86	16.99	1.64	1704.60	40.80	9.11	6.29
San Diego CA	5.95	2.47	4.95	1.09	1306.00	402519.00	87.52	24.57	1.63	1245.00	37.00	7.71	7.24
Visalia CA	5.83	1.73	5.10	1.35	586.00	215600.00	177.68	16.25	1.48	750.00	27.60	8.70	7.18
Colorado Springs CO	6.71	2.00	4.43	1.06	802.00	250336.00	88.26	26.83	1.33	684.50	22.16	6.61	4.46
Denver CO	7.31	2.09	4.73	1.01	891.00	260503.00	82.79	23.10	1.34	761.11	28.00	8.66	4.85
Fort Collins CO	4.61	2.02	4.43	1.04	802.00	251055.00	124.30	23.95	1.36	730.00	25.80	10.67	4.59
Grand Junction CO	6.29	1.89	3.90	1.25	678.00	197934.00	123.32	23.56	1.44	616.95	19.29	8.24	5.07
Greeley CO	6.83	2.03	4.45	1.03	738.00	236384.00	91.49	26.17	1.39	815.00	36.00	10.30	4.82
Gunnison CO	6.74	2.24	3.98	1.59	725.00	294000.00	136.39	22.39	1.50	825.00	22.00	9.95	5.24
Loveland CO	8.23	2.09	5.10	1.48	885.00	232694.00	115.71	25.42	1.39	1036.00	22.67	8.05	5.02
Pueblo CO	6.65	2.11	3.93	1.09	463.00	193900.00	134.73	24.14	1.36	576.00	17.60	6.95	4.52
Hartford CT	8.99	1.89	4.28	1.17	896.00	363350.00	158.25	22.39	1.42	1054.93	23.21	11.98	5.66
New Haven CT	7.82	1.87	4.13	1.19	981.00	354920.00	174.70	22.39	1.46	1209.95	37.33	9.46	5.37
Dover DE	6.99	1.72	4.95	0.92	710.00	219888.00	159.38	18.17	1.33	518.00	24.20	8.34	6.15
Wilmington DE	7.32	1.78	4.95	0.93	824.00	232945.00	141.05	18.23	1.33	590.00	25.40	7.30	6.33
Bradenton FL	6.25	1.88	4.20	1.31	731.00	226808.00	121.42	21.79	1.42	708.75	22.80	8.31	5.20

City													
Fort Myers-Cape Coral FL	6.13	1.90	4.43	1.17	808.00	199927.00	126.93	18.91	1.46	487.00	26.29	8.56	5.29
Fort Walton Beach FL	7.25	1.83	4.28	1.14	594.00	197233.00	114.02	19.06	1.43	960.00	29.00	8.76	5.68
Jacksonville FL	6.97	1.97	4.43	1.08	810.00	202995.00	98.66	20.31	1.42	530.40	27.17	7.17	5.69
Orlando FL	6.05	1.85	4.43	1.05	678.00	201387.00	128.65	23.07	1.45	754.30	33.10	7.37	5.57
Panama City FL	7.42	1.92	4.43	1.13	669.00	186240.00	119.32	21.02	1.45	717.50	30.00	7.05	5.87
Pensacola FL	6.76	1.96	4.05	1.19	648.00	197233.00	128.11	19.56	1.41	703.67	29.25	8.03	6.01
Punta Gorda/Charlotte Co FL	6.60	1.86	4.43	1.17	534.00	200400.00	132.61	18.07	1.42	623.67	25.78	8.49	5.22
St Petersburg-Clearwater FL	6.42	1.81	4.13	1.04	722.00	188295.00	134.66	20.25	1.41	652.76	24.30	8.28	5.16
Sarasota FL	5.99	1.89	4.43	0.84	750.00	249667.00	125.46	19.20	1.45	753.00	34.00	6.83	5.19
Tampa FL	6.55	1.88	4.28	0.98	858.00	203650.00	107.69	18.83	1.41	722.50	29.17	7.49	5.08
Vero Beach/Indian River FL	6.39	1.91	4.80	1.18	612.00	197242.00	129.94	18.95	1.40	469.50	28.33	10.05	5.66
West Palm Beach FL	7.49	1.90	4.65	1.25	834.00	210876.00	123.51	20.54	1.52	448.00	37.75	8.30	5.78
Albany GA	7.14	1.87	4.20	1.12	476.00	188100.00	110.40	24.08	1.26	346.75	28.60	7.02	6.29
Atlanta GA	8.22	1.70	4.20	1.14	744.00	229632.00	108.58	24.90	1.28	484.38	26.43	6.89	6.09
Augusta-Aiken GA-SC	7.77	1.97	4.43	1.53	609.00	169342.00	105.43	21.98	1.27	383.60	26.33	6.39	5.89
Bainbridge GA	6.32	2.02	3.98	1.22	538.00	182750.00	111.08	20.54	1.17	351.00	24.25	6.83	6.28
Douglas GA	5.59	2.07	3.30	1.09	513.00	179054.00	111.81	27.57	1.26	447.00	19.00	5.97	6.20
LaGrange/Troup County GA	6.99	1.86	4.13	1.10	622.00	176067.00	111.39	23.50	1.23	409.20	17.50	9.08	5.93
Marietta GA	6.76	1.85	3.60	1.11	658.00	207500.00	108.58	25.50	1.19	440.00	23.75	6.88	6.94
Tifton GA	6.81	1.83	4.05	1.16	539.00	182400.00	112.43	20.54	1.26	345.00	22.50	6.50	6.80
Valdosta GA	7.56	1.89	4.43	1.19	593.00	210417.00	110.60	18.83	1.28	345.00	28.17	7.63	6.18
Boise ID	5.74	1.35	3.75	0.92	745.00	214298.00	112.54	25.13	1.37	605.75	25.61	8.09	5.41
Twin Falls ID	5.82	1.58	4.80	1.06	613.00	196333.00	111.37	25.25	1.36	475.75	21.00	7.50	5.99
Champaign-Urbana IL	6.72	1.61	3.88	1.02	773.00	202925.00	130.57	19.03	1.43	667.00	23.79	8.73	4.29
Danville IL	5.35	1.60	3.80	1.16	600.00	188500.00	157.57	28.25	1.36	597.00	17.50	6.95	7.49
DeKalb IL	5.25	1.83	3.79	1.31	604.00	224640.00	92.96	26.17	1.48	649.50	24.20	7.18	5.24
Joliet/Will County IL	5.92	2.12	4.40	1.16	968.00	217148.00	108.81	31.37	1.59	637.50	31.50	6.38	4.75
Peoria IL	6.45	1.51	3.93	1.07	618.00	228400.00	113.44	21.91	1.37	465.50	21.50	10.05	4.65
Quincy IL	6.98	1.57	3.81	1.09	508.00	242400.00	123.38	21.74	1.44	455.00	23.00	7.97	4.74
Springfield IL	6.97	1.85	4.06	1.13	630.00	218000.00	100.04	19.33	1.33	649.33	28.40	7.80	4.49
Bloomington IN	7.69	1.96	4.28	1.39	678.00	211667.00	104.30	19.95	1.40	569.50	23.00	8.83	3.83
Elkhart-Goshen IN	7.15	1.66	4.73	1.19	580.00	196900.00	100.02	24.19	1.40	650.00	22.60	9.32	4.54
Fort Wayne/Allen County IN	6.68	1.61	3.65	1.17	716.00	208250.00	108.59	24.49	1.30	610.50	23.71	7.11	4.64
Indianapolis/Marion County IN	6.48	1.72	3.99	1.15	717.00	204458.00	108.32	20.41	1.42	519.00	27.60	8.73	4.10
Lafayette IN	5.68	1.51	3.58	1.16	690.00	195333.00	125.79	26.68	1.45	570.00	24.00	9.70	4.69
South Bend IN	6.52	1.56	3.67	1.15	747.00	252980.00	100.58	27.74	1.35	785.00	31.20	8.14	3.90
Terre Haute IN	6.79	1.29	3.77	1.17	580.00	206000.00	110.09	28.00	1.31	492.50	18.32	8.15	4.24
Ames IA	5.99	1.29	4.07	0.99	645.00	269000.00	105.85	20.00	1.36	455.00	19.00	6.74	5.52
Burlington IA	5.15	1.24	3.90	0.97	689.00	204000.00	129.83	22.50	1.39	540.00	19.94	8.68	5.84
Cedar Rapids IA	6.34	1.39	3.85	1.01	595.00	208670.00	121.54	19.14	1.34	425.12	20.70	8.91	5.70
Davenport-Moline-Rock Is IA-IL	6.26	1.61	4.19	1.32	555.00	225754.00	102.88	19.34	1.37	435.50	18.16	8.46	5.64
Des Moines IA	6.27	1.34	3.80	1.01	586.00	215000.00	112.94	19.95	1.40	601.67	24.06	8.56	5.37
Dubuque IA	6.28	1.48	3.73	1.01	632.00	301583.00	120.17	19.77	1.43	540.00	18.17	7.74	5.69
Waterloo-Cedar Falls IA	6.34	1.29	3.79	1.11	585.00	215617.00	95.72	21.80	1.38	508.33	17.25	7.33	5.56
Dodge City KS	5.06	1.68	4.35	1.10	550.00	235200.00	145.46	24.21	1.37	555.00	21.00	6.70	6.92
Garden City KS	7.22	1.32	4.05	1.21	459.00	195250.00	125.99	30.25	1.35	527.00	20.00	7.63	5.91

City													
Hays KS	7.55	9.63	21.00	550.00	1.35	20.53	122.89	228333.00	475.00	1.30	4.13	1.66	6.20
Lawrence KS	6.24	9.53	24.25	480.00	1.35	26.26	96.00	230385.00	658.00	1.04	3.98	1.52	6.95
Manhattan KS	6.85	8.03	18.67	525.00	1.40	26.54	114.38	244267.00	551.00	1.10	4.13	1.53	6.94
Salina KS	6.55	6.75	19.33	458.00	1.31	22.33	104.94	186333.00	525.00	1.15	3.53	1.44	6.45
Wichita KS	7.09	8.69	26.13	702.33	1.35	27.27	127.47	206547.00	560.00	1.20	4.65	1.59	7.99
Bowling Green KY	5.82	7.83	29.33	452.00	1.27	27.60	107.41	198450.00	585.00	1.16	4.47	1.89	7.79
Covington KY	5.07	7.96	20.20	564.00	1.38	27.63	100.79	205993.00	658.00	1.26	3.67	1.79	6.15
Hopkinsville KY	7.13	7.97	27.00	452.00	1.28	24.44	116.23	185000.00	495.00	0.92	4.00	1.67	8.74
Lexington KY	5.49	7.86	28.20	460.46	1.38	29.93	110.06	226294.00	712.00	1.37	4.64	1.73	6.67
Louisville KY	7.15	8.00	19.20	455.00	1.30	27.67	106.98	193433.00	665.00	1.04	3.72	1.65	6.91
Somerset KY	7.49	7.50	20.25	562.50	1.39	24.99	90.23	215000.00	550.00	1.19	3.11	1.69	6.25
Alexandria LA	5.96	8.08	27.50	503.50	1.31	31.96	125.24	207074.00	589.00	1.07	4.13	1.95	6.05
Lafayette LA	5.61	8.27	28.00	412.50	1.36	21.65	93.17	242896.00	748.00	0.99	4.35	1.85	8.03
Lake Charles LA	5.66	7.75	21.99	474.63	1.31	20.36	108.35	239037.00	695.00	1.12	3.90	1.68	5.94
Monroe LA	6.13	8.54	20.80	384.00	1.34	19.20	107.29	233296.00	545.00	1.13	3.53	1.68	7.08
New Orleans LA	6.56	7.85	28.83	553.40	1.35	26.06	128.29	197225.00	798.00	1.10	3.23	1.86	5.87
Shreveport-Bossier City LA	5.79	7.99	27.50	463.33	1.37	21.48	96.18	235200.00	601.00	1.09	3.90	1.83	5.39
Baltimore MD	5.79	4.48	28.60	615.80	1.41	22.48	122.73	204492.00	633.00	0.89	4.28	1.58	9.19
Boston (MA Part)	6.03	8.49	37.20	828.40	1.41	24.41	204.06	427172.00	1248.00	1.50	4.43	1.79	6.32
Fitchburg-Leominster MA	7.61	8.25	23.00	585.00	1.36	25.96	168.74	401832.00	975.00	1.19	3.23	1.76	6.67
Grand Rapids MI	6.27	10.08	26.30	467.50	1.50	22.40	112.05	240000.00	678.00	1.31	3.89	2.01	5.99
Lansing MI	6.09	7.50	19.50	551.00	1.46	24.66	87.49	206067.00	615.00	1.13	3.60	1.49	6.84
Marquette MI	5.24	10.92	18.00	585.00	1.38	21.08	93.63	254485.00	550.00	1.46	4.73	1.64	6.82
Minneapolis MN	5.79	9.68	31.20	1094.40	1.37	25.16	108.46	227540.00	815.00	1.05	4.43	2.01	7.33
St Cloud MN	4.74	8.23	21.50	640.00	1.56	26.03	120.80	202780.00	580.00	1.39	3.94	1.80	7.21
St Paul MN	5.59	9.70	29.40	1084.80	1.32	24.05	104.17	222720.00	822.00	1.07	4.35	2.02	6.41
Biloxi-Gulfport-Pascagoula	6.84	7.35	29.20	423.29	1.34	24.91	116.08	198250.00	672.00	0.96	4.35	1.90	5.31
Hattiesburg MS	6.53	7.31	25.40	360.00	1.31	24.36	116.68	226000.00	632.00	1.13	4.95	1.86	5.02
Jackson MS	6.08	6.54	29.30	302.20	1.27	28.23	107.43	209750.00	676.00	1.10	3.83	1.81	7.20
Columbia MO	6.57	7.56	30.67	624.67	1.39	21.80	111.88	233090.00	553.00	1.25	3.87	1.54	6.55
Joplin MO	6.15	6.65	18.40	642.00	1.26	17.59	113.42	188300.00	479.00	1.06	4.34	1.47	6.26
St Joseph MO	4.61	7.01	18.70	595.00	1.30	15.26	107.19	218200.00	603.00	1.23	4.04	1.58	6.73
St Louis MO-IL	6.19	8.52	26.10	522.00	1.37	24.99	127.12	231672.00	658.00	1.25	3.98	1.91	5.82
Springfield MO	6.67	6.41	22.00	539.00	1.31	18.25	90.50	189900.00	568.00	1.21	4.74	1.63	6.37
Billings MT	5.52	8.78	17.24	473.00	1.45	30.45	93.80	215000.00	550.00	1.14	4.96	1.81	6.09
Bozeman MT	6.52	10.43	25.33	466.00	1.47	31.92	94.83	261166.00	650.00	1.39	4.95	1.89	5.91
Great Falls MT	5.52	8.32	24.00	365.00	1.37	33.14	120.63	187300.00	525.00	1.11	4.70	1.75	6.89
Helena MT	6.10	9.70	18.00	508.00	1.30	23.00	94.79	190197.00	603.00	1.32	4.28	1.85	6.63
Kalispell MT	5.61	9.19	22.75	484.50	1.36	26.41	104.50	215904.00	588.00	1.16	4.70	1.76	6.39
Missoula MT	6.44	7.80	24.63	555.00	1.54	27.65	89.81	205333.00	729.00	1.42	4.58	1.79	5.92
Hastings NE	5.36	10.28	19.50	420.00	1.46	23.25	98.79	185000.00	523.00	1.19	4.27	1.49	6.70
Lincoln NE	6.07	9.43	24.40	486.67	1.42	28.00	102.95	212690.00	564.00	0.98	4.20	1.50	6.97
Omaha NE	5.23	7.93	19.60	493.71	1.35	28.54	96.47	189740.00	705.00	1.06	4.03	1.39	6.71
Reno-Sparks NV	5.12	8.69	28.20	767.00	1.54	16.74	178.96	234675.00	840.00	1.23	4.23	1.51	8.62
Hunterdon County NJ	7.24	8.51	27.11	1043.00	1.30	32.99	176.34	345200.00	923.00	1.09	4.80	1.48	6.11
Albuquerque NM	5.75	6.99	27.36	550.67	1.31	28.19	113.64	239974.00	740.00	1.21	4.90	1.81	

Farmington NM	6.32	1.81	4.73	1.31	498.00	199008.00	109.27	19.16	1.18	400.00	19.60	8.68	7.33
Las Cruces NM	5.59	1.76	4.89	1.55	567.00	201334.00	111.48	19.35	1.34	395.00	20.15	6.98	6.29
Los Alamos NM	6.69	1.49	5.29	0.99	767.00	421710.00	103.46	18.95	1.35	583.65	24.12	8.93	5.49
Rio Rancho NM	6.47	1.69	4.90	1.14	670.00	209380.00	113.46	21.24	1.33	554.42	24.00	6.53	5.67
Santa Fe NM	6.52	1.75	4.35	1.14	793.00	322500.00	102.33	24.32	1.50	450.00	27.79	8.83	6.24
Buffalo NY	6.99	1.34	4.05	1.14	714.00	224647.00	172.86	33.71	1.41	667.20	21.50	7.10	6.35
Glens Falls NY	8.74	1.38	4.13	0.99	690.00	189500.00	155.95	29.82	1.40	496.00	20.00	8.00	5.99
Syracuse NY	6.39	1.29	3.98	1.30	590.00	191683.00	145.33	29.68	1.41	726.67	20.00	9.42	6.57
Watertown/Jefferson County NY	6.19	1.29	4.35	1.27	496.00	186000.00	147.67	25.42	1.44	575.33	19.60	9.00	5.99
Asheville NC	7.83	2.03	4.28	1.11	724.00	249217.00	109.95	24.02	1.40	425.00	21.00	8.57	5.91
Burlington NC	7.11	1.92	4.20	1.19	698.00	250000.00	108.43	18.00	1.36	370.00	26.75	9.43	5.67
Chapel Hill NC	7.04	2.04	4.43	1.34	769.00	274671.00	108.44	30.00	1.32	800.00	28.33	9.59	5.39
Charlotte NC	7.05	1.84	3.90	1.19	540.00	210900.00	108.16	21.07	1.36	473.00	22.86	6.87	6.07
Durham NC	7.12	2.01	4.28	1.16	717.00	210000.00	108.44	19.51	1.37	672.00	29.75	7.57	6.23
Gastonia NC	7.45	1.90	4.13	1.17	591.00	219389.00	108.16	21.79	1.35	410.00	23.80	7.18	5.84
Goldsboro NC	6.25	1.83	4.58	1.09	600.00	203825.00	116.28	17.61	1.35	378.00	20.24	5.63	6.18
Greenville NC	7.09	1.85	3.83	1.11	588.00	224667.00	108.74	23.37	1.36	320.00	23.00	8.85	5.82
Jacksonville NC	7.66	1.97	4.43	1.29	582.00	200000.00	114.13	20.19	1.36	345.00	28.00	7.14	6.26
Raleigh NC	7.21	1.98	4.13	1.25	860.00	237843.00	114.47	17.44	1.38	550.00	29.86	8.14	7.31
Wilmington NC	7.05	2.01	4.13	1.21	681.00	222326.00	114.17	18.73	1.30	343.00	25.71	7.74	6.07
Winston-Salem NC	7.39	1.97	3.98	1.17	578.00	208580.00	109.00	18.48	1.35	292.50	29.20	7.60	6.09
Bismarck-Mandan ND	6.81	1.44	4.11	1.22	503.00	205000.00	90.56	22.68	1.46	365.50	23.00	8.20	4.87
Minot ND	6.09	1.47	4.00	0.99	475.00	196333.00	102.91	15.23	1.50	385.00	16.25	8.08	4.96
Akron OH	7.97	1.63	4.62	1.25	686.00	195880.00	131.52	21.16	1.52	700.40	22.06	7.30	6.59
Cincinnati OH	7.54	1.84	4.35	1.09	718.00	196522.00	123.44	23.00	1.33	472.40	24.60	8.29	5.99
Cleveland OH	8.13	1.62	4.80	1.31	852.00	216624.00	157.06	21.56	1.36	727.20	23.67	8.70	5.82
Dayton-Springfield OH	6.75	1.55	3.53	1.14	593.00	212822.00	117.42	20.71	1.41	689.00	26.50	9.08	6.19
Findlay OH	7.16	1.68	4.35	1.21	540.00	226667.00	143.90	19.27	1.45	440.00	23.50	10.18	6.12
Lima OH	5.69	1.81	4.35	1.47	551.00	240000.00	135.66	25.65	1.43	473.00	19.30	6.70	7.74
Toledo OH	6.65	1.75	4.88	1.02	516.00	195180.00	132.72	19.36	1.40	509.71	30.40	9.15	5.99
Troy/Miami County OH	6.79	1.79	3.92	1.25	537.00	245562.00	118.54	15.23	1.45	606.00	26.20	9.03	6.16
Youngstown-Warren OH	6.27	1.29	3.90	1.07	538.00	185000.00	146.86	20.44	1.37	411.13	23.40	7.03	6.09
Bartlesville OK	6.22	1.68	4.43	1.10	531.00	189000.00	104.26	19.31	1.19	405.00	26.00	8.65	6.12
Edmond OK	6.22	1.75	3.99	1.13	574.00	226200.00	109.08	19.99	1.30	405.00	20.67	8.78	6.49
Enid OK	6.39	1.59	4.44	1.14	528.00	183629.00	118.36	20.54	1.33	616.50	24.19	6.64	7.89
Lawton OK	6.25	1.48	4.10	1.32	520.00	189950.00	121.35	22.90	1.34	387.00	24.60	6.42	6.67
Oklahoma City OK	6.09	1.52	4.32	0.97	579.00	183960.00	114.59	23.04	1.31	423.30	23.90	7.96	6.00
Pryor Creek OK	6.39	1.46	4.31	1.18	515.00	180667.00	95.03	21.67	1.30	429.00	15.00	7.17	6.49
Stillwater OK	6.08	1.66	4.26	0.98	515.00	196000.00	113.61	23.87	1.24	365.00	22.20	7.70	6.72
Tulsa OK	6.54	1.62	3.68	0.96	622.00	206016.00	110.78	26.43	1.29	460.40	27.00	8.81	6.23
Corvallis OR	6.59	1.95	4.59	1.29	677.00	295500.00	145.76	20.07	1.33	635.00	27.00	9.98	5.33
Lincoln County OR	6.29	1.99	4.67	1.24	518.00	247916.00	99.97	22.45	1.46	679.00	24.67	9.25	6.35
Portland OR	6.97	2.11	4.35	1.31	753.00	233635.00	123.98	20.92	1.40	573.80	31.40	7.98	7.19
Salem OR	6.21	2.03	4.43	1.09	561.00	217109.00	125.28	22.95	1.34	550.00	27.33	9.88	4.93
Harrisburg PA	8.67	1.25	3.90	1.17	666.00	224933.00	122.82	22.00	1.31	795.00	25.00	9.08	7.00
Johnstown PA	7.04	1.38	4.58	1.29	489.00	204629.00	137.81	23.34	1.20	514.00	20.15	7.43	6.99

Location													
Philadelphia PA	7.49	1.48	4.95	1.06	1282.00	316319.00	165.47	21.12	1.36	1724.25	44.00	7.04	6.99
Pittsburgh PA	7.95	1.43	4.88	1.23	599.00	208250.00	131.86	16.95	1.25	614.60	26.40	8.29	6.99
Williamsport/Lycoming Co PA	7.85	1.32	3.90	1.26	550.00	205400.00	135.26	18.53	1.33	556.50	20.90	8.78	6.99
York County PA	7.39	1.44	3.70	1.16	693.00	261950.00	140.04	25.60	1.34	365.50	22.58	8.44	6.99
Charleston-N Charleston SC	7.87	1.99	4.05	1.01	734.00	240947.00	110.26	21.35	1.28	577.00	36.00	8.08	6.21
Columbia SC	8.19	1.85	4.85	0.97	731.00	219492.00	116.34	24.03	1.28	461.25	19.00	8.70	6.19
Hilton Head Island SC	7.94	1.89	4.43	1.48	823.00	241475.00	107.53	19.30	1.33	450.00	38.75	8.52	6.23
Myrtle Beach SC	8.45	1.94	4.20	1.35	593.00	230150.00	129.86	21.82	1.30	514.00	26.20	7.77	6.32
Sumter SC	6.99	2.02	4.05	1.20	561.00	205368.00	111.13	25.32	1.30	468.00	24.40	7.90	6.35
Sioux Falls SD	6.59	1.44	4.33	1.39	600.00	203134.00	94.80	27.33	1.38	548.50	21.40	7.71	6.50
Vermillion SD	5.99	1.53	4.81	1.46	600.00	235333.00	98.13	24.59	1.36	556.00	21.17	7.75	5.43
Chattanooga TN	8.22	1.85	4.28	0.86	641.00	193713.00	111.76	20.37	1.31	534.80	34.40	7.15	6.19
Clarksville TN	6.25	1.65	3.98	1.08	576.00	186340.00	101.27	18.85	1.27	368.00	25.19	7.08	5.49
Cleveland TN	5.74	1.59	4.43	0.87	505.00	205000.00	96.52	19.50	1.35	480.00	22.50	6.75	6.17
Dyersburg TN	5.77	1.64	3.83	0.98	529.00	178750.00	107.90	19.29	1.36	302.40	17.29	8.56	6.15
Johnson City TN	6.29	1.58	3.90	1.08	546.00	198000.00	100.22	22.32	1.34	417.00	22.80	7.23	5.99
Kingsport TN	6.72	1.73	3.98	1.15	558.00	193562.00	80.50	21.43	1.34	366.50	22.20	6.85	5.88
Knoxville TN	7.17	1.72	4.20	1.11	588.00	190567.00	107.72	21.08	1.30	506.40	23.80	7.80	6.99
Memphis TN	6.69	1.51	4.35	1.01	639.00	184900.00	95.79	18.10	1.33	433.80	33.80	8.37	5.99
Murfreesboro-Smyrna TN	7.32	1.76	4.43	1.36	647.00	181291.00	94.50	23.00	1.28	336.00	28.60	7.25	5.01
Abilene TX	6.32	1.58	4.35	1.27	560.00	177000.00	143.14	18.71	1.35	534.75	26.79	5.90	5.38
Amarillo TX	5.84	1.54	3.98	1.30	601.00	210300.00	88.88	17.39	1.33	415.00	23.78	6.49	4.86
Arlington TX	7.31	1.58	4.28	1.15	710.00	194533.00	121.73	20.00	1.35	525.00	20.19	6.80	5.92
Austin TX	5.08	1.69	3.90	1.50	1025.00	231366.00	127.82	19.20	1.30	510.20	36.17	6.60	5.94
Beaumont TX	5.11	1.71	4.35	1.26	629.00	220000.00	116.08	21.00	1.33	472.50	24.25	6.15	6.52
Brazoria TX	5.89	1.90	3.90	1.01	549.00	223865.00	123.90	23.00	1.34	391.67	21.70	6.21	5.31
Bryan-College Station TX	5.59	1.74	3.83	1.21	779.00	207050.00	117.66	20.30	1.31	500.00	30.57	6.59	5.56
Conroe TX	7.18	1.58	4.05	1.19	734.00	181413.00	115.55	15.74	1.35	493.50	32.00	6.78	5.00
Dallas TX	6.76	1.32	4.43	1.21	902.00	198428.00	116.89	19.22	1.40	638.38	48.60	7.75	6.43
El Paso TX	5.50	1.90	4.80	1.34	640.00	194647.00	108.03	19.75	1.39	720.00	28.30	6.71	5.49
Harlingen TX	3.99	1.69	3.62	1.50	618.00	185000.00	109.61	17.86	1.36	548.00	28.67	7.79	5.20
Houston TX	5.59	1.69	3.98	1.23	745.00	177074.00	113.90	21.37	1.36	583.10	30.00	5.90	5.45
Longview TX	5.59	1.57	3.98	1.24	523.00	181764.00	92.62	17.47	1.33	714.50	17.66	6.78	5.98
Lubbock TX	6.64	1.39	3.83	1.22	599.00	182544.00	88.18	16.91	1.31	606.00	21.80	7.11	5.38
McAllen TX	6.34	1.69	4.18	1.50	494.00	208750.00	109.45	21.95	1.35	592.50	32.00	4.68	5.67
Midland TX	4.68	1.60	4.35	1.35	538.00	174160.00	119.51	16.01	1.30	329.00	31.83	7.09	5.06
Odessa TX	4.86	1.36	4.05	1.31	544.00	183320.00	121.55	18.14	1.33	398.00	25.20	6.69	6.83
Paris TX	5.86	1.88	3.98	1.09	513.00	165682.00	131.04	17.75	1.35	450.00	24.50	5.00	6.01
Plano TX	6.71	1.37	3.80	1.11	970.00	201000.00	104.23	24.63	1.39	604.50	36.00	7.44	7.67
San Angelo TX	4.68	1.62	4.20	1.31	519.00	173864.00	89.04	20.29	1.33	454.00	24.80	7.00	5.12
San Antonio TX	4.22	1.63	3.60	1.41	611.00	186042.00	95.44	18.40	1.29	497.40	31.40	7.73	5.28
San Marcos TX	2.97	1.75	3.38	1.26	743.00	229200.00	93.37	19.16	1.31	515.00	27.67	7.62	5.74
Seguin TX	4.58	1.77	3.45	1.50	648.00	200000.00	107.86	18.90	1.27	409.00	18.67	6.33	7.35
Sherman-Denison TX	6.22	1.72	3.90	0.95	523.00	180750.00	110.72	23.69	1.35	567.50	21.60	5.82	6.32
Temple TX	4.53	1.66	3.75	1.12	763.00	213750.00	129.68	18.53	1.30	545.00	32.00	8.41	6.97
Texarkana TX-AR	6.02	1.62	4.05	1.11	550.00	176700.00	109.20	16.41	1.36	465.25	28.20	7.49	

Tyler TX	6.11	1.64	3.90	1.13	639.00	225250.00	118.10	19.50	1.33	799.67	32.20	6.78	5.92
Victoria TX	5.82	1.46	3.68	1.09	544.00	198960.00	142.38	17.96	1.38	380.00	21.13	7.83	5.63
Waco TX	5.16	1.69	3.98	1.15	665.00	191759.00	122.32	19.00	1.32	416.50	26.60	7.93	5.97
Weatherford TX	5.69	1.19	4.05	1.17	649.00	175500.00	112.01	16.63	1.36	395.00	22.39	5.96	6.39
Cedar City UT	7.02	1.74	4.72	1.39	515.00	162500.00	97.85	21.29	1.38	571.00	18.25	9.48	5.95
Logan UT	6.19	1.79	4.75	1.06	593.00	192600.00	108.26	20.85	1.36	517.20	19.56	8.13	4.95
St George UT	7.42	1.72	4.71	1.29	540.00	192600.00	105.99	21.02	1.32	612.80	20.33	9.98	5.95
Charlottesville VA	7.82	1.99	4.28	1.35	840.00	268133.00	126.29	27.17	1.33	511.50	33.20	8.54	6.47
Hampton Roads/SE Virginia VA	7.65	1.85	4.13	1.03	692.00	212033.00	135.75	25.99	1.32	395.00	24.10	8.72	5.52
Lynchburg VA	6.77	1.83	3.90	0.92	586.00	196417.00	83.01	18.42	1.31	675.00	22.60	8.47	5.37
Martinsville/Henry County VA	7.21	1.82	4.20	1.17	495.00	250805.00	82.52	22.43	1.33	375.00	20.00	7.60	6.13
Northern VA	9.30	1.86	4.82	1.34	1471.00	309882.00	138.95	22.30	1.42	688.17	37.60	7.82	6.49
Richmond VA	5.99	1.95	4.53	1.57	769.00	214871.00	127.21	26.15	1.32	499.00	26.30	7.08	5.61
Roanoke VA	6.92	1.61	4.20	1.07	582.00	203129.00	83.89	26.05	1.28	400.33	26.17	8.05	5.53
Bellingham WA	7.44	1.81	4.61	1.22	716.00	244000.00	115.10	23.92	1.36	364.50	25.33	8.40	5.29
Olympia WA	6.53	2.07	4.76	1.17	715.00	203364.00	109.12	20.01	1.36	1033.00	34.80	9.67	6.03
Richland-Kennewick-Pasco WA	6.57	1.95	4.02	1.18	767.00	231491.00	114.58	23.19	1.34	676.67	27.20	9.01	5.04
Spokane WA	6.80	2.19	4.93	1.18	593.00	244468.00	114.14	18.49	1.31	617.00	22.40	9.42	5.27
Tacoma WA	6.86	1.94	4.85	1.12	757.00	218453.00	112.06	20.91	1.39	754.71	23.18	8.82	5.74
Vancouver WA	6.51	2.04	4.57	1.69	644.00	242425.00	119.91	22.24	1.34	610.00	24.40	8.97	6.11
Wenatchee WA	6.99	1.69	4.58	1.22	650.00	267000.00	48.68	18.08	1.24	700.00	23.33	8.67	5.42
Yakima WA	6.29	1.37	4.87	1.33	692.00	219800.00	95.46	20.79	1.32	894.50	23.50	7.38	5.79
Charleston WV	7.09	1.18	4.65	1.21	606.00	220000.00	94.79	27.08	1.42	344.00	23.40	8.19	6.89
Huntington WV	6.49	1.70	5.03	1.31	500.00	188500.00	85.38	29.95	1.40	546.00	20.40	9.26	5.21
Appleton-Neenah-Menasha WI	6.26	1.35	4.43	1.11	500.00	208900.00	118.47	16.31	1.45	356.67	25.00	10.22	4.69
Eau Claire WI	6.17	1.58	4.15	1.30	593.00	236233.00	127.71	18.47	1.53	395.00	22.00	8.90	5.39
Green Bay WI	5.78	1.45	4.42	1.03	574.00	230779.00	123.33	18.79	1.46	570.25	21.00	8.89	4.73
Marshfield WI	6.32	1.51	3.85	1.08	650.00	206230.00	97.98	34.53	1.49	448.00	17.65	8.40	4.49
Sheboygan WI	6.08	1.53	4.12	1.18	705.00	205000.00	139.95	24.47	1.50	415.00	23.00	9.75	4.35
Stevens Point-Plover WI	7.42	1.75	4.54	0.99	820.00	225000.00	118.47	17.93	1.49	700.00	25.50	9.25	4.99
Wausau WI	5.75	1.41	4.05	1.15	638.00	204300.00	107.91	27.16	1.44	409.00	18.40	9.50	4.39
Cheyenne WY	7.07	1.74	5.19	1.42	586.00	252333.00	100.01	37.03	1.30	648.00	22.60	7.28	6.22

Data Set II: Data on States

C1	C2	C3	C4	C5	C6	C7	C8	C9	C10
1. Alabama	24589	9.8	56.9	5.3	87.6	39268	2131	20.4	3885
2. Alaska	30939	6.3	67.8	6.3	1.1	49418	138	28.1	189
3. Arizona	25872	7.0	56.6	4.7	45.2	36966	2057	24.6	2938
4. Arkansas	22887	8.4	56.1	5.1	51.3	35389	1619	18.4	2083
5. California	32702	5.5	59.1	5.3	217.2	53870	16297	27.5	23621
6. Colorado	33470	6.5	65.5	3.7	41.5	40222	1701	34.6	2338
7. Connecticut	42435	6.5	62.9	3.7	702.9	54300	2057	31.6	3291
8. Delaware	32472	8.8	63.8	3.5	401.1	48363	461	24.0	430
9. Florida	28947	7.2	55.7	4.8	296.4	38719	10603	22.8	19221
10. Georgia	28733	8.3	63.3	4.0	141.4	44073	3041	23.1	4111
11. Hawaii	29002	7.4	62.6	4.6	188.6	43951	326	26.3	622
12. Idaho	24621	7.2	61.9	5.0	15.6	37482	479	20.0	639
13. Illinois	33023	8.5	63.1	5.4	87.9	50000	7054	27.1	7309
14. Indiana	27783	7.8	59.8	4.4	169.5	44195	3656	17.1	4720
15. Iowa	27331	6.2	65.7	3.3	52.4	38230	1546	25.5	1453
16. Kansas	28565	7.0	65.7	4.3	32.9	36673	1468	27.3	1915
17. Kentucky	24923	7.4	57.9	5.5	101.7	27847	2596	20.5	3153
18. Louisiana	24535	9.1	54.2	6.0	102.6	35437	2238	22.5	4383
19. Maine	26723	5.4	63.9	4.0	41.3	37100	862	24.1	793
20. Maryland	35188	8.1	64.3	4.1	541.9	46200	2701	32.3	3998
21. Massachusetts	38907	5.0	61.4	3.7	809.8	49054	4097	32.7	5466
22. Michigan	29788	8.1	31.5	5.3	175.0	52037	5616	23.0	6269
23. Minnesota	33101	5.9	70.3	3.7	61.8	43330	1965	31.2	3109
24. Mississippi	21750	10.3	57.0	5.5	60.6	32800	1433	18.7	2248
25. Missouri	28226	7.5	64.3	4.7	81.2	37695	3278	26.2	4274
26. Montana	23963	6.8	64.3	4.6	6.2	34379	497	23.8	575
27. Nebraska	28886	7.0	69.0	3.1	22.3	36236	734	24.6	1225
28. Nevada	29897	6.7	63.0	5.3	18.2	41524	1006	19.3	1069
29. New Hampshire	34138	5.4	66.7	3.5	137.8	38911	664	30.1	629
30. New Jersey	38509	6.4	58.4	4.2	1134.4	54575	5056	30.1	6767
31. New Mexico	23155	6.9	57.2	4.8	15.0	36490	673	23.6	854
32. New York	36019	6.3	56.1	4.9	401.9	53081	12287	28.7	18653
33. North Carolina	27514	9.0	61.6	5.5	165.2	41991	3530	23.2	5942
34. North Dakota	25902	8.0	67.0	2.8	9.3	31709	247	22.6	501

35. Ohio	28816	7.9	60.9	4.3	277.3	44492	7128	24.6	9310
36. Oklahoma	25071	8.5	57.3	3.8	50.3	35412	2106	22.5	2137
37. Oregon	28165	5.6	62.2	6.3	35.6	43886	1924	27.2	1853
38. Pennsylvania	30720	7.2	57.1	4.7	274.0	50599	8673	24.3	13257
39. Rhode Island	30215	6.4	60.6	4.7	1003.2	49758	665	26.4	1075
40. South Carolina	24886	9.5	59.5	5.4	133.2	38943	1782	19.0	2947
41. South Dakota	26664	7.8	67.7	3.3	9.9	31295	338	25.7	564
42. Tennessee	26988	8.4	59.1	4.5	138.0	38554	2907	22.0	4907
43. Texas	28581	6.0	59.4	4.9	79.6	39293	8544	23.9	14538
44. Utah	24180	5.3	62.7	4.4	27.2	37414	298	26.4	918
45. Vermont	28594	6.3	65.3	3.6	65.8	38802	335	28.8	315
46. Virginia	32431	7.2	61.3	3.5	178.8	41262	3382	31.9	4038
47. Washington	32025	5.3	62.6	6.4	88.6	43438	2953	28.6	2843
48. West Virginia	22881	7.6	51.3	4.9	75.1	36751	1510	15.3	1656
49. Wisconsin	29270	6.9	68.3	4.6	98.8	43114	2692	23.8	3498
50. Wyoming	29416	7.0	65.1	3.9	5.1	37841	250	20.6	247

Data Set III: NBA Data

Name	Height	Weight
Abdul-Wahad, Tariq	78	225
Abdur-Rahim, Shareef	81	240
Alexander, Courtney	77	205
Allen, Malik	82	255
Allen, Ray	77	205
Amaechi, John	82	270
Anderson, Chris	82	220
Anderson, Derek	77	195
Anderson, Kenny	73	168
Anderson, Shandon	78	210
Archibald, Robert	83	250
Arenas, Gilbert	75	191
Armstrong, Brandon	77	185
Armstrong, Darrell	73	180
Arroyo, Carlos	74	202
Artest, Ron	79	246
Atkins, Chucky	71	160
Augmon, Stacey	80	213
Bagaric, Dalibor	85	290
Bailey, Toby	78	213
Baker, Vin	83	250
Barry, Brent	78	203
Barry, Jon	77	210
Bateer, Mengke	83	290
Battie, Tony	83	240
Battier, Shane	80	220
Baxter, Lonny	80	260
Bell, Raja	77	205
Bender, Jonathan	84	219
Best, Travis	71	184
Bibby, Mike	73	190
Billups, Chauncey	75	202
Blount, Corie	82	242
Blount, Mark	84	250
Booth, Calvin	83	231
Boozer, Carlos	81	258
Borchardt, Curtis	84	240
Boumtje-Boumtje, Ruben	84	257
Bowen, Bruce	79	200
Bowen, Ryan	81	220
Bradley, Michael	82	225
Bradley, Shawn	90	275
Brand, Elton	80	265
Brandon, Terell	71	173
Bremer, J.R.	74	185
Brewer, Jamison	76	184
Brezec, Primoz	85	252
Brown, Kedrick	79	222
Brown, Kwame	80	248

Brown, P.J.	83	239
Brown, Randy	74	190
Brunson, Rick	76	190
Bryant, Kobe	78	220
Bryant, Mark	81	250
Buckner, Greg	76	210
Bullard, Matt	82	229
Burke, Pat	83	250
Butler, Caron	79	217
Butler, Rasual	79	205
Caffey, Jason	80	256
Camby, Marcus	83	225
Campell, Elden	84	279
Cardinal, Brian	75	245
Carter, Anthony	74	195
Carter, Vince	78	225
Cassell, Sam	75	185
Cato, Kelvin	83	275
Chandler, Tyson	85	235
Cheaney, Calbert	79	217
Christie, Doug	78	205
Clancy, Sam	79	240
Clark, Keon	83	221
Claxton, Speedy	71	166
Cleaves, Mateen	74	205
Coleman, Derrick	82	270
Coles, Bimbo	74	182
Collier, Jason	84	260
Collins, Jarron	83	252
Collins, Jason	84	260
Crawford, Chris	81	235
Crawford, Jamal	77	190
Croshere, Austin	82	242
Curry, Eddy	83	285
Curry, Michael	77	210
Dalembert, Samuel	83	250
Dampier, Erick	83	265
Daniels, Antonio	76	205
Davis, Antonio	81	230
Davis, Baron	75	223
Davis, Dale	83	252
Davis, Emanual	77	195
Davis, Hubert	77	183
Davis, Ricky	79	195
DeClercq, Andrew	82	255
Delk, Tony	74	189
Dickau, Dan	72	190
Dickerson, Michael	77	190
Diop, DeSagana	84	300
Divac, Vlade	85	260
Dixon, Juan	75	164
Doleac, Michael	83	262

Dooling, Keyon	75	196
Drew, Bryce	75	184
Drobnja, Predrag	83	272
Dudley, Chris	83	260
Duncan, Tim	84	260
Dunleavy Jr., Mike	81	221
Eisley, Howard	74	180
Ellis, LaPhonso	80	240
Ely, Melvin	82	260
Eschmeyer, Evan	83	255
Evans, Reggie	80	245
Feick, Jamie	80	255
Ferry, Danny	82	235
Finley, Michael	79	225
Fisher, Derek	73	205
Fizer, Marcus	80	260
Ford, Alton	81	280
Forte, Joseph	76	194
Fortson, Danny	80	260
Foster, Jeff	83	242
Fowlkes, Tremaine	80	220
Fox, Rick	79	235
Foyle, Adonal	82	265
Francis, Steve	75	195
Funderburke, Lawrence	81	230
Gadzuric, Dan	83	240
Garnett, Kevin	83	220
Garrity, Pat	81	238
Gasol, Pau	84	227
George, Devean	80	240
Gill, Kendall	77	215
Ginobili, Emanuel	78	210
Giricek, Gordan	78	216
Glover, Dion	77	228
Gooden, Drew	82	230
Grant, Brian	81	254
Grant, Horace	82	245
Griffin, Andrian	77	230
Griffin, Eddie	82	232
Gugliotta, Tom	82	250
Haislip, Marcus	82	230
Ham, Darvin	79	230
Hamilton, Richard	79	193
Hardaway, Anfernee	79	215
Harpring, Matt	79	231
Harrington, Adam	77	200
Harrington, Al	81	250
Harrington, Lorinza	76	190
Harrington, Othella	81	235
Harris, Lucious	77	205
Harvey, Antonio	83	250
Harvey, Donnell	80	220

Hassell, Trenton	77	200
Haston, Kirk	81	242
Hawkins, Juaquin	79	205
Haywood, Brendan	84	268
Henderson, Alan	81	240
Hilario, Nene	83	260
Hill, Grant	80	225
Hill, Tyrone	81	250
Holiberg, Fred	77	210
Horry, Robert	82	240
House, Eddie	73	180
Houston, Allan	78	200
Howard, Juwan	81	260
Hudson, Troy	73	170
Huffman, Nate	85	245
Hughes, Larry	77	184
Humphrey, Ryan	80	235
Hunter, Lindsey	74	195
Hunter, Steven	84	220
Ilgauskas, Zydrunas	87	260
Iverson, Allen	72	165
Jackson, Bobby	73	185
Jackson, Marc	82	270
Jackson, Mark	75	195
Jackson, Stephen	80	220
Jacobsen, Casey	78	215
James, Jerome	85	272
James, Mike	74	190
Jamison, Antawn	81	223
Jannero, Pargo	74	175
Jaric, Marko	79	198
Jefferies, Chris	80	215
Jefferson, Richard	79	222
Jeffries, Jared	83	230
Johnson, Anthony	75	190
Johnson, Avery	71	180
Johnson, DerMarr	81	201
Johnson, Ervin	83	255
Johnson, Joe	79	235
Johnson, Ken	82	240
Jones, Damon	75	185
Jones, Eddie	78	210
Jones, Fred	74	218
Jones, Jumaine	80	218
Jones, Popeye	80	250
Jordan, Michael	78	216
Kemp, Shawn	82	280
Kerr, Steve	73	180
Kidd, Jason	76	212
Kirilenko, Andrei	82	225
Kittles, Kerry	77	180
Knight, Brevin	70	170

Knight, Travis	84	235
Kukoc, Toni	83	235
Laettner, Christian	83	245
LaFrentz, Raef	83	245
Lampley, Sean	82	227
Langhi, Dan	83	220
Lenard, Voshon	76	205
Lewis, Rashard	82	215
Long, Art	81	250
Lopez, Felipe	77	195
Lopez, Raul	72	160
Lue, Tyronn	72	178
Lynch, George	80	235
MacCulloch, Todd	84	280
Maddox, Tito	75	200
Madsen, Mark	81	245
Maggette, Corey	78	228
Malone, Karl	81	256
Marbury, Stephon	74	205
Marion, Shawn	79	215
Marks, Sean	82	250
Martin, Kenyon	81	234
Mashburn, Jamal	80	247
Mashell, Donyell	81	230
Mason Jr., Roger	77	200
Mason, Anthony	80	255
Mason, Desmond	77	222
Massenburg, Tony	81	250
Mayloire, Jamaal	83	259
McCarty, Walter	82	230
McCoy, Jelani	82	255
McDyess, Antonio	81	245
McGrady, Tracey	80	210
McInnis, Jeff	76	179
McKie, Aaron	77	209
Medvedenko, Stanislav	82	250
Mercer, Ron	79	210
Mickeal, Pete	77	222
Mihm, Chris	84	265
Miles, Darius	81	210
Miller, Andre	74	200
Miller, Brad	84	261
Miller, Mike	80	218
Miller, Reggie	79	195
Mills, Chris	79	220
Ming, Yao	89	296
Mobley, Cuttino	76	210
Mohammed, Nazr	82	250
Moiso, Jerome	82	232
Montross, Eric	84	270
Morris, Terence	81	221
Mourning, Alonzo	82	261

Murphy, Troy	83	245
Murray, Lamond	79	235
Murray, Ronald	76	190
Murray, Tracy	79	230
Mutombo, Dikembe	86	265
Nachbar, Bostjan	81	221
Najera, Eduardo	80	235
Nash, Steve	75	195
N'diaye, Mamadou	84	255
Nesterovic, Radoslav	84	248
Newble, Ira	79	220
Norris, Moochie	74	185
Nowitzki, Dirk	84	245
Oakley, Charles	81	245
Odom, Lamar	82	221
Okur, Mehmet	83	249
Olajuwon, Hakeem	84	255
Oliver, Dean	71	175
Ollie, Kevin	74	195
Olowokandi, Michael	84	270
O'Neal, Jermaine	83	242
O'Neal, Shaquille	85	338
Ostertag, Greg	86	280
Outlaw, Bo	80	220
Owens, Chris	79	245
Oyedeji, Olumide	82	240
Padgett, Scott	81	240
Palacio, Milt	75	195
Parker, Smush	76	180
Parker, Tony	74	180
Parks, Cherokee	83	240
Patterson, Ruben	77	224
Payton, Gary	76	180
Peeler, Anthony	76	208
Person, Wesley	78	200
Peterson, Morris	79	215
Piatkowski, Eric	78	215
Pierce, Paul	78	246
Pippen, Scottie	80	228
Pollard, Scot	83	265
Pope, Mark	82	235
Posey, James	80	215
Postell, Lavor	77	215
Potapenko, Vitaly	82	285
Price, Brent	73	185
Prince, Tayshaun	81	215
Przybilla, Joel	85	255
Radmanovic, Vladimir	82	234
Rakocevic, Igor	74	184
Randolph, Zach	81	253
Ratliff, Theo	82	230
Rebraca, Zeljko	84	257

Redd, Michael	78	220
Reid, Don	80	250
Rentzias, Efthimios	83	250
Rice, Glen	80	220
Richardson, Jason	78	220
Richardson, Quentin	78	236
Robinson, Clifford	82	225
Robinson, David	85	250
Robinson, Eddie	81	210
Robinson, Glenn	79	230
Rogers, Rodney	79	255
Rooks, Sean	82	260
Rose, Jalen	80	217
Rose, Malik	79	255
Rush, Kareem	78	215
Russell, Bryon	79	225
Sabonis, Arvydas	87	292
Salmons, John	79	210
Samake, Soumaila	82	245
Sampson, Jamal	83	235
Sanchez, Pepe	76	195
Sasser, Jeryl	78	200
Savovic, Predrag	78	225
Scalabrine, Brian	81	240
Scatterfield, Kenny	74	170
Sesay, Ansu	81	225
Shaw, Brain	78	205
Skinner, Brian	81	265
Slater, Reggie	79	255
Slay, Tamar	80	215
Smith, Charles	76	200
Smith, Joe	82	225
Smith, Steve	80	221
Snow, Eric	75	204
Sprewell, Latrell	77	195
Stackhouse, Jerry	78	218
Stepania, Vladimir	85	255
Stevenson, DeShawn	77	210
Stewart, Michael	82	230
Stockton, John	73	175
Stojakovic, Predrag	82	229
Stoudamire, Damon	70	171
Stoudemire, Amare	82	245
Strickland, Eric	75	210
Strickland, Rod	75	185
Sundov, Bruno	86	245
Sura, Bob	77	200
Swift, Stromile	81	225
Szczerbiak, Wally	79	244
Taylor, Maurice	81	260
Terry, Jason	74	180
Thomas, Etan	82	260

Thomas, Kenny	79	245
Thomas, Kurt	81	235
Thomas, Tim	82	240
Tinsley, Jamaal	75	195
Traylor, Robert	80	284
Trent, Gary	80	250
Trybanski, Cezary	86	240
Tsakalidis, Jake	86	290
Tskitishvili, Nikoloz	84	225
Turkoglu, Hidayet	82	220
Van Exel, Nick	73	195
Van Horn, Eric	82	255
Varda, Ratko	85	265
Vaughn, Jacque	73	190
Voskuhl, Jake	83	245
Wagner, Dajuan	74	200
Walker, Antonio	81	220
Walker, Samaki	81	255
Wallace, Ben	81	240
Wallace, Gerald	79	215
Wallace, Rasheed	83	230
Wang, Zhizhi	85	275
Ward, Charlie	74	185
Watson, Earl	73	190
Weatherspoon, Clarence	79	270
Webber, Chris	82	245
Wells, Bonzi	77	210
Welsch, jiri	79	208
Wesley, David	73	203
White, Jahidi	81	290
White, Rodney	81	230
Whitney, Chris	72	175
Wilcox, Chris	82	221
Williams, Aaron	81	240
Williams, Alvin	77	185
Williams, Eric	80	220
Williams, Frank	75	205
Williams, Jason	73	190
Williams, Jay	74	195
Williams, Jerome	81	206
Williams, Monty	80	225
Williams, Scott	82	260
Williams, Shammond	73	201
Williamson, Corliss	79	245
Willims, Walt	80	230
Willis, Kevin	84	245
Wolkowyski, Ruben	82	275
Woods, Loren	85	246
Woods, Qyntel	80	221
Wright, Lorenzen	83	240
Yarbrough, Vincent	79	210

Data Set IV: Manchester (Connecticut) Road Race

**Data Set IV is too large to be printed here.
It is available on the textbook's website.
See the Preface of the textbook.**

Data Set V: Sample of 500 Observations Selected
From Data Set IV

22.08	25.08	27.05	27.07	27.15	27.20	27.57	27.58	28.27
28.85	29.13	29.17	29.43	29.52	29.55	29.72	29.87	30.37
30.43	30.58	31.05	31.10	31.10	31.68	31.80	31.88	31.92
32.13	32.15	32.42	32.45	32.48	32.55	32.67	32.77	32.80
32.83	33.47	33.63	33.65	33.80	33.95	34.05	34.08	34.23
34.47	34.67	34.75	34.77	34.78	35.07	35.15	35.57	35.68
35.87	35.98	36.03	36.30	36.32	36.40	36.47	36.48	36.50
36.52	36.53	36.67	36.73	36.80	36.93	36.97	36.98	37.00
37.08	37.25	37.33	37.38	37.42	37.50	37.63	37.75	37.82
37.90	38.22	38.23	38.30	38.33	38.43	38.45	38.57	38.58
38.67	38.70	38.88	38.93	38.95	39.02	39.37	39.48	39.65
39.73	39.80	39.85	39.90	39.95	40.07	40.23	40.28	40.32
40.37	40.42	40.43	40.50	40.58	40.62	40.83	41.02	41.05
41.15	41.20	41.23	41.28	41.33	41.42	41.52	41.67	41.72
41.72	41.82	41.83	41.88	42.00	42.02	42.15	42.20	42.27
42.37	42.43	42.47	42.50	42.57	42.57	42.58	42.62	42.67
42.67	42.82	42.87	42.88	42.90	42.95	43.08	43.13	43.18
43.22	43.25	43.27	43.42	43.48	43.60	43.67	43.77	43.80
43.82	43.90	43.92	43.93	43.95	44.25	44.25	44.28	44.32
44.32	44.48	44.62	44.68	44.73	44.75	44.85	44.95	45.02
45.12	45.27	45.32	45.35	45.38	45.43	45.45	45.45	45.45
45.65	45.75	45.77	45.93	45.98	46.02	46.03	46.03	46.32
46.35	46.42	46.42	46.53	46.57	46.57	46.58	46.65	46.72
46.72	46.72	46.77	46.80	46.85	46.87	46.88	46.95	46.97
46.98	47.07	47.10	47.13	47.18	47.22	47.27	47.28	47.45
47.47	47.62	47.62	47.63	47.63	47.65	47.83	47.88	47.88
47.90	48.03	48.07	48.13	48.42	48.45	48.47	48.48	48.53
48.53	48.55	48.55	48.57	48.60	48.70	48.97	49.05	49.15
49.18	49.20	49.23	49.27	49.28	49.32	49.32	49.33	49.38
49.40	49.43	49.43	49.55	49.57	49.62	49.63	49.72	49.82
49.82	49.83	49.85	49.90	49.97	50.13	50.13	50.20	50.30
50.35	50.37	50.45	50.67	50.68	50.68	50.80	50.82	50.95
51.13	51.15	51.17	51.27	51.38	51.55	51.62	51.77	51.83
51.93	51.98	52.20	52.20	52.22	52.33	52.33	52.35	52.35
52.37	52.42	52.50	52.53	52.60	52.63	52.65	52.78	52.83
52.85	52.88	52.95	53.07	53.13	53.17	53.25	53.28	53.40
53.43	53.45	53.45	53.47	53.50	53.50	53.50	53.50	53.73
53.90	54.03	54.08	54.20	54.27	54.32	54.33	54.40	54.42
54.60	54.73	54.82	54.82	54.93	54.97	54.98	55.03	55.08
55.15	55.22	55.35	55.43	55.60	55.62	55.70	55.70	55.88
55.90	55.92	56.00	56.22	56.25	56.27	56.32	56.32	56.33
56.33	56.42	56.45	56.50	56.50	56.52	56.55	56.55	56.63
56.65	56.65	56.73	57.15	57.48	57.53	57.55	57.68	57.82
58.12	58.25	58.27	58.35	58.37	58.38	58.43	58.48	58.55
58.60	58.67	58.87	58.93	59.05	59.07	59.07	59.23	59.25
59.28	59.30	59.53	59.63	59.68	59.70	59.77	59.78	59.80
59.82	60.78	61.00	61.20	61.27	61.40	61.40	61.43	61.53
61.65	61.73	61.73	61.88	61.98	62.00	62.32	62.78	63.03
63.15	63.17	63.48	63.93	64.75	64.93	65.23	65.27	66.50
68.50	69.07	69.47	69.93	69.98	70.17	70.25	72.00	72.07
72.32	72.73	73.65	73.70	73.95	74.13	74.92	76.12	76.18
76.52	76.67	77.17	77.50	78.17	79.05	79.08	79.28	80.22
80.33	80.85	81.10	81.20	81.53	81.72	81.97	82.10	82.15
82.17	82.18	82.22	82.35	82.82	83.65	84.13	84.90	85.23
85.25	85.38	85.68	86.52	86.62	87.52	87.57	88.52	88.85
89.10	89.83	90.80	94.02	96.37				

NOTES

NOTES

NOTES

NOTES

NOTES

NOTES

NOTES